Industrial Applications of Biosurfactants and Microorganisms

Progress in Biochemistry and Biotechnology

Industrial Applications of Biosurfactants and Microorganisms

Green Technology Avenues From Lab to Commercialization

Edited by

RUBY ASLAM
School of Civil Engineering and Architecture, Chongqing University of Science and Technology, Chongqing, P.R. China

JEENAT ASLAM
Department of Chemistry, College of Science, Taibah University, Yanbu, Al-Madina, Saudi Arabia

CHAUDHERY MUSTANSAR HUSSAIN
Department of Chemistry and Environmental Sciences, New Jersey Institute of Technology, Newark, NJ, United States

Academic Press is an imprint of Elsevier
125 London Wall, London EC2Y 5AS, United Kingdom
525 B Street, Suite 1650, San Diego, CA 92101, United States
50 Hampshire Street, 5th Floor, Cambridge, MA 02139, United States
The Boulevard, Langford Lane, Kidlington, Oxford OX5 1GB, United Kingdom

Copyright © 2024 Elsevier Inc. All rights reserved.

No part of this publication may be reproduced or transmitted in any form or by any means, electronic or mechanical, including photocopying, recording, or any information storage and retrieval system, without permission in writing from the publisher. Details on how to seek permission, further information about the Publisher's permissions policies and our arrangements with organizations such as the Copyright Clearance Center and the Copyright Licensing Agency, can be found at our website: www.elsevier.com/permissions.

This book and the individual contributions contained in it are protected under copyright by the Publisher (other than as may be noted herein).

Notices

Knowledge and best practice in this field are constantly changing. As new research and experience broaden our understanding, changes in research methods, professional practices, or medical treatment may become necessary.

Practitioners and researchers must always rely on their own experience and knowledge in evaluating and using any information, methods, compounds, or experiments described herein. In using such information or methods they should be mindful of their own safety and the safety of others, including parties for whom they have a professional responsibility.

To the fullest extent of the law, neither the Publisher nor the authors, contributors, or editors, assume any liability for any injury and/or damage to persons or property as a matter of products liability, negligence or otherwise, or from any use or operation of any methods, products, instructions, or ideas contained in the material herein.

ISBN: 978-0-443-13288-9

For Information on all Academic Press publications
visit our website at https://www.elsevier.com/books-and-journals

Publisher: Stacy Masucci
Acquisitions Editor: Michelle Fisher
Editorial Project Manager: Barbara Makinster
Production Project Manager: Jayadivya Saiprasad
Cover Designer: Matthew Limbert

Typeset by MPS Limited, Chennai, India

Contents

List of contributors xiii
About the editors xvii
Preface xix

1. Biosurfactants: introduction and classification   1
Irfan Ali, Asif Jamal, Zafeer Saqib, Muhammad Ishtiaq Ali and Aetsam Bin Masood

    1.1 Introduction   1
    1.2 Fundaments aspects of biosurfactants   1
    1.3 Ecological significance of biosurfactants   4
    1.4 Production of biosurfactants   4
    1.5 Applications of biosurfactants   5
    1.6 Structural diversity of microbial surfactants   6
    1.7 Classification of biosurfactants   7
    1.8 Conclusion and future prospects   14
    References   15

2. Biosurfactants: basic properties and characterizations   25
Renata Raianny da Silva, Yali Alves da Silva, Leonie Asfora Sarubbo and Juliana Moura de Luna

    2.1 Introduction   25
    2.2 The classification of biosurfactants   26
    2.3 Physiochemical properties of biosurfactants   27
    2.4 Factors affecting biosurfactant production   30
    2.5 Conclusions and future perspectives   34
    References   34

3. Biosurfactants production utilizing microbial resources   39
Ruby Aslam, Mohammad Mobin, Saman Zehra and Jeenat Aslam

    3.1 Introduction   39
    3.2 Types of biosurfactants   40
    3.3 Sources of production of biosurfactants   41
    3.4 Fermentation process   42
    3.5 Low-cost byproducts and waste as feedstock   42
    3.6 Production of biosurfactants using microorganisms   43
    3.7 Factors affecting biosurfactants production   51

	3.8 Challenges and future research directions	51
	3.9 Conclusions	51
	Acknowledgments	52
	References	52

4. **Biosurfactant production by utilizing waste products of the food industry** — 59

   Oluwaseun Ruth Alara, Nour Hamid Abdurahman and Hassan Alsaggaf Ali

4.1 Introduction	59
4.2 Biosurfactants from different food wastes	60
4.3 Conclusion and future perspectives	71
Acknowledgments	72
References	72

5. **Factors affecting biosurfactants production** — 79

   Arif Nissar Zargar and Preeti Srivastava

5.1 Introduction	79
5.2 Factors affecting biosurfactant production	80
5.3 Factors that affect large-scale production and commercialization of biosurfactants	90
5.4 Possible approaches to improve biosurfactant production	91
5.5 Conclusions and future outlook	98
References	98

6. **Crude oil storage tank clean-up using biosurfactants** — 107

   Mohammad Mobin, Kanika Cial, Ruby Aslam and Mosarrat Parveen

6.1 Introduction	107
6.2 Biosurfactants production strategy	107
6.3 Factors affecting biosurfactants production	108
6.4 Production methods of biosurfactants	108
6.5 Classification of biosurfactants	109
6.6 Oil storage tanks in the industry	111
6.7 Use of biosurfactants in cleaning of crude oil storage tanks	114
6.8 Challenges and future outlook	116
6.9 Conclusions	117
Acknowledgments	117
References	117

7. Pollution mitigation utilizing biosurfactants — 121
Asif Jamal, Muhammad Ishtiaq Ali, Aetsam Bin Masood, Maryam Khan Wazir, Ahsan Ullah and Ramla Rehman

- 7.1 Introduction — 121
- 7.2 Sources of pollution and their impact on the ecosystem — 122
- 7.3 Natural role of biosurfactants — 123
- 7.4 Polycyclic aromatic hydrocarbons — 124
- 7.5 Biosurfactants-mediated biodegradation of polycyclic aromatic hydrocarbons — 125
- 7.6 Polychlorinated biphenyls (PCBs) — 128
- 7.7 Biosurfactants-mediated degradation of polychlorinated biphenyls — 130
- 7.8 Organopesticides — 132
- 7.9 Surfactants enhanced degradation of organopesticides — 133
- 7.10 Plant-based surfactants and their role in bioremediation — 137
- 7.11 Conclusion and future directions — 139
- References — 140

8. Strategic biosurfactant-advocated bioremediation technologies for the removal of petroleum derivatives and other hydrophobic emerging contaminants — 151
Swathi Krishnan Venkatesan, Raja Rajeswari Devi Mandava, Venkat Ramanan Srinivasan, Megha Prasad and Ramani Kandasamy

- 8.1 Introduction — 151
- 8.2 Environmental prevalence of hydrophobic contaminants/petroleum derivatives — 152
- 8.3 Toxic impacts of petroleum hydrocarbons — 154
- 8.4 Conventional treatment technologies for the mitigation of hydrophobic contaminants/petroleum derivatives — 155
- 8.5 Green surfactants: bio-based surfactants and biosurfactants — 155
- 8.6 Omics approaches for the biosurfactants—case studies in filed level applications toward the remediation of hydrophobic pollutants/petroleum derivatives — 174
- 8.7 Recent advancements in enhancing the specificity and functional properties of biosurfactants — 175
- 8.8 Bottlenecks in the real-time application of biosurfactants on the removal of hydrophobic pollutants/petroleum derivatives — 178
- 8.9 Nanotechnology—engineering of biosurfactants for the removal of hydrophobic pollutants/petroleum derivatives—challenges and future perspectives — 181
- 8.10 Challenges and future prospects — 182
- 8.11 Conclusion — 183
- References — 183

9. Removal of hydrophobic contaminant/petroleum derivate utilizing biosurfactants — 193
Chiamaka Linda Mgbechidinma and Chunfang Zhang

- 9.1 Introduction — 193
- 9.2 Hydrophobic contaminant/petroleum derivate: sources, occurrence, fate, and implications — 195
- 9.3 Biosurfactants for hydrophobic contaminant/petroleum derivate removal — 201
- 9.4 Removal mechanisms of hydrophobic contaminant/petroleum derivate in the presence of biosurfactants — 204
- 9.5 Impact of biosurfactant-mediated hydrophobic contaminant/petroleum derivate removal on microbial community — 208
- 9.6 Eco-sustainable biosurfactant-based hydrophobic contaminant/petroleum derivate removal approach toward achieving some selected United Nations Sustainable Development Goals — 209
- 9.7 Conclusion and possible future outlooks — 211
- References — 212

10. Role of biosurfactants in improving target efficiency of drugs and designing novel drug delivery systems — 217
Ramla Rehman, Asif Jamal, Irfan Ali, Munira Quddus and Aziz ur Rehman

- 10.1 Introduction — 217
- 10.2 Unique self-assembly features of biosurfactants and their suitability for drug adaptation and target improvement — 218
- 10.3 Solubility and emulsion formation by biosurfactants for hydrophobic drug bioavailability — 221
- 10.4 Interaction of biosurfactants with bio-interfaces — 226
- 10.5 Biosurfactants as delivery carriers for DNA-/RNA-based drug vehicles — 229
- 10.6 New biosurfactants with improved drug target efficiency — 231
- 10.7 Patents concerning the applications of biosurfactants for the pharmaceutical industry — 236
- 10.8 Unknown aspects of biosurfactants and future directions — 238
- 10.9 Conclusion and future outlook of biosurfactants in drug development — 239
- References — 239

11. Recent advancements in biosurfactant-aided adsorption technologies for the removal of pharmaceutical drugs — 249
Jagriti Jha Sanjay, Swathi Krishnan Venkatesan and Ramani Kandasamy

- 11.1 Introduction — 249
- 11.2 Insight on properties pertaining to ecotoxicological impact of pharmaceutical drugs — 250
- 11.3 A comprehensive account of the biosurfactants in terms of their types, characteristics, sources, and applications for removing toxic pharmaceutical compounds — 256

11.4	Elucidating biosurfactant drug adsorption properties and mechanisms	262
11.5	Recent advancements in biosurfactant-aided adsorption technologies for removal of drugs from the environment	263
11.6	Limitations preventing for extensive application of biosurfactant for drug removal from environment and peculiar advantages associated	268
11.7	Future prospects—planning possible strategies to overcome the limited application of biosurfactant over a wide spectrum	268
11.8	Conclusion	270
	References	270

12. Potential of biosurfactants in corrosion inhibition — 277

Qihui Wang and Zhitao Yan

12.1	Introduction	277
12.2	Sources and classification of biosurfactants	279
12.3	Corrosion inhibition mechanism of biosurfactants	286
12.4	Practical application of biosurfactants	291
12.5	Characterization methods for biosurfactants	296
12.6	Conclusion and outlook	299
	References	300

13. Antimicrobial and anti-biofilm potentials of biosurfactants — 307

John Adewole Alara and Oluwaseun Ruth Alara

13.1	Introduction	307
13.2	Biosurfactants as an antimicrobial agent and their mechanisms of action	309
13.3	Antimicrobial properties of biosurfactants	314
13.4	Microbial formation of biofilm	319
13.5	Biosurfactants as antibiofilm agent and their mechanisms of action	321
13.6	Current industrial and medical applications and commercialization of biosurfactant compounds with anti-biofilm and antimicrobial property	329
13.7	Future trends and conclusions	331
13.8	Conclusion	332
	Conflict of interest	332
	References	332

14. Insecticidal potential of biosurfactants — 341

Natalia Andrade Teixeira Fernandes, Luara Aparecida Simões, Angelica Cristina Souza and Disney Ribeiro Dias

14.1	Introduction	341
14.2	Chemical pesticides	342
14.3	Biosurfactants as agricultural biopesticides	342

14.4	Biosurfactant as biocontrol for organic agriculture	345
14.5	Conclusion and future direction	350
	References	351

15. Potential of biosurfactants as antiadhesive biological coating — 355
John Adewole Alara

15.1	Introduction	355
15.2	Microbial adhesion and biofilms	358
15.3	The antiadhesive coating property of biosurfactants	359
15.4	Antiadhesive property of glycolipids biosurfactants	360
15.5	Antiadhesive property of lipopeptides biosurfactants	362
15.6	Production of antiadhesive and antiineffective biomaterials	363
15.7	Results of some patents related to antiadhesive biological coating property of biosurfactants	364
15.8	Challenges and future perspective	365
15.9	Conclusion	366
	Reference	366

16. Advantages of biosurfactants over petroleum-based surfactants — 371
Angelica Cristina de Souza, Monique Suela Silva, Luara Aparecida Simões, Natalia Andrade Teixeira Fernandes, Rosane Freitas Schwan and Disney Ribeiro Dias

16.1	Introduction	371
16.2	Chemical classification of biosurfactants	373
16.3	Microbial biosurfactant production	375
16.4	Renewable natural resources used in biosurfactant production	377
16.5	Properties and advantages of biosurfactants	379
16.6	Production using renewable raw materials	382
16.7	Challenges that limit the production of biosurfactants	383
16.8	Concluding remarks	384
	References	385

17. Commercialization of biosurfactants — 395
Ruby Aslam, Jeenat Aslam and C.M. Hussain

17.1	Introduction	395
17.2	Global biosurfactant market and their impact on the COVID-19 pandemic	396
17.3	Factors affecting scale up of biosurfactants	397
17.4	Scale-up studies of biosurfactant production	398
17.5	Patents related to the biosurfactants	399
17.6	Challenges and future outlook	404
17.7	Conclusion	404

References		404
Further reading		406

18. Biosurfactants for environmental health and safety — 407

Luara Aparecida Simões, Natalia Andrade Teixeira Fernandes, Angelica Cristina de Souza and Disney Ribeiro Dias

18.1	Introduction	407
18.2	Environmental effect of synthetic surfactants	408
18.3	Role of biosurfactants in environmental pollution	412
18.4	Biosurfactants and sustainability	415
18.5	Beneficial effects on plants	415
18.6	Concluding remarks and future perspectives	418
	References	419

19. Biosurfactants: sustainable alternatives to chemical surfactants — 425

Arif Nissar Zargar and Preeti Srivastava

19.1	Introduction	425
19.2	Drive for global sustainability	426
19.3	Chemical surfactants and their production	428
19.4	Sustainability assessment of chemical surfactants	429
19.5	Biosurfactants: sustainable alternatives to chemical surfactants	430
19.6	Concluding remarks and future outlook	433
	References	434

20. Biosurfactants for sustainability — 437

Oluwaseun Ruth Alara, Nour Hamid Abdurahman and Hassan Alsaggaf Ali

20.1	Introduction	437
20.2	Production of biosurfactants from wastes and renewable materials for sustainability	439
20.3	Methods for enhancing sustainability in biosurfactant production	440
20.4	Conclusion	447
	Acknowledgments	448
	References	448

Index — *455*

List of contributors

Nour Hamid Abdurahman
Centre for Research in Advanced Fluid and Processes (Fluid Centre), Universiti Malaysia Pahang, Gambang, Pahang, Malaysia; Faculty of Chemical and Process Engineering Technology, Universiti Malaysia Pahang, Gambang, Pahang, Malaysia

John Adewole Alara
Department of Chemical Engineering, College of Engineering, University Malaysia Pahang, Gambang, Pahang, Malaysia; Oyo State Primary Healthcare Board, Ogbomoso, Oyo State, Nigeria; St. John of God Accord, Greensborough, VIC, Australia

Oluwaseun Ruth Alara
Centre for Research in Advanced Fluid and Processes (Fluid Centre), Universiti Malaysia Pahang, Gambang, Pahang, Malaysia; Faculty of Chemical and Process Engineering Technology, Universiti Malaysia Pahang, Gambang, Pahang, Malaysia; School of Property, Construction and Project Management, RMIT University, Melbourne, VIC, Australia

Hassan Alsaggaf Ali
Eastern Unity Technology, Kuala Lumpur, Malaysia

Irfan Ali
Centre of Agricultural Biochemistry and Biotechnology (CABB), University of Agriculture Faisalabad, Faisalabad, Punjab, Pakistan

Muhammad Ishtiaq Ali
Faculty of Biological Sciences, Department of Microbiology, Quaid-i-Azam University, Islamabad, Pakistan

Jeenat Aslam
Department of Chemistry, College of Science, Taibah University, Yanbu, Al-Madina, Saudi Arabia

Ruby Aslam
School of Civil Engineering and Architecture, Chongqing University of Science and Technology, Chongqing, P.R. China

Kanika Cial
Corrosion Research Laboratory, Department of Applied Chemistry, Faculty of Engineering and Technology, Aligarh Muslim University, Aligarh, Uttar Pradesh, India

Renata Raianny da Silva
Northeast Biotechnology Network (RENORBIO), Federal Rural University of Pernambuco, Recife, Pernambuco, Brazil

Yali Alves da Silva
Masters degree in Development of Environmental Processes, Catholic University of Pernambuco, Recife, Brazil

Juliana Moura de Luna
Advanced Institute of Technology and Innovation (IATI), Recife, Pernambuco, Brazil; School of Health and Life Sciences, Catholic University of Pernambuco, Recife, Pernambuco, Brazil

Angelica Cristina de Souza
Department of Biology, Federal University of Lavras, Lavras, Minas Gerais, Brazil

Disney Ribeiro Dias
Department of Food Science, Federal University of Lavras, Lavras, Minas Gerais, Brazil

Natalia Andrade Teixeira Fernandes
Department of Chemistry, University of California, Davis, CA, United States

C.M. Hussain
Department of Chemistry and Environmental Sciences, New Jersey Institute of Technology, Newark, NJ, United States

Asif Jamal
Faculty of Biological Sciences, Department of Microbiology, Quaid-i-Azam University, Islamabad, Pakistan

Ramani Kandasamy
Industrial and Environmental Sustainability Laboratory, Department of Biotechnology, School of Bioengineering, SRM Institute of Science and Technology, Kattankulathur, Chengalpattu, Tamil Nadu, India

Raja Rajeswari Devi Mandava
Industrial and Environmental Sustainability Laboratory, Department of Biotechnology, School of Bioengineering, SRM Institute of Science and Technology, Kattankulathur, Chengalpattu, Tamil Nadu, India

Aetsam Bin Masood
Faculty of Biological Sciences, Department of Microbiology, Quaid-i-Azam University, Islamabad, Pakistan

Chiamaka Linda Mgbechidinma
Institute of Marine Biology and Pharmacology, Ocean College, Zhejiang University, Zhoushan, Zhejiang, P.R. China

Mohammad Mobin
Corrosion Research Laboratory, Department of Applied Chemistry, Faculty of Engineering and Technology, Aligarh Muslim University, Aligarh, Uttar Pradesh, India

Mosarrat Parveen
Corrosion Research Laboratory, Department of Applied Chemistry, Faculty of Engineering and Technology, Aligarh Muslim University, Aligarh, Uttar Pradesh, India

Megha Prasad
Industrial and Environmental Sustainability Laboratory, Department of Biotechnology, School of Bioengineering, SRM Institute of Science and Technology, Kattankulathur, Chengalpattu, Tamil Nadu, India

Munira Quddus
Faculty of Biological Sciences, Department of Microbiology, Quaid-i-Azam University, Islamabad, Pakistan

Aziz ur Rehman
Faculty of Biological Sciences, Department of Microbiology, Quaid-i-Azam University, Islamabad, Pakistan

Ramla Rehman
Institute of Industrial Biotechnology, Government College University, Lahore, Punjab, Pakistan

Jagriti Jha Sanjay
Industrial and Environmental Sustainability Laboratory, Department of Biotechnology, SRM Institute of Science and Technology, Kattankulathur, Chengalpattu, Tamil Nadu, India

Zafeer Saqib
Department of Environmental Sciences, International Islamic University, Islamabad, Pakistan

Leonie Asfora Sarubbo
Icam Tech School, Catholic University of Pernambuco, Recife, Pernambuco, Brazil; Advanced Institute of Technology and Innovation (IATI), Recife, Pernambuco, Brazil

Rosane Freitas Schwan
Department of Biology, Federal University of Lavras, Lavras, Minas Gerais, Brazil

Monique Suela Silva
Department of Food Science, Federal University of Lavras, Lavras, Minas Gerais, Brazil

Luara Aparecida Simões
Biology Department, University of Porto, Porto, Portugal

Angelica Cristina Souza
Department of Biology, Federal University of Lavras, Lavras, Minas Gerais, Brazil

Venkat Ramanan Srinivasan
Industrial and Environmental Sustainability Laboratory, Department of Biotechnology, School of Bioengineering, SRM Institute of Science and Technology, Kattankulathur, Chengalpattu, Tamil Nadu, India

Preeti Srivastava
Department of Biochemical Engineering and Biotechnology, Indian Institute of Technology Delhi, Hauz Khas, New Delhi, India

Ahsan Ullah
Faculty of Biological Sciences, Department of Microbiology, Quaid-i-Azam University, Islamabad, Pakistan

Swathi Krishnan Venkatesan
Industrial and Environmental Sustainability Laboratory, Department of Biotechnology, School of Bioengineering, SRM Institute of Science and Technology, Kattankulathur, Chengalpattu, Tamil Nadu, India

Qihui Wang
School of Civil Engineering and Architecture, Chongqing University of Science and Technology, Chongqing, P.R. China

Maryam Khan Wazir
Faculty of Biological Sciences, Department of Microbiology, Quaid-i-Azam University, Islamabad, Pakistan

Zhitao Yan
School of Civil Engineering and Architecture, Chongqing University of Science and Technology, Chongqing, P.R. China

Arif Nissar Zargar
Department of Biochemical Engineering and Biotechnology, Indian Institute of Technology Delhi, Hauz Khas, New Delhi, India

Saman Zehra
Corrosion Research Laboratory, Department of Applied Chemistry, Faculty of Engineering and Technology, Aligarh Muslim University, Aligarh, Uttar Pradesh, India

Chunfang Zhang
Institute of Marine Biology and Pharmacology, Ocean College, Zhejiang University, Zhoushan, Zhejiang, P.R. China

About the editors

Dr. Ruby Aslam is a Postdoctoral fellow in the School of Civil Engineering and Architecture at Chongqing University of Science and Technology, Chongqing, China. She received her M.Sc., M. Phil., and Ph.D. degrees from Aligarh Muslim University, India. She has authored/coauthored several research papers in international peer reviewed journals of wide readership, including critical reviews and book chapters. She has edited many books for American Chemical Society, Elsevier, Springer, Wiley, Walter de Gruyter, and Taylor & Francis.

Dr. Jeenat Aslam works as an associate professor at the Department of Chemistry, College of Science, Taibah University, Yanbu, Al-Madina, Saudi Arabia. She earned her PhD in surface science/chemistry from the Aligarh Muslim University, Aligarh, India. Materials and corrosion, nanotechnology, and surface chemistry are the primary areas of her research. She has published several research and review articles in peer-reviewed international journals such as ACS, Wiley, Elsevier, Springer, Taylor & Francis, and Bentham Science. She has authored more than 30 book chapters and edited more than 20 books for the American Chemical Society, Elsevier, Springer, Wiley, Walter de Gruyter, and Taylor & Francis.

Dr. Chaudhery Mustansar Hussain is an adjunct professor and director of laboratories in the Department of Chemistry & Environmental Sciences at the New Jersey Institute of Technology (NJIT), Newark, New Jersey, United States. His research is focused on the applications of nanotechnology and advanced materials, environmental management, analytical chemistry, and other industries. He has authored numerous papers in peer-reviewed journals and is a prolific author and editor of nearly 150 books, including scientific monographs and handbooks, in his research areas. He has published with Elsevier, the American Chemical Society, the Royal Society of Chemistry, John Wiley & Sons, CRC Press, and Springer.

Preface

Microorganisms produce biosurfactants, which are surface-active molecules either secreted extracellularly or on the surface of cells. Biosurfactants produce a thin layer on the surface of microorganisms that assists in their adhesion or dissociation from other cell surfaces. Due to the growing global need for sustainable solutions, biosurfactants derived from microorganisms have been investigated as a potential alternative to synthetic surfactants in various industrial processes, including food, medicine, petroleum biotechnology, oil recovery, biomedical and therapeutic, and bioremediation. The book covers the most recent academic developments, significant applications, and implementation studies from around the world.

The book is separated into three parts, with each part consisting of several chapters, to capture a comprehensive picture of fundamental, industrial applications, and greener avenues of biosurfactants and offer readers a rational and impressive design of the topic and concentrated up-to-date references. The fundamentals of biosurfactants are examined in PART 1. Introduction and classification, basic properties and characterizations, production using microbial resources and waste products of the food industry, and factors affecting biosurfactant production are the topics covered in Chapters 1–5. PART 2 examines the industrial applications of biosurfactants. Chapters 6–16 cover topics such as crude oil storage tank cleanup using biosurfactants, pollution mitigation using biosurfactants, application of biosurfactants on the remediation of hydrophobic pollutants/petroleum derivatives, the role of biosurfactants in improving target efficiency of drugs and designing novel drug delivery systems, the role of biosurfactants in drug adsorption, the potential of biosurfactants in corrosion inhibition, antimicrobial and antibiofilm potentials of biosurfactants, insecticidal potential of biosurfactants, potential of biosurfactants as an antiadhesive biological coating, and advantages of biosurfactants over petroleum-based surfactants. PART 3 explores the greener avenues of biosurfactants. Commercialization of biosurfactants, biosurfactants for environmental health and safety, biosurfactants as sustainable alternatives to chemical surfactants, and biosurfactants for sustainability are discussed in Chapters 17–20.

This book aims to present the most recent developments in the field of biosurfactants for use in industrial applications. This book is written for a highly diverse audience that works in surface chemistry, colloids and interface chemistry, and other related subjects. This book will be a priceless resource for libraries in academic and professional settings, government and nonprofit organizations, solitary research groups, and scientists. This book is intended to be a resource for scientists, researchers, and advanced undergraduate and graduate students seeking biosurfactants for industrial applications to meet current research demands.

All chapters were authored by renowned academic and professional researchers, scientists, and subject matter specialists. We would like to express our gratitude to all chapter authors on behalf of Elsevier for their extraordinary and sincere efforts in producing this book. For their unwavering support and assistance throughout this project, we are extremely grateful to Dr. Linda Buschman (Senior Acquisition Editor), Ms. Barbara Makinster (Senior Editorial Project Manager), and the editorial team of Elsevier. In the end, Elsevier deserves all praise for releasing the book.

Ruby Aslam
Jeenat Aslam
Chaudhery Mustansar Hussain

CHAPTER 1

Biosurfactants: introduction and classification

Irfan Ali[1], Asif Jamal[2], Zafeer Saqib[3], Muhammad Ishtiaq Ali[2] and Aetsam Bin Masood[2]

[1]Centre of Agricultural Biochemistry and Biotechnology (CABB), University of Agriculture Faisalabad, Faisalabad, Punjab, Pakistan
[2]Faculty of Biological Sciences, Department of Microbiology, Quaid-i-Azam University, Islamabad, Pakistan
[3]Department of Environmental Sciences, International Islamic University, Islamabad, Pakistan

1.1 Introduction

Surfactants are used in almost every industrial sector of the modern society. With an increasing global population, the market for surfactants in different technological fields and commercial applications is expected to reach $66 408 million by the end of 2025 (Wieczorek & Kwaśniewska, 2020). They find applications in household detergents, soaps, personal care and cosmetics, food, beverages, agriculture formulations, textile, electrochemical, oilfield, plastic, and pharmaceutical industries with an annual component growth rate of 5.4% (Shaban et al., 2020). Currently, surfactants are being used extensively as emulsification, solubilizing, stabilizing, flocculating, and wetting agents in emulsion formulations, improving the solubility of hydrophobic drugs and drug-permeability enhancers (Teng et al., 2021). Besides having huge implications, the applications of chemical surfactants, mostly derived from petrochemicals, have been associated with cellular and ecological toxicities (Drobeck, 2019). During the past few decades, the sustainability drive has considerably pushed green synthesis and applications of natural products. Microbial surfactants or biosurfactants have gained much interest and dendritic growth recently because of their ecological significance, multi-faceted properties, and biotechnological applications (Bhadani et al., 2020).

1.2 Fundaments aspects of biosurfactants

Biosurfactants are organic molecules produced by bacteria, yeast, and filamentous fungi that are either released extracellularly or displayed on cell surfaces. Like synthetic surfactants, biosurfactants are amphiphilic chemicals with distinct hydrophilic and hydrophobic termini, allowing them to partition at liquid–liquid interfaces (Markande et al., 2021). The amphiphilic chemical structure of biosurfactants determines their chemical, physical, and biological properties (Crouzet et al., 2020). With the amphiphilic structure, surfactants molecules are adsorbed at the biotic and abiotic interfaces and thereby reduce Gibbs free energy of two phases and alter surface and interfacial tensions to have a stabilizing effect (Zdziennicka et al., 1934). Furthermore, biosurfactant monomers exhibit self-assembling properties in the solution, which is quite a remarkable feature of these wonder biomolecules. In solution, once saturated, surfactant monomers start to self-assemble in the form of very fine, thermodynamically stable supramolecular aggregates known as micelles (Baccile et al., 2021). In a surfactant micelle, molecules are arranged such that their hydrophobic tails form the micelle core, whereas the hydrophilic heads are oriented toward the aqueous environment. The morphology of the surfactant micelles is governed by critical packing parameters and surfactant concentration, technically termed critical micelle concentration (cmc) (Durval et al., 2019). The cmc value has great implications for surfactant use and efficiency in any specific reaction condition. Notably, the biosurfactants' solubilization action is a function of surfactant concentration above its cmc value (Ribeiro et al., 2020). The cmc depends on the length of the hydrophobic tail of the surfactant molecule and is influenced by the presence of counterions, pH, and temperature. At cmc, a rapid transition in phase behavior and properties of the surfactant can be observed, such as reduction of surface tension, adsorption capacity, detergency, and electric conductivity of the system in which surfactants are employed (Yea et al., 2019). Many scientists have reported that biosurfactants are powerful natural surfactants that can reduce the surface tension of water to 29 mN/m and the interfacial tension of an oil–water emulsion to 1 mN/m. They improve the aqueous solubility of hydrophobic substrates, form microemulsion, and enhance their bioavailability (Rehman et al., 2021). Fig. 1.1 depicts the surfactants' cmc and micelle formation.

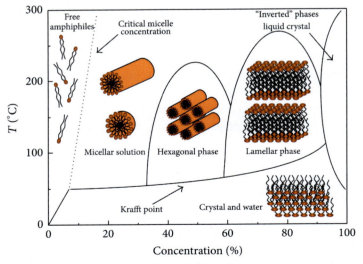

Figure 1.1 Critical micelle concentration and micelle formation (Lombardo et al., 2015).

Other important properties of the biosurfactants are emulsification, wettability, desorption, and partitioning efficiency at the air—water and water—oil interfaces. These properties result from the surfactant self-aggregation phenomenon (Markande et al., 2021). It is reported that biosurfactants form different supramolecular aggregates allowing encapsulation of hydrophobic contaminants and drugs in their micelle core. This process is called microemulsion formation, which enhances the solubility and bioavailability of less soluble hydrophobic substrates (Gudiña & Rodrigues, 2019). In technical terms, microemulsions produced by microbial surfactants have ultralow interfacial tension and higher solubilization efficiencies, allowing their applications in enhanced oil recovery and drug delivery systems (Nasiri & Biria, 2020). Biosurfactants form different types of micelles, lamellar sheets, and crystalline structures in the solution; every geometrical arrangement corresponds to different properties. Microorganisms produce chemically diverse surfactants, which can make stable emulsions under varying reaction conditions. Both glycolipids and lipopeptides secreted by *Pseudomonas aeruginosa and Bacillus subtilis* have been known to produce stable emulsions of crude oil, vegetable oil, kerosene, diesel oil, n-hexane and many other hydrophobic hydrocarbons and permit their use in the bioremediation of hydrophobic contaminants (Kaczorek et al., 2018). It is important to note that the stability of the emulsion is greatly influenced by the hydrophilic—lipophilic balance of the surfactants. As a principle, biosurfactants with low HLB values, between 3 and 6, form W/O microemulsions.

In contrast, biosurfactants having higher HLB values (8—18) perform better for making O/W microemulsions (Ohadi et al., 2020). The details of the surfactants' structure—function relationship suggested that biosurfactants show significant variability in their chemical structures as compared to the synthetic surfactants, creating more complex and dynamic micelle systems with a high degree of uncertainty of their phase behavior (Oliva et al., 2020). As biosurfactants are produced in the form of a complex mixture of isomers, they can generate micelles with different geometrical arrangements, posing great difficulty in evaluating experimental data. Understanding structure—function relationships and phase behavior of biosurfactants in various chemical environments has thus become a new frontier of surfactant science and technology (Manga et al., 2021). Biosurfactants have the following fundamental properties:
- Biodegradability.
- Structural diversity.
- Surface and interfacial tension reduction.
- Microemulsion formation.
- Self-assembly and aggregation.
- Critical micelle concentration (cmc) and micelle formation.
- Adsorption and desorption.
- Partitioning and dispersion efficiency.

1.3 Ecological significance of biosurfactants

Biosurfactants-producing microbes are naturally present in diverse ecological sources, including soil, water, and marine habitats. The synthesis of biosurfactants is an important biochemical feature of the microbial cells associated with obvious physiological advantages over non-surfactant producers (Mohanty et al., 2021). It has been widely recognized that biosurfactant production helps bacteria in substrate accessibility, colonization, swarming mobility, and cell defense (Hou et al., 2019). In soil systems, biosurfactants can emulsify hydrophobic organic compounds, making them biologically available for the cell by improving cell surface hydrophobicity. On the other hand, the architecture of biofilm and cell-to-cell communication is also supported by the production of biosurfactants (Jahan et al., 2020). It has been demonstrated that rhamnolipids are involved in the swarming mobility of *P. aeruginosa*, help their colonization on solid surfaces, and facilitate nutritional supply within the biofilm (Qi & Christopher, 2019). These processes are driven by complex communication networks of the bacteria called Quorum Sensing (QS) (Victor et al., 2019). Biosurfactants have enormous biological significance in agricultural soil. They help establish biofilm at the rhizosphere and thereby facilitate the bioavailability of nutrients for the plants. Owing to their strong antimicrobial activity, biosurfactants are antagonistic against several plant pathogens (Naughton et al., 2019). The isolation of biosurfactant-producing bacteria from crude petroleum and hydrocarbons contaminated sites has been frequently cited. Basically, in these soils, the production of biosurfactants is known to decrease the surface and interfacial tension of petroleum hydrocarbons and enhance their solubility in the aqueous phase (Baccile et al., 2021). Subsequently, the bioavailability of these hydrophobic contaminants is increased, leading to their uptake by microbial cells. This highlights the potential role of biosurfactant-producing microbes in pollution control and global carbon cycle regulation (Osman et al., 2019). Biosurfactant-producing microbes have also been isolated from forests soil, marine environment, sediments, and mangroves settings, making them ubiquitous and nature's favorite chemicals (Dikit et al., 2019). Biosurfactants play various important ecological roles, including:

- Solubilizing hydrophobic contaminants by emulsification.
- Promoting substrate bioavailability.
- Enhancing the mobility of nutrients and metals in the soil matrix.
- Regulating the attachment and colonization of microbes to surfaces.
- Protecting plants from pathogens with their strong antimicrobial activity, and
- Regulating cell-to-cell communication and cell defense.

1.4 Production of biosurfactants

One of the most striking features of the biosurfactants is their production through the fermentation process using cheap carbon substrates such as biodiesel waste glycerol, olive oil

(Shakeri et al., 2020), mill waste, milk whey, corn steep liquor, molasses, animal fats, waste frying oil, oil refinery waste, cassava waste, soap stock, and distillery waste (Hentati et al., 2019). In addition, simple carbon compounds such as glucose, sucrose, alkanes, and glycerol have also been cited for biosurfactant production (Retnaningrum & Wilopo, 2018). Multiple cultivation strategies have been used in pursuit of biosurfactant production at laboratory and commercial scales. In most cases, the production of biosurfactants relies on various process-specific parameters, including media composition, pH, temperature, multivalent ions, and agitation speed (Mohanty et al., 2021). Macronutrients such as carbon, nitrogen, and phosphate play a significant role in the biosynthesis of these molecules (Kashif et al., 2022). The media composition also affects the chemical composition of the biosurfactants. For example, *P. aeruginosa* and *Myerozyma* sp. produce a complex mixture of rhamnolipids and sophorolipids, respectively, with varying molar ratios of these congeners (Rehman et al., 2021). The biosynthesis of BS is also influenced by the ratios of nutrients used in the fermentation media. It has been reported that a low C:N ratio promotes the cellular production of biosurfactants from *P. aeruginosa* (Hrůzová et al., 2020). Currently, the pursuit of gaining maximum product yield through bioprocess optimization has been the most critical aspect of the high-volumetric biosurfactant production. Computer-aided optimization methods, in particular, response surface methodology (RSM), have been among the most efficient tools for improving the efficacy of a biological system, including biosurfactants (Datta et al., 2018). For instance, Eslami et al., 2020; reported improved rhamnolipids production from *P. aeruginosa* EMS1 using the RSM design (Eswari et al., 2013) achieved 18.07 g/dm^{-3} of the rhamnolipids using multi-objective optimization method.

1.5 Applications of biosurfactants

Biosurfactants have enormous applications in the food, cosmetics, petroleum, and pharmaceutical industries as wetting agents, dispersants, detergents, and emulsifying agents (Fiechter, 1992). In the food industry, biosurfactants produced by *Enterobacter cloacae* have been employed as viscosity enhancers (Swidsinski et al., 2007). Similarly, rhamnolipid derived from *P. aeruginosa* can improve dough stability and texture and volume of bakery products (Nitschke et al., 2010). In the petroleum industry, surfactin, lichenysin, and trehaloselipids produced by *B. subtilis*, *B. lichniformis*, and *Rhodococcus erythropolis*, respectively, are used for spill management (Phulpoto et al., 2020). Pertaining to high stability at extreme conditions, biosurfactants like rhamnolipid and surfactin improved the yield of abundant wells by mobilizing heavy crude oil from the reservoirs. The biosurfactants show excellent potential for application in the agriculture and pharmaceutical industries as antibacterial, antiviral, anticancer, and antifungal agents (Ramalingam et al., 2019). They have been applied in various agriculture formulations to control fungal phytopathogens, such as *Aspergillus flavus*, *Colletotrichum*, and *Fusarium oxysporum* (Krishnan et al., 2019). Recently, owing to their lipid solubilization activity, biosurfactants have been

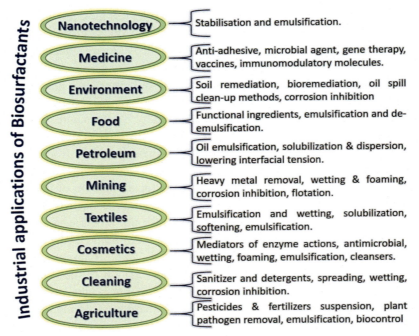

Figure 1.2 Applications of biosurfactants in different fields (Rehman et al., 2021)

investigated for the treatment of SARS-CoV-2. Biosurfactants act as immunomodulators and attenuate cytokine storm during SARS-CoV-2 infection, thus limiting the progression of the virus (Sen et al., 2022). In microbial fuel cell technology, biosurfactants enhance the bioavailability of hydrophobic substrates, improve biofilm formation, and facilitate electron shuttle for better power performance of the system. Based on their adsorption pattern, micellization and interactions with bio-interfaces, biosurfactants are used in drug delivery systems (Philippova & Molchanov, 2019). The applications of the biosurfactants in different fields are summarized in Fig. 1.2.

1.6 Structural diversity of microbial surfactants

Microorganisms produce structurally diverse surfactants with excellent surface and interfacial properties. The glycolipids and lipopeptides are low-molecular-weight biosurfactants. The lipoproteins, lipopolysaccharides—protein complexes, and polysaccharides—protein—fatty acid complexes are classified as high-molecular-weight biosurfactants (Drakontis & Amin, 2020). Biosurfactants, like chemical surfactants, are bipolar molecules with hydrophobic and hydrophilic ends. The hydrophilic part could be a sugar, an amino acid, or a short-chain peptide. The lipidic or hydrophobic portion may comprise saturated or unsaturated fatty acids with varying chain lengths, isopreniod, or hydrophobic amino acids (Carolin et al., 2021). Generally, five classes of biosurfactants have been reported in the literature, that is, glycolipids, lipopeptides, polymeric biosurfactants, particulate biosurfactants, and phospholipids and natural lipids (Singh et al., 2019).

1.7 Classification of biosurfactants
1.7.1 Glycolipids

Glycolipids, sugar—lipid conjugates, are the most prominent microbial surfactants produced as a series of structurally related amphiphiles, each with a minor modification in the parent molecule (Abdel-Mawgoud & Stephanopoulos, 2018). Glycolipids contain a sugary hydrophilic part linked with a hydrophobic portion, which could be a fatty acid molecule of varying chain length (Kirschbaum et al., 2021). Although there are many different types of glycolipids, we will only discuss three main biosurfactants, that is, rhamnolipids, sophorolipids, and trehaloselipids (Kareem, 2020).

1.7.2 Rhamnolipids

Rhamnolipids (RLs) are a best-known class of glycolipid biosurfactants first reported by Bergström and coworkers in 1946 as an extracellular product of *P. pyocyanea* (*P. aeruginosa*) (Platel et al., 2022). Later on, the works of Jarvis and Johnson and Edwards and Hayashi described the chemical structure of RL molecules having two hydroxydecanoic acids (hydrophobic part) and two L-rhamnose moieties (hydrophilic part) linked through a 1,2-glycosidic bond (Kumar & Das, 2018). Because of their fascinating properties, RLs have been a subject of extensive academic and commercial interest. So far, *P. aeruginosa* is considered a potent microbial resource for producing RLs. The non-*Pseudomonas* RL producers include *B. mallei*, *B. pseudomallei*, *B. thailandensis*, *Acinetobacter calcoaceticus*, and *Pantoea stewartii*. The microbiology of RLs biosynthesis is fascinating. It has been reported that a single strain of the bacterium can produce a complex mixture of different RL congeners (Sidrim et al., 2020). The variation of rhamnose units and carbon chain length of fatty acid creates structural diversity and complexity in the RL molecules, making them the most versatile natural amphiphiles. The composition of fermentation media, specifically the carbon source, has been linked to RL diversity (Varjani et al., 2021). So far, more than 60 structurally distinct rhamnolipid variants have been discovered under different cultivation conditions and carbon substrates (Pirog et al., 2019). In general, *P. aeruginosa* strains produce four types of RLs, including mono-rhamno-mono-lipid, di-rhamno-mono-lipid, mono-rhamno-di-lipid, and di-rhamno-di-lipid, each with distinctive physical, chemical, and biological properties (Eslami et al., 2020). The chemical structures of various RL types are depicted in Fig. 1.3.

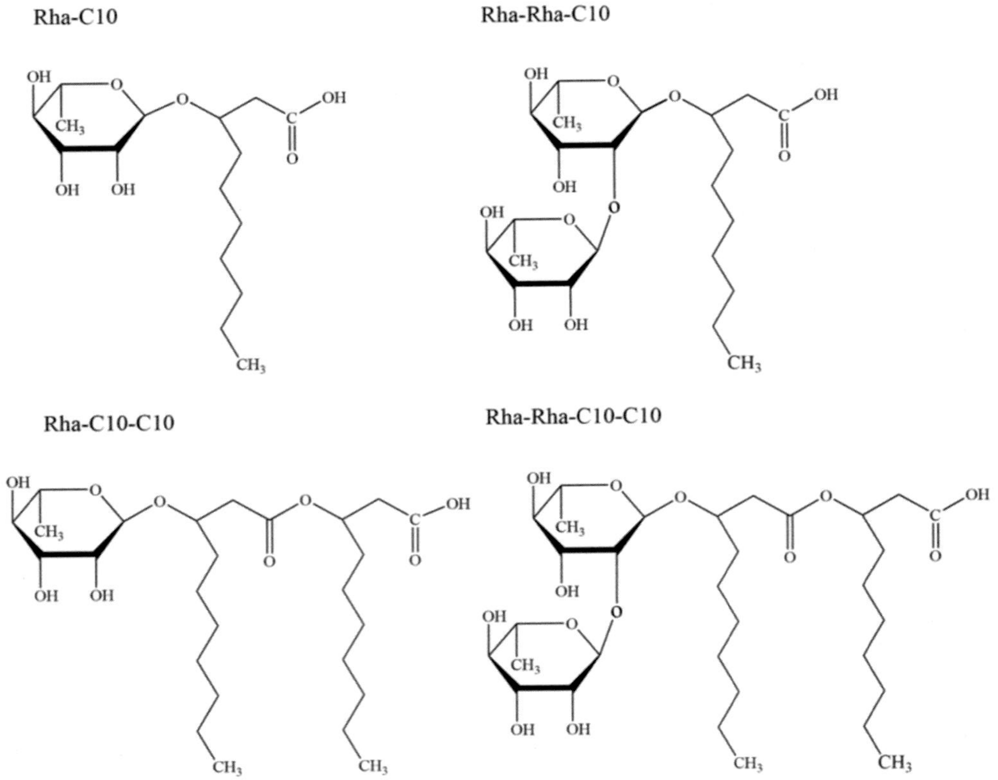

Figure 1.3 Structure of different rhamnolipids types (RL-1 to RL-4) (Shu et al., 2021)

Rhamnolipids display a broad spectrum of physical, chemical, and biological properties. RLs have low molecular weight, are slightly acidic anionic glycolipids, with pKa values ranging from 4.1 to 5.6 (Bai et al., 1998). The average molecular mass of RLs ranges between 302 and 989 Da (Hauser & Karnovsky, 1957). Rhamnolipids alter the surface tension of the water to 25 mN/m and the interfacial tension of the oil−water system to <1 mN/m (Penfold et al., 2011). The cmc value of standard di-rhamnolipid is 110 mg/L (Li et al., 2019). Because of their excellent surface activity, RLs are applied extensively to rehabilitate crude oil and hydrocarbons contaminated sites. In aqueous and soil systems, RLs improved the transformation of recalcitrant hydrocarbons, such as hexadecane, octadecane, and phenanthrene, better than synthetic surfactants (Wei et al., 2020). The addition of RLs produced by *P. aeruginosa* along with co-substrates has also given promising results when tested on soil contaminated with polyaromatic hydrocarbons and pesticides. Due to high stability in extreme conditions, RLs are used extensively in microbial-enhanced oil recovery (Zhao et al., 2019). They are equally important for the remediation of metal-contaminated sites (Liu et al., 2018). As they producemixed micelle system, RLs perform better in treating sites contaminated with a complex mixture of hydrocarbons. In the food industry, RLs are used to enhance the shelf life and quality of bakery products (Dobler et al., 2020). RLs have been added into animal fed formulations to prevent inflammatory diseases (Crouzet et al., 2020). As, they are the source of rhamnose, they can serve as a precursor for producing high-quality flavor products (Adetunji et al., 2018). Similarly, the use of RLs as a food emulsifier has been suggested (Nitschke & Silva, 2018). In the agriculture industry, RLs have been used to control plant pathogens such as Plasmospora, *P. caspsici*, and *P. aphanidermatum* (Kim et al., 2000). RLs show excellent potential for biomedical applications as antibacterial, antifungal, antiviral, antiadhesive, and antiproliferative agents (Niaz et al., 2019). Recently, the application of RLs has been suggested for making nanoparticles of different materials (Ma et al., 2020). Besides, many decades have passed in understanding the properties and applications, and the scientific knowledge of RL molecules in certain areas is quite limited. Further insight into the phase transition and molecular behavior of RL molecules will likely expand their applications in various innovative fields.

1.7.3 Sophorolipids

Sophorolipids (SLs) are among the second most important glycolipid biosurfactants. A typical SL molecule consists of two sophorose units connected by a glycosidic bond to a long hydroxy fatty acid chain (Prasad et al., 2021). Sophorolipids are usually produced in two chemical forms: (1) an acidic form, in which the fatty acid tail is free and (2) a lactonic form, where the carboxyl group of the fatty acid chain is associated with the hydroxyl group of sophorose sugar (Borsanyiova et al., 2016). SLs are reported from various nonpathogenic fungal strains, including *Candida bombicola*, *C. apicola*, *Myerozyma* sp., and *Starmerella bombicola* (Wang et al., 2019). The biosynthesis of SLs is regulated through production-specific genes, including ugtA1, cyp52, and ugtB1 (Van-Bogaert et al., 2013). Like other glycolipids, strain type, medium composition, and carbon source affect yield and chemical composition of the SLs (Rau et al., 2001). Different types of isomers of both lactonic and acidic SLs are identified from different fungi. The variation in the chemical structures of sophorolipids emerges due to acetylation, the addition of hydroxyl groups, and the chain length of fatty acids found in different SLs congeners (Nuñez et al., 2001). Other variations may include linkages within SLs molecules because of an additional number of carbon atoms in lipid moiety (Jiménez-Peñalver et al., 2020). SLs are produced under resting culture strategy, yielding 120 g/L after 8 days (Casas & García-Ochoa, 1999). SLs can reduce surface tension up to 34 mN/m, which implies their great industrial potential (Daverey et al., 2021). The surface/interfacial properties with excellent antimicrobial activity offered by SLs spur broad-spectrum applications of these biological amphiphiles in health, agriculture, and biomedical industries (Adu et al., 2022). SLs show

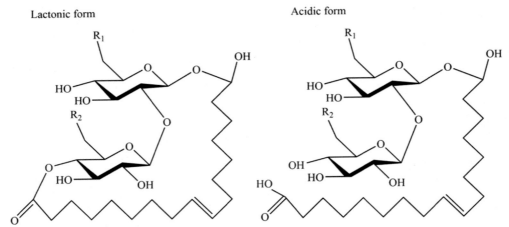

Figure 1.4 Structures of acidic and lactonic forms of sophorolipids (Shu et al., 2021).

higher process efficiency at a very low cmc value of 27.17 mg/L, engraving their marketplace at a steady pace (Rajkhowa & Sarma, 2022).

SLs molecules exhibit a strong antimicrobial action against bacteria, fungi, and viruses (de O Caretta et al., 2022; Pontes et al., 2016). The acidic SLs showed effective antagonistic activity against nosocomial bacterial agents, including *Escherichia coli* and *P. aeruginosa*, with an effective dose of 5 mg/L. SLs can solubilize the lipid membranes of viruses by causing perturbation in the viral structure, leading to viral death (Borsanyiova et al., 2016; Sun et al., 2004). In a recent study, SLs showed a high antiviral effect against SARS-CoV-2 because of their lipid-solubilizing potential (Daverey et al., 2021). SLs are employed in the cosmetic industry to make deodorant formulations, skin protection agents, hair conditioners, and antidandruff agents (Morya et al., 2013). Commercially available products containing SLs include Sopholiance (Kaga et al., 2022). Besides the cosmetics industry, SLs also find applications in cancer treatment and metastasis prevention because of their tumor-suppressing prospects (Miceli et al., 2022; Mohamed et al., 2019). With their remarkable surface-active properties, SLs are being used for the bioremediation of hydrocarbons and metals contaminated sites. With the emergence of exciting properties, the number of SL products and industrial applications are growing, securing their place in the international market as biocompatible surfactants. Fig. 1.4 represents the structure of two different forms of SLs.

1.7.4 Trehalolipids

Trehalolipids (TLs), discovered in 1933, are representative of the microbial glycolipids produced by various strains, such as *Corynebacteria, Mycobacteria,* and *Nocardia* sp. (de Sousa-D'Auria et al., 2022). Generally, two molecular forms of TLs are produced under varying fermentation conditions, including mono-corynomycolate (TL-1) and di-corynomycolate (TL-2) (Lang et al., 1989). They are also acclaimed for reducing surface tensions in the 32—36 mN/m range, and the interfacial tension is same as in other members of the glycolipids family (Janek et al., 2018; Mortita et al., 2016). The calculated cmc value for TLs molecules is 0.140 mg/mL (Janek et al., 2018) and their biosynthesis is regulated by multiple genes (de Paula et al., 2022). The obtained production titer of TLs ranges from 1.56 g/L (Ruhal & Choudhury, 2012) to 10.9 g/L (Mutalik et al., 2008). The structural diversity of TLs emerges due to the interaction of mycolic acid and disaccharide trehalose units, creating different sizes, shapes, and fascinating chemical properties that can be used in various applications (Franzetti et al., 2010). The commercial arena of the TLs is expanding, owing to their pH, temperature, and chemical tolerance under different reaction conditions. Rhodococcus is well known for producing heat and pH-stable trehalolipid emulsions with remarkable stability between 20°C and 100°C, pH; 2%—10% and 5%—25% salt concentrations (Kundu et al., 2013). TLs show antimicrobial and emulsification properties for cosmetic, food, and bioremediation applications. One of the commercial formulations containing TLs is Lucentis (Luyckx & Baudouin, 2011). TLs are also used for the biodegradation and bioremediation of hydrophobic contaminants owing to their solubilization properties. They can enhance mobilization and bioavailability of complex water-insoluble substrates in agriculture and contaminated soils (Feofilova et al., 2014). These wonder molecules are pushing the boundaries of science and technology with their micelle-forming properties and antiviral and antimicrobial actions as a sustainable alternative to synthetic surfactants (Wu et al., 2015). Nonetheless, further research could open new avenues for their possible use in various innovative applications.

1.7.5 Lipopeptides

Lipopeptides (LPs) are chemically diverse and unique biosurfactants class produced by various soil microorganisms. They are synthesized by nonribosomal peptide synthetase (NRPS) platform. LPs are produced by different bacterial strains under the effect of varying carbon compounds (Baltz, 2014). The common types of lipopeptides include surfactin, iturin, fengycin, and lichenysin, which are produced through multimodular enzyme complexes associated with their assembly lines (Carolin et al., 2021). Surfactin is a lipopeptide biosurfactant which is studied extensively in structural, functional, and commercial aspects (Ongena & Jacques, 2008). There are several different variants of LP biosurfactants; however, surfactin and fengycin will be discussed in this chapter.

1.7.6 Surfactin

Surfactin production was first reported in 1968 by Arima et al. from *Bacillus subtilis* as a powerful bacterial surfactant. Later, in the following years, Kakinuma and colleagues elucidated its structure as a lipopeptide containing cyclic heptapeptide (hydrophilic portion) attached to a hydroxyl fatty acid chain (hydrophobic portion). Surfactin shows exceptional chemical diversity because of the immense variation in the amino acids and lipid composition. Surfactin is produced as an extracellular amphiphile of *B. subtilis*. Because of great structural diversity, surfactin is further classified into surfactin A (containing L-leucine in the structure), surfactin B (L-valine), and surfactin C (L-isoleucine) (Ahimou et al., 2000). These isoforms are produced due to variations in growth conditions and genetics of the bacterium (Davis et al., 1999). The surfactin biosynthesis is attributed to the involvement of the srfA operon in *B. subtilis* (Nakano et al., 1991). Surfactin synthesis follows a unique nonribosomal pathway that consists of a multimodular mega-enzyme system. This system catalyzes the synthesis of peptide products using proteinogenic and nonproteinogenic amino acids.

Surfactin has been produced at a laboratory scale using a high cell density fermentation approach up to 23.7 g/L by *B. subtilis* (Klausmann et al., 2021). Surfactin and related lipopeptides with surface and biological activities are commonly used in the environment, petroleum, agricultural, pharmaceutical, and health sectors as emulsification, dispersing, wetting, chelating, and antimicrobial agents (Falk, 2019; Zhu et al., 2021). Owing to their strong surface tension reduction (up to 27 mN/m) and interfacial performance, surfactin shows high potential for enhanced oil recovery, increased biodegradation of insoluble aromatics, and the removal of heavy metal contaminants from soil. Recently, surfactin has been used for making microemulsions for to improve drug delivery and target efficiency (Ohadi et al., 2020). In pursuit of its biomedical potential, the role of surfactin as a blood clot inhibitor, antifungal, antiviral, anticancer, antiinflammatory and antibacterial agent is also cited in the literature (Hisham et al., 2019). The antimicrobial activity of surfactin is evident from its ability to form pores in biological membranes, leading to bacterial cell death and virulence reduction (Chen et al., 2022).

1.7.7 Fengycin

Fengycin is an LP biosurfactant containing a heptapeptide moiety with a carbon chain length of 14–17 (Wu et al., 2019). Different types of fengycins have been identified based on variations in their molecular structures, including fengycin A and fengycin B (Khedher et al., 2021). Fengycin exhibits excellent antifungal activity because of its ability to change membrane permeability and unstabilize ergosterols, leading to fluid leakage and loss of physiological functions of the fungal cells (Sur et al., 2018). Fengycin is derived from different strains of Bacillus (Peñaranda-López et al., 2020). These molecules possess antimicrobial properties (Talón et al., 2019) and are well known for inhibiting biofilm and promoting plant growth by preventing disease progression (Wu et al., 2018). The antimicrobial action of fengycin against bacterial pathogens is not well documented and needs further scientific attention.

1.7.8 Polymeric microbial surfactants

Polymeric biosurfactants are high-molecular-weight BS consisting of emulsan, alasan, biodispersan, and polysaccharide−protein complexes (Mujumdar et al., 2019). These BS can be produced by various microorganisms, including *Acinetobacter* sp. The production of emulsan up to 20 g/L was reported by (Shabtai & Wang, 1990). Chemically, it is a polysaccharide−protein complex and the most widely studied bacterial bioemulsifier. The purified emulsan shows emulsification properties at a remarkably low concentration of 0.01%−0.001%. It improves the bioavailability of less-soluble substrates and promotes the biodegradation of hydrophobic materials by encasing them in their micelle core. In this process, the entrapped contaminants can be delivered back to the producing bacterium for efficient transport through the cell membrane. The emulsification action of emulsan is mainly associated with its fatty acid part, which acts as a binding site for different hydrophobic phases (Kaplan et al., 1987; Lukondeh et al., 2003).

Alasan is an anionic alanine-containing bioemulsifier produced by *Acinetobacter radioresistens*. Chemically, it is a complex of poly sugars, proteins, and alanine found in both cell-bound and free states. Through its emulsification properties, it can increase the biodegradation rate (up to 20 times) of a wide range of alkanes, aromatics, poly aromatics, and crude oil. The alasan production titer of 2.2 g/L is reported under submerged fermentation conditions. Various genes, such as *alanA*, *alanB*, and *alanC*, are involved in the biosynthesis of alasan (Navon-Venezia et al., 1995). It should be noted that the protein component of alasan is responsible for its emulsification properties. The 45 kDa protein component of the alasan contains folded hydrophobic regions that cause solubilization and emulsification of the substrates. Biodispersan is a polymeric biosurfactant composed of lipids, carbohydrates, and proteins. Its production has been reported from *Acinetobacter baumanii* with a yield up to 5 g/L (Hyder, 2015). Owing to their high surface-active properties, the applications of polymeric biosurfactants continue to increase in various fields (Ezzat et al., 2018). Recently, polymeric surfactants have been investigated for their promising role in stimulating plant growth, nutrient mobilization, and hydrocarbon bioavailability (Hafiane & Fatimi, 2022).

1.7.9 Particulate biosurfactants

Particulate biosurfactants are unique among microbial surfactants. They are produced as vesicles on the cell surface and form microemulsions. These vesicles mainly consist of proteins, phospholipids, and polysaccharides (Gayathiri et al., 2022), which show unique properties, including adherence to surfaces and hydrocarbons, contact angle changes, and salting out aggregation kinetics. Particulate BS can enclose drugs and other materials and thus are potential candidates for drug delivery applications (Dutta & Bhatnagar, 2022). One of the fascinating facts includes the ability of a microbial cell to act as a vesicle and function as a BS. The bacteria possessing these unique characteristics may be classified as BS producers or BS (Chen et al., 2020). They have been used in various industrial and agricultural applications (Farjami & Madadlou, 2019). The uptake of hydrocarbons via vesicular interactions, alkane uptake, transference, and solubilization of complex substrates are among the promising properties associated with particulate biosurfactants (Sharma & Sharma, 2020; Singh et al., 2020). These BS normally range from 20 to 50 mm in diameter and contain a membrane on the exterior for material entrapment (Susanti et al., 2021). The emerging data on bacterial cell structures and surface hydrophobicity provide insight into the structure–function relationship of particulate biosurfactants; however, many critical aspects of their biology and chemistry remain unclear (Kaur & Bakshi, 2020).

1.7.10 Fatty acids, phospholipids, and neutral lipids

Some bacterial and yeast strains are known to produce high quantities of fatty acid and phospholipid biosurfactants, such as corynomycolic acid, spiculisporic acid, and phosphotidylethanolamine (Appanna et al., 1995; Jang et al., 2002), during their growth on different hydrocarbon substrates (Fenibo et al., 2019; Santos et al., 2016). For example, the production of spiculisporic acid was carried out from *Talaromyces trachyspermus* using a fed-batch culture technique with a production rate of 6.6 g/L/day. The strain produced spiculisporic acid from different sugary substrates; however, glucose and sucrose were the most appropriate for optimum BS production (Moriwaki-Takano et al., 2021). *Acinetobacter* sp. produces phosphatidylethanolamine, a type of phospholipid, using hexadecane for its growth (Muthusamy et al., 2008). Because of environmental and public health safety and excellent properties, tricarboxylic-type surfactants have gained considerable attention in recent years. Particularly, the application of spiculisporic acid has been suggested for the preparation of innovative emulsions, bioactive materials, and superfine microcapsules (Moriwaki-Takano et al., 2021).

1.8 Conclusion and future prospects

With an increasing number of patents and their inclusion in commercial formulations, biosurfactants are quickly becoming one of the most important products of the modern biotechnological industry. Fascinating details of the structure–function relationship continue to emerge, enabling a significant growth of these molecules in the international market. Structurally, biosurfactants are diverse, from simple glycolipids to the most complex particulate biosurfactants. The remarkable properties, chemical diversity, and biological action of biosurfactants is expanding the boundaries of surfactant sciences and their technological applications. On the other side, the market potential of biosurfactants is not fully realized owing to low production, a higher process cost, a laborious purification process, and limited knowledge of their phase behavior. The transition from lab to industry needs extensive research on their production and purification processes. Applying innovative BS production and purification techniques could promote their further growth at an industrial scale in the future.

References

Abdel-Mawgoud, A. M., & Stephanopoulos, G. (2018). Simple glycolipids of microbes: Chemistry, biological activity and metabolic engineering. *Synthetic and Systems Biotechnology (Reading, Mass.)*, *3*(1), 3–19. Available from https://doi.org/10.1016/j.synbio.2017.12.001.

Adetunji, C. O., Adejumo, I. O., Afolabi, I. S., Adetunji, J. B., & Ajisejiri, E. S. (2018). Prolonging the shelf life of 'Agege Sweet' orange with chitosan–rhamnolipid coating. *Horticulture Environment and Biotechnology*, *59*(5), 687–697. Available from https://doi.org/10.1007/s13580-018-0083-2.

Adu, S., Twigg, M., Naughton, P., Marchant, R., & Banat, I. M. (2022). *Microbial sophorolipids: Natural surfactants for topical chemotherapy and cosmetic applications*. Cardiff, United Kingdom: Early Career Scientist Research Symposium.

Ahimou, F., Jacques, P., & Deleu, M. (2000). *Surfactin and iturin A effects on* Bacillus subtilis *surface hydrophobicity, Enzyme and microbial technology* (Vol. 27, pp. 749–754). Elsevier Science Inc Issue 10. Available from https://doi.org/10.1016/S0141-0229(00)00295-7.

Appanna, V. D., Finn, H., & Pierre, M. S. (1995). Exocellular phosphatidylethanolamine production and multiple-metal tolerance in *Pseudomonas fluorescens*. *FEMS Microbiology Letters*, *131*(1), 53–56. Available from https://doi.org/10.1016/0378-1097(95)00234-V.

Baccile, N., Seyrig, C., Poirier, A., Alonso-De Castro, S., Roelants, S. L. K. W., & Abel, S. (2021). Self-assembly, interfacial properties, interactions with macromolecules and molecular modelling and simulation of microbial bio-based amphiphiles (biosurfactants). A tutorial review. *Green Chemistry*, *23*(11), 3842–3944. Available from https://doi.org/10.1039/d1gc00097g.

Bai, G., Brusseau, M. L., & Miller, R. M. (1998). Influence of cation type, ionic strength, and pH on solubilization and mobilization of residual hydrocarbon by a biosurfactant. *Journal of Contaminant Hydrology*, *30*(3–4), 265–279. Available from https://doi.org/10.1016/S0169-7722(97)00043-0.

Baltz, R. H. (2014). Combinatorial biosynthesis of cyclic lipopeptide antibiotics: A model for synthetic biology to accelerate the evolution of secondary metabolite biosynthetic pathways. *ACS Synthetic Biology*, *3*(10), 748–758. Available from https://doi.org/10.1021/sb3000673.

Bhadani, A., Kafle, A., Ogura, T., Akamatsu, M., Sakai, K., Sakai, H., & Abe, M. (2020). Current perspective of sustainable surfactants based on renewable building blocks. *Current Opinion in Colloid and Interface Science*, *45*, 124–135. Available from https://doi.org/10.1016/j.cocis.2020.01.002.

Borsanyiova, M., Patil, A., Mukherji, R., Prabhune, A., & Bopegamage, S. (2016). Biological activity of sophorolipids and their possible use as antiviral agents. *Folia Microbiologica*, *61*(1), 85–89. Available from https://doi.org/10.1007/s12223-015-0413-z.

Carolin C, F., Kumar, P. S., & Ngueagni, P. T. (2021). A review on new aspects of lipopeptide biosurfactant: Types, production, properties and its application in the bioremediation process. *Journal of Hazardous Materials*, *407*, 124827. Available from https://doi.org/10.1016/j.jhazmat.2020.124827.

Casas, J. A., & García-Ochoa, F. (1999). Sophorolipid production by Candida bombicola: Medium composition and culture methods. *Journal of Bioscience and Bioengineering*, *88*(5), 488–494. Available from https://doi.org/10.1016/S1389-1723(00)87664-1.

Chen, A., Wang, F., Zhou, Y., & Xu, J. H. (2020). In situ measurements of interactions between switchable surface-active colloid particles using optical tweezers. *Langmuir: The ACS Journal of Surfaces and Colloids*, *36*(17), 4664–4670. Available from https://doi.org/10.1021/acs.langmuir.0c00398.

Chen, X., Chen, X., Zhu, L., Liu, W., & Jiang, L. (2022). Programming an orthogonal self-assembling protein cascade based on reactive peptide-protein pairs for in vitro enzymatic trehalose production. *Journal of Agricultural and Food Chemistry*, *70*(15), 4690–4700. Available from https://doi.org/10.1021/acs.jafc.2c01118.

Crouzet, J., Arguelles-Arias, A., Dhondt-Cordelier, S., Cordelier, S., Pršić, J., Hoff, G., Mazeyrat-Gourbeyre, F., Baillieul, F., Clément, C., Ongena, M., & Dorey, S. (2020). Biosurfactants in plant protection against diseases: Rhamnolipids and lipopeptides case study. *Frontiers in Bioengineering and Biotechnology*, *8*, 1014. Available from https://doi.org/10.3389/fbioe.2020.01014.

Datta, P., Tiwari, P., & Pandey, L. M. (2018). Isolation and characterization of biosurfactant producing and oil degrading *Bacillus subtilis* MG495086 from formation water of Assam oil reservoir and its suitability for enhanced oil recovery. *Bioresource Technology*, *270*, 439–448. Available from https://doi.org/10.1016/j.biortech.2018.09.047.

Daverey, A., Dutta, K., Joshi, S., & Daverey, A. (2021). Sophorolipid: A glycolipid biosurfactant as a potential therapeutic agent against COVID-19. *Bioengineered*, *12*(2), 9550–9560. Available from https://doi.org/10.1080/21655979.2021.1997261.

Davis, D. A., Lynch, H. C., & Varley, J. (1999). The production of surfactin in batch culture by *Bacillus subtilis* ATCC 21332 is strongly influenced by the conditions of nitrogen metabolism. *Enzyme and Microbial Technology*, *25*(3–5), 322–329. Available from https://doi.org/10.1016/S0141-0229(99)00048-4.

de O Caretta, T., I Silveira, V. A., Andrade, G., Macedo, F., & P C Celligoi, M. A. (2022). Antimicrobial activity of sophorolipids produced by *Starmerella bombicola* against phytopathogens from cherry tomato. *Journal of the Science of Food and Agriculture*, *102*(3), 1245–1254. Available from https://doi.org/10.1002/jsfa.11462.

de Paula, F., Vieira, N. V., da Silva, G. F., Delforno, T. P., & Duarte, I. C. S. (2022). A comparison of microbial communities of mango and orange residues for bioprospecting of biosurfactant producers. *Ecologies*, *3*(2), 120–130. Available from https://doi.org/10.3390/ecologies3020010.

de Sousa-D'Auria, C., Constantinesco, F., Bayan, N., Constant, P., Tropis, M., Daffé, M., Graille, M., & Houssin, C. (2022). Cg1246, a new player in mycolic acid biosynthesis in *Corynebacterium glutamicum*. *Microbiology (United Kingdom)*, *168*, 4. Available from https://doi.org/10.1099/mic.0.001171.

Dikit, P., Maneerat, S., & Saimmai, A. (2019). Production and application of biosurfactant produced by *Agrobacterium rubi* L5 isolated from mangrove sediments. *Applied Mechanics and Materials*, *886*, 98–104. Available from https://doi.org/10.4028/http://www.scientific.net/amm.886.98.

Dobler, L., Ferraz, H. C., Araujo de Castilho, L. V., Sangenito, L. S., Pasqualino, I. P., Souza dos Santos, A. L., Neves, B. C., Oliveira, R. R., Guimarães Freire, D. M., & Almeida, R. V. (2020). Environmentally friendly rhamnolipid production for petroleum remediation. *Chemosphere*, *252*, 126349. Available from https://doi.org/10.1016/j.chemosphere.2020.126349.

Drakontis, C. E., & Amin, S. (2020). Biosurfactants: Formulations, properties, and applications. *Current Opinion in Colloid and Interface Science*, *48*, 77–90. Available from https://doi.org/10.1016/j.cocis.2020.03.013.

Drobeck, H. P. (2019). *Current topics on the toxicity of cationic surfactants* (pp. 61–94). Informa UK Limited. Available from https://doi.org/10.1201/9780429270376-5.

Durval, I. J. B., Resende, A. H. M., Figueiredo, M. A., Luna, J. M., Rufino, R. D., & Sarubbo, L. A. (2019). Studies on biosurfactants Produced using *Bacillus cereus* isolated from seawater with biotechnological potential for marine oil-spill bioremediation. *Journal of Surfactants and Detergents*, *22*(2), 349–363. Available from https://doi.org/10.1002/jsde.12218.

Dutta, N., & Bhatnagar, A. (2022). *Biosurfactants current trends and applications* (pp. 241–252). Informa UK Limited. Available from https://doi.org/10.1201/9781003260165-14.

Eslami, P., Hajfarajollah, H., & Bazsefidpar, S. (2020). Recent advancements in the production of rhamnolipid biosurfactants by *Pseudomonas aeruginosa*. *RSC Advances*, *10*(56), 34014–34032. Available from https://doi.org/10.1039/d0ra04953k.

Eswari, J. S., Anand, M., & Venkateswarlu, C. (2013). Optimum culture medium composition for rhamnolipid production by *Pseudomonas aeruginosa* AT10 using a novel multi-objective optimization method. *Journal of Chemical Technology and Biotechnology*, *88*(2), 271–279. Available from https://doi.org/10.1002/jctb.3825.

Ezzat, A. O., Atta, A. M., Al-Lohedan, H. A., & Hashem, A. I. (2018). Synthesis and application of new surface active poly (ionic liquids) based on 1,3-dialkylimidazolium as demulsifiers for heavy petroleum crude oil emulsions. *Journal of Molecular Liquids*, *251*, 201–211. Available from https://doi.org/10.1016/j.molliq.2017.12.081.

Falk, N. A. (2019). Surfactants as antimicrobials: A brief overview of microbial interfacial chemistry and surfactant antimicrobial activity. *Journal of Surfactants and Detergents*, *22*(5), 1119–1127. Available from https://doi.org/10.1002/jsde.12293.

Farjami, T., & Madadlou, A. (2019). An overview on preparation of emulsion-filled gels and emulsion particulate gels. *Trends in Food Science and Technology*, *86*, 85–94. Available from https://doi.org/10.1016/j.tifs.2019.02.043.

Fenibo, E. O., Douglas, S. I., & Stanley, H. O. (2019). A review on microbial surfactants: Production, classifications, properties and characterization. *Journal of Advances in Microbiology*, *18*(3), 1–22. Available from https://doi.org/10.9734/jamb/2019/v18i330170.

Feofilova, E. P., Usov, A. I., Mysyakina, I. S., & Kochkina, G. A. (2014). Trehalose: Chemical structure, biological functions, and practical application. *Microbiology (Reading, England)*, *83*(3), 184–194. Available from https://doi.org/10.1134/s0026261714020064.

Fiechter, A. (1992). Biosurfactants: Moving towards industrial application. *Trends in Biotechnology*, *10*(C), 208–217. Available from https://doi.org/10.1016/0167-7799(92)90215-H.

Franzetti, A., Gandolfi, I., Bestetti, G., Smyth, T. J. P., & Banat, I. M. (2010). Production and applications of trehalose lipid biosurfactants. *European Journal of Lipid Science and Technology*, *112*(6), 617–627. Available from https://doi.org/10.1002/ejlt.200900162.

Gayathiri, E., Prakash, P., Karmegam, N., Varjani, S., Awasthi, M. K., & Ravindran, B. (2022). Biosurfactants: Potential and eco-friendly material for sustainable agriculture and environmental safety—A review. *Agronomy*, *12*(3), 662. Available from https://doi.org/10.3390/agronomy12030662.

Gudiña, E. J., & Rodrigues, L. R. (2019). *Research and production of biosurfactants for the food industry* (pp. 125–143). Wiley. Available from https://doi.org/10.1002/9781119434436.ch6.

Hafiane, F. Z., & Fatimi, A. (2022). An emulsion-based formulation for increasing the resistance of plants to salinity stress: US20160302416A1 patent evaluation. *Environmental Sciences Proceedings*, *16*(1), 4.

Hauser, G., & Karnovsky, M. L. (1957). Rhamnose and rhamnolipid biosynthesis by *Pseudomonas aeruginosa*. *Journal of Biological Chemistry*, *224*(1), 91–105. Available from https://doi.org/10.1016/s0021-9258(18)65013-6.

Hentati, D., Chebbi, A., Hadrich, F., Frikha, I., Rabanal, F., Sayadi, S., Manresa, A., & Chamkha, M. (2019). Production, characterization and biotechnological potential of lipopeptide biosurfactants from a novel marine *Bacillus stratosphericus* strain FLU5. *Ecotoxicology and Environmental Safety*, *167*, 441–449. Available from https://doi.org/10.1016/j.ecoenv.2018.10.036.

Hisham, N. H., Ibrahim, M. F., Ramli, N., & Abd-Aziz, S. (2019). Production of biosurfactant produced from used cooking oil by *Bacillus* sp. HIP3 for heavy metals removal. *Molecules (Basel, Switzerland)*, *24*(14), 617. Available from https://doi.org/10.3390/molecules24142617.

Hou, L., Debru, A., Chen, Q., Bao, Q., & Li, K. (2019). AmrZ regulates swarming motility through cyclic di-GMP-dependent motility inhibition and controlling Pel Polysaccharide production in *Pseudomonas aeruginosa* PA14. *Frontiers in Microbiology*, *10*, 1847. Available from https://doi.org/10.3389/fmicb.2019.01847.

Hrůzová, K., Patel, A., Masák, J., Maťátková, O., Rova, U., Christakopoulos, P., & Matsakas, L. (2020). A novel approach for the production of green biosurfactant from *Pseudomonas aeruginosa* using renewable forest biomass. *Science of the Total Environment*, *711*, 135099. Available from https://doi.org/10.1016/j.scitotenv.2019.135099.

Hyder, N. H. (2015). Production, characterization and antimicrobial activity of a bioemulsifier produced by *Acinetobacter baumanii* AC5 utilizing edible oils. *Iraqi Journal of Biotechnology*, *14*, 2.

Jahan, R., Bodratti, A. M., Tsianou, M., & Alexandridis, P. (2020). Biosurfactants, natural alternatives to synthetic surfactants: Physicochemical properties and applications. *Advances in Colloid and Interface Science*, *275*, 102061. Available from https://doi.org/10.1016/j.cis.2019.102061.

Janek, T., Krasowska, A., Czyżnikowska, Ż., & èukaszewicz, M. (2018). Trehalose lipid biosurfactant reduces adhesion of microbial pathogens to polystyrene and silicone surfaces: An experimental and computational approach. *Frontiers in Microbiology*, *9*, 2441. Available from https://doi.org/10.3389/fmicb.2018.02441.

Jang, K. H., Park, Y. I., & Britz, M. L. (2002). Quantitative analysis of corynomycolic acids in fermentation broth. *Journal of Microbiology and Biotechnology*, *12*(5), 793–800.

Jiménez-Peñalver, P., Koh, A., Gross, R., Gea, T., & Font, X. (2020). Biosurfactants from waste: Structures and interfacial properties of sophorolipids produced from a residual oil cake. *Journal of Surfactants and Detergents*, *23*(2), 481–486. Available from https://doi.org/10.1002/jsde.12366.

Kaczorek, E., Pacholak, A., Zdarta, A., & Smułek, W. (2018). The impact of biosurfactants on microbial cell properties leading to hydrocarbon bioavailability increase. *Colloids and Interfaces*, *2*(3), 35. Available from https://doi.org/10.3390/colloids2030035.

Kaga, H., Nakamura, A., Orita, M., Endo, K., Akamatsu, M., Sakai, K., & Sakai, H. (2022). Removal of a model biofilm by sophorolipid solutions: A QCM-D Study. *Journal of Oleo Science*, *71*(5), 663–670. Available from https://doi.org/10.5650/jos.ess21360.

Kaplan, N., Zosim, Z., & Rosenberg, E. (1987). Reconstitution of emulsifying activity of *Acinetobacter calcoaceticus* BD4 emulsan by using pure polysaccharide and protein. *Applied and Environmental Microbiology*, *53*(2), 440–446. Available from https://doi.org/10.1128/aem.53.2.440-446.1987.

Kareem, M. K. A. (2020). A review of biosurfactants (Glycolipids): The characteristics, composition and application. *International Journal of Psychosocial Rehabilitation*, *24*(5), 3795–3807. Available from https://doi.org/10.37200/ijpr/v24i5/pr202088.

Kashif, A., Rehman, R., Fuwad, A., Shahid, M. K., Dayarathne, H. N. P., Jamal, A., Aftab, M. N., Mainali, B., & Choi, Y. (2022). Current advances in the classification, production, properties and applications of microbial biosurfactants – A critical review. *Advances in Colloid and Interface Science*, *306*, 102718. Available from https://doi.org/10.1016/j.cis.2022.102718.

Kaur, R., & Bakshi, M. S. (2020). Mechanistic aspects of simultaneous extraction of silver and gold nanoparticles across aqueous–organic interfaces by surface active iron oxide nanoparticles. *Langmuir: The ACS Journal of Surfaces and Colloids*, *36*(26), 7505–7516. Available from https://doi.org/10.1021/acs.langmuir.0c01102.

Khedher, S., Mejdoub-Trabelsi, B., & Tounsi, S. (2021). Biological potential of *Bacillus subtilis* V26 for the control of Fusarium wilt and tuber dry rot on potato caused by Fusarium species and the promotion of plant growth. *Biological Control*, *152*, 104444. Available from https://doi.org/10.1016/j.biocontrol.2020.104444.

Kim, B. S., Lee, J. Y., & Hwang, B. K. (2000). In vivo control andin vitro antifungal activity of rhamnolipid B, a glycolipid antibiotic, against *Phytophthora capsici* and *Colletotrichum orbiculare*. *Pest Management Science*, *56*(12), 1029–1035, https://doi.org/10.1002/1526-4998(200012)56:12 < 1029::aid-ps238 > 3.0.co;2-q.

Kirschbaum, C., Greis, K., Mucha, E., Kain, L., Deng, S., Zappe, A., Gewinner, S., Schöllkopf, W., von Helden, G., Meijer, G., Savage, P. B., Marianski, M., Teyton, L., & Pagel, K. (2021). Unravelling the structural complexity of glycolipids with cryogenic infrared spectroscopy. *Nature Communications*, *12*(1), 1. Available from https://doi.org/10.1038/s41467-021-21480-1.

Klausmann, P., Hennemann, K., Hoffmann, M., Treinen, C., Aschern, M., Lilge, L., Morabbi Heravi, K., Henkel, M., & Hausmann, R. (2021). *Bacillus subtilis* high cell density fermentation using a sporulation-deficient strain for the production of surfactin. *Applied Microbiology and Biotechnology*, *105*(10), 4141–4151. Available from https://doi.org/10.1007/s00253-021-11330-x.

Krishnan, N., Velramar, B., & Velu, R. K. (2019). Investigation of antifungal activity of surfactin against mycotoxigenic phytopathogenic fungus *Fusarium moniliforme* and its impact in seed germination and mycotoxicosis. *Pesticide Biochemistry and Physiology*, *155*, 101–107. Available from https://doi.org/10.1016/j.pestbp.2019.01.010.

Kumar, R., & Das, A. J. (2018). *Rhamnolipid biosurfactants and their properties* (pp. 1–13). Springer Science and Business Media LLC. Available from https://doi.org/10.1007/978-981-13-1289-2_1.

Kundu, D., Hazra, C., Dandi, N., & Chaudhari, A. (2013). Biodegradation of 4-nitrotoluene with biosurfactant production *by Rhodococcus pyridinivorans* NT2: Metabolic pathway, cell surface properties and toxicological characterization. *Biodegradation*, *24*(6), 775–793. Available from https://doi.org/10.1007/s10532-013-9627-4.

Lang, S., Katsiwela, E., & Wagner, F. (1989). Antimicrobial effects of biosurfactants. *Lipid/Fett*, *91*(9), 363–366. Available from https://doi.org/10.1002/lipi.19890910908.

Li, Z., Zhang, Y., Lin, J., Wang, W., & Li, S. (2019). High-yield di-rhamnolipid production by *Pseudomonas aeruginosa* YM4 and its potential application in MEOR. *Molecules (Basel, Switzerland)*, *24*(7), 1433. Available from https://doi.org/10.3390/molecules24071433.

Liu, G., Zhong, H., Yang, X., Liu, Y., Shao, B., & Liu, Z. (2018). Advances in applications of rhamnolipids biosurfactant in environmental remediation: A review. *Biotechnology and Bioengineering*, *115*(4), 796–814. Available from https://doi.org/10.1002/bit.26517.

Lombardo, D., Kiselev, M. A., Magazù, S., & Calandra, P. (2015). Amphiphiles self-assembly: basic concepts and future perspectives of supramolecular approaches. *Advances in Condensed Matter Physics*, *2015*.

Lukondeh, T., Ashbolt, N. J., & Rogers, P. L. (2003). Evaluation of *Kluyveromyces marxianus* FII 510700 grown on a lactose-based medium as a source of a natural bioemulsifier. *Journal of Industrial Microbiology and Biotechnology*, *30*(12), 715–720. Available from https://doi.org/10.1007/s10295-003-0105-6.

Luyckx, J., & Baudouin, C. (2011). Trehalose: An intriguing disaccharide with potential for medical application in ophthalmology. *Clinical Ophthalmology*, *5*(1), 577–581. Available from https://doi.org/10.2147/OPTH.S18827.

Ma, Y., Chen, S., Liao, W., Zhang, L., Liu, J., & Gao, Y. (2020). Formation, physicochemical stability, and redispersibility of curcumin-loaded rhamnolipid nanoparticles using the pH-driven method. *Journal of Agricultural and Food Chemistry*, *68*(27), 7103–7111. Available from https://doi.org/10.1021/acs.jafc.0c01326.

Manga, E. B., Celik, P. A., Cabuk, A., & Banat, I. M. (2021). Biosurfactants: Opportunities for the development of a sustainable future. *Current Opinion in Colloid and Interface Science*, *56*, 1514. Available from https://doi.org/10.1016/j.cocis.2021.101514.

Markande, A. R., Patel, D., & Varjani, S. (2021). A review on biosurfactants: Properties, applications and current developments. *Bioresource Technology*, *330*, 124963. Available from https://doi.org/10.1016/j.biortech.2021.124963.

Miceli, R. T., Corr, D. T., Barroso, M., Dogra, N., & Gross, R. A. (2022). Sophorolipids: Anti-cancer activities and mechanisms. *Bioorganic & Medicinal Chemistry*, *65*, 116787. Available from https://doi.org/10.1016/j.bmc.2022.116787.

Mohamed, S. K., Asif, M., Nazari, M. V., Baharetha, H. M., Mahmood, S., Yatim, A. R. M., Majid, A. S. A., & Majid, A. M. S. A. (2019). Antiangiogenic activity of sophorolipids extracted from refined bleached deodorized palm olein. *Indian Journal of Pharmacology*, *51*(1), 45–54. Available from https://doi.org/10.4103/ijp.IJP_312_18.

Mohanty, S. S., Koul, Y., Varjani, S., Pandey, A., Ngo, H. H., Chang, J. S., Wong, J. W. C., & Bui, X. T. (2021). A critical review on various feedstocks as sustainable substrates for biosurfactants production: A way towards cleaner production. *Microbial Cell Factories*, *20*(1), 3. Available from https://doi.org/10.1186/s12934-021-01613-3.

Moriwaki-Takano, M., Asada, C., & Nakamura, Y. (2021). Production of spiculisporic acid by *Talaromyces trachyspermus* in fed-batch bioreactor culture. *Bioresources and Bioprocessing*, *8*(1), 1. Available from https://doi.org/10.1186/s40643-021-00414-1.

Morya, V. K., Ahn, C., Jeon, S., & Kim, E. K. (2013). Medicinal and cosmetic potentials of sophorolipids. *Mini-Reviews in Medicinal Chemistry*, *13*(12), 1761–1768. Available from https://doi.org/10.2174/13895575113139990002.

Mujumdar, S., Joshi, P., & Karve, N. (2019). Production, characterization, and applications of bioemulsifiers (BE) and biosurfactants (BS) produced by Acinetobacter spp.: A review. *Journal of Basic Microbiology*, *59*(3), 277–287. Available from https://doi.org/10.1002/jobm.201800364.

Mutalik, S. R., Vaidya, B. K., Joshi, R. M., Desai, K. M., & Nene, S. N. (2008). Use of response surface optimization for the production of biosurfactant from Rhodococcus spp. MTCC 2574. *Bioresource Technology*, *99*(16), 7875–7880. Available from https://doi.org/10.1016/j.biortech.2008.02.027.

Muthusamy, K., Gopalakrishnan, S., Ravi, T. K., & Sivachidambaram, P. (2008). Biosurfactants: Properties, commercial production and application. *Current Science*, *94*(6), 736–747. Available from http://www.ias.ac.in/currsci/mar252008/736.pdf.

Nakano, M. M., Magnuson, R., Myers, A., Curry, J., Grossman, A. D., & Zuber, P. (1991). srfA is an operon required for surfactin production, competence development, and efficient sporulation in *Bacillus subtilis*. *Journal of Bacteriology*, *173*(5), 1770–1778. Available from https://doi.org/10.1128/jb.173.5.1770-1778.1991.

Nasiri, M. A., & Biria, D. (2020). Extraction of the indigenous crude oil dissolved biosurfactants and their potential in enhanced oil recovery. *Colloids and Surfaces A: Physicochemical and Engineering Aspects*, *603*, 125216. Available from https://doi.org/10.1016/j.colsurfa.2020.125216.

Naughton, P. J., Marchant, R., Naughton, V., & Banat, I. M. (2019). Microbial biosurfactants: Current trends and applications in agricultural and biomedical industries. *Journal of Applied Microbiology*, *127*(1), 12–28. Available from https://doi.org/10.1111/jam.14243.

Navon-Venezia, S., Zosim, Z., Gottlieb, A., Legmann, R., Carmeli, S., Ron, E. Z., & Rosenberg, E. (1995). Alasan, a new bioemulsifier from *Acinetobacter radioresistens*. *Applied and Environmental Microbiology*, *61*(9), 3240–3244. Available from https://doi.org/10.1128/aem.61.9.3240-3244.1995.

Niaz, T., Shabbir, S., Noor, T., & Imran, M. (2019). Antimicrobial and antibiofilm potential of bacteriocin loaded nano-vesicles functionalized with rhamnolipids against foodborne pathogens. *LWT, 116*, 108583. Available from https://doi.org/10.1016/j.lwt.2019.108583.

Nitschke, M., & Silva, S. S. e (2018). Recent food applications of microbial surfactants. *Critical Reviews in Food Science and Nutrition, 58*(4), 631–638. Available from https://doi.org/10.1080/10408398.2016.1208635.

Nitschke, M., Costa, S. G. V. A. O., & Contiero, J. (2010). Structure and applications of a rhamnolipid surfactant produced in soybean oil waste. *Applied Biochemistry and Biotechnology, 160*(7), 2066–2074. Available from https://doi.org/10.1007/s12010-009-8707-8.

Nuñez, A., Ashby, R., Foglia, T. A., & Solaiman, D. K. Y. (2001). Analysis and characterization of sophorolipids by liquid chromatography with atmospheric pressure chemical ionization. *Chromatographia, 53*(11–12), 673–677. Available from https://doi.org/10.1007/BF02493019.

Ohadi, M., Shahravan, A., Dehghannoudeh, N., Eslaminejad, T., Banat, I. M., & Dehghannoudeh, G. (2020). Potential use of microbial surfactant in microemulsion drug delivery system: A systematic review. *Drug Design, Development and Therapy, 14*, 541–550. Available from https://doi.org/10.2147/DDDT.S232325.

Oliva, A., Teruel, J. A., Aranda, F. J., & Ortiz, A. (2020). Effect of a dirhamnolipid biosurfactant on the structure and phase behaviour of dimyristoylphosphatidylserine model membranes. *Colloids and Surfaces B: Biointerfaces, 185*, 576. Available from https://doi.org/10.1016/j.colsurfb.2019.110576.

Ongena, M., & Jacques, P. (2008). Bacillus lipopeptides: Versatile weapons for plant disease biocontrol. *Trends in Microbiology, 16*(3), 115–125. Available from https://doi.org/10.1016/j.tim.2007.12.009.

Osman, M. S., Ibrahim, Z., Japper-Jaafar, A., & Shahir, S. (2019). Biosurfactants and its prospective application in the petroleum industry. *Journal of Sustainability Science and Management, 14*(3), 125–140. Available from http://jssm.umt.edu.my/.

Peñaranda-López, A. L., Brito-de la Fuente, E., & Torrestiana-Sánchez, B. (2020). Fractionation of hydrolysates from concentrated lecithin free egg yolk protein dispersions by ultrafiltration. *Food and Bioproducts Processing, 123*, 209–216. Available from https://doi.org/10.1016/j.fbp.2020.07.001.

Penfold, J., Chen, M., Thomas, R. K., Dong, C., Smyth, T. J. P., Perfumo, A., Marchant, R., Banat, I. M., Stevenson, P., Parry, A., Tucker, I., & Grillo, I. (2011). Solution self-assembly of the sophorolipid biosurfactant and its mixture with anionic surfactant sodium dodecyl benzene sulfonate. *Langmuir: The ACS Journal of Surfaces and Colloids, 27*(14), 8867–8877. Available from https://doi.org/10.1021/la201661y.

Philippova, O. E., & Molchanov, V. S. (2019). Enhanced rheological properties and performance of viscoelastic surfactant fluids with embedded nanoparticles. *Current Opinion in Colloid and Interface Science, 43*, 52–62. Available from https://doi.org/10.1016/j.cocis.2019.02.009.

Phulpoto, I. A., Yu, Z., Hu, B., Wang, Y., Ndayisenga, F., Li, J., Liang, H., & Qazi, M. A. (2020). Production and characterization of surfactin-like biosurfactant produced by novel strain Bacillus nealsonii S2MT and it's potential for oil contaminated soil remediation. *Microbial Cell Factories, 19*(1), 4. Available from https://doi.org/10.1186/s12934-020-01402-4.

Pirog, T. P., Lutsay, D. A., Kliuchka, L. V., & Beregova, K. A. (2019). Antimicrobial activity of surfactants of microbial origin. *Biotechnologia Acta, 12*(1), 39–57.

Platel, R., Lucau-Danila, A., Baltenweck, R., Maia-Grondard, A., Chaveriat, L., Magnin-Robert, M., Randoux, B., Trapet, P., Halama, P., Martin, P., Hilbert, J. L., Höfte, M., Hugueney, P., Reignault, P., & Siah, A. (2022). Bioinspired rhamnolipid protects wheat against *Zymoseptoria tritici* through mainly direct antifungal activity and without major impact on leaf physiology. *Frontiers in Plant Science, 13*, 878272. Available from https://doi.org/10.3389/fpls.2022.878272.

Pontes, C., Alves, M., Santos, C., Ribeiro, M. H., Gonçalves, L., Bettencourt, A. F., & Ribeiro, I. A. C. (2016). Can Sophorolipids prevent biofilm formation on silicone catheter tubes? *International Journal of Pharmaceutics, 513*(1–2), 697–708. Available from https://doi.org/10.1016/j.ijpharm.2016.09.074.

Prasad, R. V., Kumar, R. A., Sharma, D., Sharma, A., & Nagarajan, S. (2021). Sophorolipids and rhamnolipids as a biosurfactant: Synthesis and applications. In *Green sustainable process for chemical and environmental engineering and science: Microbially-derived biosurfactants for improving sustainability in industry* (pp. 423–472). Elsevier. Available from https://doi.org/10.1016/B978-0-12-823380-1.00014-9.

Qi, L., & Christopher, G. F. (2019). Role of flagella, Type IV pili, biosurfactants, and extracellular polymeric substance polysaccharides on the formation of pellicles by *Pseudomonas aeruginosa*. *Langmuir: The ACS Journal of Surfaces and Colloids*, *35*(15), 5294–5304. Available from https://doi.org/10.1021/acs.langmuir.9b00271.

Rajkhowa, S., & Sarma, J. (2022). *Biosurfactant: An alternative towards sustainability. In innovative bio-based technologies for environmental remediation* (pp. 377–402). CRC Press. Available from https://doi.org/10.1201/9781003004684-20.

Ramalingam, V., Varunkumar, K., Ravikumar, V., & Rajaram, R. (2019). Production and structure elucidation of anticancer potential surfactin from marine actinomycete *Micromonospora marina*. *Process Biochemistry*, *78*, 169–177. Available from https://doi.org/10.1016/j.procbio.2019.01.002.

Rau, U., Hammen, S., Heckmann, R., Wray, V., & Lang, S. (2001). Sophorolipids: A source for novel compounds. *Industrial Crops and Products*, *13*(2), 85–92. Available from https://doi.org/10.1016/S0926-6690(00)00055-8.

Rehman, R., Ali, M. I., Ali, N., Badshah, M., Iqbal, M., Jamal, A., & Huang, Z. (2021). Crude oil biodegradation potential of biosurfactant-producing *Pseudomonas aeruginosa* and Meyerozyma sp. *Journal of Hazardous Materials*, *418*, 126276. Available from https://doi.org/10.1016/j.jhazmat.2021.126276.

Retnaningrum, E., & Wilopo, W. (2018). Production and characterization of biosurfactants produced by *Pseudomonas aeruginosa* B031 isolated from a hydrocarbon phytoremediation field. *Biotropia*, *25*(2), 130–139. Available from https://doi.org/10.11598/btb.2018.25.2.808.

Ribeiro, B. G., Guerra, J. M. C., & Sarubbo, L. A. (2020). Biosurfactants: Production and application prospects in the food industry. *Biotechnology Progress*, *36*(5), 3030. Available from https://doi.org/10.1002/btpr.3030.

Ruhal, R., & Choudhury, B. (2012). Improved trehalose production from biodiesel waste using parent and osmotically sensitive mutant of *Propionibacterium freudenreichii* subsp. shermanii under aerobic conditions. *Journal of Industrial Microbiology and Biotechnology*, *39*(8), 1153–1160. Available from https://doi.org/10.1007/s10295-012-1124-y.

Santos, D. K. F., Rufino, R. D., Luna, J. M., Santos, V. A., & Sarubbo, L. A. (2016). Biosurfactants: Multifunctional biomolecules of the 21st century. *International Journal of Molecular Sciences*, *17*(3). Available from https://doi.org/10.3390/ijms17030401.

Sen, S., Banerjee, S., Ghosh, A., & Sarkar, K. (2022). Selected lipopeptides of *Bacillus* as plausible inhibitors of SARS-COV-2 chymotrypsin-like protease 3CLPro: An in-silico analysis. Authorea.

Shaban, S. M., Kang, J., & Kim, D. H. (2020). Surfactants: Recent advances and their applications. Composites. *Communications*, *22*, 537. Available from https://doi.org/10.1016/j.coco.2020.100537.

Shabtai, Y., & Wang, D. I. C. (1990). Production of emulsan in a fermentation process using soybean oil (SBO) in a carbon–nitrogen coordinated feed. *Biotechnology and Bioengineering*, *35*(8), 753–765. Available from https://doi.org/10.1002/bit.260350802.

Shakeri, F., Babavalian, H., Amoozegar, M. A., Ahmadzadeh, Z., Zuhuriyanizadi, S., & Afsharian, M. P. (2020). Production and application of biosurfactants in biotechnology. *Biointerface Research in Applied Chemistry*, *11*(3), 10446–10460. Available from https://doi.org/10.33263/BRIAC113.1044610460.

Sharma, P., & Sharma, N. (2020). Microbial biosurfactants-an ecofriendly boon to industries for green revolution. *Recent Patents on Biotechnology*, *14*(3), 169–183. Available from https://doi.org/10.2174/1872208313666191212094628.

Shu, Q., Lou, H., Wei, T., Liu, X., & Chen, Q. (2021). Contributions of glycolipid biosurfactants and glycolipid-modified materials to antimicrobial strategy: A review. *Pharmaceutics*, *13*(2), 227. Available from https://doi.org/10.3390/pharmaceutics13020227.

Sidrim, J. J. C., Ocadaque, C. J., Amando, B. R., De M Guedes, G. M., Costa, C. L., Brilhante, R. S. N., A Cordeiro, R. D., Rocha, M. F. G., & Scm Castelo-Branco, D. (2020). Rhamnolipid enhances *Burkholderia pseudomallei* biofilm susceptibility, disassembly and production of virulence factors. *Future Microbiology*, *15*(12), 1109–1121. Available from https://doi.org/10.2217/fmb-2020-0010.

Singh, P., Patil, Y., & Rale, V. (2019). Biosurfactant production: Emerging trends and promising strategies. *Journal of Applied Microbiology*, *126*(1), 2–13. Available from https://doi.org/10.1111/jam.14057.

Singh, S., Kumar, V., Singh, S., Dhanjal, D. S., Datta, S., Sharma, D., Singh, N. K., & Singh, J. (2020). *Biosurfactant-based bioremediation* (pp. 333–358). Elsevier BV. Available from https://doi.org/10.1016/b978-0-12-819025-8.00016-8.

Sun, X. X., Choi, J. K., & Kim, E. K. (2004). A preliminary study on the mechanism of harmful algal bloom mitigation by use of sophorolipid treatment. *Journal of Experimental Marine Biology and Ecology*, *304*(1), 35–49. Available from https://doi.org/10.1016/j.jembe.2003.11.020.

Sur, S., Romo, T. D., & Grossfield, A. (2018). Selectivity and mechanism of fengycin, an antimicrobial lipopeptide, from molecular dynamics. *Journal of Physical Chemistry B*, *122*(8), 2219–2226. Available from https://doi.org/10.1021/acs.jpcb.7b11889.

Susanti, M. E., Maftuch, M., & Prihanto, A. A. (2021). Screening of potential biosurfactant producing bacteria from Tanjung Perak Port, Surabaya. *The Journal of Experimental Life Sciences*, *11*(3), 100–105. Available from https://doi.org/10.21776/ub.jels.2021.011.03.05.

Swidsinski, A., Sydora, B. C., Doerffel, Y., Loening-Baucke, V., Vaneechoutte, M., Lupicki, M., Scholze, J., Lochs, H., & Dieleman, L. A. (2007). Viscosity gradient within the mucus layer determines the mucosal barrier function and the spatial organization of the intestinal microbiota. *Inflammatory Bowel Diseases*, *13*(8), 963–970. Available from https://doi.org/10.1002/ibd.20163.

Talón, E., Lampi, A. M., Vargas, M., Chiralt, A., Jouppila, K., & González-Martínez, C. (2019). Encapsulation of eugenol by spray-drying using whey protein isolate or lecithin: Release kinetics, antioxidant and antimicrobial properties. *Food Chemistry*, *295*, 588–598. Available from https://doi.org/10.1016/j.foodchem.2019.05.115.

Teng, Y., Stewart, S. G., Hai, Y. W., Li, X., Banwell, M. G., & Lan, P. (2021). Sucrose fatty acid esters: Synthesis, emulsifying capacities, biological activities and structure-property profiles. *Critical Reviews in Food Science and Nutrition*, *61*(19), 3297–3317. Available from https://doi.org/10.1080/10408398.2020.1798346.

Van-Bogaert, I. N. A., Holvoet, K., Roelants, S. L. K. W., Li, B., Lin, Y. C., Van de Peer, Y., & Soetaert, W. (2013). The biosynthetic gene cluster for sophorolipids: A biotechnological interesting biosurfactant produced by *Starmerella bombicola*. *Molecular Microbiology*, *88*(3), 501–509. Available from https://doi.org/10.1111/mmi.12200.

Varjani, S., Rakholiya, P., Yong Ng, H., Taherzadeh, M. J., Hao Ngo, H., Chang, J. S., Wong, J. W. C., You, S., Teixeira, J. A., & Bui, X. T. (2021). Bio-based rhamnolipids production and recovery from waste streams: Status and perspectives. *Bioresource Technology*, *319*, 124213. Available from https://doi.org/10.1016/j.biortech.2020.124213.

Victor, I. U., Kwiencien, M., Tripathi, L., Cobice, D., McClean, S., Marchant, R., & Banat, I. M. (2019). Quorum sensing as a potential target for increased production of rhamnolipid biosurfactant in *Burkholderia thailandensis* E264. *Applied Microbiology and Biotechnology*, *103*(16), 6505–6517. Available from https://doi.org/10.1007/s00253-019-09942-5.

Wang, H., Roelants, S. L. K. W., To, M. H., Patria, R. D., Kaur, G., Lau, N. S., Lau, C. Y., Van Bogaert, I. N. A., Soetaert, W., & Lin, C. S. K. (2019). *Starmerella bombicola*: Recent advances on sophorolipid production and prospects of waste stream utilization. *Journal of Chemical Technology and Biotechnology*, *94*(4), 999–1007. Available from https://doi.org/10.1002/jctb.5847.

Wei, Z., Wang, J. J., Gaston, L. A., Li, J., Fultz, L. M., DeLaune, R. D., & Dodla, S. K. (2020). Remediation of crude oil-contaminated coastal marsh soil: Integrated effect of biochar, rhamnolipid biosurfactant and nitrogen application. *Journal of Hazardous Materials*, *396*, 122595. Available from https://doi.org/10.1016/j.jhazmat.2020.122595.

Wieczorek, D., & Kwaśniewska, D. (2020). *8. Novel trends in technology of surfactants* (pp. 223–250). Walter de Gruyter GmbH. Available from https://doi.org/10.1515/9783110656367-008.

Wu, D., Wan, J., Lu, J., Wang, X., Zhong, S., Schwarz, P., Chen, B., & Rao, J. (2018). Chitosan coatings on lecithin stabilized emulsions inhibit mycotoxin production by Fusarium pathogens. *Food Control*, *92*, 276–285. Available from https://doi.org/10.1016/j.foodcont.2018.05.009.

Wu, Q., Jiang, D., Huang, C., Van Dyk, L. F., Li, L., & Chu, H. W. (2015). Trehalose-mediated autophagy impairs the anti-viral function of human primary airway epithelial cells. *PLoS One*, *10*(4), 124524. Available from https://doi.org/10.1371/journal.pone.0124524.

Wu, S., Liu, G., Zhou, S., Sha, Z., & Sun, C. (2019). Characterization of antifungal lipopeptide biosurfactants produced by marine Bacterium Bacillus sp. CS30. *Marine Drugs*, *17*(4), 199. Available from https://doi.org/10.3390/md17040199.

Yea, D., Jo, S., & Lim, J. (2019). *Synthesis of eco-friendly nano-structured biosurfactants from vegetable oil sources and characterization of their interfacial properties for cosmetic applications, MRS advances* (Vol. 4, pp. 377–384). Materials Research Society Issue 7. Available from https://doi.org/10.1557/adv.2018.619.

Zdziennicka, A., Krawczyk, J., Szymczyk, K., & Jańczuk, B. (1934). Macroscopic and microscopic properties of some surfactants and biosurfactants. *International Journal of Molecular Sciences*, *19*(7), 1934.

Zhao, F., Jiang, H., Sun, H., Liu, C., Han, S., & Zhang, Y. (2019). Production of rhamnolipids with different proportions of mono-rhamnolipids using crude glycerol and a comparison of their application potential for oil recovery from oily sludge. *RSC Advances*, *9*(6), 2885–2891. Available from https://doi.org/10.1039/c8ra09351b.

Zhu, Z., Zhang, B., Cai, Q., Cao, Y., Ling, J., Lee, K., & Chen, B. (2021). A critical review on the environmental application of lipopeptide micelles. *Bioresource Technology*, *339*, 125602. Available from https://doi.org/10.1016/j.biortech.2021.125602.

CHAPTER 2

Biosurfactants: basic properties and characterizations

Renata Raianny da Silva[1], Yali Alves da Silva[2], Leonie Asfora Sarubbo[3,4] and Juliana Moura de Luna[4,5]

[1]Northeast Biotechnology Network (RENORBIO), Federal Rural University of Pernambuco, Recife, Pernambuco, Brazil
[2]Masters degree in Development of Environmental Processes, Catholic University of Pernambuco, Recife, Brazil
[3]Icam Tech School, Catholic University of Pernambuco, Recife, Pernambuco, Brazil
[4]Advanced Institute of Technology and Innovation (IATI), Recife, Pernambuco, Brazil
[5]School of Health and Life Sciences, Catholic University of Pernambuco, Recife, Pernambuco, Brazil

2.1 Introduction

Biosurfactants are molecules with amphiphilic ends, composed of hydrophobic and hydrophilic moieties, measuring surface interactions at the interface through their ends. The nonpolar end is composed of fatty acids and saturated or unsaturated hydrocarbon chains and exhibits water repulsion property. The polar end is composed of mono, di or polysaccharides, acid, peptides, anions, or cations and has the property of affinity with the water produced by a specific microorganism. These characteristics give surfactants surface-active properties, such as the reduction of surface and interfacial tensions in aqueous solutions and mixtures of hydrocarbons (Ambaye et al., 2021; Jahan, Bodratti, et al., 2020).

The most produced surfactants today are chemically derived from petroleum, generally have high toxicity, and are difficult to break down through biodegradation. The use of these compunds in the environment results in an extra source of contamination due to this factor, and because it is potentially toxic, the use of these agents is limited (Mandalenaki et al., 2021). In contrast, biosurfactants are biomolecules naturally synthesized by microorganisms, such as bacteria, yeasts, and filamentous fungi (Drakontis & Amin, 2020).

Natural and synthetic surfactants share many properties, such as surface tension reduction, foaming ability, emulsification, stabilizing capacity, solubility, and detergency. However, biosurfactants demonstrate better performance and have additional properties that make them more attractive than chemical surfactants (Sarubbo et al., 2022).

Biosurfactants are a diversified group of compounds whose characteristics, such as low toxicity, biodegradability, biocompatibility, and even production involving renewable and low-cost sources, make them an eco-friendly alternative for many industrial processes in terms of reduced release of chemicals pollutants and contaminants in the environment (Barrantes et al., 2021). These surface-active agents are used in a wide range of industrial, domestic, and biological processes, acting in various roles, such as emulsifiers, wetting agents, phase-suspending and -dispersing agents, and lubricants (Barbosa et al., 2022).

These surface-active compounds are classified according to their molecular structure. High-molecular-weight compounds, better known as bioemulsifiers, include proteins, lipoproteins, lipopolysaccharides, and heteropolysaccharides. Low-molecular-weight compounds, biosurfactants, include glycolipids and lipopeptides. Such high- and low-molecular-weight compounds have a variety of applications, such as in personal and household care, bioremediation processes and enhanced oil recovery, agrochemistry, textile industries, and biomedical areas (Twigg et al., 2021).

The microorganisms used in the production of low- and high-molecular-weight biosurfactants belong to different genera, such as *Clostridium, Brevibacterium, Pseudomonas, Rhodococcus, Acinetobacter, Thiobacillus, Bacillus, Leuconostoc, Lactobacillus, Enterobacter, Saccharomyces, Aspergillus, Ustilago, Penicillium, Corynebacterium, Citrobacter, Candida*, and *Paenibacillus* (Ahamed & Lichtfouse, 2021; Fooladi et al., 2016; Magalhães et al., 2018; Singh et al., 2019). The products obtained from these microorganisms used in the production of biosurfactants can be glycolipids, phospholipids, lipopeptides, fatty acids, saponins, and alkyl poly glycosides. All biosurfactants under different conditions and other systems demonstrate the ability to reduce surface and interfacial tensions of oil/water mixtures (Bjerk et al., 2021).

Accordingly, this chapter aims to highlight the properties and structures of microbial biosurfactants, as well as the criteria for their classification and the factors that influence their production.

2.2 The classification of biosurfactants

The term biosurfactant refers to surface-acting agents capable of improving surface—surface interactions by forming naturally produced micelles. In addition, these surfactants are used to reduce the interfacial surface tension between the solution and surface or air/water or oil/water interfaces. The addition of surfactants in these systems reduces the surface tension to the point that the surfactants form structures such as micelles, vesicles, and bilayers; when this occurs, the critical point, known as critical micelle concentration (CMC), is reached. When added in the mixture of water and oil, the surfactant remains at the oil/water interface and forms emulsions, providing an excellent emulsifying, foaming, and dispersing capacity, resulting in a very versatile bioproduct for industrial processes (Kumar & Singh, Kant, et al., 2021).

Surfactants can be classified according to the nature of the charge in the individual polar regions. They can be anionic, with a negative charge because of a sulfonate or sulfur group. Nonionic, that is, they do not have an ionic component, and most nonionic ones are polymerization products of 1,2-epoxyethane. Cationic, that is, defined by a quaternary ammonium group, has a positive charge. Amphoterics have both positively and negatively charged portions in the same biomolecule. In addition, surfactants can be classified according to their molecular weight (Firdosbanu & Hajoori, 2022).

Surfactants are also divided according to their molecular weight. Low-molecular-weight surfactants present high efficiency in reducing surface and interfacial tensions, and high-molecular-weight surfactants are agents with the capacity and efficiency to stabilize emulsions. Regarding chemical composition, low-molecular-weight biosurfactants can be classified as glycolipids, phospholipids, fatty acids, lipopeptides, polyketideglycosides, and spiculisporic acid. On the other hand, high-molecular-weight biosurfactants are classified as polymeric and particulate surfactants (Vieira et al., 2021).

According to Sajid et al. (2020), the family of natural surfactants includes glycolipids (cellobiolipids, trehalolipids, rhamnolipid, sophorolipid, mannosylerythritol lipid, and others); phospholipids (phosphatidylethanolamine), lipopeptides (surfactin, subtilisin, viscosine, amphisine, putisolvin, serrawettin, polymyxins, ornithine/taurine/lysine-containing lipids, and pseudofactin-II), fatty acids (corinocholic acids, spiculisporic acids), neutral, polymeric lipids (emulsion, liposan, biodispersan, mannoprotein, and polysaccharide complex), and particulate chemical substances (M protein, lipoteichoic acid, protein-A, layer-A, prodigiosin, fine fimbriae, and whole cells). Table 2.1 presents the chemical structure of some surfactants achieved through microbial production, and some producing species (Elsoud & Ahmed, 2021).

2.3 Physiochemical properties of biosurfactants

Both synthetic and natural surfactants share many properties in common, such as reducing surface and interfacial tensions, foaming ability, emulsification, stabilizing capacity, solubility, and detergency; these properties make these surfactants useful in a wide range of industrial activities (Ribeiro et al., 2020a, 2020b).

Many different techniques are used characterize and identify the biosurfactants' properties produced from different microorganisms. In recent years, techniques such as high-performance liquid chromatography (HPLC), thin-layer chromatography (TLC), liquid chromatography—mass spectrometry (LC-MS), gas chromatography—mass spectrometry (GC-MS), nuclear magnetic resonance (NMR), and matrix-assisted laser desorption/ionization—time of flight mass spectroscopy (MALDI-TOF) have been applied for this purpose (Ambaye et al., 2021).

According to Sarubbo et al. (2022), compared with chemical surfactants, biosurfactants demonstrate better performance due to their additional properties. This occurs because biosurfactants are natural products achieved through the production of microorganisms in which they occupy various environmental niches and therefore manage to maintain their effectiveness even under adverse conditions. These natural surfactants are stable and maintain their physicochemical activities at high temperatures and

Table 2.1 Chemical structure of some surfactants achieved through microbial production.

Group	Class	Chemical structure	Producing species. (According group)	References
Glycolipids	Rhamnolipids Sophorolipids Trehalolipids	Composed of 3-hydroxy fatty acids in hydrophobic moiety and rhamnose, a disaccharide hydrophilic moiety. Consist of a hydroxyl fatty acid and sophorose, a dimeric sugar. These two moieties are linked to each other by a β-glycosidic bond. Two types of sophorolipids can be differentiated: lactonic and nonlactonic. Composed of trehalose (hydrophilic disaccharide) linked to β-hydroxy fatty acid chain with long α-branches mycolic acid	Pseudomonas aeruginosa Candida sphaerica Candida guilliermondii Torulopsis apicola, Rhodococcus erythropolis	Pacwa-Plociniczak et al. (2011) Santos et al. (2021) Lira et al. (2022) Pacwa-Plociniczak et al. (2011) Pacwa-Plociniczak et al. (2011)
Neutral lipids Phospholipids Fatty Acids	Corynomycolic Acid Spiculisporic Acid Phosphatidylethanolamine	Have hydrophilic phosphate group on one end and hydrophobic fatty acids on the other. These fatty acids can be either straight-chain or complex with alkyl branches and hydroxyl groups. The balance between hydrophilicity and hydrophobicity of the fatty acid is highly related to the chain length of hydrocarbon and its degree of complexity. Most surface-active fatty acids have a length between 12 and 14 carbon atoms	Corynebacterium lepus Penicillium spiculisporum Rhodococcus erythropolis, Acinetobacter sp	Pacwa-Plociniczak et al. (2011) Pacwa-Plociniczak et al. (2011) Pacwa-Plociniczak et al. (2011) Pacwa-Plociniczak et al. (2011)

Polymeric Biosurfactants	Liposan	Are exopolysaccharide (EPS) in nature	*Candida lipolytica*	Pacwa-Plociniczak et al. (2011)
Lipopeptide	Surfactin	A peptide loop represents the hydrophilic moiety of surfactin, while a fatty acid chain of 13–15 carbons long represents the hydrophobic moiety. The peptide loop is composed of seven successive amino acids, namely, L-aspartic acid, L-leucine, glutamic acid, L-leucine, L-valine, and two D-leucines	*Bacillus cereus*	Durval et al. (2020)

Source: Adapted from Elsoud, A., & Ahmed, M. M. (2021). *Classification and production of microbial surfactants* (pp. 65–89). Springer Science and Business Media LLC. https://doi.org/10.1007/978-981-15-6607-3_4.

within a wide pH range (3–12). In addition, biosurfactants can tolerate salt concentrations of up to 10% (w/v), while chemical surfactants are inactivated with concentrations equal to or greater than 2% NaCl.

The origin, as well as the production and purification processes of biosurfactants, significantly influence their molecular characteristics and interfacial behavior. Therefore, surface and interfacial tensions, self-assembly, solubilization, and emulsification influence the production of a biosurfactant (Jahan, Bodratti, et al., 2020).

The main property of a surfactant is the reduction of surface and interfacial tensions because it plays a fundamental role in formulations. Biosurfactants adsorb at air–liquid, liquid–liquid, and solid–liquid interfaces because of their amphiphilic nature. The surfactant reduces the intermolecular force between the biomolecules and, thus, decreases surface and interfacial tensions (Kumar & Ngueagni, 2021).

Biosurfactants can self-assemble and form micelles, which allows them to have morphologically different structures from each other, increasing their specificity (Sarubbo et al., 2022). Amphiphilic self-assembly in aqueous media results from a balance between the hydrophobic interaction of hydrocarbon chains in water and the electrostatic repulsion of the main group of surfactant molecules. The counter-ions contained in the aqueous medium reduce the electrostatic repulsion between the fundamental groups of ionic surfactants, thus decreasing the entropy of the surfactant molecules and neutralizing the formation of micelles. The CMC is the concentration for the initiation of surfactant self-assembly, forming micelles (Jahan, 2020).

After self-assembly in aqueous solutions, surfactants can solubilize hydrophobic compounds. The solubility of hydrophobic organic compounds in the presence of the surfactant takes into account their relative concentrations, as well as the pH range and salinity concentration of aqueous solutions. On the other hand, emulsions are unbalanced systems, kinetically stabilized. The structural stability and appearance characteristics of the surfactant depend on both its components and its composition, as well as the preparation conditions, such as temperature and pressure, and the process, such as input energy, mixing time and type of equipment (Jahan, Bodratti, et al., 2020).

2.4 Factors affecting biosurfactant production

In the last decades, the interest in biosurfactants has increased due to their diversified forms of application in a wide range of industries due to characteristics that highlight them, such as low toxicity, biodegradability, high emulsifying capacity, and low CMC, making them a new generation of surfactants called "green surfactants," as these compounds are seen to have low environmental impacts (Santos et al. 2016; Farias et al. 2021).

One of the great challenges of biosurfactant production is to obtain a substantial yield as many factors directly influence microbial growth and metabolism during its production. Thus, investigations regarding the best and most promising combinations of culture medium for the production of these compounds have been carried out. The main factors to be considered for optimizing the production of biosurfactant are, above all, the cultivation conditions, which include carbon and nitrogen sources, and the availability of micronutrients for microbial growth, in addition to pH, temperature, aeration, agitation, and the inoculum (Sarubbo et al., 2022).

Furthermore, according to Sarubbo et al. (2022), fermentation processes can be optimized using statistical methods, making it possible to study the effects of interactions between multiple variables, enabling the search for better and ideal cultivation conditions aiming to obtain maximum production of the desired biosurfactant at more affordable costs.

Several microorganisms produce biosurfactants when cultivated in different culture media with different sources of carbon and nitrogen. The carbon source plays an important role in microbial growth and, consequently, in the structure of the biosurfactant produced. Thus, it becomes possible to select culture media to obtain specific properties desired for a given application (Farias et al. 2021).

The carbon source is extremely important for biosurfactant production, as it directly influences the growth of the microorganism, in addition to the yield of biosurfactant production, in terms of quantity and quality (Jimoh & Lin, 2019; Sarubbo et al., 2022). In Table 2.2 are listed some works produced in recent years in which various carbon sources were used to produce biosurfactants.

In addition to carbon, the nitrogen source is also essential for cell growth and development and can be from the organic or inorganic origin, such as urea, amino acids, and yeast extract or ammonium nitrates and sulfates, respectively (Sarubbo et al., 2022; Vieira et al., 2021). Nitrogen is an important nutrient for influencing microbial growth, as well as protein and enzyme synthesis (Fenibo et al., 2019). Table 2.3 reports some works conducted using nitrogen sources responsible for influencing the growth of microorganisms and the production of biosurfactants.

Other nutrients such as phosphorus, manganese, sulfur, and iron are also essential for microbial growth as they also affect the fermentation processes for biosurfactant production (Jimoh & Lin, 2019; Sarubbo et al., 2022).

Environmental variables such as temperature, pH, incubation time, and aeration also directly affect microbial growth and biosurfactant production. According to Auhim and Mohamed (2013), temperatures between 25°C and 37°C influence the microbial growth of biosurfactant producers, a physical factor that deserves attention as it significantly influences production. Culture media with a pH around 6−8 have been reported to be optimal for biosurfactant production, and, as a rule, bacteria tend to do better at alkaline pH, while yeast and fungi thrive better in acid conditions (Jimoh & Lin, 2019). The aeration and agitation of the culture media are also strong influencers of the production of biosurfactants as both facilitate the oxygenation of the culture medium (Roy, 2017), thus favoring higher production with better yields and a high surfactant activity. In the literature, it is reported that aeration speeds between 50 and 250 rpm directly

Table 2.2 Studies conducted using different carbon sources for the production of biosurfactants by different microorganisms.

Carbon source (Cs)	Classification	Microorganism	(Cs)	Biosurfactant (g/L)	References
Restaurant food waste hydrolysate	Sophorolipid	*Starmerella bombicola* (previously *Candida bombicola*)	100 g/L	115.2	Kaur et al. (2019)
Glucose	Surfactin	*Bacillus subtili*	8 g/L	1.1	Willenbacher et al. (2015)
Molasses	Lipopeptide	*Kurthia gibsonii* KH2	10% (w/v)	2	Nor et al. (2021)
Sucrose	Surfactin	*Micromonospora marina*	10 g/L	1.22	Ramalingam et al. (2019)
Sucrose	Lipopeptide	*Fusarium sp.*	2% (w/v)	1.21	Qazi et al. (2014)
Brown sugar	Lipopeptide	*Bacillus atrophaeus*	10 g/L	0.77	Zhang et al. (2016)
Palm fatty acid distillate	Rhamnolipid	*Pseudomonas aeruginosa*	100 g/L	0.43	Radzuan et al. (2017)
Glucose	Sophorolipid	*Starmerella bombicola* (previously *Candida bombicola*)	10% (w/v)	51.5	Jadhav et al. (2019)
Animal fat	Rhamnolipid	*Pseudomonas aeruginosa*	40 g/L	25.8	Jia et al. (2020)
Crude oil	Rhamnolipid	*Pseudomonas aeruginosa* PG1	2% (v/v)	2.26	Patowary et al. (2017)
Soybean waste frying oil	Lipoprotein	*Streptomyces sp.*	10 g/L	1.9	Santos et al. (2019)
Biodiesel	Sophorolipid	*Starmerella bombicola* (previously *Candida bombicola*)	10% (w/v)	224.2	Kim et al. (2020)

Source: Adapted from Vieira, I. M. M., Santos, B. L. P., Ruzene, D. S. & Silva, D. P. (2021). An overview of current research and developments in biosurfactants. *Journal of Industrial and Engineering Chemistry*, 100, 1–18. https://doi.org/10.1016/j.jiec.2021.05.017.

Table 2.3 Studies conducted using different nitrogen sources for the production of biosurfactants by different microorganisms.

Nitrogen (N)	Classification	Microorganism	(N)	Biosurfactant (g/L)	References
KNO$_3$	Lipopeptide	*Brevibacillus* sp.	0.5% (w/v)	1.28	Vigneshwaran et al. (2018)
Peptone	Lipopeptide	*Bacillus cereus* UCP 1615	0.12% (w/v)	4.70	Durval et al. (2021)
Yeast extract	Glycolipid	*Pseudomonas taiwanensis*	1 g/L	1.2	Liu et al. (2017)
Tryptone	Rhamnolipid	*Pseudomonas aeruginosa*	1% (w/v)	1.12	Shi et al. (2021)
Yeast extract (NH$_4$)$_2$SO$_4$	Mannosylerythritol lipid (MEL-A)	*Ceriporia lacerate*	1 g/L	129.64	Niu et al. (2017)
NH$_4$NO$_3$	Rhamnolipid	*Pseudomonas fluorescens*	1 g/L	2	Abouseoud et al. (2008)
NaNO$_3$	Glycolipid	*Stenotrophomonas acidaminiphila*	3 g/L	2.31	Onlamool et al. (2020)
NaNO$_3$	Surfactin	*Bacillus subtilis*	5 g/L	1.12	Abdel-Mawgoud et al. (2008)
(NH$_4$)$_2$SO$_4$	Surfactin	*Bacillus subtilis*	50 mM	1.1	Willenbacher et al. (2015)
Corn steep liquor	Glycolipid	*Candida sphaerica*	9% (w/v)	10	Santos et al. (2021)
Corn steep liquor and NaNO$_3$	Glycolipid	*Pseudomonas cepacia* CCT6659	3% (w/v) and 0.20%	25	Silva et al. (2021)
Corn steep liquor	Glycolipid	*Candida bombicola*	5%	221.9	Pinto et al. (2022)
Corn steep liquor	Glycolipid	*Candida guilliermondii*	5%	21	Lira et al. (2022)

Source: Adapted from Vieira, I. M. M., Santos, B. L. P., Ruzene, D. S. & Silva, D. P. (2021). An overview of current research and developments in biosurfactants. *Journal of Industrial and Engineering Chemistry*, 100, 1–18. https://doi.org/10.1016/j.jiec.2021.05.017.

affect the production of biosurfactants. The incubation period also varies, depending on the microbial growth phase, ranging from a few hours to several days (Sarubbo et al., 2022). In the literature, incubation periods of 24 h for *Yarrowia lipolytica* (Fontes et al., 2010), 120 h for *Saccharomyces cerevisiae* (Ribeiro et al., 2020a, 2020b) and 96 h for *Azotobacter chrococcum* (Auhim and Mohamed 2013) have been reported, demonstrating different incubation times for fungal and bacterial species.

Another necessary factor to be considered is the size of the inoculum and the growth phase of the inoculant culture, as these have a direct influence on the production and biomass yield of the biosurfactant because many physiological processes are dependent on cell density (Jimoh & Lin, 2019; Twigg et al., 2021).

2.5 Conclusions and future perspectives

Biosurfactants are surface-active agents with great industrial potential as they can be applied in a highly diversified spectrum of areas. Thus, it is important to investigate the optimization methods and cost reduction to obtain these compounds on an industrial scale. The use of industrial waste as a source of carbon and nitrogen has already shown to be promising for production, in addition to being an alternative for the sustainable destination of these wastes from various industries.

The main qualities of biosurfactants, such as low toxicity, high degradability, emulsifying properties, and efficient surface activity, are attractive and demanded aspects in recent years, which has increased the interest in these agents. Biosurfactants have the potential to be used in various industrial sectors, such as the oil, pharmaceutical, chemical, food, agriculture, and cosmetics industries. Due to a wide range of applications, it is important to know the details of production minutely to optimize the entire process, resulting in a quality product, economically viable, and eco-friendly.

In this chapter, therefore, the basic and physicochemical properties of biosurfactants, in addition to the factors that can influence the production of these compounds, were discussed.

More and more studies have shown the diversity of applications of biosurfactants. Thus, it is necessary to continue investigating ways to optimize all production processes, from the materials used in the culture medium to the purification process, to make costs less onerous, making this production viable on a large scale. Only then will be possible to significantly replace chemical surfactants with biological ones so that environmental impacts on human health are reduced for the use of using green agents.

References

Abdel-Mawgoud, A. M., Aboulwafa, M. M., & Hassouna, N. A. H. (2008). Optimization of surfactin production by bacillus subtilis isolate BS5. *Applied Biochemistry and Biotechnology*, *150*, 305–BS325. Available from https://doi.org/10.1007/s12010-008-8155-x.

Abouseoud, M., Maachi, R., Amrane, A., Boudergua, S., & Nabi, A. B. (2008). Evaluation of different carbon and nitrogen sources in production of biosurfactant by *Pseudomonas fluorescens*. *Desalination*, *223*(1−3), 143−151. Available from https://doi.org/10.1016/j.desal.2007.01.198.

Ahamed, M. I., & Lichtfouse, E. (Eds.), (2021). *Water pollution and remediation: Photocatalysis*. Springer International Publishing.

Ambaye, T. G., Vaccari, M., Prasad, S., & Rtimi, S. (2021). Preparation, characterization and application of biosurfactant in various industries: A critical review on progress, challenges and perspectives. *Environmental Technology & Innovation*, *24*, 102090. Available from https://doi.org/10.1016/j.eti.2021.102090.

Barbosa, F. G., Ribeaux, D. R., Rocha, T., Costa, R. A., Guzmán, R. R., Marcelino, P. R., & Silva, S. S. D. (2022). Biosurfactants: Sustainable and versatile molecules. *Journal of the Brazilian Chemical Society*, *33*, 870−893. Available from https://doi.org/10.21577/0103-5053.20220074.

Barrantes, K., Araya, J. J., Chacón, L., Procupez-Schtirbu, R., Lugo, F., Ibarra, G., & Soto, V. H. (2021). Antiviral, antimicrobial, and antibiofilm properties of biosurfactants: Sustainable use in food and pharmaceuticals. *Biosurfactants for a sustainable future: Production and applications in the environment and biomedicine* (pp. 245−268). Wiley. Available from https://doi.org/10.1002/9781119671022.ch11.

Bjerk, T. R., Severino, P., Jain, S., Marques, C., Silva, A. M., Pashirova, T., & Souto, E. B. (2021). Biosurfactants: Properties and applications in drug delivery, biotechnology and ecotoxicology. *Bioengineering*, *8*(8), 115. Available from https://doi.org/10.3390/bioengineering8080115.

Drakontis, C. E., & Amin, S. (2020). Biosurfactants: Formulations, properties, and applications. *Current Opinion in Colloid & Interface Science*, *48*, 77−90. Available from https://doi.org/10.1016/j.cocis.2020.03.013.

Durval, I. J. B., Mendonça, A. H. R., Rocha, I. V., Luna, J. M., Rufino, R. D., Converti, A., & Sarubbo, L. A. (2020). Production, characterization, evaluation and toxicity assessment of a *Bacillus cereus* UCP 1615 biosurfactant for marine oil spills bioremediation. *Marine Pollution Bulletin*, *157*, 111357. Available from https://doi.org/10.1016/j.marpolbul.2020.111357.

Durval, I. J. B., Ribeiro, B. G., Aguiar, J. S., Rufino, R. D., Converti, A., & Sarubbo, L. A. (2021). Application of a biosurfactant produced by *Bacillus cereus* UCP 1615 from waste frying oil as an emulsifier in a cookie formulation. *Fermentation*, *7*(3), 189. Available from https://doi.org/10.3390/fermentation7030189.

Elsoud, A., & Ahmed, M. M. (2021). *Classification and production of microbial surfactants. Microbial biosurfactants* (pp. 65−89). Singapore: Springer. Available from http://doi.org/10.1007/978-981-15-6607-3_4.

Farias, C. B. B., Almeida, F. C., Silva, I. A., Souza, T. C., Meira, H. M., Rita de Cássia, F., Luna, J. M., Santos, V. A., Converti, A., Banat, I. M., & Sarubbo, L. A. (2021). Production of green surfactants: Market prospects. *Electronic Journal of Biotechnology*, *51*, 28−39. Available from https://doi.org/10.1016/j.ejbt.2021.02.002.

Fenibo, E. O., Ijoma, G. N., Selvarajan, R., & Chikere, C. B. (2019). Microbial surfactants: The next generation multifunctional biomolecules for applications in the petroleum industry and its associated environmental remediation. *Microorganisms*, *7*(11), 1−29. Available from https://doi.org/10.3390/microorganisms7110581.

Firdosbanu, P., & Hajoori, M. (2022). Microbial biosurfactant: An intermediate with enhanced degradative potential. *International Journal of Research in Engineering and Science*, *10*(81), 445−453.

Fontes, G. C., Amaral, F., Filomena, P., Nele, M., Coelho, Z., & Alice, M. (2010). Factorial design to optimize biosurfactant production by Yarrowia lipolytica. *BioMed Research International*, *821306*, 1−8. Available from https://doi.org/10.1155/2010/821306.

Fooladi, T., Moazami, N., Abdeshahian, P., Kadier, A., Ghojavand, H., Yusoff, W. M. W., & Hamid, A. A. (2016). Characterization, production and optimization of lipopeptide biosurfactant by new strain *Bacillus pumilus* 2IR isolated from an Iranian oil field. *Journal of Petroleum Science and Engineering*, *145*, 510−519. Available from https://doi.org/10.1016/j.petrol.2016.06.015.

Jahan, R. (2020). *Aqueous self-assembly properties of novel surfactants* (Doctoral dissertation). State University of New York at Buffalo.

Jadhav, J. V., Pratap, A. P., & Kale, S. B. (2019). Evaluation of sunflower oil refinery waste as feedstock for production of sophorolipid. *Process Biochemistry*, *78*, 15−24. Available from https://doi.org/10.1016/j.procbio.2019.01.015.

Jahan, R., Bodratti, A. M., Tsianou, M., & Alexandridis, P. (2020). Biosurfactants, natural alternatives to synthetic surfactants: Physicochemical properties and applications. *Advances in Colloid and Interface Science, 275*, 102061. Available from https://doi.org/10.1016/j.cis.2019.102061.

Jia, L., Zhou, J., Cao, J., Wu, Z., Liu, W., & Yang, C. (2020). Foam fractionation for promoting rhamnolipids production by *Pseudomonas aeruginosa* D1 using animal fat hydrolysate as carbon source and its application in intensifying phytoremediation. *Chemical Engineering and Processing - Process Intensification, 158*, 108177. Available from https://doi.org/10.1016/j.cep.2020.108177.

Jimoh, A. A., & Lin, J. (2019). Biosurfactant: A new frontier for greener technology and environmental sustainability. *Ecotoxicology and Environmental Safety, 184*, 109607. Available from https://doi.org/10.1016/j.ecoenv.2019.109607.

Kaur, G., Wang, H., To, M. H., Roelants, S. L. K. W., Soetaert, W., Lin, C. S. K., & Clean, J. (2019). Efficient sophorolipids production using food waste. *Journal of Cleaner Production, 232*, 1–11. Available from https://doi.org/10.1016/j.jclepro.2019.05.326.

Kim, J.-H., Oh, Y.-R., Hwang, J., Jang, Y.-A., Lee, S. S., Hong, S. H., & Eom, G. T. (2020). Value-added conversion of biodiesel into the versatile biosurfactant sophorolipid using *Starmerella bombicola*. *Cleaner Engineering and Technology, 1*, 100027. Available from https://doi.org/10.1016/j.clet.2020.100027.

Kumar, A., Singh, S. K., Kant, C., Verma, H., Kumar, D., Singh, P. P., & Kumar, M. (2021). Microbial biosurfactant: A new frontier for sustainable agriculture and pharmaceutical industries. *Antioxidants, 10*(9), 1472. Available from https://doi.org/10.3390/antiox10091472.

Kumar, P. S., & Ngueagni, P. T. (2021). A review on new aspects of lipopeptide biosurfactant: Types, production, properties and its application in the bioremediation process. *Journal of Hazardous Materials, 407*, 124827. Available from https://doi.org/10.1016/j.jhazmat.2020.124827.

Lira, I. C. A. S., Santos, E. M. S., Guerra, J. M. C., Meira, H. M., Sarubbo, L. A., & Luna, J. M. (2022). Microbial biosurfactant: Production, characterization and application as a food emulsions. *Research, Society and Development, 11*(5), 28339, e44111528339. Available from https://doi.org/10.33448/rsd-v11i5.28339.

Liu, C., You, Y., Zhao, R., Sun, D., Zhang, P., Jiang, J., Zhu, A., & Liu, W. (2017). Biosurfactant production from *Pseudomonas taiwanensis* L1011 and its application in accelerating the chemical and biological decolorization of azo dyes. *Ecotoxicology and Environmental Safety, 145*, 8–15. Available from https://doi.org/10.1016/j.ecoenv.2017.07.012.

Magalhães, E. R. B., Silva, F. L., Sousa, M. A. D. S. B., & Dos Santos, E. S. (2018). Use of different agroindustrial waste and produced water for biosurfactant production. *Biosciences Biotechnology Research Asia, 15*(1), 17–26. Available from https://doi.org/10.13005/bbra/2604.

Mandalenaki, A., Kalogerakis, N., & Antoniou, E. (2021). Production of high purity biosurfactants using heavy oil residues as carbon source. *Energies, 14*(12), 3557. Available from https://doi.org/10.3390/en14123557.

Niu, Y., Fan, L., Gu, D., Wu, J., & Chen, Q. (2017). Characterization, enhancement and modelling of mannosylerythritol lipid production by fungal endophyte *Ceriporia lacerate* CHZJU. *Food Chemistry, 228*, 610–617. Available from https://doi.org/10.1016/j.foodchem.2017.02.042.

Nor, F. H. M., Abdullah, S., Yuniarto, A., Ibrahim, Z., Nor, M. H. M., & Hadibarata, T. (2021). Production of lipopeptide biosurfactant by *Kurthia gibsonii* KH2 and their synergistic action in biodecolourisation of textile wastewater. *Environmental Technology & Innovation, 22*, 101533. Available from https://doi.org/10.1016/j.eti.2021.101533.

Onlamool, T., Saimmai, A., Meeboon, N., & Maneerat, S. (2020). Enhancement of glycolipid production by Stenotrophomonas acidaminiphila TW3 cultivated in low cost substrate. *Biocatalysis and Agricultural Biotechnology (Reading, Mass.), 26*, 101628. Available from https://doi.org/10.1016/j.bcab.2020.101628.

Pacwa-Płociniczak, M., Płaza, G. A., Piotrowska-Seget, Z., & Cameotra, S. S. (2011). Environmental applications of biosurfactants: Recent advances. *International Journal of Molecular Sciences, 12*(1), 633–654. Available from https://doi.org/10.3390/ijms12010633.

Patowary, K., Patowary, R., Kalita, M. C., & Deka, S. (2017). Characterization of biosurfactant produced during degradation of hydrocarbons using crude oil as sole source of carbon. *Frontiers in Microbiology*, *8*, 279. Available from https://doi.org/10.3389/fmicb.2017.00279.

Pinto, M. I. S., Guerra, J. M. C., Meira, H. M., Sarubbo, L. A., & Luna, J. M. (2022). A Biosurfactant from Candida bombicola: Its synthesis, characterization, and its application as a food emulsions. *Foods*, *11*(4), 561. Available from https://doi.org/10.3390/foods11040561.

Qazi, M. A., Kanwal, T., Jadoon, M., Ahmed, S., & Fatima, N. (2014). Isolation and characterization of a biosurfactant-producing Fusarium sp. BS-8 from oil contaminated soil. *Biotechnology Progress*, *30*(5), 1065−1075. Available from https://doi.org/10.1002/btpr.1933.

Radzuan, M. N., Banat, I. M., & Winterburn, J. (2017). Production and characterization of rhamnolipid using palm oil agricultural refinery waste. *Bioresource Technology*, *225*, 99−105. Available from https://doi.org/10.1016/j.biortech.2016.11.052.

Ramalingam, V., Varunkumar, K., Ravikumar, V., & Rajaram, R. (2019). Production and structure elucidation of anticancer potential surfactin from marine actinomycete Micromonospora marina. *Process Biochemistry*, *78*, 169−177. Available from https://doi.org/10.1016/j.procbio.2019.01.002.

Ribeiro, B. G., Guerra, J. M. C., & Sarubbo, L. A. (2020a). Potential food application of a biosurfactant produced by *Saccharomyces cerevisiae* URM 6670. *Frontiers in Bioengineering and Biotechnology*, *8*(434), 1−13. Available from https://doi.org/10.3389/fbioe.2020.00434.

Ribeiro, B. G., Guerra, J. M. C., & Sarubbo, L. A. (2020b). Biosurfactants: Production and application prospects in the food industry. *Biotechnology Progress*, *36*(5), e3030. Available from https://doi.org/10.1002/btpr.3030.

Roy, A. A. (2017). Review on the biosurfactants: Properties, types and its application. *Journal of Fundamentals of Renewable Energy and Applications*, *8*(1), 1−14. Available from https://doi.org/10.4172/2090-4541.1000248.

Sajid, M., Khan, M. S. A., Cameotra, S. S., & Al-Thubiani, A. S. (2020). Biosurfactants: Potential applications as immunomodulator drugs. *Immunology Letters*, *223*, 71−77. Available from https://doi.org/10.1016/j.imlet.2020.04.003.

Santos, D. K. F., Rufino, R. D., Luna, J. M., Santos, V. A., & Sarubbo, L. A. (2016). Biosurfactants: multifunctional biomolecules of the 21st century. *International journal of molecular sciences*, *17*(3), 401. Available from https://doi.org/10.3390/ijms17030401.

Santos, E. F., Teixeira, M. F. S., Converti, A., Porto, A. L. F., & Sarubbo, L. A. (2019). Production of a new lipoprotein biosurfactant by Streptomyces sp. DPUA1566 isolated from lichens collected in the Brazilian Amazon using agroindustry wastes. *Biocatalysis and Agricultural Biotechnology (Reading, Mass.)*, *142*, 142−150. Available from https://doi.org/10.1016/j.bcab.2018.10.014.

Santos, E. M. S., Lira, I. C. A. S., Meira, H. M., Aguiar, J. S., Rufino, R. D., Almeida, D. G., Casazza, A. A., Converti, A., Sarubbo, L. A., & Luna, J. M. (2021). Enhanced oil removal by a non-toxic biosurfactant formulation. *Energies*, *14*(2), 467. Available from https://doi.org/10.3390/en14020467.

Sarubbo, L. A., Maria da Gloria, C. S., Durval, I. J. B., Bezerra, K. G. O., Ribeiro, B. G., Silva, I. A., & Banat, I. M. (2022). Biosurfactants: Production, properties, applications, trends, and general perspectives. *Biochemical Engineering Journal*, 108377. Available from https://doi.org/10.1016/j.bej.2022.108377.

Shi, J., Chen, Y., Liu, X., & Li, D. (2021). Rhamnolipid production from waste cooking oil using newly isolated halotolerant *Pseudomonas aeruginosa* M4. *Journal of Cleaner Production*, *278*(123879). Available from https://doi.org/10.1016/j.jclepro.2020.123879.

Silva, R. C. F. S., Luna, J. M., Rufino, R. D., & Sarubbo, L. A. (2021). Ecotoxicity of the formulated biosurfactant from Pseudomonas cepacia CCT 6659 and application in the bioremediation of terrestrial and aquatic environments impacted by oil spills. *Process Safety and Environmental Protection*, *154*, 338−347. Available from https://doi.org/10.1016/j.psep.2021.08.038.

Singh, P., Patil, Y., & Rale, V. (2019). Biosurfactant production: Emerging trends and promising strategies. *Journal of Applied Microbiology*, *126*(1), 2−13. Available from https://doi.org/10.1111/jam.14057.

Twigg, M. S., Baccile, N., Banat, I. M., Déziel, E., Marchant, R., Roelants, S., & Van Bogaert, I. N. (2021). Microbial biosurfactant research: Time to improve the rigour in the reporting of synthesis, functional characterization and process development. *Microbial Biotechnology*, *14*(1), 147–170. Available from https://doi.org/10.1111/1751-7915.13704.

Vieira, I. M. M., Santos, B. L. P., Ruzene, D. S., & Silva, D. P. (2021). An overview of current research and developments in biosurfactants. *Journal of Industrial and Engineering Chemistry*, *100*, 1–18. Available from https://doi.org/10.1016/j.jiec.2021.05.017.

Vigneshwaran, C., Sivasubramanian, V., Vasantharaj, K., Krishnanand, N., & Jerold, M. (2018). Potential of Brevibacillus sp. AVN 13 isolated from crude oil contaminated soil for biosurfactant production and its optimization studies. *Journal of Environmental Chemical Engineering*, *6*(4), 4347–4356. Available from https://doi.org/10.1016/j.jece.2018.06.036.

Willenbacher, J., Yeremchuk, W., Mohr, T., Syldatk, C., & Hausmann, R. (2015). Enhancement of surfactin yield by improving the medium composition and fermentation process. *AMB Express*, *5*(57). Available from https://doi.org/10.1186/s13568-015-0145-0.

Zhang, J., Xue, Q., Gao, H., Lai, H., & Wang, P. (2016). Production of lipopeptide biosurfactants by *Bacillus atrophaeus* 5-2a and their potential use in microbial enhanced oil recovery. *Microbial Cell Factories*, *15*(168). Available from https://doi.org/10.1186/s12934-016-0574-8.

CHAPTER 3

Biosurfactants production utilizing microbial resources

Ruby Aslam[1], Mohammad Mobin[2], Saman Zehra[2] and Jeenat Aslam[3]

[1]School of Civil Engineering and Architecture, Chongqing University of Science and Technology, Chongqing, P.R. China
[2]Corrosion Research Laboratory, Department of Applied Chemistry, Faculty of Engineering and Technology, Aligarh Muslim University, Aligarh, Uttar Pradesh, India
[3]Department of Chemistry, College of Science, Taibah University, Yanbu, Al-Madina, Saudi Arabia

3.1 Introduction

The term surfactant (SURFace-ACTive AgeNTS) includes a wide variety of compounds, both synthetic and biological, having similar surface-active properties (Aslam et al., 2021; Mobin et al., 2017). Biosurfactants (biological surface-active compounds or microbial surface-active agents) are the biomolecules produced by living cells. They are amphiphilic biochemical molecules containing hydrophobic and hydrophilic groups that allow them to exist at the interface between polar and nonpolar media. They are produced on microbial cell surfaces or secreted as extracellular products. Structurally, they contain a hydrophilic moiety, which can be a carbohydrate, amino acid, cyclic peptide, phosphate, carboxylic acid or an alcohol, and a hydrophobic moiety, which can be a long-chain fatty acid, hydroxyl fatty acid, or α-alkyl β-hydroxy fatty acid.

Worldwide interest in biosurfactants significantly increased in recent years due to their ability to mitigate the drawbacks of chemical surfactants. Advances in the era of industrial globalization have increasingly directed several industries toward biotechnology. While the world market for biotechnology products was US $1.7 billion in 1992, it has increased beyond US $500 billion. Surfactants constitute an integral part of chemical feedstock inventory to many industries and are mainly synthesized from petrochemicals. Their worldwide production was estimated to exceed 4 million tons and by $9–10 billion per year (Makkar & Cameotra, 2002). The market for biosurfactants is estimated to be USD 1.2 billion in 2022, and it will grow from 2022 to 2027 at a rate of 11.2%. A huge global surfactant market is opened to generate great revenues (Surekha et al., 2017).

Microorganisms such as yeast, bacteria, or fungi can produce biosurfactants-surface-active compounds using different substrates such as oils, glycerol, alkanes, sugars, and agricultural and industrial wastes (Shekhar et al., 2015; Amaral et al., 2010). Biosurfactants are biodegradable, making them an attractive alternative to chemically synthesized surfactants that are normally petroleum-based and environmentally hazardous

(Maikudi Usman et al., 2016). Biosurfactants have special advantages over their chemically manufactured counterparts because of their lower toxicity, biodegradable nature, effectiveness at extreme temperatures, pH, and salinity (Gaur et al., 2022), and ease of synthesis. In addition, they possess surface-active properties differing in many cases from synthetic surfactants. Several reviews and monographs have appeared in the literature on the properties, chemistry, biosynthesis, and applications of biosurfactants. In this chapter, efforts are made to give a brief account of the sources and properties of various classes of biosurfactants and to review the present status of biosurfactant production.

3.2 Types of biosurfactants

Biosurfactants produced by microorganisms are classified into five major groups, including glycolipids, lipopeptides, phospholipids, polymeric surfactants, and the particulate type (Sobrinho et al., 2014).

3.2.1 Glycolipids

Glycolipids long-chain aliphatic acids or hydroxy aliphatic acids attached to carbohydrates via an ester group. The majority of biosurfactants are glycolipids. The most well-known glycolipids are sophorolipids, trehalolipids, and rhamnolipids.

3.2.2 Lipopeptides or lipoprotein

A lipid and a polypeptide chain are combined to form a type of protein known as a lipopeptide. These molecules can reduce interfacial and surface tension and are characterized by their structural variety. Two primary types of molecules are involved in this process: acyl tails and linear oligopeptide sequences with an amide bond (Freitas de Oliveira et al., 2013). The biosurfactant's hydrophobic tail and hydrophilic head are made up of a combination of components, including a peptide sequence and a hydrocarbon chain. The peptide component is also equipped with anionic and cationic residues. The lipopeptide's potential as an anticancer and antibacterial agent has been studied in various studies. Due to their unique structural and functional characteristics, these molecules are commonly used in various sectors. Based on their structural differences, the various groups of lipopeptide surfactants include isoforms with various D- and L-amino acids. Some of these include the viscosine and iturin from *Bacillus subtilis*, the serravettin from *B. licheniformis*, the gramicidin from *B. fluorescens*, and the polymyxin from *B. polymyxa* (Dhasayan et al., 2015).

3.2.3 Phospholipids and fatty acids (mycolic acids)

Phospholipids and fatty acids are the byproducts of microbial oxidation resulting from alkanes (Fawzy et al., 2018). While fatty acids are commonly utilized in the food industry, gene carrier systems have started using phospholipids due to their membranous nature. Some of the commonly used biosurfactants for these types of lipids include lysolecithin and lecithin (Mulligan et al., 2014; Müller et al., 2012). In addition to straight-chain fatty acids, microbes can also create complex fatty acids with alkyl branching and form hydroxyl groups. Examples of such complex acids (Karpenko et al., 2009) include Corynomucolic acids, which are used as surfactant. The ratio of fatty acids that are either hydrophilic or lipophilic, depending on the length of the hydrocarbon chain, is related to surface and interfacial tensions. The carbon atom with a 12–14 range is the best choice for reducing interfacial and surface tensions.

3.2.4 Polymeric surfactants

Polysaccharide—protein biosurfactant Alasan, emulsan, and lipomannan have been studied and are widely used in various industries. Emulsan is a water-soluble emulsifier that is useful in reducing the viscosity of water. It can also be utilized as an effective oil-soluble emulsifier in the food and cosmetics industries. The other biosurfactant, produced by *Candida lipolytica*, is composed of 17% carbohydrate and 83% protein.

3.2.5 The particulate type

The extracellular membrane is made up of various protein and lipid structures designed to help microbes transport alkanes. The thick and vibrant vesicles of acinetobacter species strain H1N are 20–50 nm thick and have a buoyancy of 1.158 g/cm^3. They play an important role in the uptake of alkane by the cells by partitioning hydrocarbons. Phospholipid, protein, and lipopolysaccharide are the main components of the purified vesicles. Compared to the outer membrane of an organism, the purified vesicles have a 360-fold higher polysaccharide content.

3.3 Sources of production of biosurfactants

3.3.1 Microorganisms and growing media

Various microorganisms, such as Corynebacterium, Bacillus, and Candida, can be used to create biosurfactants. Among these, one of the most common is *P. aeruginosa*, as it is used to produce rhamnolipids. The rapid emergence and evolution of new technological tools and the high demand for biosurfactants are some of the factors expected to drive the market for this product in the future. In addition to these, the microorganisms can also use other compounds to grow (Eras-Muñoz et al., 2022). One of the most common sources of carbon for these organisms is glucose and glycerol. As glucose is an industrial feedstock, this can increase the cost of production (Abdul Hamid, Mohd Shamzi and Lai Yee Phang, 2018).

3.4 Fermentation process

The two main types of bioprocessing methods used to produce biosurfactants are solid-state fermentation (SSF) and submerged fermentation. Compared to conventional methods, SSF offers various advantages. It is energy-efficient and can use industrial residue and agro-industrial waste without inhibiting the substrates. It is also more rentable (Lizardi-Jiménez and Hernández-Martínez, 2017). However, it has drawbacks, such as the complexity of the downstream processing and operational monitoring (Ajila et al., 2011).

3.5 Low-cost byproducts and waste as feedstock

Fig. 3.1 shows the various sources of biosurfactant production. Biosurfactants can be made from various industrial wastes such as animal fat, starch, whey, corn steep liquor, and molasses. These materials, typically cheaper than raw materials, can be used to produce different biosurfactants (Soumen, Palashpriya and Ramkrishna, 2006). The energy, carbon, and nitrogen used by microbes to grow on low-cost substrates can help increase biosurfactants. Using starch, glucose, and other agro-industrial products can also help boost the number of biosurfactants in the cell (Khadydja et al., 2016). L-amino acids, which include glutamic acid, levulinic acid, and -valine, were selected as the ideal source of nitrogen for the production of biosurfactants (Abdul Hamid, Mohd Shamzi and Lai Yee Phang, 2018). Different types of fungi, bacteria, and yeast can also be used to create biosurfactants. Most commonly, they are produced by the *Candida* and *Pseudomonas* species. Biosurfactants are usually synthesized during exponential or stationary phases.

Researchers are looking for new substrates for the production of biosurfactant products due to the increasing demand. These low-cost substrates can be utilized to reduce the manufacturing costs and provide waste management. The following section

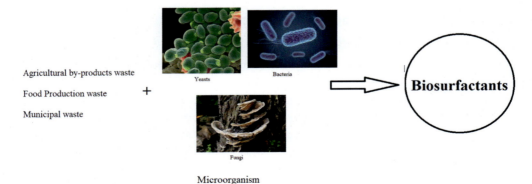

Figure 3.1 Various sources of biosurfactant production.

will provide an overview of the various feedstocks used in the process. This section covers the various types of solid waste that can be utilized for biosurfactant production. Various low-cost substrates, such as food waste, industrial waste, and agricultural products, can be used to address this issue. It also provides an overview of their advantages. Unfortunately, there is a lack of coverage in the literature regarding this subject. This section also explores the various problems related to the production of biosurfactants (Bueno et al., 2019; Singh et al., 2019). These include the high cost of processing, the lack of suitable raw materials, and the effects of batch processing on the quality of the product. Despite the technological advancements that have occurred in the field, many challenges remain to be resolved (Patel et al., 2019). To address the issue of solid waste management, processes that utilize waste as substrates can be developed. This approach can help minimize pollution and improve the efficiency of the process (Mohanty et al., 2021).

Due to the high cost of bioemulsifiers and biosurfactants, these products should be produced on inexpensive and renewable substrates using various carbon and nitrogen sources and water-soluble and -insoluble compounds. The production location within a cell can also vary depending on the carbon source and the substrate. For instance, different types of carbon sources, such as ethanol, hexadecane, and gasoline, can be utilized as water-based compounds.

3.6 Production of biosurfactants using microorganisms
3.6.1 Bacteria

In the production of biosurfactants, bacteria play a significant role. Table 3.1 shows the various species that are involved in this process. For instance, the dominant genus in biosurfactant production is the bacterium pseudomonas. In a study, Coelho et al. (2003) reported that the degradation of marine *Pseudomonas* sp. strain GU 104 by quinoline-degrading bacteria resulted in the production of biosurfactant products. Another study showed that the same bacterium produced polymeric biosurfactants (Husain et al., 1997). Emulsifying activity was also observed in the extracellular biosurfactants produced by the pseudomonas isolated from a coastal zone of the Mediterranean Sea. Depending on the type of hydrocarbons and carbon sources, various types of emulsifiers can be produced by microorganisms. The study conducted by Desai et al. (1988) revealed that the hydrocarbon-degrading *Pseudomonas fluorescens* can lead to the production of trehalose lipid-*o*-dialkyl monoglyceride—protein emulsifiers.

The bacteria known as *Bacillus subtilis* is mainly known for its ability to produce various nutrients, such as lipids, subtilisins, and lichenysin. In a study conducted on wastes from the production of biodiesel, Santos et al. (2014) discovered that subtilis grew on beet peel, corn steep liquor, and glycerin. A statistical model was then utilized to analyze the various factors affecting the production of biosurfactants (Ostendorf et al., 2019).

Table 3.1 Summary of various bacterium used to produce biosurfactants.

Biosurfactant	Bacterium	Carbon source	References
Rhamnolipid	Pseudomonas aeruginosa TGC01	Glycerol 4% (w/v)	Bezerra et al. (2019)
Rhamnolipid	Pseudomonas aeruginosa Dr1	Mango oil and glucose 1% (w/v)	Sathi Reddy et al. (2016)
Rhamnolipid		Mango oil 1% (w/v)	
Rhamnolipid	Pseudomonas aeruginosa	Glucose 6% (w/v)	Sahebnazar et al. (2018)
Rhamnolipid	Pseudomonas aeruginosa SR17	Glucose 2% (w/v) and paneer whey waste Crude oil 2% (w/v)	Patowary et al. (2016)
Rhamnolipid	Pseudomonas aeruginosa	Glycerol, glucose, mannitol, molasses, and n-hexadecane at 2% (w/v)	Patowary et al. (2018)
Rhamnolipid	Pseudomonas aeruginosa PBS	Glucose 5% (w/v) and kerosene 3% (w/v)	Sharma et al. (2018a)
Rhamnolipid	Pseudomonas aeruginosa PA1	Glycerol 2% (w/v)	Tomar et al. (2019)
Rhamnolipid	Pseudomonas LSH-7	Crude oil 0.5% (w/v)	Pi et al. (2017)
Mono- and di-rhamnolipid	Bacillus algicola 003-Phe1, Rhodococcus soli 102-Na5, Isoptericola chiayiensis 103-Na4, and Pseudoalteromonas agarivorans SDRB-Py1	Mannitol 2%, glucose, glycerol, starch, and crude oil 1% (w/v)	Lee et al. (2018)
Rhamnolipid	Pseudomonas aeruginosa	Glycerol 2% (v/v)	Huang et al. (2020)
Rhamnolipid	Pseudomonas aeruginosa UCP 0992	Corn steep liquor 0.5% and vegetable oil residue 4% (v/v)	Huang et al. (2020)
Rhamnolipid	Pseudomonas aeruginosa	Glucose 1% (w/v)	Silva et al. (2018)
Surfactin	Bacillus velezensis MHNK1	Residual frying oil 2% (v/v)	Jakinala et al. (2019)
Surfactin	Bacillus subtilis ATCC 21332	Glucose 4% (w/v)	Long et al. (2017)
Surfactin	Bacillus methylotrophicus	Soybean oil 2% (v/v) and whey	Machado et al. (2020)
Surfactin	Bacillus amyloliquefaciens SAS-1 and Bacillus subtilis BR-15	Glycerol 5% (w/v) and glucose 2.17% (w/v)	Ben Ayed et al. (2015)

Surfactin and others	Bacillus lichenimorfis L20	Glucose, sucrose, and lactose 10 g/L	Liu et al. (2021)
Crude lipopeptide	Bacillus amyloliquefaciensAn6	Glucose 20 g/L	Sharma et al. (2018b)
Crude lipopeptide	Bacillus subtilis CN2	Glycerol 4% (v/v)	Bezza & Chirwa (2015)
Crude lipopeptide	Paenibacillus dendritiformis	Sunflower oil 3% (v/v) and anthracene 0.01% (w/v)	Bezza & Chirwa (2017)
Crude lipopeptide	Bacillus subtilis AS2, Bacillus licheniformis AS3 and Bacillus velezensis AS4	Crude oil 2% (w/v)	Prakash et al. (2021)
Crude lipopeptide	Bacillus cereus UCP 1615	Waste frying soybean oil 2% (w/v)	Durval et al. (2020)
Crude lipopeptide	Bacillus subtilis	Glucose 15 g/L and crude oil 1% (w/v)	Bezza & Nkhalambayausi Chirwa (2016)
Crude lipopeptide	Pseudomonas aeruginosa	Glycerol 60 g/L	Tian et al. (2016)
Crude lipopeptide	Rhodococcus sp. TW53 R. ruber MP4	n-hexadecane 2% (w/v) Petroleum + glucose	Peng et al. (2008) Yalaoui-Guellal et al. (2020)
Syringafactin	Rhodococcus sp. ADL36 Pseudomonas putida	Diesel Glucose 2.5 g/L	Syahir et al. (2020) Zouari et al. (2019)
Other biosurfactants			
Biosurfactant extract	Serratia marcescens and Serratia nematodiphila	Kerosene 2% (v/v)	Borah et al. (2019)
Biosurfactant extract	Enterobacteriaceae, Pseudomonas, Microbacterium and Rhodanobacteraceae	Colza oil and glucose 20 g/L	Cazals et al. (2020)

In this study, the researchers used wastes from the agro-industrial sector, as well as molasses and waste frying oil, to create a biosurfactant that can contain a lipopeptide. The biosurfactant was produced in concentrations up to 2.05 g/L with 2% molasses and 1% corn steep liquor using *B. cereus*. The presence of marine algae, *Artemia salina*, exhibited low toxicity and the potential to disperse and desorption of motor oil. A study (Moshtagh et al., 2017) conducted on the production of subtilis biosurfactants from industrial wastewater revealed that the process reduced surface tension by 36% and enabled the emulsification of motor oil and soybean oil. In 2016, Nalini et al. (2016) conducted a study on the production of a protein known as lipopeptide from *B. cereus* using different oily residue food (SSF) substrates. They found that the low-cost substrates, such as peanut oil cake, were suitable for producing the protein. The researchers noted that the concentration of substrates plays a vital role in producing bio-based substances (BS) in the SSF process. In addition to providing a high level of nutrients, the availability of substrates also plays a significant role in the development of sustainable processes. In 2013, Zhu and colleagues (2013) investigated how surfactin was produced by *Bacillus amyloliquefaciens* in soybean flour and rice straw substrates. They also studied the effects of different sources of nitrogen and carbon, such as glucose, maltose, and sucrose. The researchers found that the addition of various sources of nitrogen did not affect surfactin production. They also noted that the increase in surfactin concentration was observed on adding 2.0% (w/w) of maltose and glycerol. The economics of the production of the peptide were determined by *B. mojavenis* A21 versus *Fusarium* sp. The substance production by *B. subtilis* (Moshtagh et al., 2019) was evaluated by analyzing the brewery waste. A response surface methodology was utilized to improve the efficiency of the process. The lowest surface tension was recorded at 27.3 mN/m, and the maximum was 107 mg/L. The researchers found that the optimal concentration of 0.66 mg/L of biosurfactant was reached at 6.22 g/L of waste. In another study, Paraszkiewicz and colleagues (Paraszkiewicz et al., 2018) investigated how *B. subtilis* can produce a type of peptide from renewable substrates. They also analyzed the effects of different sources of sugar, including beets, carrots, and apples.

In 2013, Cormack et al. (1997) isolated a bacterium, namely, Acinetobacter sp., from the soil samples taken near the shoreline in Jubany Station (King George Island, South Shetland Islands). It plays a vital role in the degradation of hydrocarbons (*n*-dodecane and *n*-hexadecane) and is also known to be involved in bioremediation processes. In Dongying, China, Bao and colleagues (2014) isolated the bacteria Acinetobacter sp. D3–2 from an oil-contaminated area. The ideal environmental conditions for the growth of the bacteria were set at pH 8.0 and a 3% (w/v) NaCl around 30°C. The bacteria could potentially use different hydrocarbon substrates for their energy and carbon needs. The researchers noted that the strain could produce biosurfactants at a production rate of 0.52 g/Ls. The researchers also noted that the culture broth's surface tension decreased from 48.02 to 26.30 mN/m.

The surface-active molecules produced by the *Rhodococcus* sp. bacterium are known to be involved in the production of various types of lipids. A few members of the group are known to produce liposomes, but none of them can produce hydrophobic peptides. The characteristics of the Rhodococcus species, such as their wide catabolic diversity and high persistence in low oxygen conditions, make them an excellent candidate for the remediation of various types of petroleum hydrocarbons (Kuyukina & Ivshina, 2010a). They also resist environmental stresses well. Some of the common Rhodococcus species known to produce biosurfactants are *Rhodococcus erythropolis*, *Rhodococcus opacus*, *Rhodococcus rhodochrous*, *Rhodococcus ruber*, and *Rhodococcus fascians* (Kuyukina & Ivshina, 2010b; Malavenda et al., 2015). However, to date, the exact number of strains capable of producing biosurfactants is still unclear. Their structural and characterization studies are still sparse. In a study (Peng et al., 2007) conducted on the soil of the Island of Xiamen in China, researchers identified a strain of the *Rhodococcus erythropolis* strain 3C-9 bacterium that can degrade oil.

The bacteria called *Halomonas* sp. (Juni, 1978) is mainly known for its ability to produce exopolysaccharides that have emulsifying properties. There are a few reports about its ability to create surface-active agents. However, a strain of Halomonas ANT-3b was isolated from the Ross Sea's Terra Nova Bay station. Emulsifying glycolipids are produced by this bacterium. Bioemulsifiers based on glycoproteins produced by *Halomonas* sp. have been characterized by Pepi et al. (2005) and Gutierrez et al. (2007).

Table 3.1 shows the various sources of bacterium used to produce biosurfactants (Bezerra et al., 2019; Patowary et al., 2016; Sahebnazar et al., 2018; Patowary et al., 2016; Patowary et al., 2018; Sharma et al., 2018a; Tomar et al., 2019; Pi et al., 2017; Lee et al., 2018; Huang et al., 2020; Silva et al., 2018; Sun et al., 2019; Jakinala et al., 2019; Long et al., 2017; Machado et al., 2020; Ben Ayed et al., 2015; Abdul Hamid, Mohd Shamzi and Lai Yee Phang, 2018; Liu et al., 2021; Sharma et al., 2018b; Bezza & Chirwa, 2015; Bezza & Chirwa, 2017; Prakash et al., 2021; Durval et al., 2020; Bezza & Nkhalambayausi Chirwa, 2016; Al-Kashef et al., 2018; Peng et al., 2008; Yalaoui-Guellal et al., 2020; Syahir et al., 2020; Zouari et al., 2019; Borah et al., 2019; Cazals et al., 2020).

3.6.2 Fungi

Due to the limited availability of commercially available fungi for biosurfactant production, only a few reports on this subject have been published. Ascomycetes are a type of anamorphs that produce asexual spores on branching structures known as conidiophores. These are septate fungi with filaments partitioned by septa cells (McConnaughey, 2014). The secondary metabolites of ascomycetes can be biosynthesized through exchange or absorption. Some of these are known to be biosurfactant producers, such as Penicillium, Aspergillus, and Fusarium (Ron & Rosenberg, 2001). Compared to yeasts, ascomycetes with filaments exhibited significant advantages in terms of their production of biosurfactants (Bhardwaj, 2013). These fungi are known to be excellent producers of stable emulsifiers and biosurfactants. They can also reduce interfacial and surface tension (Laine et al., 1972; Reis et al., 2018).

The study conducted by Minucelli et al. (2017) analyzed the sophorolipid by *Candida bombicola* ATCC 22214 production process using sugar cane molasses, glucose, and sucrose as hydrophilic sources, as well as chicken fat or sunflower oil as a hydrophobic one. The results indicated that the production of sophorolipid was 39.81 g/L. The optimal condition of 75 g of chicken fat, 77.5 g of glucose, 2.5 g of yeast extract, and no urea was achieved. The sophorolipid produced was characterized by their surface tension properties, which exhibited a surface tension of 35 mN/m and CMC 65 mg/L, and showed an emulsification index of greater than 10 in toluene, *n*-heptane, and lubricant oil.

Table 3.2 shows the various types of biosurfactants produced by various fungi (Pele et al., 2019; Kannahi & Sherley, 2012; Al-Kashef et al., 2018; Silva et al., 2021; Qazi et al., 2014; Gautam, 2014; Pitocchi et al., 2020; Cicatiello et al., 2019).

Table 3.2 Summary of various fungi used to produce biosurfactants.

Biosurfactant	Fungi	Carbon source	References
Biosurfactant extract	Rhizopus arrhizus UCP1607	Crude glycerol and corn steep liquor	Pele et al. (2019)
Glycolipid	Aspergillus niger	Mineral salt broth with crude oil, olive oil, groundnut oil, and coconut oil	Kannahi & Sherley (2012)
	Fusarium sp.	sunflower oil cake and pineapple waste mixture	Al-Kashef et al. (2018)
	Aspergillus niger	-	Silva et al. (2021)
Lipopeptide	Fusarium SP BS-8	sucrose	Qazi et al. (2014)
Lipopeptide	Penicillium chrysogenum SNP5		Gautam (2014)
Cerato-platanins	Aspergillus terreus MUT 271	crude oil	Pitocchi et al. (2020)
Sap-Pc protein	Penicillium chrysogenum MUT 5039	-	Cicatiello et al. (2019)

3.6.3 Yeast

Compared to bacteria, yeasts are known to produce higher-concentration biosurfactants. However, they are not widely used yet due to the high cost of obtaining them. This makes them uneconomical for industrial-scale production. One of the ways to reduce production costs is by using alternative substrates (de Lima & Contiero, 2009). This can be done using agricultural waste or food industry products containing high levels of lipids and carbohydrates (Benincasa, 2007). Examples of residue substrates include preused soap stock, waste products from the production of sugar cane and molasses, and various types of oils and fats (Desai & Banat, 1997). These can be used in the process of obtaining biosurfactant-producing materials. Compared to using conventional substrates, the use of residual ones can significantly reduce production costs.

An unusual yeast strain isolated from a region of the Earth that has high emulsification index (80%) and crude oil (76%)-removal capabilities has been reported to produce biosurfactant or bioemulsifier (Oloke & Glick, 2005). This strain, which is known as *Pseudozyma* sp., was mainly utilized to produce biosurfactants.

To create a biosurfactant from Antarctic soil, the researchers isolated 68 yeast strains (Bueno et al., 2019). Eleven of them were able to effectively create biosurfactants after passing the collapse and emulsification index tests, with Strain 1_4.0, which was later determined to be *Candida glaebosa*, being the top producer. For the study, the researchers fed yeast extract into a medium containing glycerol for 120 h to determine the kinetics of the process. The increase in the C/N ratio significantly affected the yeast growth and production. Cooper and Paddock (1983) isolated *Torulopsis petrophilum* to produce sophorolipids. Kakugawa et al. (2002) isolated *Kurtzmanomyces* spp. from soybean oil as a sole carbon source.

The study conducted by Pratap et al. (Amit et al., 2021) utilized waste sirup from the jaggery plant, oleic acid, and corn oil for the synthesis of sophorolipids. The resulting products were then purified through column chromatography and solvent extraction. At an optimal pH control level, the *Starmerella bombicola* (ATCC 22214) produced a maximum yield of biosurfactant of 42.9 g/L. The sophorolipid was then characterized and identified through various techniques, such as thin-layer chromatography (TLC), Fourier-transform infrared spectroscopy (FTIR), high-performance liquid chromatography (HPLC), and liquid chromatography-mass spectroscopy (LC-Ms). The surface tension of the surface of water decreased from 72.28 to 28.83 mN/m. The sophorolipid produced from nonedible jaggery plants had a total CMC of 76.12 mg/L. The highest concentration of sophorolipid was found in oleic acid and sunflower oil. On the other hand, the products produced from waste sirup and corn oil had an emulsification index of 48.44%.

In a study conducted by Jiménez-Pealver et al. (2018), *S. bombicola* produced sophorolipids using stearic acid as a low-cost substrate. Due to its high melting point and chemical properties, stearic acid is difficult to work with in submerged fermentation. The researchers monitored the process for 16 days. They achieved the highest yield of sophorolipids on day 13 (0.211 g/g). They also observed a gradual increase in the consumption of the substrate. This shows the importance of using hydrophobic substrates in the production of biosurfactants.

A study (Rufino et al., 2014) conducted on the yeast *C. lipolyticum* UCP 0988 revealed that the fermentation time significantly increases biosurfactant production. It was also associated with cell growth. However, the concentrations of these substances in the stationary phase increased during the 72-h fermentation. In this study, biosurfactant production was also associated with yeast growth. The emulsification rates during the stationary phase ($\geq$30% E.I. at 120 h) were higher than during the exponential phase.

Table 3.3 lists various biosurfactants-producing yeast (Gaur et al., 2019; Daniel et al., 1998; Hommel et al., 1994; Rodríguez et al., 2021; Frautz et al., 1986; Katemai et al., 2008; Luna et al., 2015; da Rocha Junior et al., 2019; Sajna et al., 2015; Teixeira Souza et al., 2018).

Table 3.3 Summary of various yeasts used to produce biosurfactants.

Biosurfactant	Yeasts	Carbon Source	References
Sophorolipid	*Candida albicans* and *Candida glabrata*	Glucose 2% (w/v)	Gaur et al. (2019)
Sophorolipid	*Candida bombicola* ATCC 22214	Whey, rapeseed oil	Daniel et al. (1998)
Sophorolipid	*Candida apicola* IMET 43147	Glucose, sunflower oil	Hommel et al. (1994)
Sophorolipid	*Starmerella bombicola* ATCC 22214	Winterization oil cake (WOC) and molasses	Rodríguez et al. (2021)
Cellulose lipid	*Ustilago maydis* ATCC 14826	Coconut oil	Frautz et al. (1986)
Fatty acid	*Issatchenkia orientalis* SR4	Glucose, weathered crude oil, xylene	Katemai et al. (2008)
Sophorolipid	*Candida sphaerica* UCP 0995	Corn steep liquor 9% and groundnut oil refinery residue 9% (v/v)	Luna et al. (2015)
Sophorolipid	*Candida tropicalis*	Corn liqueur 4%, molasses 2.5% and canola frying oil 2.5% (v/v)	da Rocha Junior et al. (2019)
MELs	*Pseudozyma* sp. NII 08165	Soybean oil, diesel, kerosene, and petroleum	Sajna et al. (2015)
Biosurfactant extract	*Wickerhamomyces anomalus* CCMA 0358	Glucose 2 g/L and olive oil 20 g/L	Teixeira Souza et al. (2018)

3.7 Factors affecting biosurfactants production

Various factors such as glucose, mannitol, carbon source, and oil can affect the process of biosurfactant production. Other factors that can affect the process include agitation speed, pH, nitrogen source, and the presence of a certain type of group (De et al., 2015).

3.8 Challenges and future research directions

Biosurfactants have many advantages, but they also have certain drawbacks. One of the disadvantages is their high production costs. This is especially true for applications in the pharmaceutical, cosmetic, and food industries. In addition, the process of diluting the broths can take several sequential steps. One of the most important factors that can affect the production of biosurfactants is the presence of overproducing bacteria. This issue can be prevented by implementing the O_2-limitation process. The addition of waste substrates to the process, which can counter the effects of pollution, can help overcome the challenges associated with the production of biosurfactants. A comprehensive economic study should also be conducted to analyze the various steps involved in the purification process and their potential applications, which would help improve the efficiency of the biosurfactants manufacturing process. Considering the various factors that affect the composition of end products, it is important to find the right biosurfactants for industrial scale-up. For instance, regarding environmental remediation, the purity of the biosurfactants used should not be compromised. Due to the increasing cost of microbial cultivation, the production of industrial biosurfactants is still in its early stages. Due to the complexity of the process, it has been suggested that instead of using pure substrates, waste, and byproducts can be used as a feedstock for the production of sophorolipid. Due to the complexity of the process involved in recovering biosurfactant, the cost of production during the downstream stages is often significantly higher than the cost of production. Therefore, the government and private sectors must work together to address the financial crises in the biosurfactant industry. Currently, various substrates such as wheat straw, cassava, and cane sugar can be used for the production of biosurfactants.

3.9 Conclusions

Biosurfactants have low toxicity under different environmental conditions, such as salinity, temperature, and pH. Compared to synthetic chemicals, these are safer and more sustainable due to their high biodegradability and low toxicity. They can be made from various bio-resources wastes and renewable substrates, significantly lowering their production cost.

The production of biosurfactants requires a comprehensive understanding of their life cycle. New molecular techniques for identifying microbial strains are needed to make biosurfactants from wastes at high yields. To improve the yields of biosurfactants, further research is needed on the composition of wastes. It is also important to improve the purification and recovery procedures. A deeper understanding of the product purity can help in determining the ideal level of biosurfactant for different applications. To achieve a full-scale production process, studies on the production of biosurfactants in an in-place environment are needed. This method can help minimize the need for additional surfactants.

Acknowledgments

The authors gratefully acknowledge the Council of Scientific and Industrial Research (CSIR), New Delhi, India, for the funding under the research project [File number: 22(0832)/20/EMR-II].

References

Abdul Hamid, N., Mohd Shamzi, M., & Lai Yee, P. (2018). Culture medium development for microbial-derived surfactants production—An overview. *Molecules, 23*((5), 1049.

Ajila, C. M., Brar, S. K., Verma, M., Tyagi, R. D., & Valéro, J. R. (2011). Solid-state fermentation of apple pomace using Phanerocheate chrysosporium — Liberation and extraction of phenolic antioxidants. *Food Chemistry, 126*(3), 1071−1080. Available from https://doi.org/10.1016/j.foodchem.2010.11.129.

Al-Kashef, A., Shaban, S., Nooman, M., & Rashad, M. (2018). Effect of fungal glycolipids produced by a mixture of sunflower oil cake and pineapple waste as green corrosion inhibitors. *Journal of Environmental Science and Technology, 11*(3), 119−131. Available from https://doi.org/10.3923/jest.2018.119.131.

Amaral, P. F. F., Coelho, M. A. Z., Marrucho, I. M. J., & Coutinho, J. A. P. (2010). Biosurfactants from yeasts: Characteristics, production and application. *Advances in Experimental Medicine and Biology, 672*, 236−249. Available from https://doi.org/10.1007/978-1-4419-5979-9_18.

Amit, P. P., Rohan, S. M., & Suraj, N. M. (2021). Waste derived-green and sustainable production of Sophorolipid. *Current Research in Green and Sustainable Chemistry, 4*, 100209.

Aslam, R., Mobin, M., Aslam, J., Aslam, A., Zehra, S., & Masroor, S. (2021). Application of surfactants as anticorrosive materials: A comprehensive review. *Advances in Colloid and Interface Science, 295*, 102481. Available from https://doi.org/10.1016/j.cis.2021.102481.

Bao, M., Pi, Y., Wang, L., Sun, P., Li, Y., & Cao, L. (2014). Lipopeptide biosurfactant production bacteria Acinetobacter sp. D3-2 and its biodegradation of crude oil. *Environmental Sciences: Processes and Impacts, 16*(4), 897−903. Available from https://doi.org/10.1039/C3EM00600J.

Ben Ayed, H., Jemil, N., Maalej, H., Bayoudh, A., Hmidet, N., & Nasri, M. (2015). Enhancement of solubilization and biodegradation of diesel oil by biosurfactant from Bacillus amyloliquefaciens An6. *International Biodeterioration and Biodegradation, 99*, 8−14. Available from https://doi.org/10.1016/j.ibiod.2014.12.009.

Benincasa, M. (2007). Rhamnolipid produced from agroindustrial wastes enhances hydrocarbon biodegradation in contaminated soil. *Current Microbiology, 54*(6), 445−449. Available from https://doi.org/10.1007/s00284-006-0610-8.

Bezerra, K. G. O., Gomes, U. V. R., Silva, R. O., Sarubbo, L. A., & Ribeiro, E. (2019). The potential application of biosurfactant produced by *Pseudomonas aeruginosa* TGC01 using crude glycerol on the enzymatic hydrolysis of lignocellulosic material. *Biodegradation, 30*(4), 351−361. Available from https://doi.org/10.1007/s10532-019-09883-w.

Bezza, F. A., & Chirwa, E. M. N. (2015). Production and applications of lipopeptide biosurfactant for bioremediation and oil recovery by Bacillus subtilis CN2. *Biochemical Engineering Journal, 101*, 168−178. Available from https://doi.org/10.1016/j.bej.2015.05.007.

Bezza, F. A., & Chirwa, E. M. N. (2017). The role of lipopeptide biosurfactant on microbial remediation of aged polycyclic aromatic hydrocarbons (PAHs)-contaminated soil. *Chemical Engineering Journal, 309*, 563−576. Available from https://doi.org/10.1016/j.cej.2016.10.055.

Bezza, F. A., & Nkhalambayausi Chirwa, E. M. (2016). Biosurfactant-enhanced bioremediation of aged polycyclic aromatic hydrocarbons (PAHs) in creosote contaminated soil. *Chemosphere*, *144*, 635–644. Available from https://doi.org/10.1016/j.chemosphere.2015.08.027.

Bhardwaj, G. (2013). Biosurfactants from Fungi: A review. *Journal of Petroleum & Environmental Biotechnology*, *04*(06), 160. Available from https://doi.org/10.4172/2157-7463.1000160.

Borah, D., Agarwal, K., Khataniar, A., Konwar, D., Gogoi, S. B., & Kallel, M. (2019). A newly isolated strain of Serratia sp. from an oil spillage site of Assam shows excellent bioremediation potential. *3 Biotech*, *9*(7). Available from https://doi.org/10.1007/s13205-019-1820-7.

Bueno, J. L., Santos, P. A. D., da Silva, R. R., Moguel, I. S., Pessoa, A., Vianna, M. V., Pagnocca, F. C., Sette, L. D., & Gurpilhares, D. B. (2019). Biosurfactant production by yeasts from different types of soil of the South Shetland Islands (Maritime Antarctica). *Journal of Applied Microbiology*, *126*(5), 1402–1413. Available from https://doi.org/10.1111/jam.14206.

Cazals, F., Huguenot, D., Crampon, M., Colombano, S., Betelu, S., Galopin, N., Perrault, A., Simonnot, M.-O., Ignatiadis, I., & Rossano, S. (2020). Production of biosurfactant using the endemic bacterial community of a PAHs contaminated soil, and its potential use for PAHs remobilization. *Science of The Total Environment*, *709*, 136143. Available from https://doi.org/10.1016/j.scitotenv.2019.136143.

Cicatiello, P., Stanzione, I., Dardano, P., De Stefano, L., Birolo, L., De Chiaro, A., Monti, D. M., Petruk, G., D'errico, G., & Giardina, P. (2019). Characterization of a surface-active protein extracted from a marine strain of penicillium chrysogenum. *International Journal of Molecular Sciences*, *20*(13), 133242. Available from https://doi.org/10.3390/ijms20133242.

Coelho, P. A., Queiroz-Machado, J., & Sunkel, C. E. (2003). Condensin-dependent localisation of topoisomerase II to an axial chromosomal structure is required for sister chromatid resolution during mitosis. *Journal of Cell Science*, *116*(23), 4763–4776. Available from https://doi.org/10.1242/jcs.00799.

Cooper, D. G., & Paddock, D. A. (1983). Torulopsis petrophilum and surface activity. *Applied and Environmental Microbiology*, *46*(6), 1426–1429. Available from https://doi.org/10.1128/aem.46.6.1426-1429.1983.

Cormack, W. P. M., & Fraile, E. R. (1997). Characterization of a hydrocarbon degrading psychrotrophic Antarctic bacterium. *Antarctic Science*, *9*(2), 150–155. Available from https://doi.org/10.1017/s0954102097000199.

da Rocha Junior, R. B., Meira, H. M., Almeida, D. G., Rufino, R. D., Luna, J. M., Santos, V. A., & Sarubbo, L. A. (2019). Application of a low-cost biosurfactant in heavy metal remediation processes. *Biodegradation*, *30*(4), 215–233. Available from https://doi.org/10.1007/s10532-018-9833-1.

Daniel, H. J., Reuss, M., & Syldatk, C. (1998). Production of sophorolipids in high concentration from deproteinized whey and rapeseed oil in a two stage fed batch process using Candida bombicola ATCC 22214 and Cryptococcus curvatus ATCC 20509. *Biotechnology Letters*, *20*(12), 1153–1156. Available from https://doi.org/10.1023/A:1005332605003.

de Lima, C. J. B., & Contiero, J. (2009). Use of soybean oil fry waste for economical biosurfactant production by isolated *Pseudomonas aeruginosa* using response surface methodology. *Current Trends in Biotechnology and Pharmacy*, *3*(2), 162–171. Available from http://www.abap.co.in/files/journal/2009-3-vol-2-pdf/Paper-6.pdf.

De, S., Malik, S., Ghosh, A., Saha, R., & Saha, B. (2015). A review on natural surfactants. *RSC Advances*, *5*(81), 65757–65767. Available from https://doi.org/10.1039/c5ra11101c.

Desai, A. J., Patel, K. M., & Desai, J. D. (1988). Emulsifier production by *Pseudomonas fluorescens* during the growth on hydrocarbons. *Curr. Sci*, *57*(9), 500–501.

Desai, J. D., & Banat, I. M. (1997). Microbial production of surfactants and their commercial potential. *Microbiology and Molecular Biology Reviews*, *61*(1), 47–64. Available from https://doi.org/10.1128/0.61.1.47-64.1997.

Dhasayan, A., Selvin, J., & Kiran, S. (2015). Biosurfactant production from marine bacteria associated with sponge *Callyspongia diffusa*. *3 Biotech*, *5*(4), 443–454. Available from https://doi.org/10.1007/s13205-014-0242-9.

Durval, I. J. B., Mendonça, A. H. R., Rocha, I. V., Luna, J. M., Rufino, R. D., Converti, A., & Sarubbo, L. A. (2020). Production, characterization, evaluation and toxicity assessment of a Bacillus cereus UCP 1615 biosurfactant for marine oil spills bioremediation. *Marine Pollution Bulletin*, *157*, 111357. Available from https://doi.org/10.1016/j.marpolbul.2020.111357.

Eras-Muñoz, E., Farré, A., Sánchez, A., Font, X., & Gea, T. (2022). Microbial biosurfactants: A review of recent environmental applications. *Bioengineered*, *13*(5), 12365−12391. Available from https://doi.org/10.1080/21655979.2022.2074621.

Fawzy, A., Abdallah, M., Zaafarany, I. A., Ahmed, S. A., & Althagafi, I. I. (2018). Thermodynamic, kinetic and mechanistic approach to the corrosion inhibition of carbon steel by new synthesized amino acids-based surfactants as green inhibitors in neutral and alkaline aqueous media. *Journal of Molecular Liquids*, *265*, 276−291. Available from https://doi.org/10.1016/j.molliq.2018.05.140.

Frautz, B., Lang, S., & Wagner, F. (1986). Formation of cellobiose lipids by growing and resting cells of *Ustilago maydis*. *Biotechnology Letters*, *8*(11), 757−762. Available from https://doi.org/10.1007/BF01020817.

Freitas de Oliveira, D. W., Lima França, I. W., Nogueira Félix, A. K., Lima Martins, J. J., Aparecida Giro, M. E., Melo, V. M. M., & Gonçalves, L. R. B. (2013). Kinetic study of biosurfactant production by *Bacillus subtilis* LAMI005 grown in clarified cashew apple juice. *Colloids and Surfaces B: Biointerfaces*, *101*, 34−43. Available from https://doi.org/10.1016/j.colsurfb.2012.06.011.

Gaur, V. K., Sharma, P., Sirohi, R., Varjani, S., Taherzadeh, M. J., Chang, J.-S., Yong, Ng, H., Wong, J. W. C., & Kim, S.-H. (2022). Production of biosurfactants from agro-industrial waste and waste cooking oil in a circular bioeconomy: An overview. *Bioresource Technology*, *343*, 126059. Available from https://doi.org/10.1016/j.biortech.2021.126059.

Gaur, V. K., Bajaj, A., Regar, R. K., Kamthan, M., Jha, R. R., Srivastava, J. K., & Manickam, N. (2019). Rhamnolipid from a *Lysinibacillus sphaericus* strain IITR51 and its potential application for dissolution of hydrophobic pesticides. *Bioresource Technology*, *272*, 19−25. Available from https://doi.org/10.1016/j.biortech.2018.09.144.

Gautam, G. (2014). A cost effective strategy for production of bio-surfactant from locally isolated *Penicillium chrysogenum* SNP5 and its applications. *Journal of Bioprocessing & Biotechniques*, *04*(06). Available from https://doi.org/10.4172/2155-9821.1000177.

Gutiérrez, T., Mulloy, B., Black, K., & Green, D. H. (2007). Glycoprotein emulsifiers from two marine *Halomonas* species: Chemical and physical characterization. *Journal of Applied Microbiology*, *103*(5), 1716−1727. Available from https://doi.org/10.1111/j.1365-2672.2007.03407.x.

Hommel, R. K., Weber, L., Weiss, A., Himmelreich, U., Rilke, O., & Kleber, H. P. (1994). Production of sophorose lipid by Candida (Torulopsis) apicola grown on glucose. *Journal of Biotechnology*, *33*(2), 147−155. Available from https://doi.org/10.1016/0168-1656(94)90107-4.

Huang, X., Zhou, H., Ni, Q., Dai, C., Chen, C., Li, Y., & Zhang, C. (2020). Biosurfactant-facilitated biodegradation of hydrophobic organic compounds in hydraulic fracturing flowback wastewater: A dose−effect analysis. *Environmental Technology & Innovation*, *19*, 100889. Available from https://doi.org/10.1016/j.eti.2020.100889.

Husain, D. R., Goutx, M., Acquaviva, M., Gilewicz, M., & Bertrand, J. C. (1997). Short communication: The effect of temperature on eicosane substrate uptake modes by a marine bacterium *Pseudomonas nautica* strain 617: Relationship with the biochemical content of cells and supernatants. *World Journal of Microbiology and Biotechnology*, *13*(5), 587−590. Available from https://doi.org/10.1023/A:1018581829320.

Jakinala, P., Lingampally, N., & Kyama, a. (2019). Ecotoxicology and environmental safety enhancement of atrazine biodegradation by marine isolate Bacillus velezensisMHNK1 in presence of surfactin lipopeptide. *Ecotoxicol Environ Saf*, *182*, 109372.

Jiménez-Peñalver, P., Castillejos, M., Koh, A., Gross, R., Sánchez, A., Font, X., & Gea, T. (2018). Production and characterization of sophorolipids from stearic acid by solid-state fermentation, a cleaner alternative to chemical surfactants. *Journal of Cleaner Production*, *172*, 2735−2747. Available from https://doi.org/10.1016/j.jclepro.2017.11.138.

Juni, E. (1978). Genetics and physiology of Acinetobacter. *Annual Review of Microbiology*, *32*, 349−371. Available from https://doi.org/10.1146/annurev.mi.32.100178.002025.

Kakugawa, K., Tamai, M., Imamura, K., Miyamoto, K., Miyoshi, S., Morinaga, Y., Suzuki, O., & Miyakawa, T. (2002). Isolation of yeast kurtzmanomyces sp. I-11, novel producer of mannosylerythritol lipid. *Bioscience, Biotechnology and Biochemistry*, *66*(1), 188−191. Available from https://doi.org/10.1271/bbb.66.188.

Kannahi, M., & Sherley, M. (2012). Biosurfactant production by *Pseudomonas putida* and *Aspergillus niger* from oil contamined site. *International Journal of Chemistry and Pharmaceutical Sciences*, *3*, 37−42.

Karpenko, E. V., Pokin'Broda, T. Y., Makitra, R. G., & Pal'Chikova, E. Y. (2009). Optimal methods of isolation of biogenic ramnolipid surfactants. *Russian Journal of General Chemistry*, *79*(12), 2637−2640. Available from https://doi.org/10.1134/S1070363209120135.

Katemai, W., Maneerat, S., Kawai, F., Kanzaki, H., Nitoda, T., & H-Kittikun, A. (2008). Purification and characterization of a biosurfactant produced by Issatchenkia orientalis SR4. *Journal of General and Applied Microbiology*, *54*(1), 79−82. Available from https://doi.org/10.2323/jgam.54.79.

Khadydja, D., Raquel, D. R., Santos, V. A., Juliana, M. L., & Luna, S. (2016). Biosurfactants: Multifunctional biomolecules of the 21st century. *International Journal of Molecular Sciences*, *17*(3), 401.

Kuyukina, M. S., & Ivshina, I. B. (2010a). *Application of rhodococcus in bioremediation of contaminated environments* (pp. 231−262). Springer Science and Business Media LLC. Available from https://doi.org/10.1007/978-3-642-12937-7_9.

Kuyukina, M. S., & Ivshina, I. B. (2010b). *Rhodococcus biosurfactants: Biosynthesis, properties, and potential applications* (pp. 291−313). Springer Science and Business Media LLC. Available from https://doi.org/10.1007/978-3-642-12937-7_11.

Laine, R. A., Griffin, P. F. S., Sweeley, C. C., & Brennan, P. J. (1972). Monoglucosyloxyoctadecenoic Acid-a Glycolipid from *Aspergillus niger*. *Biochemistry*, *11*(12), 2267−2271. Available from https://doi.org/10.1021/bi00762a009.

Lee, D. W., Lee, H., Kwon, B. O., Khim, J. S., Yim, U. H., Kim, B. S., & Kim, J. J. (2018). Biosurfactant-assisted bioremediation of crude oil by indigenous bacteria isolated from Taean beach sediment. *Environmental Pollution*, *241*, 254−264. Available from https://doi.org/10.1016/j.envpol.2018.05.070.

Liu, Q., Niu, J., Yu, Y., Wang, C., Lu, S., Zhang, S., Lv, J., & Peng, B. (2021). Production, characterization and application of biosurfactant produced by *Bacillus licheniformis* L20 for microbial enhanced oil recovery. *Journal of Cleaner Production*, *307*, 127193. Available from https://doi.org/10.1016/j.jclepro.2021.127193.

Lizardi-Jiménez, M. A., & Hernández-Martínez, R. (2017). Solid state fermentation (SSF): Diversity of applications to valorize waste and biomass. *3 Biotech*, *7*, 44.

Long, X., He, N., He, Y., Jiang, J., & Wu, T. (2017). Biosurfactant surfactin with pH-regulated emulsification activity for efficient oil separation when used as emulsifier. *Bioresource Technology*, *241*, 200−206. Available from https://doi.org/10.1016/j.biortech.2017.05.120.

Luna, J. M., Rufino, R. D., Jara, A. M. A. T., Brasileiro, P. P. F., & Sarubbo, L. A. (2015). Environmental applications of the biosurfactant produced by Candida sphaerica cultivated in low-cost substrates. *Colloids and Surfaces A: Physicochemical and Engineering Aspects*, *480*, 413−418. Available from https://doi.org/10.1016/j.colsurfa.2014.12.014.

Machado, T. S., Decesaro, A., Cappellaro, Â. C., Machado, B. S., van Schaik Reginato, K., Reinehr, C. O., Thomé, A., & Colla, L. M. (2020). Effects of homemade biosurfactant from Bacillus methylotrophicus on bioremediation efficiency of a clay soil contaminated with diesel oil. *Ecotoxicology and Environmental Safety*, *201*, 110798. Available from https://doi.org/10.1016/j.ecoenv.2020.110798.

Maikudi Usman, M., Dadrasnia, A., Tzin Lim, K., Fahim Mahmud, A., & Ismail, S. (2016). Application of biosurfactants in environmental biotechnology; Remediation of oil and heavy metal. *AIMS Bioengineering*, *3*(3), 289−304. Available from https://doi.org/10.3934/bioeng.2016.3.289.

Makkar, R., & Cameotra, S. (2002). An update on the use of unconventional substrates for biosurfactant production and their new applications. *Applied Microbiology and Biotechnology*, *58*(4), 428−434. Available from https://doi.org/10.1007/s00253-001-0924-1.

Malavenda, R., Rizzo, C., Michaud, L., Gerçe, B., Bruni, V., Syldatk, C., Hausmann, R., & Lo Giudice, A. (2015). Biosurfactant production by Arctic and Antarctic bacteria growing on hydrocarbons. *Polar Biology*, *38*(10), 1565−1574. Available from https://doi.org/10.1007/s00300-015-1717-9.

McConnaughey, M. (2014). *Physical chemical properties of fungi* ☆. Elsevier BV. Available from https://doi.org/10.1016/b978-0-12-801238-3.05231-4.

Minucelli, T., Ribeiro-Viana, R. M., Borsato, D., Andrade, G., Cely, M. V. T., de Oliveira, M. R., Baldo, C., & Celligoi, M. A. P. C. (2017). Sophorolipids production by *Candida bombicola* ATCC

22214 and its potential application in soil bioremediation. *Waste and Biomass Valorization, 8*(3), 743−753. Available from https://doi.org/10.1007/s12649-016-9592-3.

Mobin, M., Aslam, R., & Aslam, J. (2017). Non toxic biodegradable cationic gemini surfactants as novel corrosion inhibitor for mild steel in hydrochloric acid medium and synergistic effect of sodium salicylate: Experimental and theoretical approach. *Materials Chemistry and Physics, 191*, 151−167. Available from https://doi.org/10.1016/j.matchemphys.2017.01.037.

Mohanty, S. S., Koul, Y., Varjani, S., Pandey, A., Ngo, H. H., Chang, J. S., Wong, J. W. C., & Bui, X. T. (2021). A critical review on various feedstocks as sustainable substrates for biosurfactants production: A way towards cleaner production. *Microbial Cell Factories, 20*(1). Available from https://doi.org/10.1186/s12934-021-01613-3.

Moshtagh, B., Hawboldt, K., & Zhang, B. (2019). Optimization of biosurfactant production by *Bacillus subtilis* N3-1P using the brewery waste as the carbon source. *Environmental Technology, 40*(25), 3371−3380. Available from https://doi.org/10.1080/09593330.2018.1473502.

Müller, M. M., Kügler, J. H., Henkel, M., Gerlitzki, M., Hörmann, B., Pöhnlein, M., Syldatk, C., & Hausmann, R. (2012). Rhamnolipids-next generation surfactants? *Journal of Biotechnology, 162*(4), 366−380. Available from https://doi.org/10.1016/j.jbiotec.2012.05.022.

Mulligan, C. N., Sharma, S. K., & Mudhoo, A. (2014). *Biosurfactants: Research trends and applications*. Routledge.

Nalini, S., Parthasarathi, R., & Prabudoss, V. (2016). Production and characterization of lipopeptide from Bacillus cereus SNAU01 under solid state fermentation and its potential application as anti-biofilm agent. *Biocatalysis and Agricultural Biotechnology, 5*, 123−132. Available from https://doi.org/10.1016/j.bcab.2016.01.007.

Oloke, J. K., & Glick, B. R. (2005). Production of bioemulsifier by an unusual isolate of salmon/red melanin containing *Rhodotorula glutinis*. *African Journal of Biotechnology, 4*(2), 164−171. Available from http://www.academicjournals.org/AJB/PDF/Pdf2005/Feb/Oloke%20and%20Glick.pdf.

Ostendorf, T. A., Silva, I. A., Converti, A., & Sarubbo, L. A. (2019). Production and formulation of a new low-cost biosurfactant to remediate oil-contaminated seawater. *Journal of Biotechnology, 295*, 71−79. Available from https://doi.org/10.1016/j.jbiotec.2019.01.025.

Paraszkiewicz, K., Bernat, P., Kuśmierska, A., Chojniak, J., & Płaza, G. (2018). Structural identification of lipopeptide biosurfactants produced by *Bacillus subtilis* strains grown on the media obtained from renewable natural resources. *Journal of Environmental Management, 209*, 65−70. Available from https://doi.org/10.1016/j.jenvman.2017.12.033.

Patel, S., Homaei, A., Patil, S., & Daverey, A. (2019). Microbial biosurfactants for oil spill remediation: Pitfalls and potentials. *Applied Microbiology and Biotechnology, 103*(1), 27−37. Available from https://doi.org/10.1007/s00253-018-9434-2.

Patowary, R., Patowary, K., Kalita, M. C., & Deka, S. (2018). Application of biosurfactant for enhancement of bioremediation process of crude oil contaminated soil. *International Biodeterioration and Biodegradation, 129*, 50−60. Available from https://doi.org/10.1016/j.ibiod.2018.01.004.

Patowary, R., Patowary, K., Kalita, M. C., & Deka, S. (2016). Utilization of paneer whey waste for cost-effective production of rhamnolipid biosurfactant. *Applied Biochemistry and Biotechnology, 180*(3), 383−399. Available from https://doi.org/10.1007/s12010-016-2105-9.

Pele, M. A., Ribeaux, D. R., Vieira, E. R., Souza, A. F., Luna, M. A. C., Rodríguez, D. M., Andrade, R. F. S., Alviano, D. S., Alviano, C. S., Barreto-Bergter, E., Santiago, A. L. C. M. A., & Campos-Takaki, G. M. (2019). Conversion of renewable substrates for biosurfactant production by Rhizopus arrhizus UCP 1607 and enhancing the removal of diesel oil from marine soil. *Electronic Journal of Biotechnology, 38*, 40−48. Available from https://doi.org/10.1016/j.ejbt.2018.12.003.

Peng, F., Liu, Z., Wang, L., & Shao, Z. (2007). An oil-degrading bacterium: Rhodococcus erythropolis strain 3C-9 and its biosurfactants. *Journal of Applied Microbiology, 102*(6), 1603−1611. Available from https://doi.org/10.1111/j.1365-2672.2006.03267.x.

Peng, F., Wang, Y., Sun, F., Liu, Z., Lai, Q., & Shao, Z. (2008). A novel lipopeptide produced by a Pacific Ocean deep-sea bacterium, *Rhodococcus* sp. TW53. *Journal of Applied Microbiology, 105*(3), 698−705. Available from https://doi.org/10.1111/j.1365-2672.2008.03816.x.

Pepi, M., Cesàro, A., Liut, G., & Baldi, F. (2005). An antarctic psychrotrophic bacterium *Halomonas* sp. ANT-3b, growing on n-hexadecane, produces a new emulsyfying glycolipid. *FEMS Microbiology Ecology*, *53*(1)), 157−166. Available from https://doi.org/10.1016/j.femsec.2004.09.013.

Pi, Y., Bao, M., Liu, Y., Lu, T., & He, R. (2017). The contribution of chemical dispersants and biosurfactants on crude oil biodegradation by Pseudomonas sp. LSH-7′. *Journal of Cleaner Production*, *153*, 74−82. Available from https://doi.org/10.1016/j.jclepro.2017.03.120.

Pitocchi, R., Cicatiello, P., Birolo, L., Piscitelli, A., Bovio, E., Cristina Varese, G., & Giardina, P. (2020). Cerato-platanins from marine fungi as effective protein biosurfactants and bioemulsifiers. *International Journal of Molecular Sciences*, *21*(8). Available from https://doi.org/10.3390/ijms21082913.

Prakash, A. A., Prabhu, N. S., Rajasekar, A., Parthipan, P., AlSalhi, M. S., Devanesan, S., & Govarthanan, M. (2021). Bio-electrokinetic remediation of crude oil contaminated soil enhanced by bacterial biosurfactant. *Journal of Hazardous Materials*, *405*, 124061. Available from https://doi.org/10.1016/j.jhazmat.2020.124061.

Qazi, M. A., Kanwal, T., Jadoon, M., Ahmed, S., & Fatima, N. (2014). Isolation and characterization of a biosurfactant-producing Fusarium sp. BS-8 from oil contaminated soil. *Biotechnology Progress*, *30*(5), 1065−1075. Available from https://doi.org/10.1002/btpr.1933.

Reis, C. B. L. D., Morandini, L. M. B., Bevilacqua, C. B., Bublitz, F., Ugalde, G., Mazutti, M. A., & Jacques, R. J. S. (2018). First report of the production of a potent biosurfactant with α,β-trehalose by Fusarium fujikuroi under optimized conditions of submerged fermentation. *Brazilian Journal of Microbiology*, *49*, 185−192. Available from https://doi.org/10.1016/j.bjm.2018.04.004.

Rodríguez, A., Gea, T., & Font, X. (2021). Sophorolipids production from oil cake by solid-state fermentation. inventory for economic and environmental assessment. *Frontiers in Chemical Engineering*, *3*, 632752. Available from https://doi.org/10.3389/fceng.2021.632752.

Ron, E. Z., & Rosenberg, E. (2001). Natural roles of biosurfactants. Minireview. *Environmental Microbiology*, *3*(4), 229−236. Available from https://doi.org/10.1046/j.1462-2920.2001.00190.x.

Rufino, R. D., de Luna, J. M., de Campos Takaki, G. M., & Sarubbo, L. A. (2014). Characterization and properties of the biosurfactant produced by Candida lipolytica UCP 0988. *Electronic Journal of Biotechnology*, *17*(1), 34−38. Available from https://doi.org/10.1016/j.ejbt.2013.12.006.

Sahebnazar, Z., Mowla, D., Karimi, G., & Yazdian, F. (2018). Zero-valent iron nanoparticles assisted purification of rhamnolipid for oil recovery improvement from oily sludge. *Journal of Environmental Chemical Engineering*, *6*(1), 917−922. Available from https://doi.org/10.1016/j.jece.2017.11.043.

Sajna, K. V., Sukumaran, R. K., Gottumukkala, L. D., & Pandey, A. (2015). Crude oil biodegradation aided by biosurfactants from Pseudozyma sp. NII 08165 or its culture broth. *Bioresource Technology*, *191*, 133−139. Available from https://doi.org/10.1016/j.biortech.2015.04.126.

Sathi Reddy, K., Yahya Khan, M., Archana, K., Gopal Reddy, M., & Hameeda, B. (2016). Utilization of mango kernel oil for the rhamnolipid production by *Pseudomonas aeruginosa* DR1 towards its application as biocontrol agent. *Bioresource Technology*, *221*, 291−299. Available from https://doi.org/10.1016/j.biortech.2016.09.041.

Secato, J. F. F., dos Santos, B. F., & Ponezi, A. N. (2017). Optimization techniques and development of neural models applied in biosurfactant production by bacillus subtilis using alternative substrates. *Advances in Bioscience and Biotechnology*, *10*, 343−360.

Sharma, R., Singh, J., & Verma, N. (2018a). Optimization of rhamnolipid production from Pseudomonas aeruginosa PBS towards application for microbial enhanced oil recovery. *3 Biotech*, *8*(1). Available from https://doi.org/10.1007/s13205-017-1022-0.

Sharma, R., Singh, J., & Verma, N. (2018b). Production, characterization and environmental applications of biosurfactants from Bacillus amyloliquefaciens and Bacillus subtilis. *Biocatalysis and Agricultural Biotechnology*, *16*, 132−139. Available from https://doi.org/10.1016/j.bcab.2018.07.028.

Shekhar, S., Sundaramanickam, A., & Balasubramanian, T. (2015). Biosurfactant producing microbes and their potential applications: A review. *Critical Reviews in Environmental Science and Technology*, *45*(14), 1522−1554. Available from https://doi.org/10.1080/10643389.2014.955631.

Silva, E. J., Correa, P. F., Almeida, D. G., Luna, J. M., Rufino, R. D., & Sarubbo, L. A. (2018). Recovery of contaminated marine environments by biosurfactant-enhanced bioremediation. *Colloids and Surfaces B: Biointerfaces*, *172*, 127−135. Available from https://doi.org/10.1016/j.colsurfb.2018.08.034.

Silva, M. E. T., Duvoisin, S., Oliveira, R. L., Banhos, E. F., Souza, A. Q. L., & Albuquerque, P. M. (2021). Biosurfactant production of Piper hispidum endophytic fungi. *Journal of Applied Microbiology*, *130*(2), 561−569. Available from https://doi.org/10.1111/jam.14398.

Singh, P., Patil, Y., & Rale, V. (2019). Biosurfactant production: Emerging trends and promising strategies. *Journal of Applied Microbiology*, *126*(1), 2−13. Available from https://doi.org/10.1111/jam.14057.

Sobrinho, H., Luna, J., Rufino, R., Porto, A., & Sarubbo, L. A. (2014). Biosurfactants: Classification, properties and environmental applications. *Biotechnology*, *11*, 1−29.

Syahir, H., Siti Aqlima, A., Wan Lutfi, W. J., Mohd Yunus, A. S., Siti Aisyah, A., Jerzy, S., Nurul Hani, S., Nur Syafiqah, A. R., & Nur Adeela, Y. (2020). Production of lipopeptide biosurfactant by a hydrocarbon-degrading antarctic rhodococcus. *International Journal of Molecular Sciences*, *21*(17), 6138.

Santos, B. F., Ponezib, A. N., & Filetia, A. M. F. (2014). Strategy of using waste for biosurfactant production through fermentation by *Bacillus subtilis*. *Chemical Engineering Transactions*, *37*, 727−732.

Soumen, M., Palashpriya, D., & Ramkrishna, S. (2006). Towards commercial production of microbial surfactants. *Trends in Biotechnology*, *24*(11), 509−515.

Sun, S., Wang, Y., Zang, T., Wei, J., Wu, H., Wei, C., Qiu, G., & Li, F. (2019). A biosurfactant-producing Pseudomonas aeruginosa S5 isolated from coking wastewater and its application for bioremediation of polycyclic aromatic hydrocarbons. *Bioresource Technology*, *281*, 421−428. Available from https://doi.org/10.1016/j.biortech.2019.02.087.

Surekha, K., Satpute, G. A., Płaza, A. G., Banpurkar., Production, B., From., & Resources, N. (2017). Example of innovative and smart technology in circular bioeconomy. *Management Systems in Production Engineering*, *25*, 46−54.

Teixeira Souza, K. S., Gudiña, E. J., Schwan, R. F., Rodrigues, L. R., Dias, D. R., & Teixeira, J. A. (2018). Improvement of biosurfactant production by Wickerhamomyces anomalus CCMA 0358 and its potential application in bioremediation. *Journal of Hazardous Materials*, *346*, 152−158. Available from https://doi.org/10.1016/j.jhazmat.2017.12.021.

Tian, W., Yao, J., Liu, R., Zhu, M., Wang, F., Wu, X., & Liu, H. (2016). Effect of natural and synthetic surfactants on crude oil biodegradation by indigenous strains. *Ecotoxicology and Environmental Safety*, *129*, 171−179. Available from https://doi.org/10.1016/j.ecoenv.2016.03.027.

Tomar, S., Lal, M., Khan, M. A., Singh, B. P., & Sharma, S. (2019). Characterization of glycolipid biosurfactant from *Pseudomonas aeruginosa* PA 1 and its efficacy against Phytophtora infestans. *Journal of Environmental Biology*, *40*(4), 725−730. Available from https://doi.org/10.22438/jeb/40/4/MRN-910.

Yalaoui-Guellal, D., Fella-Temzi, S., Djafri-Dib, S., Brahmi, F., Banat, I. M., & Madani, K. (2020). Biodegradation potential of crude petroleum by hydrocarbonoclastic bacteria isolated from Soummam wadi sediment and chemical-biological proprieties of their biosurfactants. *Journal of Petroleum Science and Engineering*, *184*, 106554. Available from https://doi.org/10.1016/j.petrol.2019.106554.

Zhu, Z., Zhang, F., Wei, Z., Ran, W., & Shen, Q. (2013). The usage of rice straw as a major substrate for the production of surfactin by *Bacillus amyloliquefaciens* XZ-173 in solid-state fermentation. *Journal of Environmental Management*, *127*, 96−102. Available from https://doi.org/10.1016/j.jenvman.2013.04.017.

Zouari, O., Lecouturier, D., Rochex, A., Chataigne, G., Dhulster, P., Jacques, P., & Ghribi, D. (2019). Bio-emulsifying and biodegradation activities of syringafactin producing Pseudomonas spp. strains isolated from oil contaminated soils. *Biodegradation*, *30*(4), 259−272. Available from https://doi.org/10.1007/s10532-018-9861-x.

CHAPTER 4

Biosurfactant production by utilizing waste products of the food industry

Oluwaseun Ruth Alara[1,2], Nour Hamid Abdurahman[1,2] and Hassan Alsaggaf Ali[3]
[1]Centre for Research in Advanced Fluid and Processes (Fluid Centre), Universiti Malaysia Pahang, Gambang, Pahang, Malaysia
[2]Faculty of Chemical and Process Engineering Technology, Universiti Malaysia Pahang, Gambang, Pahang, Malaysia
[3]Eastern Unity Technology, Kuala Lumpur, Malaysia

4.1 Introduction

Globally, waste generation and by-products have significant effects on the social, economic, and environmental sectors. The generation of waste is becoming worrisome because of the progressive side effects on human health and the environment. Using these generated wastes in producing valuable products has created novel avenues for environmental sustainability. Growing world population and globalization have enhanced the evolution of complex food supply chains. Food products, especially from agricultural products, generate greater quantities of waste and by-products. More than 50% of fruit by-products, including stems, bagasse, trimmings, peels, shells, seeds, and bran, are obtained from fresh fruits; these by-products contain different functional and nutritional contents (Ayala-Zavala et al., 2011). However, diverse food wastes and by-products are not being used and eventually end up in municipal landfills, where they cause unexplainable danger to the environment.

The use of food waste is associated with significant quantities of carbon sources that are viable substitutes in the production of biosurfactants. Biosurfactants are mobile surface molecules comprising both hydrophobic and hydrophilic components that empower them to assemble at the interface between nonpolar and polar phases; thus, they modify the surface and interface effects and improve the solubilities of polar substances in nonpolar substances and otherwise (Aguirre-Ramírez et al., 2021). Biosurfactants are generally produced from bacteria, filamentous fungi, and yeasts. They are categorized into fatty acids and phospholipids, glycolipids, lipopeptides and lipoproteins, particulate surfactants, and polymeric surfactants (Chen et al., 2015). The ability to minimize the surface tension shows the efficiency of a biosurfactant; an adequate biosurfactant can minimize water surface tension from 72 to 35 mN/m (Santos et al., 2016). Another factor in determining a biosurfactant's efficiency is critical micellular concentration value; an effective biosurfactant will have a value between 10 and 40 times lower than synthetic surfactants (Sharma, 2016).

Biosurfactants are majorly produced from carbon feedstocks, including fats and oils, hydrocarbons, and carbohydrates, through aerobic microorganisms in the aqueous media. They are discharged into the culture media to enhance microorganism growth by facilitating insoluble substrates over cell membranes (Silva et al., 2010). Furthermore, biosurfactants possess minimal toxicity, higher foaming ability, higher level of biodegradability, and optimal performance at the highest conditions of temperature, salinity, and pH levels (Neboh & Abu, 2015). Different sectors that use biosurfactants are finding ways to replace the use of chemical-based surfactants with sustainable biosurfactants. However, the higher cost of biosurfactants is the main setback (Silva et al., 2010).

Biosurfactants are categorized into lipoproteins and lipopeptides, glycolipids, fatty acids and phospholipids, particulates, and polymeric surfactants. The lipopeptides can further be grouped into linear or cyclic compounds; they comprise fatty acids in combination with the peptide residue (Mnif et al., 2013); examples of lipoproteins include fengysin, iturin, and surfactin. Some microorganisms can grow in hydrophobic nutrients that include fatty acids, alkanes, neutral lipids, and phospholipids; these enhance the consumption and absorption of nutrients (Sharma, 2016). The glycolipids comprise diverse sugars that are joined by an ester group to branched or linear alkyl groups (Otzen, 2017); examples of glycolipids include sophorolipids, trehalolipids, and rhamnolipids. In the case of particulate biosurfactants, they are generated as extracellular membrane cells to create microemulsions that impact the alkane absorption into the microbial cells (Vijayakumar & Saravanan, 2015). Furthermore, polymeric biosurfactants are biopolymers of higher molecular weights; they account for the proteins, lipoproteins, lipopolysaccharides, polysaccharides, or a mixture of these compounds (Vijayakumar & Saravanan, 2015).

Until now, different commercial substances have been utilized to produce biosurfactants; these include different concentrations of glucose and diverse kinds of organic nitrogen sources, including a pancreatic digest of casein, urea, yeast extracts, and beef extracts (Shilpa, 2021). Using these commercial substances is not environment-friendly; these have propelled the need to seek alternative renewable substrates that are eco-friendly and cheaper. Moreover, some of the substrates used to produce biosurfactants have a higher cost, foam during the batch processes, and affect downstream processing and purification system; these have limited the scale-up production of biosurfactants. The use of waste, especially food waste, has seen greater attention due to its availability, ability to reduce overall production costs, and achieve higher yields of biosurfactants. Therefore, this chapter explores the use of wastes from the food industry in the production of biosurfactants.

4.2 Biosurfactants from different food wastes

Different sources of waste from food can potentially be used to produce biosurfactants. These sources include agri-food wastes such as starch waste, vegetable and fruit waste, wastes from cooking oil, animal fats, distillery wastes, and sugar industry wastes. Table 4.1

Table 4.1 Some of the biosurfactants produced from food wastes.

Sr. no.	Food waste	Microorganism	Biosurfactant type	Process conditions	Mode of cultivation	Yield (g/L)	Applications	References
1.	Animal fats, animal wastes, and corn steep liquor	*Candida bombicola* ATCC 22214	Sophorolipids	30°C, 150 rpm, 24 h	Batch shake flask	39.81	Bioremediation of lubricating-oil-contaminated soils	Minucelli et al. (2017)
2.		*Candida lipolytica* UCP0988	Glycolipids	28°C, pH of 5.3, 200 rpm	Batch bioreactor	0.99	Bioremediation processes	Santos et al. (2014)
3.		*Pseudomonas aeruginosa* ATCC 9027 and ATCC 10145	Rhamnolipids	121°C, pH of 7.0,	Batch bioreactor	3.84	Bioremediation of soils polluted by oily compounds	Borges et al. (2012)
4.	Banana peel	*Halobacteriaceae archaeon* AS65	Lipopeptides	200 rpm for 24 h at 30°C	Batch shake flask	5.30	Safe and effective therapeutic agents	Chooklin et al. (2014)
5.	Carrot peels, brewery wastewaters, apple peels, beet molasses	*Bacillus subtilis* I′-1a	Lipopeptides (Iturin)	28°C, 120 rpm	Batch shake flask	428.70 mg/L	Removal of hydrocarbon contaminants from sand	Paraszkiewicz et al. (2018)
6.	Cashew apple juice	*Bacillus subtilis*	Lipopeptides	30°C, 180 rpm	Batch shake flask	–	Removal of hydrocarbon contaminants from sand	Nogueira Felix et al. (2019)
7.		*Bacillus subtilis* LAMI005	Surfactin	30°C, 180 rpm	Batch bioreactor	0.123	Oil recovery and environmental usage	Giro et al. (2009)
8.		*Bacillus subtilis* LAMI008	Surfactin	30°C, 180 rpm	Batch shake flask	0.0035	Oil recovery and environmental usage	Rocha et al. (2007)
9.		*Bacillus subtilis* LAMI005	Surfactin	30°C, 180 rpm	Batch shake flask	319.30 mg/L	In the food and pharmaceutical industries	Freitas de Oliveira et al. (2013)
10.		*Pseudomonas aeuroginosa* ATCC 10145	Rhamnolipids	30°C, 150 rpm	Batch shake flask	3.80	Bioremediation processes	Rocha et al. (2007)
11.		*Yarrowia lipolytica*	Biosurfactants	28°C, 250 rpm	Batch shake flask	6.90	–	Fontes et al. (2012)
12.	Cassava flour wastewater	*Bacillus subtilis* LB5a	mannosylerylthritol lipids	30°C, 150 rpm	Batch shake flask	1.26	Effective topical moisturizers	Almeida et al. (2017)
13.		*subtilis* LB5a	Surfactin	35°C, 150 rpm	Batch bioreactor	2.40	Good tenso-active agent	Barros et al. (2008)

(*Continued*)

Table 4.1 (Continued)

Sr. no.	Food waste	Microorganism	Biosurfactant type	Process conditions	Mode of cultivation	Yield (g/L)	Applications	References
14.		*Bacillus subtilis* LB5a	Surfactin	121°C, 8000 g	Batch shake flask	3.00	Many industrial applications	Nitschke & Pastore (2006)
15.		*Pseudomonas aeruginosa*	Rhamnolipids	30°C, 200 rpm	Batch shake flask	0.66	Bioremediation processes	Costa et al. (2009)
16.	Corn steep liquor, sugarcane molasses, and canola waste frying oil	*Pseudomonas cepacia* CCT6659	Glycolipid	28°C, 120 h, 200 rpm	Batch shake flask	4.19	Surface cleaning and oil recovery	Almeida et al. (2017)
17.	Corn steep liquor and soybean cooking oil	*Streptomyces* sp. DPUA1566	Lipoprotein (Bioelan)	28°C, 96 h, 150 rpm	Batch shake flask	1.90	Bioremediation	Santos et al. (2019)
18.	Corn waste oil and cassava flour wastewater	*Serratia marcescens*	Biosurfactants	28°C, 150 rpm	Batch shake flask	–	To remove hydrophobic pollutants from contaminated sand	Araújo et al. (2019)
19.	Crude glycerol and palm oil effluent	*Bacillus subtilis* TD4	Lipopeptides	30°C, 200 rpm	Batch shake flask	1.18	Biodegradation of oil industry wastes	Louhasakul et al. (2020)
20.	Frying oil waste	*Bacillus stratosphericus* FLU5	Lipopeptides	37°C, 180 rpm	Batch shake flask	0.05	Environmental remediation processes	Hentati et al. (2019)
21.	Frying oil waste	*Burkholderia thailandensis* E264	Rhamnolipids	37°C, 200 rpm	Batch bioreactor	2.2	Efficient emulsifiers	Kourmentza et al. (2018)
22.		*Candida bombicola*	Sophorolipids	30 ± 1°C, 200 rpm	Batch shake flask	34	Cleaning of hard surface	Shah et al. (2007)
23.		*Candida bombicola* ATCC22214	Sophorolipids	30°C, 130 rpm	Batch shake flask	50	Cleaning of hard surface	Fleurackers (2006)
24.		*Mucor circinelloides*	Glycolipids	28°C, 150 rpm	Batch shake flask	12.30	Bioremediation of oil slicks, oil-contaminated fields, and MEOR processes	Hasanizadeh et al. (2018)
25.		*Pseudomonas aeruginosa* DG30	Rhamnolipids	100°C, 8000 rpm	Batch bioreactor	15.56	Agriculture, bioremediation, cosmetics, food processing, petroleum, and pharmaceuticals	Zheng et al. (2011)

		Microorganism	Biosurfactant type	Conditions	Mode	Yield	Application	References
26.		*Pseudomonas aeruginosa* OG1	Rhamnolipids	80°C, 10000 rpm	Batch shake flask	13.31	Agriculture, bioremediation, cosmetics, food processing, petroleum, and pharmaceuticals	Ozdal et al. (2017)
27.		*Pseudomonas aeruginosa* PACL	Rhamnolipids	30 ± 1°C, 170 rpm	Batch bioreactor	3.3	Agriculture, bioremediation, cosmetics, food processing, petroleum, and pharmaceuticals	De Lima et al. (2009)
28.		*Pseudomonas aeruginosa* zju.ul M	Rhamnolipids	35 °C, 300 rpm	Batch bioreactor	20.00	Applicable industrially as the sole carbon source	Zhu et al. (2007)
29.		*Pseudomonas cepacia* CCT6659	Biosurfactants	28°C, 250 rpm	Batch bioreactor	40.50	In the petroleum industry	Soares da Silva et al. (2019)
30.		*Streptomyces* sp. DPUA 1559	Glycoproteins	28°C, 200 rpm	Batch shake flask	1.74	Industrial biotechnological area	Santos et al. (2018)
31.		*Streptomyces* sp. DPUA1559	Lipoproteins	28°C, 150 rpm	Batch shake flask	1.90	Bioremediation processes, cosmetic, and pharmaceutical industries	Santos et al. (2019)
32.	Lignocellulose hydrolysates	*Bacillus tequilensis* ZSB10	Biosurfactants	48.5°C, 150 rpm	Batch bioreactor	1.52	—	Cortés-Camargo et al. (2016)
33.		*Candida bombicola* ATCC 22214	Sophorolipids	30°C, 200 rpm	Batch shake flask	120	Food and cosmetic industries	Deshpande & Daniels (1995)
34.		*Candida bombicola*	Sophorolipids	25°C, 120 rpm	Batch shake flask	84.60	Bactericidal agent	Samad et al. (2015)
35.		*Candida bombicola* ATCC 22214	Sophorolipids	25°C, 120 rpm	Batch shake flask	52.10	Bactericidal agent	Samad et al. (2017)
36.		*Latobacillus pentosus*	Intracellular biosurfactants	31°C, 150 rpm	Batch shake flask	0.0048	Production of pharmaceuticals and cosmetics	Konishi et al. (2015)
37.		*Starmerella bombicola* NBRC 10243	Sophorolipids	28°C, 500–1000 rpm	Batch bioreactor	49.20	Bactericidal agent	Konishi et al. (2015)
38.	Molasses	*Bacillus subtilis* RSL-2	Lipopeptides	37°C, 180 rpm	Batch shake flask	12.34	Applicability in the food, environmental, and medicinal industries	Verma et al. (2020)

(*Continued*)

Table 4.1 (Continued)

Sr. no.	Food waste	Microorganism	Biosurfactant type	Process conditions	Mode of cultivation	Yield (g/L)	Applications	References
39.	Orange peel	*Bacillus licheniformis* KC710973	Lipopeptides	30°C, 10000 rpm	Batch shake flask	1.80	Applicability in the food, environmental, and medicinal industries	Kumar et al. (2016)
40.		*Pseudomonas aeuroginosa* MTCC 2297	Rhamnolipids	37°C, 10000 rpm	Batch shake flask	9.18	Mitigate the waste management challenges	George & Jayachandran (2009)
41.	Potato peel powder	*Klebsiella sp* RJ-03	Biosurfactants	32°C, 120 rpm	Batch shake flask	15.40	Bioremediation	Jain et al. (2013)
42.	Potato peels and bagasse	*Pseudomonas azotoformans* AJ15	Rhamnolipids	30°C, 180 rpm	Batch shake flask	1.16	Oil recovery from the sand matrix	Das & Kumar (2018)
43.	Potato processing effluent	*Bacillus subtilis* 21332	Surfactin	30°C, 150 rpm	Batch shake flask	0.44	Oil recovery or environmental remediation	Thompson et al. (2000)
44.			Surfactin	30°C, 250 rpm	Batch bireactor	0.90	Oil recovery or environmental remediation	Noah et al. (2005)
45.	Rice husk	*Mucor indicus*	Glycolipids	40°C, 4000 rpm	Batch shake flask	7.80	Spilled oil remediation	Oje et al. (2016)
46.	Sesame peel flour and tuna fish cooking residue	*Bacillus subtilis* SPB1	Lipopeptides	30°C, 150 rpm	Batch shake flask	4.5	Solubilization of diesel oil	Mnif et al. (2013)
47.	Soy molasses	*Candida bombicola*	Sophorolipids	28°C, 700 rpm	Batch shake flask	55	Food and cosmetic industries	Solaiman et al. (2007)
48.	Sugarcane baggase	*Bacillus safensis* J2	Surfactin	30°C, 150 rpm	Batch shake flask	0.92	Repair of diesel-contaminated soil and efficiency of oil recovery	Das & Kumar (2019)
49.	Sugarcane molasses, oil mill wastewater, and corn steep liquor	*Pseudomonas aeruginosa*	Rhamnolipids	37°C, 180 rpm, 1 vvm	Batch shake flask	5.10	Mitigate the waste management challenges	Gudiña et al. (2016)
50.	Vegetable oil processing waste and by-products	*Bacillus subtilis* 3–10	Iturin A	28°C, 600 rpm, 2 vvm	Batch shake flask	0.60	*Application* in biomedicines	Jin et al. (2014)

51.		Bacillus subtilis KB1	Surfactin	37°C, 150 rpm	Solid-state fermentation	4–16 mg/mL	Oil recovery or environmental remediation	Jajor et al. (2016)
52.		Bacillus subtilis SPB1	Lipopeptides	37°C, 10000 rpm	Solid-state fermentation	30.67 mg/g	Antibiotics, antitumors and antifungal agents	Zouari et al. (2014)
53.		Candida sphaerica UCP0995	Glycolipids	28°C, 150 rpm	Batch shake flask	4.50	Oil recovery from sand	Sobrinho et al. (2008)
54.		Pseudomonas aeruginosa 47T2	Rhamnolipids	30°C, 150 rpm	Batch shake flask	2.70	Removal of heavy metals	Haba et al. (2000)
55.		Pseudomonas aeruginosa AB4	Rhamnolipids	40°C, 120 rpm	Batch shake flask	40	Removal of heavy metals	Hazra et al. (2011)
56.		Pseudomonas aeruginosa LBI	Rhamnolipids	30°C, 200 rpm	Batch shake flask	11.70	Removal of heavy metals	Nitschke et al. (2005)
57.		Pseudomonas aeruginosa PAO1	Rhamnolipids	37°C, 160 rpm	Batch shake flask	191.46 mg/L	Removal of heavy metals	Moya-Ramírez et al. (2016)
58.		Starmerella bombicola ATCC 22,214	Sophorolipids	30°C, 180 rpm	Batch shake flask	0.179 g/g	Used in the pharmaceutical sector	Jiménez-Peñalver et al. (2016)
59.		S. bombicola MTCC1910	Sophorolipids	30°C, 200 rpm	Batch bioreactor	51.50	Used in the pharmaceutical sector	Jadhav et al. (2019)
60.	Vineyard pruning waste	Lactobacillus paracasei	Glycolipopeptide	25°C, 150 rpm	Fermenter	5	Emulsifying agent	Ferreira et al. (2017)

presents some of the biosurfactants produced from food waste. Fig. 4.1 presents two major sources of substrates from the food sector, which are agri-food and oily wastes. The use of these wastes in the presence of adequate microorganisms (some are provided in Table 4.1) can produce biosurfactants. The produced biosurfactants are rich with various merits, including cost-effectiveness, nontoxicity, and environmental-friendly. However, some demerits still abound when using food wastes, including downstream processing, availability of raw materials, and large-scale production.

4.2.1 Agric-food wastes

It is believed that expanded production of biosurfactants is not economically viable because of the challenges associated with increased cost of fermentation. Different researchers are now considering the potential of using low-cost substrates to achieve mass production of biosurfactants. Given this, the use of agri-food wastes is being examined as feedstocks in the production of biosurfactants (Gaur et al., 2022; De Lima et al., 2009). These agri-food wastes are organic, renewable substrates comprising hemicellulose, cellulose, and lignin that can be used for fermentation media. The conversion of these wastes biologically to obtain valuable products can significantly reduce environmental problems associated with disposal and improve energy costs. Moreover, the industrial processing of vegetables and fruits, including pineapple, banana, lime, carrot, and orange, can generate greater quantities of waste, which are currently considered substrates to produce biosurfactants.

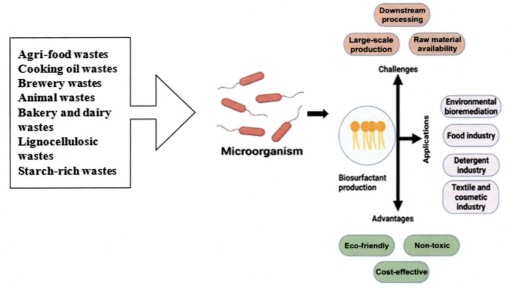

Figure 4.1 Biosurfactant production using wastes from the food industry, the merits, demerits, and applications.

Annually, food waste is generated in billions of tons across the globe. The use of vegetables and fruits in the food sector, including cooking and other eatables, generates much including household and food industry waste. Minimizing these generated wastes keep challenging the environment globally. These wastes can be converted from food sources to resolve several environmental and improve cost-effectiveness. Different studies have considered the use of wastes from vegetable and fruit processing wastes as substrates to produce biosurfactants. A study considered the use of food-sourced wastes, including *Citrus medica* peels, *Citrus lambiri* peels, banana peels, potato peels, and orange peels, to generate 2% glycerol when using *Bacillus licheniformis* (Kumar et al., 2016). The outcome of the study showed that a maximum yield of 1.295 g/L was achieved with the use of orange peels, followed by 1.116 g/L using banana peels, and 1.058 g/L using potato peels. The bagasse produced from the sugarcane industry had been reportedly used as substrates in the production of biosurfactants. The rhamnolipid produced from *Pseudomonas azotoformans* AJ15 was assessed by utilizing a mixture of potato peels and bagasse through a fermentation process (Sindhu et al., 2016). The rhamnolipids produced was better in terms of activity when a mixture of 2% bagasse powder and 2% potato peels was used. Another study confirmed the prospect of bagasse and potato peels was more productive than dextrose. The surfactin produced from *B. safensis* J2 was 0.92 g/L when 15 g of bagasse was used (Das & Kumar, 2018). A study established the potential of *Fusarium oxysporum* LM5634 (a fungal strain) in the production of biosurfactants using the extracts from fruit peels of *Bactris gasipaes* Kunth, *Astrocaryum aculeatum*, *Musa paradisiaca*, and *Theobroma grandiflorum* Schumann. The highest productivity of biosurfactants was observed in the extract from the peels of *Bactris gasipaes* Kunth (Sanches et al., 2018).

Most importantly, the cost of fermentation can be minimized by optimizing the use of industry wastes. An optimization study was conducted on the use of date molasse in the production of biosurfactants from the *B. subtilis* strain Al-Dhabi-130 (Al-Dhabi et al., 2020). The study's outcome revealed that the biosurfactants produced were 74 mg/g substrates. Considering the recent studies, the effluents generated from the potato processing industry had been considered suitable for recovering surfactin and the cultivation of *B. subtilis* 21332 (Rivera et al., 2007). A study by Vieira et al. (2021) presented the prospect of pineapple peels in the preparation of culture media to produce biosurfactants. The results showed that pineapple peels could potentially substitute salt and glucose nutrients partially. This can reduce the cost of exogenous minerals as supplementary. In another study, agro-industrial wastes such as barley bran, wheat straw, cassava flour waste, rice paddy, de-oiled cakes, and potato wastes were reported as potential substitutes for the synthetic media used to produce biosurfactants by utilizing *Candida* sp., *Pseudomonas sp.*, and *Bacillus* sp. (Rajasimman et al., 2021). Using the wastes from the food industry in the fermentation processes through green chemistry to generate biosurfactants is an essential tool to improve innovations that can keep up with the current market (de Freitas Ferreira et al., 2019).

4.2.2 Wastes from cooking oils

The recent reports from FAO (Food and Agriculture Organization) and OECD (Organization for Economic Cooperation and Development) on vegetable oils as a commodity market indicated that about 210 million tons are produced and consumed annually. For this, larger quantities of waste are generated from using vegetable oils to cook. It was estimated that about 42 million tons of waste from cooking oils are produced globally annually (OECD-FAO, 2019). By-products generated from the processing of vegetable oils are rich in oils, fats, and other compounds such as oil seed cakes, soap stocks, fatty acid residues, water-soluble effluents, and semisolid effluents (Dumont & Narine, 2007). Importantly, the residues as mentioned earlier are sources of contamination to soil and water because of the reduced degradability of lipids in them (Cammarota & Freire, 2006). *Candida sphaerica* UCP 0995 was cultured alongside refinery wastes from groundnut oil (5%) and corn steep liquid (2.5%) in distilled water (Sobrinho et al., 2008). The results from this study showed a yield of 4.5 g/L biosurfactants and the ability to recover about 65% oils from the sand. A similar study reported the recovery of 9 g/L of biosurfactants using the same materials, such as *Candida sphaerica* UCP 0995, wastes from groundnut oil (9%), and corn steep liquid (9%) (Luna et al., 2013).

Moreover, a previous study optimized humidity, inoculum, and agro-industrial residue concentration to produce biosurfactants using *B. subtilis* SPB1 and by-products from olive oil in solid-state fermentation. The biosurfactants produced were 30.67 mg/g utilizing 4 g of olive cake flour and 6 g of olive leaf residue flour (Zouari et al., 2014). Another study considered using mahua oil cake to produce biosurfactants through the culturing of *Serratia rubidaea* SNAU02 under solid-state fermentation conditions. The recovered biosurfactants showed an appreciable quantity of rhamnolipids; this exhibited no toxicity against *Artemia salina* and *Brassica oleracea* seeds with antifungal activity. This positioned the produced rhamnolipids as a potential biocontrol agent against pathogens of plants (Nalini & Parthasarathi, 2014). Olive mill waste was used as the main carbon source to produce surfactin from *B. subtilis* N1 and rhamnolipids from *P. aeruginosa* PAO1. Optimum surfactin (3.1 mg/L) was produced using 2% of olive mill wastes, while the optimum rhamnolipids (191.4 mg/L) were produced using 10% of olive mill wastes. Nevertheless, the hydrolysis pretreatment of olive mill wastes yielded 26.5 mg/L of surfactin and 299 mg/L of rhamnolipids using 5% and 2% of olive mill wastes, respectively. This study presented that enzymatic hydrolysis effectively improves biosurfactant production (Moya-Ramírez et al., 2016). Under solid-state fermentation conditions, sophorolipids (179 mg/g) were produced by utilizing sugar beet molasses, winterization oil cake, and *S. bombicola* ATCC 22214 (Jiménez-Peñalver et al., 2016).

Recently, some studies have also studied the efficacy of using wastes of cooking oils. Wastes from the sunflower oil refinery were employed as feedstocks to produce sophorolipids of the highest yield of 41.60 g/L through the culture with *Starmerella bombicola*

MTCC1910 at the shake flask level, while the use of a fermenter produced a higher yield of sophorolipids (51.50 g/L) with 10% glucose and 10% of waste oil (Jadhav et al., 2019). Another study employed soybean oil, waste cooking oil, and *Pseudomonas aphidis* ZJUDM34 to produce mannosylerythritol lipids (MELs) with a yield of 61.50 g/L (Nui et al., 2019). The culture of 2% used cooking oil with *Bacillus sp.* yielded 9.50 g/L of surfactin within the 7 days of incubation (Md Badrul Hisham et al., 2019). Likewise, rice bran oil with *P. aeruginosa* CR1 yielded rhamnolipids of 19.22 g/L; however, with the addition of Luria Bertani and glycerol, the yield of rhamnolipids increased to 21.77 g/L (Sood et al., 2020). Waste from frying oil was used as a carbon source to produce biosurfactants using *Exiguobacterium profundum*; this process yielded 8.20 g/L of biosurfactant (Mardiah et al., 2022). Another study produced rhamnolipid biosurfactants (16 g/L) with 94% of waste cooking oil biodegradation capability of *P. aeruginosa* P7815 (Sharma et al., 2022). A patent was published on the production of biosurfactants using green processes through the utilization of wastes (US20110151100A1). The patent reported a yield of 0.05% biosurfactant using wastes.

4.2.3 Brewery wastes

Brewery is another source of waste from the food industry, which includes the production of beer; the first phase (lauter and mash tun) of beer production generates much waste, referred to as brewery waste. These wastes are sources of carbon that can serve as substrates for the microorganisms in the production of biosurfactants. A study used brewery waste with *B. subtilis* N3−1P strains and ammonium nitrate to produce biosurfactants, with an yield of 657 mg/L (Moshtagh et al., 2019). In another study, brewery substrates with *B. subtilis* ATCC6051 yielded 210.11 mg/L of surfactin after 28 h (Nazareth et al., 2020). Thus, this shows the potential of using brewery wastes as substrates in the production of biosurfactants.

4.2.4 Animal wastes from the food industry

Diverse wastes are generated from animal sources in the food industry. A study considered using fish wastes such as fish livers and heads to produce biosurfactants. It was reported that the wastes from fish were good sources of ash, total nitrogen, and organic carbon; the fish liver had a carbon/nitrogen ratio of 3:10 and fish head had a carbon/nitrogen ratio of 4:12 (Zhu et al., 2020). Another study produced rhamnolipid using fish meal media in addition to different strains of *P. aeruginosa*, such as SY1, H1, and ATCC. This study reported rhamnolipid yields as 10.30, 9.30, and 12.30 g/L, respectively (Kaskatepe et al., 2015). Fish wastes had also been used as the substrates to produce lipopeptides by *B. subtilis* N3−1P through the optimization study. The critical micelle dilution values of the produced biosurfactants from the fish head and fish liver wastes were 47.59 and 54.72, respectively (Zhu et al., 2020).

Apart from the wastes from fish, poultry also produced varieties of wastes, including generated wastes as effluents from where pigs and chickens are slaughtered. These effluents were used as substrates in the production of biosurfactants. A study investigated the impacts of diverse concentrations of ammonium nitrate, brewery residual yeast, and fat on the production of biosurfactants by *P. aeruginosa* ATCC 10145. The outcome of the study provided that the optimal concentrations of ammonium nitrate, brewery residual yeast, and fat were 0, 15, and 12 g/L, respectively, when boosted with the extracts from meat. Hence, the biosurfactant produced had an emulsification activity of 100% and a surface tension of 27.50 dyne/cm (Borges et al., 2012).

4.2.5 Bakery and dairy wastes

The biochemical oxygen demand (BOD) in dairy wastes is high, and disposing of these wastes is a major environmental problem. It has been reported that about 50% of dairy waste cannot be recycled (Shilpa, 2021). Hence, using these wastes as substrates to produce biosurfactants will benefit the environment and economy. The wastes from dairy, including buttermilk, whey, and other by-products, are utilized as substrates to produce biosurfactants. For instance, curd whey has been studied as a suitable substrate that is rich in nitrogen, vitamins such as pantothenic acids and riboflavin, organic carbon, and minerals including potassium, calcium, phosphorus, and calcium. Curd whey has been used with *Kocuria turfanesis* J and *P. aeruginosa* PP2 to produce biosurfactants. More yields were obtained from *P. aeruginosa* PP2 than *Kocuria turfanesis* J (Dubey et al., 2012). The production of sophorolipids was increased using cultivated *Cryptococcus curvatus* on concentrated lactose-rich whey (Mohanty et al., 2021). Moreover, rhamnolipids were produced using discarded mixed bakery waste and *P. aeruginosa* PG1. The mineral salt media was enriched with powdered and dried bakery wastes (Shilpa, 2021).

4.2.6 Lignocellulosic wastes

One of the main sources organic carbon is lignocellulose. The main cellulosic substances are generated from plants. Sophorolipids were produced from the use of stover, sweet sorghum bagasse, and *C. bombicola*; the study established the importance of lignocellulosic feedstock as an eco-friendly substance and substrate to produce biosurfactant (Samad et al., 2015). Another study produced sophorolipids from *C. bombicola* and *Starmerella bombicola* NBRC using corn fiber, sweet sorghum bagasse, and corncob hydrolysate media (Konishi et al., 2015). Cellulosic sugar wastes (vineyard pruning waste) were used alongside *B. tequilensis* ZSB10 in the production for extracellular biosurfactant production and cell-bound. The result showed that the production of extracellular biosurfactants was better than cell-bound (Cortés-Camargo et al., 2016). A more recent study reported the use of hemicellulose hydrolysate from sugarcane bagasse and *Cutaneotrichosporon mucoides* UFMG-CM-Y6148 to produce sophorolipids; the highest yield of 12.50 g/L was achieved at 72 h (Marcelino et al., 2019).

4.2.7 Starch-rich wastes

Industrially, the extraction of starches from foods, including rice, corn, potato, cassava, and wheat, produces higher quantities of wastewater; these are rich in husks and starch and can be used as substrates in the production of biosurfactants. For instance, potato is enriched with fiber (1%—1.80%), proteins (2%—2.50%), fatty acids (0.15%), and starch (16%—20%). A potato, including its peels, comprises higher quantities of vitamins B and C, potassium, and minerals including iron, magnesium, and phosphorus (Graeme & Sansonetti, 2009). A study used the effluent from potatoes as a substrate to produce biosurfactant by *B. subtilis* ATCC 21332. The following were compared in terms of their performances: a fixed potato medium, solid and liquid potato media, and potato starch available in the market in mineral salt media. The surface tension was reported to reduce from 71.30 to 28.30 mN/m in solid media as the strains were cultivated in only potato substrate (60 g/L). The authors suggested exchanging the conventional carbon source in the production of biosurfactants with potato substrates (Fox & Bala, 2000). In another study, the effluents from potato processing and *B. subtilis* 21332 were used to produce surfactin that yielded 0.44 g/L (Thompson et al., 2000).

4.3 Conclusion and future perspectives

Adequate management of waste is the main problem facing the globe. The use of cheap natural wastes is the essential technique to minimize the cost and control environmental pollution caused by these wastes. This chapter has reviewed some of the wastes generated by the food industry in producing biosurfactants. The exploitation of fortified and unprocessed wastes from the food industry can improve novel techniques in the production of biosurfactants. Using the wastes generated from the food industry to produce biosurfactants will not only stabilize the management of wastes but also minimize production costs. Genuine implementation of these techniques can improve the production of biosurfactants and establish an adequate competitive market through bioconversion of food wastes and revenue generation. Furthermore, there are a few recommendations to help achieve this goal of producing biosurfactants, including constant availability of food waste materials and the cost of transporting these wastes to minimize the operating costs. Moreover, studies can focus on the use of viable microorganisms alongside the wastes from food industry to increase the yields of biosurfactants.

Acknowledgments

The authors thankfully acknowledge Eastern Unity Technology, Malaysia for the financial support through the grant (UIC190806).

References

Aguirre-Ramírez, M., Silva-Jiménez, H., Banat, I. M., & Díaz De Rienzo, M. A. (2021). Surfactants: Physicochemical interactions with biological macromolecules. *Biotechnology Letters*, *43*(3), 523–535. Available from https://doi.org/10.1007/s10529-020-03054-1.

Al-Dhabi, N. A., Esmail, G. A., & Arasu, M. V. (2020). Enhanced production of biosurfactant from *Bacillus subtilis* strain al-dhabi-130 under solid-state fermentation using date molasses from Saudi Arabia for bioremediation of crude-oil-contaminated soils. *International Journal of Environmental Research and Public Health*, *17*(22), 1–20. Available from https://doi.org/10.3390/ijerph17228446.

Almeida, D. G., da Silva, R. d. C. F. S., Luna, J. M., Rufino, R. D., Santos, V. A., & Sarubbo, L. A. (2017). Response surface methodology for optimizing the production of biosurfactant by *Candida tropicalis* on industrial waste substrates. *Frontiers in Microbiology*, *8*, 157. Available from https://doi.org/10.3389/fmicb.2017.00157.

Araújo, H. W. C., Andrade, R. F. S., Montero-Rodríguez, D., Rubio-Ribeaux, D., Alves Da Silva, C. A., & Campos-Takaki, G. M. (2019). Sustainable biosurfactant produced by *Serratia marcescens* UCP 1549 and its suitability for agricultural and marine bioremediation applications. *Microbial Cell Factories*, *18*(1). Available from https://doi.org/10.1186/s12934-018-1046-0.

Ayala-Zavala, J. F., Vega-Vega, V., Rosas-Domínguez, C., Palafox-Carlos, H., Villa-Rodriguez, J. A., Siddiqui, M. W., Dávila-Aviña, J. E., & González-Aguilar, G. A. (2011). Agro-industrial potential of exotic fruit byproducts as a source of food additives. *Food Research International*, *44*(7), 1866–1874. Available from https://doi.org/10.1016/j.foodres.2011.02.021.

Barros, F. F. C., Ponezi, A. N., & Pastore, G. M. (2008). Production of biosurfactant by *Bacillus subtilis* LB5a on a pilot scale using cassava wastewater as substrate. *Journal of Industrial Microbiology and Biotechnology*, *35*(9), 1071–1078. Available from https://doi.org/10.1007/s10295-008-0385-y.

Borges, W., da, S., Cardoso, V. L., & Resende, M. M. de (2012). Use of a greasy effluent floater treatment station from the slaughterhouse for biosurfactant production. *Biotechnology and Applied Biochemistry*, *59*(3), 238–244. Available from https://doi.org/10.1002/bab.1018.

Cammarota, M. C., & Freire, D. M. G. (2006). A review on hydrolytic enzymes in the treatment of wastewater with high oil and grease content. *Bioresource Technology*, *97*(17), 2195–2210. Available from https://doi.org/10.1016/j.biortech.2006.02.030.

Chen, W. C., Juang, R. S., & Wei, Y. H. (2015). Applications of a lipopeptide biosurfactant, surfactin, produced by microorganisms. *Biochemical Engineering Journal*, *103*, 158–169. Available from https://doi.org/10.1016/j.bej.2015.07.009.

Chooklin, C. S., Maneerat, S., & Saimmai, A. (2014). Utilization of banana peel as a novel substrate for biosurfactant production by *Halobacteriaceae archaeon* AS65. *Applied Biochemistry and Biotechnology*, *173* (2), 624–645. Available from https://doi.org/10.1007/s12010-014-0870-x.

Cortés-Camargo, S., Pérez-Rodríguez, N., Oliveira, R. P. d. S., Huerta, B. E. B., & Domínguez, J. M. (2016). Production of biosurfactants from vine-trimming shoots using the halotolerant strain Bacillus tequilensis ZSB10. *Industrial Crops and Products*, *79*, 258–266. Available from https://doi.org/10.1016/j.indcrop.2015.11.003.

Costa, S. G. V. A. O., Lépine, F., Milot, S., Déziel, E., Nitschke, M., & Contiero, J. (2009). Cassava wastewater as a substrate for the simultaneous production of rhamnolipids and polyhydroxyalkanoates by pseudomonas aeruginosa. *Journal of Industrial Microbiology and Biotechnology*, *36*(8), 1063–1072. Available from https://doi.org/10.1007/s10295-009-0590-3.

Das, A. J., & Kumar, R. (2018). Utilization of agro-industrial waste for biosurfactant production under submerged fermentation and its application in oil recovery from sand matrix. *Bioresource Technology*, *260*, 233–240. Available from https://doi.org/10.1016/j.biortech.2018.03.093.

Das, A. J., & Kumar, R. (2019). Production of biosurfactant from agro-industrial waste by *Bacillus safensis* J2 and exploring its oil recovery efficiency and role in restoration of diesel contaminated soil. *Environmental Technology & Innovation*, *16*, 100450. Available from https://doi.org/10.1016/j.eti.2019.100450.

de Freitas Ferreira, J., Vieira, E. A., & Nitschke, M. (2019). The antibacterial activity of rhamnolipid biosurfactant is pH dependent. *Food Research International*, *116*, 737−744. Available from https://doi.org/10.1016/j.foodres.2018.09.005.

De Lima, C. J. B., Ribeiro, E. J., Sérvulo, E. F. C., Resende, M. M., & Cardoso, V. L. (2009). Biosurfactant production by *Pseudomonas aeruginosa* grown in residual soybean oil. *Applied Biochemistry and Biotechnology*, *152*(1), 156−168. Available from https://doi.org/10.1007/s12010-008-8188-1.

Deshpande, M., & Daniels, L. (1995). Evaluation of sophorolipid biosurfactant production by *Candida bombicola* using animal fat. *Bioresource Technology*, *54*(2), 143−150. Available from https://doi.org/10.1016/0960-8524(95)00116-6.

Dubey, K. V., Charde, P. N., Meshram, S. U., Shendre, L. P., Dubey, V. S., & Juwarkar, A. A. (2012). Surface-active potential of biosurfactants produced in curd whey by *Pseudomonas aeruginosa* strain-PP2 and *Kocuria turfanesis* strain-J at extreme environmental conditions. *Bioresource Technology*, *126*, 368−374. Available from https://doi.org/10.1016/j.biortech.2012.05.024.

Dumont, M. J., & Narine, S. S. (2007). Soapstock and deodorizer distillates from North American vegetable oils: Review on their characterization, extraction and utilization. *Food Research International*, *40*(8), 957−974. Available from https://doi.org/10.1016/j.foodres.2007.06.006.

Ferreira, A., Vecino, X., Ferreira, D., Cruz, J. M., Moldes, A. B., & Rodrigues, L. R. (2017). Novel cosmetic formulations containing a biosurfactant from *Lactobacillus paracasei*. *Colloids and Surfaces B: Biointerfaces*, *155*, 522−529. Available from https://doi.org/10.1016/j.colsurfb.2017.04.026.

Fleurackers, S. J. J. (2006). On the use of waste frying oil in the synthesis of sophorolipids. *European Journal of Lipid Science and Technology*, *108*(1), 5−12. Available from https://doi.org/10.1002/ejlt.200500237.

Fontes, G. C., Ramos, N. M., Amaral, P. F. F., Nele, M., & Coelho, M. A. Z. (2012). Renewable resources for biosurfactant production by *Yarrowia lipolytica*. *Brazilian Journal of Chemical Engineering*, *29*(3), 483−493. Available from https://doi.org/10.1590/S0104-66322012000300005.

Fox, S. L., & Bala, G. A. (2000). Production of surfactant from *Bacillus subtilis* ATCC 21332 using potato substrates. *Bioresource Technology*, *75*(3), 235−240. Available from https://doi.org/10.1016/S0960-8524(00)00059-6.

Freitas de Oliveira, D. W., Lima França, I. W., Nogueira Félix, A. K., Lima Martins, J. J., Aparecida Giro, M. E., Melo, V. M. M., & Gonçalves, L. R. B. (2013). Kinetic study of biosurfactant production by *Bacillus subtilis* LAMI005 grown in clarified cashew apple juice. *Colloids and Surfaces B: Biointerfaces*, *101*, 34−43. Available from https://doi.org/10.1016/j.colsurfb.2012.06.011.

Gaur, V. K., Sharma, P., Sirohi, R., Varjani, S., Taherzadeh, M. J., Chang, J. S., Yong Ng, H., Wong, J. W. C., & Kim, S. H. (2022). Production of biosurfactants from agro-industrial waste and waste cooking oil in a circular bioeconomy: An overview. *Bioresource Technology*, *343*, 126059. Available from https://doi.org/10.1016/j.biortech.2021.126059.

George, S., & Jayachandran, K. (2009). Analysis of rhamnolipid biosurfactants produced through submerged fermentation using orange fruit peelings as sole carbon source. *Applied Biochemistry and Biotechnology*, *158*(3), 694−705. Available from https://doi.org/10.1007/s12010-008-8337-6.

Giro, M. E. A., Martins, J. J. L., Rocha, M. V. P., Melo, V. M. M., & Gonçalves, L. R. B. (2009). Clarified cashew apple juice as alternative raw material for biosurfactant production by *Bacillus subtilis* in a batch bioreactor. *Biotechnology Journal*, *4*(5), 738−747. Available from https://doi.org/10.1002/biot.200800296.

Graeme, T., & Sansonetti, G. (2009). *New light on a hidden treasure: International year of the potato 2008, an end-of-year review* (Vol. 189, pp. 243−249). Rome, Italy: Food and Agriculture Organization of the United Nations.

Gudiña, E. J., Rodrigues, A. I., de Freitas, V., Azevedo, Z., Teixeira, J. A., & Rodrigues, L. R. (2016). Valorization of agro-industrial wastes towards the production of rhamnolipids. *Bioresource Technology*, *212*, 144−150. Available from https://doi.org/10.1016/j.biortech.2016.04.027.

Haba, E., Espuny, M. J., Busquets, M., & Manresa, A. (2000). Screening and production of rhamnolipids by *Pseudomonas aeruginosa* 47T2 NCIB 40044 from waste frying oils. *Journal of Applied Microbiology, 88*(3), 379–387. Available from https://doi.org/10.1046/j.1365-2672.2000.00961.x.

Hasani zadeh, P., Moghimi, H., & Hamedi, J. (2018). Biosurfactant production by Mucor circinelloides : Environmental applications and surface-active properties. *Engineering in Life Sciences, 18*(5), 317–325. Available from https://doi.org/10.1002/elsc.201700149.

Hazra, C., Kundu, D., Ghosh, P., Joshi, S., Dandi, N., & Chaudhari, A. (2011). Screening and identification of *Pseudomonas aeruginosa* AB4 for improved production, characterization and application of a glycolipid biosurfactant using low-cost agro-based raw materials. *Journal of Chemical Technology and Biotechnology, 86*(2), 185–198. Available from https://doi.org/10.1002/jctb.2480.

Hentati, D., Chebbi, A., Hadrich, F., Frikha, I., Rabanal, F., Sayadi, S., Manresa, A., & Chamkha, M. (2019). Production, characterization and biotechnological potential of lipopeptide biosurfactants from a novel marine *Bacillus stratosphericus* strain FLU5. *Ecotoxicology and Environmental Safety, 167*, 441–449. Available from https://doi.org/10.1016/j.ecoenv.2018.10.036.

Jadhav, J. V., Pratap, A. P., & Kale, S. B. (2019). Evaluation of sunflower oil refinery waste as feedstock for production of sophorolipid. *Process Biochemistry, 78*, 15–24. Available from https://doi.org/10.1016/j.procbio.2019.01.015.

Jain, R. M., Mody, K., Joshi, N., Mishra, A., & Jha, B. (2013). Effect of unconventional carbon sources on biosurfactant production and its application in bioremediation. *International Journal of Biological Macromolecules, 62*, 52–58. Available from https://doi.org/10.1016/j.ijbiomac.2013.08.030.

Jajor, P., Piłakowska-Pietras, D., Krasowska, A., & èukaszewicz, M. (2016). Surfactin analogues produced by *Bacillus subtilis* strains grown on rapeseed cake. *Journal of Molecular Structure, 1126*, 141–146. Available from https://doi.org/10.1016/j.molstruc.2016.02.014.

Jiménez-Peñalver, P., Gea, T., Sánchez, A., & Font, X. (2016). Production of sophorolipids from winterization oil cake by solid-state fermentation: Optimization, monitoring and effect of mixing. *Biochemical Engineering Journal, 115*, 93–100. Available from https://doi.org/10.1016/j.bej.2016.08.006.

Jin, H., Zhang, X., Li, K., Niu, Y., Guo, M., Hu, C., Wan, X., Gong, Y., & Huang, F. (2014). Direct bio-utilization of untreated rapeseed meal for effective iturin a production by *Bacillus subtilis* in submerged fermentation. *PLoS One, 9*(10). Available from https://doi.org/10.1371/journal.pone.0111171.

Kaskatepe, B., Yildiz, S., Gumustas, M., & Ozkan, S. A. (2015). Biosurfactant production by *Pseudomonas aeruginosa* in kefir and fish meal. *Brazilian Journal of Microbiology, 46*(3), 855–859. Available from https://doi.org/10.1590/S1517-838246320140727.

Konishi, M., Yoshida, Y., & Horiuchi, Ji (2015). Efficient production of sophorolipids by *Starmerella bombicola* using a corncob hydrolysate medium. *Journal of Bioscience and Bioengineering, 119*(3), 317–322. Available from https://doi.org/10.1016/j.jbiosc.2014.08.007.

Kourmentza, C., Costa, J., Azevedo, Z., Servin, C., Grandfils, C., De Freitas, V., & Reis, M. A. M. (2018). Burkholderia thailandensis as a microbial cell factory for the bioconversion of used cooking oil to polyhydroxyalkanoates and rhamnolipids. *Bioresource Technology, 247*, 829–837. Available from https://doi.org/10.1016/j.biortech.2017.09.138.

Kumar, A. P., Janardhan, A., Viswanath, B., Monika, K., Jung, J. Y., & Narasimha, G. (2016). Evaluation of orange peel for biosurfactant production by *Bacillus licheniformis* and their ability to degrade naphthalene and crude oil. *3 Biotech, 6*(1), 1–10. Available from https://doi.org/10.1007/s13205-015-0362-x.

Louhasakul, Y., Cheirsilp, B., Intasit, R., Maneerat, S., & Saimmai, A. (2020). Enhanced valorization of industrial wastes for biodiesel feedstocks and biocatalyst by lipolytic oleaginous yeast and biosurfactant-producing bacteria. *International Biodeterioration & Biodegradation, 148*, 104911. Available from https://doi.org/10.1016/j.ibiod.2020.104911.

Luna, J. M., Rufino, R. D., Sarubbo, L. A., & Campos-Takaki, G. M. (2013). Characterisation, surface properties and biological activity of a biosurfactant produced from industrial waste by *Candida sphaerica* UCP0995 for application in the petroleum industry. *Colloids and Surfaces B: Biointerfaces, 102*, 202–209. Available from https://doi.org/10.1016/j.colsurfb.2012.08.008.

Marcelino, P. R. F., Peres, G. F. D., Terán-Hilares, R., Pagnocca, F. C., Rosa, C. A., Lacerda, T. M., dos Santos, J. C., & da Silva, S. S. (2019). Biosurfactants production by yeasts using sugarcane bagasse hemicellulosic hydrolysate as new sustainable alternative for lignocellulosic biorefineries. *Industrial Crops and Products, 129*, 212−223. Available from https://doi.org/10.1016/j.indcrop.2018.12.001.

Mardiah, I., Puspitaningrum, K., Hamdani, S., & Setiani, N. A. (2022). Utilization of waste frying oil as a source of carbon in the production of biosurfactant using exiguobacterium profundum. *Nusantara Science and Technology Proceedings*, 27−33. Available from https://doi.org/10.11594/nstp.2022.2104.

Md Badrul Hisham, N. H., Ibrahim, M. F., Ramli, N., & Abd-Aziz, S. (2019). Production of biosurfactant produced from used cooking oil by Bacillus sp. HIP3 for heavy metals removal. *Molecules, 24*(14). Available from https://doi.org/10.3390/molecules24142617.

Minucelli, T., Ribeiro-Viana, R. M., Borsato, D., Andrade, G., Cely, M. V. T., de Oliveira, M. R., Baldo, C., & Celligoi, M. A. P. C. (2017). Sophorolipids production by *Candida bombicola* ATCC 22214 and its potential application in soil bioremediation. *Waste and Biomass Valorization, 8*(3), 743−753. Available from https://doi.org/10.1007/s12649-016-9592-3.

Mnif, I., Ellouze-Chaabouni, S., & Ghribi, D. (2013). Economic production of *Bacillus subtilis* SPB1 biosurfactant using local agro-industrial wastes and its application in enhancing solubility of diesel. *Journal of Chemical Technology & Biotechnology, 88*(5), 779−787. Available from https://doi.org/10.1002/jctb.3894.

Mohanty, S. S., Koul, Y., Varjani, S., Pandey, A., Ngo, H. H., Chang, J. S., Wong, J. W. C., & Bui, X. T. (2021). A critical review on various feedstocks as sustainable substrates for biosurfactants production: A way towards cleaner production. *Microbial Cell Factories, 20*(1). Available from https://doi.org/10.1186/s12934-021-01613-3.

Moshtagh, B., Hawboldt, K., & Zhang, B. (2019). Optimization of biosurfactant production by *Bacillus subtilis* N3-1P using the brewery waste as the carbon source. *Environmental Technology, 40*(25), 3371−3380. Available from https://doi.org/10.1080/09593330.2018.1473502.

Moya-Ramírez, I., Altmajer-Vaz, D., Banat, I. M., Marchant, R., Jurado-Alameda, E., & García-Román, M. (2016). Hydrolysis of olive mill waste to enhance rhamnolipids and surfactin production. *Bioresource Technology, 205*, 1−6.

Nalini, S., & Parthasarathi, R. (2014). Production and characterization of rhamnolipids produced by *Serratia rubidaea* SNAU02 under solid-state fermentation and its application as biocontrol agent. *Bioresource Technology, 173*, 231−238. Available from https://doi.org/10.1016/j.biortech.2014.09.051.

Nazareth, T. C., Zanutto, C. P., Tripathi, L., Juma, A., & Maass, D. (2020). The use of low- cost brewery waste product for the production of surfactin as a natural microbial biocide. *Biotechnology Reports, 28*, 1−10.

Neboh, H., & Abu, G. (2015). *Biosurfactant production from Palm Oil Mill Effluent (POME) for application as an oil field chemical in Nigeria* (pp. 1−15). Society of Petroleum Engineers.

Nitschke, M., Costa, S. G. V. A. O., Haddad, R., Gonçalves, L. A. G., Eberlin, M. N., & Contiero, J. (2005). Oil wastes as unconventional substrates for rhamnolipid biosurfactant production by *Pseudomonas aeruginosa* LBI. *Biotechnology Progress, 21*(5), 1562−1566. Available from https://doi.org/10.1021/bp050198x.

Nitschke, M., & Pastore, G. M. (2006). Production and properties of a surfactant obtained from *Bacillus subtilis* grown on cassava wastewater. *Bioresource Technology, 97*(2), 336−341. Available from https://doi.org/10.1016/j.biortech.2005.02.044.

Noah, K. S., Bruhn, D. F., & Bala, G. A. (2005). Surfactin production from potato process effluent by *Bacillus subtilis* in a chemostat. *Applied biochemistry and biotechnology - Part A enzyme engineering and biotechnology* (122, pp. 465−473). Humana Press 1−3. Available from https://doi.org/10.1007/978-1-59259-991-2_41.

Nogueira Felix, A. K., Martins, J. J. L., Lima Almeida, J. G., Giro, M. E. A., Cavalcante, K. F., Maciel Melo, V. M., Loiola Pessoa, O. D., Ponte Rocha, M. V., Rocha Barros Gonçalves, L., & Saraiva de Santiago Aguiar, R. (2019). Purification and characterization of a biosurfactant produced by *Bacillus subtilis* in cashew apple juice and its application in the remediation of oil-contaminated soil. *Colloids and Surfaces B: Biointerfaces, 175*, 256−263. Available from https://doi.org/10.1016/j.colsurfb.2018.11.062.

Nui, Y., Wu, J., Wang, W., & Chen, Q. (2019). Production and characterization of a new glycolipid, mannosylerythritol lipid, from waste cooking oil biotransformation by Pseudozyma aphidis ZJUDM34. *Food Science & Nutrition, 7*(3), 937−948. Available from https://doi.org/10.1002/fsn3.880.

OECD-FAO. (2019). *OECD-FAO agricultural outlook*. OECD Publishing.

Oje, O. A., Okpashi, V. E., Uzor, J. C., Uma, U. O., Irogbolu, A. O., & Onwurah, I. N. E. (2016). Effect of acid and alkaline pretreatment on the production of biosurfactant from rice husk using mucor indicus. *Research Journal of Environmental Toxicology*, *10*(1), 60–67. Available from https://doi.org/10.3923/rjet.2016.60.67.

Otzen, D. E. (2017). Biosurfactants and surfactants interacting with membranes and proteins: Same but different? *Biochimica et Biophysica Acta - Biomembranes*, *1859*(4), 639–649. Available from https://doi.org/10.1016/j.bbamem.2016.09.024.

Ozdal, M., Gurkok, S., & Ozdal, O. G. (2017). Optimization of rhamnolipid production by *Pseudomonas aeruginosa* OG1 using waste frying oil and chicken feather peptone. *3 Biotech*, *7*(2). Available from https://doi.org/10.1007/s13205-017-0774-x.

Paraszkiewicz, K., Bernat, P., Kuśmierska, A., Chojniak, J., & Płaza, G. (2018). Structural identification of lipopeptide biosurfactants produced by *Bacillus subtilis* strains grown on the media obtained from renewable natural resources. *Journal of Environmental Management*, *209*, 65–70. Available from https://doi.org/10.1016/j.jenvman.2017.12.033.

Rajasimman, M., Suganya, A., Manivannan, P., & Pandian, A. M. K. (2021). Utilization of agroindustrial wastes with a high content of protein, carbohydrates, and fatty acid used for mass production of biosurfactant. *Green sustainable process for chemical and environmental engineering and science: Microbially-derived biosurfactants for improving sustainability in industry* (pp. 127–146). Elsevier. Available from https://doi.org/10.1016/B978-0-12-823380-1.00007-1.

Rivera, O. M. P., Moldes, A. B., Torrado, A. M., & Domínguez, J. M. (2007). Lactic acid and biosurfactants production from hydrolyzed distilled grape marc. *Process Biochemistry*, *42*(6), 1010–1020. Available from https://doi.org/10.1016/j.procbio.2007.03.011.

Rocha, M. V. P., Souza, M. C. M., Benedicto, S. C. L., Bezerra, M. S., Macedo, G. R., Pinto, G. A. S., & Gonçalves, L. R. B. (2007). Production of biosurfactant by *Pseudomonas aeruginosa* grown on cashew apple juice. *Applied biochemistry and biotechnology* (137–140, pp. 185–194). Humana Press 1–12. Available from https://doi.org/10.1007/s12010-007-9050-6.

Samad, A., Zhang, J., Chen, D., Chen, X., Tucker, M., & Liang, Y. (2017). Sweet sorghum bagasse and corn stover serving as substrates for producing sophorolipids. *Journal of Industrial Microbiology and Biotechnology*, *44*(3), 353–362. Available from https://doi.org/10.1007/s10295-016-1891-y.

Samad, A., Zhang, J., Chen, D., & Liang, Y. (2015). Sophorolipid production from biomass hydrolysates. *Applied Biochemistry and Biotechnology*, *175*(4), 2246–2257. Available from https://doi.org/10.1007/s12010-014-1425-x.

Sanches, M., Santos, R., Cortez, A., Mariner, R., & Souza, J. (2018). Biosurfactant production by *Fusarium oxysporum* LM 5634 using peels from the fruit of Bactris gasipaes (Kunth) as substrate. *Biotechnology Journal International*, *21*(1), 1–9. Available from https://doi.org/10.9734/bji/2018/39405.

Santos, A. P. P., Silva, M. D. S., Costa, E. V. L., Rufino, R. D., Santos, V. A., Ramos, C. S., Sarubbo, L. A., & Porto, A. L. F. (2018). Production and characterization of a biosurfactant produced by Streptomyces sp. DPUA 1559 isolated from lichens of the Amazon region. *Brazilian Journal of Medical and Biological Research*, *51*(2). Available from https://doi.org/10.1590/1414-431x20176657.

Santos, D. K. F., Brandão, Y. B., Rufino, R. D., Luna, J. M., Salgueiro, A. A., Santos, V. A., & Sarubbo, L. A. (2014). Optimization of cultural conditions for biosurfactant production from *Candida lipolytica*. *Biocatalysis and Agricultural Biotechnology*, *3*(3), 48–57. Available from https://doi.org/10.1016/j.bcab.2014.02.004.

Santos, D. K. F., Rufino, R. D., Luna, J. M., Santos, V. A., & Sarubbo, L. A. (2016). Biosurfactants: Multifunctional biomolecules of the 21st century. *International Journal of Molecular Sciences*, *17*(3). Available from https://doi.org/10.3390/ijms17030401.

Santos, E. F., Teixeira, M. F. S., Converti, A., Porto, A. L. F., & Sarubbo, L. A. (2019). Production of a new lipoprotein biosurfactant by Streptomyces sp. DPUA1566 isolated from lichens collected in the Brazilian Amazon using agroindustry wastes. *Biocatalysis and Agricultural Biotechnology*, *17*, 142–150. Available from https://doi.org/10.1016/j.bcab.2018.10.014.

Shah, V., Jurjevic, M., & Badia, D. (2007). Utilization of restaurant waste oil as a precursor for sophorolipid production. *Biotechnology Progress*, *23*(2), 512–515. Available from https://doi.org/10.1021/bp0602909.

Sharma, D. (2016). *Biosurfactants in food*. Springer.

Sharma, S., Verma, R., Dhull, S., Maiti, S. K., & Pandey, L. M. (2022). Biodegradation of waste cooking oil and simultaneous production of rhamnolipid biosurfactant by *Pseudomonas aeruginosa* P7815 in batch and fed-batch bioreactor. *Bioprocess and Biosystems Engineering*, 45(2), 309–319. Available from https://doi.org/10.1007/s00449-021-02661-0.

Shilpa, M. (2021). Use of natural wastes for biosurfactant (BS) and bioemulsifier (BE) production and their applications – A review. *Open Access Journal of Microbiology & Biotechnology*, 6(3), 1–17. Available from https://doi.org/10.23880/oajmb-16000203.

Silva, S. N. R. L., Farias, C. B. B., Rufino, R. D., Luna, J. M., & Sarubbo, L. A. (2010). Glycerol as substrate for the production of biosurfactant by *Pseudomonas aeruginosa* UCP0992. *Colloids and Surfaces B: Biointerfaces*, 79(1), 174–183. Available from https://doi.org/10.1016/j.colsurfb.2010.03.050.

Sindhu, R., Gnansounou, E., Binod, P., & Pandey, A. (2016). Bioconversion of sugarcane crop residue for value added products – An overview. *Renewable Energy*, 98, 203–215. Available from https://doi.org/10.1016/j.renene.2016.02.057.

Soares da Silva, R. d. C. F., de Almeida, D. G., Brasileiro, P. P. F., Rufino, R. D., de Luna, J. M., & Sarubbo, L. A. (2019). Production, formulation and cost estimation of a commercial biosurfactant. *Biodegradation*, 30(4), 191–201. Available from https://doi.org/10.1007/s10532-018-9830-4.

Sobrinho, H. B. S., Rufino, R. D., Luna, J. M., Salgueiro, A. A., Campos-Takaki, G. M., Leite, L. F. C., & Sarubbo, L. A. (2008). Utilization of two agroindustrial by-products for the production of a surfactant by *Candida sphaerica* UCP0995. *Process Biochemistry*, 43(9), 912–917. Available from https://doi.org/10.1016/j.procbio.2008.04.013.

Solaiman, D. K. Y., Ashby, R. D., Zerkowski, J. A., & Foglia, T. A. (2007). Simplified soy molasses-based medium for reduced-cost production of sophorolipids by *Candida bombicola*. *Biotechnology Letters*, 29(9), 1341–1347. Available from https://doi.org/10.1007/s10529-007-9407-5.

Sood, U., Singh, D. N., Hira, P., Lee, J. K., Kalia, V. C., Lal, R., & Shakarad, M. (2020). Rapid and solitary production of mono-rhamnolipid biosurfactant and biofilm inhibiting pyocyanin by a taxonomic outlier *Pseudomonas aeruginosa* strain CR1. *Journal of Biotechnology*, 307, 98–106. Available from https://doi.org/10.1016/j.jbiotec.2019.11.004.

Thompson, D. N., Fox, S. L., & Bala, G. A. (2000). *Biosurfactants from potato process effluents*, Applied biochemistry and biotechnology - Part A Enzyme engineering and biotechnology (84–86, pp. 917–930). Humana Press. Available from https://doi.org/10.1385/ABAB:84-86:1-9:917.

Verma, R., Sharma, S., Kundu, L. M., & Pandey, L. M. (2020). Experimental investigation of molasses as a sole nutrient for the production of an alternative metabolite biosurfactant. *Journal of Water Process Engineering*, 38, 101632. Available from https://doi.org/10.1016/j.jwpe.2020.101632.

Vieira, I. M. M., Santos, B. L. P., Silva, L. S., Ramos, L. C., de Souza, R. R., Ruzene, D. S., & Silva, D. P. (2021). Potential of pineapple peel in the alternative composition of culture media for biosurfactant production. *Environmental Science and Pollution Research*, 28(48), 68957–68971. Available from https://doi.org/10.1007/s11356-021-15393-1.

Vijayakumar, S., & Saravanan, V. (2015). Biosurfactants-types, sources and applications. *Research Journal of Microbiology*, 10(5), 181–192. Available from https://doi.org/10.3923/jm.2015.181.192.

Zheng, C., Luo, Z., Yu, L., Huang, L., & Bai, X. (2011). The utilization of lipid waste for biosurfactant production and its application in enhancing oil recovery. *Petroleum Science and Technology*, 29(3), 282–289. Available from https://doi.org/10.1080/10916460903117586.

Zhu, Y., Gan, J. J., Zhang, G. L., Yao, B., Zhu, W. J., & Meng, Q. (2007). Reuse of waste frying oil for production of rhamnolipids using *Pseudomonas aeruginosa* zju.u1M. *Journal of Zhejiang University: Science A*, 8(9), 1514–1520. Available from https://doi.org/10.1631/jzus.2007.A1514.

Zhu, Z., Zhang, B., Cai, Q., Ling, J., Lee, K., & Chen, B. (2020). Fish waste based lipopeptide production and the potential application as a bio-dispersant for oil spill control. *Frontiers in Bioengineering and Biotechnology*, 8, 734. Available from https://doi.org/10.3389/fbioe.2020.00734.

Zouari, R., Ellouze-Chaabouni, S., & Ghribi-Aydi, D. (2014). Optimization of *Bacillus subtilis* SPB1 biosurfactant production under solid-state fermentation using by-products of a traditional olive mill factory. *Achievements in the Life Sciences*, 8(2), 162–169. Available from https://doi.org/10.1016/j.als.2015.04.007.

CHAPTER 5

Factors affecting biosurfactants production

Arif Nissar Zargar and Preeti Srivastava
Department of Biochemical Engineering and Biotechnology, Indian Institute of Technology Delhi, Hauz Khas, New Delhi, India

5.1 Introduction

Surfactants are surface-active chemicals with a polar hydrophilic head and a nonpolar hydrophobic tail. Because of their amphipathic nature, these compounds can align at the interface of two distinct phases and reduce interfacial tension at the boundary. They can reduce liquid surface tension and hence improve surface wettability. Because of their qualities, they are widely used in various everyday products such as food, cosmetics, medications, and detergents (Gayathiri et al., 2022). Surfactants are extremely important in modern society since their safe use determines people's quality of life and health (Georgiou et al., 1992).

Chemical surfactant production began in the early twentieth century and saw significant growth following World War II, with the emergence of the petrochemical sector, which began providing raw materials for their synthesis (Georgiou et al., 1992). Since then, there has been a continuous increase in the demand, production, and application of surfactants across diverse industries such as detergent, food and beverage, cosmetics, agricultural, pharmaceutical, and petroleum. The global market for surfactants was valued at USD 41.3 billion in 2019, with a projected increase to USD 58.8 billion by 2027 (Beuker et al., 2014). However, growing sustainability agendas, public awareness of environmental toxicity, and safety concerns about the use of synthetic surfactants have prompted a search for green alternatives that can replace synthetic surfactants to protect nature and human health while avoiding environmental degradation.

Biosurfactants have evolved as effective alternatives to conventional surfactants and demonstrated significant potential to replace their chemical equivalents practically in all industries (Mulligan et al., 1989b). Biosurfactants are amphiphilic molecules of biological origin produced by various living organisms such as plants, animals, bacteria, and fungi. They serve distinct functions in different microorganisms. Biosurfactants produced by plants and animals have antimicrobial properties and, hence, play an important role in immunity (Gomaa, 2013). Biosurfactants produced by bacteria, fungi, and yeasts are among the few metabolites produced by these organisms that impart biophysical advantages to the microbes producing them. They assist these microbes in their cellular development, adjusting to osmotic pressure, biofilm formation, nutrient assimilation, and in cellular communication and motility (Gomaa, 2013; Sharma et al., 2021). Compared to biosurfactants produced by plants and animals, microbial biosurfactants are of specific importance as their production in large amounts at a commercial scale is achievable.

Biosurfactants are one of the many structurally and functionally diverse metabolites produced by microorganisms with diverse applications in various industries (Banat et al., 2010; Banat et al., 2014; De Almeida et al., 2016; Desai & Banat, 1997; Geetha et al., 2018; Ju et al., 2016; Kosaric & Sukan, 2010; Rodrigues et al., 2006). Microbial biosurfactants belong to various chemical groups, which include glycolipids, glycolipoproteins, lipopeptides, glycopeptides, lipoproteins, fatty acids, neutral acids, phospholipids, glycoglycerolipids, and lipopolysaccharides (Banat et al., 2010; Desai & Banat, 1997; Marchant & Banat, 2012; Wicke et al., 2000; Zargar et al., 2022a; 2022b). Biosurfactants have been reported to exhibit properties similar to chemical surfactants in terms of emulsification, CMC, foaming, detergency, wetting, dispersion, surface and interfacial tension reduction, antimicrobial agents, antiadhesive nature, and residual oil recovery (Vijayakumar & Saravanan, 2015). These properties, along with their degradability, environmental friendliness, lower toxicity and activity under extreme conditions, have allowed biosurfactants to be considered attractive alternatives to their chemical counterparts (Sharma et al., 2022; Zargar et al., 2022a). The production of biosurfactants from renewable substrates is another advantage compared with chemical surfactants derived from fossil fuels (Le Roes-Hill et al., 2019).

Microorganisms mainly produce biosurfactants to emulsify hydrophobic substrates to increase their bioavailability. Besides, biosurfactants are produced by microbes to enable metal binding and biofilm formation. Biosurfactants are also involved in pathogenesis, as some biosurfactants act as virulence factors and antimicrobial agents. Microbial biosurfactants facilitate pyrogenicity, lethal toxicity, immunogenicity, and mitogenicity on the interaction of microbes with eucaryotic cells (Neu, 1996). Production of biosurfactants also enables the microorganisms to change their cell surface hydrophobicity, enabling microbial adhesion to surfaces, cells, and biofilms (Wicken & Knox, 1980). Microbial biosurfactants also play a critical role in regulating cell motility, antagonism, virulence, and cellular communication (Sharma et al., 2021).

5.2 Factors affecting biosurfactant production

Microorganisms produce biosurfactants as secondary metabolites, either as growth-associated or non-growth-associated compounds, to facilitate the functions mentioned earlier. Like other secondary metabolites, their production is governed by various factors that can be intrinsic (depending on the microbial strain), like the type of

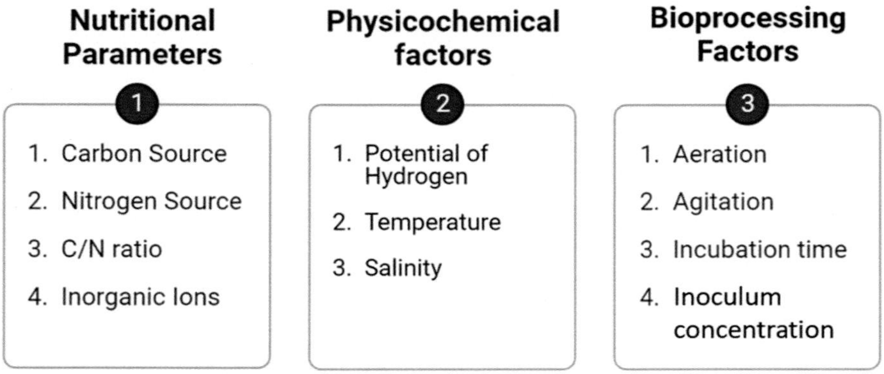

Nutritional Parameters (1)	Physicochemical factors (2)	Bioprocessing Factors (3)
1. Carbon Source 2. Nitrogen Source 3. C/N ratio 4. Inorganic Ions	1. Potential of Hydrogen 2. Temperature 3. Salinity	1. Aeration 2. Agitation 3. Incubation time 4. Inoculum concentration

Figure 5.1 Factors affecting biosurfactant production.

biosurfactant produced, or extrinsic (depending on the process conditions), like productivity and purity of the biosurfactant produced. These factors can affect biosurfactant production at three different stages of production: prefermentation, fermentation, and postfermentation. Fig. 5.1 lists the external factors that affect biosurfactant production by a particular microbial strain.

A detailed discussion of the factors and their effect on biosurfactant production follows.

5.2.1 Microbial strain

Till date, numerous microorganisms, which include bacteria, fungi, and yeasts, have been reported to produce biosurfactants. Among the three, bacteria have been majorly utilized to produce biosurfactants. Extensive reports on the production of biosurfactants by Bacillus and Pseudomonas are available in the literature, and they have been described as super-producers (Santos et al., 2016). Biosurfactant production by yeasts like Candida, Yarrowia, and Saccharomyces have also been reported by various groups (Bertrand et al., 2018; Diniz Rufino et al., 2014; Fontes et al., 2010). The main factor that determines the production of biosurfactants by microorganisms is its genetic make-up (Randhawa, 2014; Thavasi & Banat, 2019b). The presence of the genes coding for the enzymes involved in the production of the biosurfactant dictates not only whether the microbial strain can produce biosurfactants but also what type of biosurfactant it will produce and whether it will produce a single type of biosurfactant or a mixture of different congener compounds.

The growth requirements of the microbial strain dictate the biosurfactant's properties (Randhawa, 2014). Biosurfactants produced by the extremophile will usually be stable under extreme conditions. Therefore, if one needs to use biosurfactants under extreme environmental conditions, e.g., for the recovery of residual oil, it would be better to use extremophiles to produce biosurfactants. Schultz et al., in their review, have summarized various reports of biosurfactants produced by extremophiles and described the biotechnological applications of extreme biosurfactants (Schultz & Rosado, 2020). An ideal microbial strain utilized for the production of biosurfactant will depend on the biosurfactant's application, e.g., if the area of application is pharmaceuticals, then the strain utilized should produce only a specific type of biosurfactant (Fracchia et al., 2019). On the contrary, if the area of application is environmental remediation, then it may be advantageous to use a strain that produces a mixture of biosurfactants having a broad range of action (Finnerty, 1994; Pacwa-Płociniczak et al., 2011).

For commercial production of the biosurfactant using a particular strain, it would be advantageous to know the details about the molecular genetics, biochemical enzymes, and metabolic pathways involved in the production of biosurfactants. It would give greater control over the production of the biosurfactant. However, except for rhamnolipids, there is very limited information about the genetics and the biochemical pathways for the production of other types of biosurfactants (Das et al., 2008).

5.2.2 Nutritional parameters

The composition of the culture medium has a critical effect on the growth of the microorganisms and the production of the biosurfactants. Nutritional factors that affect biosurfactant production include carbon source, nitrogen source, and inorganic ions (iron, multivalent ions, phosphate concentration, and presence of salts).

5.2.2.1 Carbon source

The carbon sources present in the culture affect the growth of the microbial strains and are also utilized as building blocks for synthesizing biosurfactants (Abbasi et al., 2013). Therefore, the production of biosurfactants is highly dependent on the type and concentration of carbon source present in the culture medium (Kashif et al., 2022). The carbon source used in the bioprocesses significantly affects the type, quality, and yield of the biosurfactant (Vieira et al., 2021). Sugars such as rhamnose form the polar hydrophilic head of the biosurfactant and, therefore, their presence in the culture medium results in the formation of glycolipids (George & Jayachandran, 2013). Different carbon sources result in the synthesis of different types of biosurfactants in the same organism as well. Vecino et al. reported that utilization of lactose as a carbon source resulted in the synthesis of glycolipid, while sugars derived from vineyard pruning waste resulted in the synthesis of glycolipopeptide in the same microorganism (Vecino et al., 2017). Jain et al. reported that the type of carbon source affected yields and physical and chemical properties of the produced biosurfactant (Jain et al., 2013). Biosurfactant synthesis may be regulated by different nutrient sources by induction or catabolic repression (Desai & Desai, 1993; Pal et al., 2009). Hydrocarbons and other immiscible substrates act as inducers for the production of biosurfactants. Some bacterial strains capable of hydrocarbon degradation do not produce biosurfactants in the presence of sugars such as glucose due to catabolite repression. Three types of carbon sources have mostly been used for biosurfactant production: carbohydrates, hydrocarbons, and vegetable oils (Varjani & Upasani, 2017). Simpler carbohydrates such as glucose, glycerol, ethanol, and mannitol, have been reported to promote rhamnolipid production in microorganisms (Gudiña et al., 2016; Robert et al., 1989; Sharma et al., 2022; Varjani & Upasani, 2016). Utilization of hydrocarbons, such as crude oil, alkanes, and diesel, as carbon sources have been reported to result in lower rhamnolipid production (Ramírez et al., 2015; Ron & Rosenberg, 2014; Varjani & Upasani, 2017).

5.2.2.2 Nitrogen source

Nitrogen, similar to carbon, is another essential medium required for the production of biosurfactants. Apart from being utilized for the growth and development of microorganisms, nitrogen is required as a building block to form polar head moieties of lipopeptide and glycolipopeptide biosurfactants (Bertrand et al., 2018). Amino acids form the polar head of the lipopeptides, and their presence in the culture medium has been reported to increase surfactin and iturin production (Makkar & Cameotra, 2002). Both organic and inorganic sources of nitrogen have been utilized for the production of biosurfactants. Various groups have reported the importance of nitrogen sources for the production of biosurfactants (Jimoh & Lin, 2019a). High concentration of nitrogen has been reported to have a detrimental effect on biosurfactant production (Abouseoud et al., 2008a). Vigneshwaran et al. have also compared the effect of three inorganic nitrogen sources ($NaNO_3$, KNO_3, and NH_4NO_3) on biosurfactant production. The group reported a maximum biosurfactant production when KNO_3 was used as a nitrogen source; however, a higher concentration of KNO_3 was found harmful to the fermentation and biosurfactant production (Vigneshwaran et al., 2018). Abouseoud et al. reported higher production of biosurfactants using NH_4NO_3 as a nitrogen source. The group also concluded that a higher concentration of nitrogen is harmful to the production of biosurfactants (Abouseoud et al., 2008a). Recently, Zargar et al. also reported that a higher concentration of yeast extract inhibited the production of saponin biosurfactants (Zargar et al., 2022a). Higher carbon-to-nitrogen ratio has been reported to enhance microbial growth and biosurfactant production (Kashif et al., 2022). A C:N ratio between 16:1 and 18:1 has been reported to result in maximum rhamnolipid production (Guerra-Santos et al., 1984). The group also reported that rhamnolipid production stops below C:N ratio of less than 11:1. Various reports on the effect of organic and inorganic nitrogen sources are available in literature (Abdel-Mawgoud et al., 2008; Abushady et al., 2005; Hassan et al., 2016; Onlamool et al., 2020). The groups have reported higher biosurfactant yields using inorganic nitrogen sources. The use of nitrates as a nitrogen source has been reported to enhance biosurfactant production in *Pseudomonas aeruginosa* (Maqsood & Jamal, 2011). Nitrogen is essential for the production of lipopeptides and glycolipopeptides and plays a key role in the production of glycolipids. Efficient rhamnolipid production has been reported under nitrogen-limiting conditions (Kashif et al., 2022). The addition of a nitrogen source has been reported to inhibit rhamnolipid production in *Pseudomonas aeruginosa* (Syldatk et al., 1985). A simultaneous increase in the glutamine synthase activity and rhamnolipid production was observed under nitrogen-limiting conditions in *P. aeruginosa* (Robert et al., 1989; Toda & Itoh, 2012; Varjani & Upasani, 2016).

5.2.2.3 Inorganic ions

In addition to carbon and nitrogen sources, inorganic ions present in the medium also significantly effect the production of the biosurfactant by a microbial strain. The inorganic ions present in the medium can be either required in considerable quantities (macronutrients) or sufficient in trace amounts (micronutrients) for the growth of microbial strain (Joshi et al., 2007). Macronutrients such as potassium, magnesium, iron, manganese, cobalt, and zinc usually act as cofactors for synthesizing nucleic acids and amino acids. As a result, these factors also impact the synthesis of biosurfactants from the most diverse microorganisms. Metallic salts such as $FeSO_4$, $MnSO_4$, and $MgSO_4$ also act as cofactors for enzymes involved in the biosynthesis of surfactin, and their addition to the culture medium has been reported to increase the biosurfactant production by Gudiña et al. (2015). Similar results have been reported by Makkar and Cameotra, who concluded that adding metallic supplements (iron, magnesium, calcium, and trace elements) to the culture medium also had a substantial effect on biosurfactant production (Makkar & Cameotra, 2002). However, as these compounds mainly act as cofactors, they are required in smaller quantities. Their presence in high concentrations in the culture medium has been reported to inhibit biosurfactant production (Makkar & Cameotra, 2002). Multiple reports on increased biosurfactant production by iron supplementation at low concentrations are available in the literature (Nitschke & Pastore, 2006; Varjani & Upasani, 2017). In a study by Wei et al., the effect of K^+, Mg^{2+}, Ca^{2+}, Fe^{2+}, and Mn^{2+} was evaluated for the production of surfactin by *Bacillus subtilis* (Wei et al., 2007). The group found that surfactin production was influenced by all ions except for calcium, which did not affect the growth or biosurfactant production. An important observation of their study was that when all ions in their optimized concentrations were simultaneously added to the culture medium, the biosurfactant production decreased. The group concluded that there exists a strong interaction between these ions, which determines their overall effect on biosurfactant production. The presence of phosphates in the culture medium has also been reported to affect microbial biosurfactant production (Joshi et al., 2007; Praharyawan et al., 2013; Wei et al., 2007). The direct effect of phosphates on biosurfactant production is unknown; however, it is speculated that phosphates indirectly affect biosurfactant production by regulating the biosynthesis of nucleic acids and phospholipids required for the growth of microorganisms. These studies highlight the important role of inorganic ions in promoting bacterial growth and biosurfactant production.

5.2.3 Physicochemical factors

Similar to all other bioprocessings, biosurfactant production is also affected by various physicochemical factors, including pH, temperature, salinity, oxygen availability, and incubation time. These factors mainly affect the growth and cellular metabolism of the microorganism and, therefore, indirectly affect the biosurfactant production.

5.2.3.1 Potential of hydrogen

The potential of hydrogen (pH) of a fermentation medium impacts both cell growth and metabolite production and therefore is a key factor for biosurfactant production. Multiple statistical studies have assessed the effect of pH on the biosurfactant production by taking CMC or the biosurfactant concentration as responses (Darvishi et al., 2011; Sen & Swaminathan, 1997; Vigneshwaran et al., 2018). In a study conducted by Vigneshwaran et al., pH was found to be a statistically significant factor for the production of the biosurfactant using a Brevibacillus strain. The group reported maximum biosurfactant production at a pH of 6.755 when the range tested was from 4 to 8 (Vigneshwaran et al., 2018). Similar results were obtained by Makkar and Cameotra, who evaluated the effect of pH from 4.5 to 10.5 and reported maximum biosurfactant production at pH 7 (Makkar & Cameotra, 2002). Darvishi et al. also evaluated the effect of pH (4—10) on biosurfactant production by a consortium and reported maximum production at neutral pH (Darvishi et al., 2011). Biosurfactant production has been reported to decrease in the absence of pH control, indicating its importance in biosurfactant production. In contrast to biosurfactant production by bacteria, biosurfactant production by yeasts has been reported at lower pH values as well. Maximum glycolipid production by *Candida antarctica* and *C. apicola* has been reported to occur at pH 5.5 (Bednarski et al., 2004). Similarly, optimal biosurfactant production by *Yarrowia lipolytica* has been reported to occur at a pH value of 5.0 (Bednarski et al., 2004). Other reports on biosurfactant production by *Candida lipolytica* and *C. batistae* reported optimal pH values of 5 and 6, respectively (Santos et al., 2016). The studies have conclusively proven that maximum biosurfactant production occurs at a specific pH, which in most cases, ranges between 6.5 and 7 (Darvishi et al., 2011; Sen & Swaminathan, 1997; Vigneshwaran et al., 2018; Zargar et al., 2022a).

5.2.3.2 Temperature

The microbial growth always depends on the incubation temperature and plays a crucial role in determining the bioprocess performance. Besides, temperature has been reported to influence the composition of the biosurfactant (Abdel-Mawgoud et al., 2008). All microbes have a specific temperature where the growth rate is maximum. Most studies have reported biosurfactant production between 30°C and 37°C by mesophilic microorganisms (Abushady et al., 2005; Hassan et al., 2016; Najafi et al., 2010; Sharma et al., 2022; Silva et al., 2010; Zargar et al., 2022b). The optimal temperature for biosurfactant production is usually the same as that for microbial growth. However, Abdel-Mawgoud et al. reported optimal biosurfactant production using *Bacillus subtilis* at 30°C, while the strain showed high cellular growth at different temperatures (Abdel-Mawgoud et al., 2008). Onlamool et al. reported the highest biosurfactant production by *S. acidaminiphila* between 30°C and 37°C when the optimal temperature for the growth of the strain was 30°C (Onlamool et al., 2020). Surfactin production at different temperatures (25°C–60°C) was studied by Abushady et al., who concluded that maximum production was obtained at 30°C (Abushady et al., 2005). Almansoory et al. evaluated the effect of temperature on biosurfactant production by *S. marcescens* and reported optimal production at 30°C when the range tested was 20°C–40°C (Almansoory et al., 2017). Another study conducted by Hassan et al. showed that temperature critically affects rhamnolipid production (Hassan et al., 2016). The group observed maximum rhamnolipid production at 30°C, and reported that any variation from the optimum temperatures resulted in reduced biosurfactant production.

Depending on the growth requirement of the microbial strain, biosurfactant production has been reported to occur under awide range of temperatures. Psychrophilic and thermophilic microbes have also been reported to produce biosurfactants under extreme temperatures (Schultz & Rosado, 2020). Biosurfactant production by psychrophiles has been reported to help the microbes to interact with water, ice, hydrophobic compounds, and gases under cold conditions (Perfumo et al., 2018). Studies on psychrophiles have concluded that biosurfactant production is a common feature of the cold-loving microorganisms, as 50% of the microbes in a community in polar soils tested positive for biosurfactant production (Vasileva-Tonkova & Gesheva, 2004; Vollú et al., 2014). Konishi et al. reported biosurfactant production by Pseudomonas, Pseudoalteromonas, Marinomonas, Halomonas, Rhodococcus, and Cobetia isolated from deep polar marine sediments (Konishi et al., 2014). Biosurfactant production has also been reported from the microbes isolated from the cold habitats of the Canadian arctic, Antarctic, and Polar Arctic (Cai et al., 2014). Besides bacteria, some psychrophilic yeasts like *Moesziomyces antarcticus* and Rhodotorula have been also been reported to produce biosurfactants (Perfumo et al., 2018).

Thermophilic microorganisms obtained from hot springs, volcanoes, and deserts have not been extensively explored for biosurfactant production. However, biosurfactant production from microbes dwelling in deep oil reservoirs has been reported. Tabatabaee et al. reported biosurfactant production by microbes isolated from deep oil reservoirs in Iran where the temperature ranged from 65°C to 110°C

(Tabatabaei et al., 2005). The group reported that 23 strains of 35 isolates tested positive for biosurfactant production. Daryasafar et al. also reported biosurfactant production by *Bacillus licheniformis* isolated from an oil reservoir in Iran (Daryasafar et al., 2016). They found that the strain could grow at 50°C, and at that temperature, it was successful in reducing the surface tension from 72 to 23.8 mN/m. Another species of Bacillus isolated from the Niage field of Egypt was reported to produce biosurfactant, resulting in an emulsification index of 96% and reducing the surface tension from 72 to 36 mN/m (El-Sheshtawy et al., 2015). Coronel-León et al. reported biosurfactant production by thermophilic *Bacillus licheniformis* AL1.1 isolated from Antarctic volcano (Coronel-León et al., 2015). The biosurfactant produced by the strain was found to be stable up to 110°C and reduced the surface tension of water to 28.5°C. Another biosurfactant-producing strain (*Ochrobactrum intermedium*) was isolated by Zarinviarsagh et al. from a hot spring in Iran (Zarinviarsagh et al., 2017). Biosurfactant produced by the strain had a high emulsification index of 65% at 60°C.

5.2.3.3 Salinity

The salinity of the fermentation medium is another factor that affects biosurfactant production. A study by Almansoory et al. examined the effect of varying concentrations of salt (1%–5% NaCl) in the fermentation medium on biosurfactant production by *S. marcescens* (Almansoory et al., 2017). Limited biomass growth and biosurfactant production was observed in the absence of salt, while the highest biosurfactant yield and surface tension reduction were observed when 1% NaCl was added to the medium. The study showed that the salinity of the fermentation medium is a critical factor for the production of the biosurfactant. Similar results on limited growth and biosurfactant production by *Vibrio sp.* in the absence of salt have been reported by Hu et al. (2015). The group examined the effect of 0, 5, 10, 15, 20, and 30 g/L NaCl and reported the highest biosurfactant production when the fermentation medium was supplemented with 20 g/L NaCl. Compared with these studies, a study by Darvishi et al. on biosurfactant production by a microbial consortium showed that a higher concentration of NaCl (15%) deterred biosurfactant production (Darvishi et al., 2011). In other studies, biosurfactant production was found to be unaffected by salt concentration (Makkar & Cameotra, 2002). These studies show no clear consensus on the effect of salinity on biosurfactant production. The salinity of the medium might interact with other medium components and, therefore, affect biosurfactant production differently under different conditions. Zargar et al. have reported that increasing the salinity of the medium can increase the osmolarity of the medium when other components of the medium are in excess and, therefore, can induce salt stress on the microbial cells and, thus, limit their growth and biosurfactant production (Zargar et al., 2022a).

Various reports on biosurfactant production using halophilic microorganisms are available in the literature (Schultz & Rosado, 2020). Some halophilic microorganisms reported for biosurfactant production are *Salinibacter ruber*, *Salicola* sp. *Haloarcula argentinensis*, *Haloarcula japonica*, *Haloarcula vallismortis*, and *Halorubrum tebenquichense* (Donio et al., 2013; Llamas et al., 2012; Nercessian et al., 2015; Sarafin et al., 2014). Kebbouche-Gana et al. reported peptidoglycolipid biosurfactant production by Halovivax and Haloarcula isolated from salterns in Ain Salah, Algeria (Kebbouche-Gana et al., 2009). The strain could grow and produce stable biosurfactant in a medium containing 35% NaCl. Biosurfactant production by halophilic *Bacillus licheniformis*, *Bacillus subtilis*, *Bacillus asahii*, *B. detrensis*, *Kukuria marina* BS15, *Halomonas* sp. BS4, *Fictibacillus barbaricus*, and *Paenisporosarcina indica* (Couto et al., 2015). Kim et al. and Zinjarde et al. isolated biosurfactant-producing strains of *Yarrowia lipolytica* from hypersaline and marine locations (Kim et al., 2007; Zinjarde et al., 2008). Halophilic biosurfactant-producing strains can be used for PAH remediation, wastewater treatment, or enhanced oil recovery.

5.2.4 Bioprocessing factors
5.2.4.1 Aeration and agitation
Oxygen is a key component required for the growth of aerobic microbes, as many intracellular enzymatic reactions are regulated by oxygen. Aeration and agitation are the key factors that control the oxygen concentration inside the culture vessel. Like any other bioprocessing, the yield and productivity of biosurfactant production are critically dependent on aeration and agitation. Proper agitation not only ensures homogeneous distribution of nutrients and biomass inside the culture vessel but also plays a key role in gas—liquid and liquid—solid mass transfer. Optimum aeration and agitation prevent impeller flooding and loading inside the fermentation vessel, ensuring complete gas dispersion and exchange of oxygen from the gas to the aqueous phase. Aeration and agitation determine the dissolved oxygen concentration inside the culture medium to ensure the oxygen transfer rate (OTR) is equivalent to the oxygen uptake rate (OUR). Insufficient aeration or agitation can result in reduction in OTR, in which case OTR will become less than OUR, dissolved oxygen concentration will drop to zero, the microbes will face oxygen limitation, and their growth and biosurfactant production will be affected. Higher aeration and lower agitation also have detrimental effects on microbial growth as such a condition can lead to impeller loading and flooding; therefore, gas dispersion will be impaired, and the microbes will again face oxygen limitation.

Ghribi and Ellouze-Chaabouni have studied the effect of aeration on biosurfactant production and have concluded that it strongly influences the biosurfactant yield (Ghribi & Ellouze-Chaabouni, 2011). The group concluded that maximum biosurfactant production was obtained when 30% oxygen saturation was maintained. They also concluded that very low and very high aeration rates impair biosurfactant production. Fontes et al. also observed higher biosurfactant production when the agitation speed increased from 160 to 200 rpm (Fontes et al., 2010). In another study by Silva et al., a biosurfactant yield of 6.5 g/L was obtained at an impeller speed of 200 rpm. When the impeller speed of 150 rpm was used, only 5 g/L biosurfactant was obtained. At the same time, aeration evaluated at 60%, 80%, and 90% did not affect biosurfactant production (Silva et al., 2010). Wei et al. observed in their study that an increase in the agitation speed from 80 to 250 rpm increases the rhamnolipid production by 80% (Wei et al., 2007). Excessive agitation can also result in the formation of foam, which can either trap a considerable fraction of biomass and affect the overall fermentation process or lead to contamination of the fermentation run.

5.2.4.2 Incubation time

Biosurfactant production can be a growth-associated or non-growth-associated phenomenon, which means that the maximum accumulation of biosurfactant can either be at the end of the exponential phase (growth-associated) or start at the beginning of the stationary phase (non-growth-associated). Therefore, incubation time plays a significant role in biosurfactant production. The production of biosurfactants during the cell growth can be checked by either measuring the broth's surface tension or performing emulsion index assay of the collected broth during the fermentation run. Zargar et al. found saponin production to be a growth-associated phenomenon as biosurfactant production started as soon as the culture entered the exponential phase and stopped when the stationary phase was reached (Zargar et al., 2022a). In another study, Abouseoud et al. observed biosurfactant formation by *Pseudomonas fluorescence* started at the 36th hour of incubation and reached its maximum in the 56th hour (Abouseoud et al., 2008a).

5.2.4.3 Inoculum concentration

As the efficiency of any bioprocess depends on the cell density, the inoculum concentration is of great relevance for the production of any by-product. Biosurfactant production is also affected by the inoculum concentration as it controls cell density and biomass growth. A lower inoculum size leads to a reduction in the initial amount of microbial cells inside the culture vessel; therefore, an extended incubation time is required to achieve optimal growth for the production of biosurfactant. On the contrary, a higher-than-optimal inoculum size may decrease microbial activity due to the deregulation of the substrate inside the production vessel (Vieira et al., 2021).

Statistical studies conducted by Mnif et al. showed that age and inoculum concentration are critical factors for biosurfactant production (Mnif et al., 2013). The group observed an increase in the biosurfactant yields by changing the inoculum values. In a separate study by Jimoh and Lin, maximum biosurfactant yield was obtained using an inoculum concentration of 1.5% (Jimoh & Lin, 2019b). Abdel-Mawgoud et al.

reported that a higher inoculum size (6%) led to decrease in the growth of *Bacillus subtilis* and slow accumulation of surfactin (Abdel-Mawgoud et al., 2008). Other studies evaluating the effect of inoculum size on biosurfactant production include (Ghribi & Ellouze-Chaabouni, 2011; Nalini & Parthasarathi, 2018).

5.3 Factors that affect large-scale production and commercialization of biosurfactants

Despite the benefits of biosurfactants over chemical counterparts, commercialization is still difficult and expensive (Zargar et al., 2023). Large-scale production of biosurfactants is limited by various factors, which can broadly be categorized into three types: economic constraints, technical constraints, and safety concerns (Fig. 5.2).

5.3.1 Economic constraints

Economic constraints include the high cost of raw materials and the cost involved in the extraction and purification of biosurfactants (Petrides, 2000). Together, these two constitute up to 80% of the total production cost. Compared with chemical biosurfactants, the cost involved in downstream processing of biosurfactants is 10 times higher, making biosurfactants costlier than chemical surfactants, which has an indirect effect on their market demand (Winterburn & Martin, 2012). Most studies on biosurfactants have focused on purification and detailed characterization of the biosurfactants at the lab scale; however, limited studies have been conducted on process economics involved in large-scale production and downstream processing.

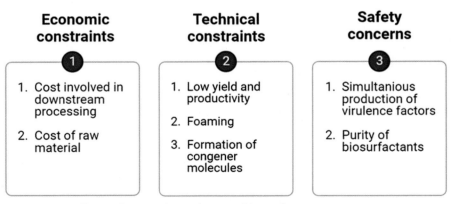

Figure 5.2 Factors affecting large-scale production of biosurfactants.

5.3.2 Technical constraints

Technical constraints include low process yields and productivity and other problems such as foaming faced during biosurfactant production (Henkel et al., 2012; Kronemberger et al., 2007; Makkar et al., 2011). In most cases, a biosurfactant concentration of less than 8 g/L has been obtained. The lower yields obtained significantly hinder the commercialization of most biosurfactants and restrict the large-scale manufacture of biosurfactants. Although some studies have shown a high production yield at the laboratory scale, most biosurfactants still fail to achieve the yield required at the industrial scale on an economically viable scale.

Foam generation during biosurfactant production leads to its accumulation in the headspace of the fermentation vessel. The accumulated foam entraps a considerable fraction of the biomass and therefore reduces biomass fraction in the bulk medium, resulting in lower biosurfactant yields. Foam generation also complicates the process control and is a major source of contamination (Winterburn & Martin, 2012).

Another technical constraint that limits the production of biosurfactants at a large scale is the purification of biosurfactants from a mixture of congener molecules produced by the microorganism of varying in shapes, chain lengths, degree of unsaturation, and physical and chemical properties (Marchant & Banat, 2012; Smyth et al., 2010). The level of the purification required is dictated by the end application of the biosurfactant; e.g., for pharmaceutical applications or cosmetic applications, the biosurfactant must be in its purest form. However, for the application in the detergent industry or petroleum industry, the level of purity required can be low. In cases of high purity, the downstream processing involved is very expensive and not economically viable on industrial scale.

5.3.3 Safety concerns

Another factor that limits large-scale production and commercialization of biosurfactants is the simultaneous secretion of virulence factors by the strains producing biosurfactants (Cha et al., 2008; Dhanya, 2021; Rikalović et al., 2015). The presence of these virulent factors has an adverse effect on the overall demand for biosurfactants in the pharmaceutical, cosmetic, food, and detergent industries. The solution is to purify the biosurfactants produced, but the steps involved in purification increase the overall cost of the final product, thus affecting its economic viability.

5.4 Possible approaches to improve biosurfactant production

As discussed in the preceding section, biosurfactant production is affected by various factors at different production stages. Proper optimizations at each step are required to maximize biosurfactant production (Fig. 5.3).

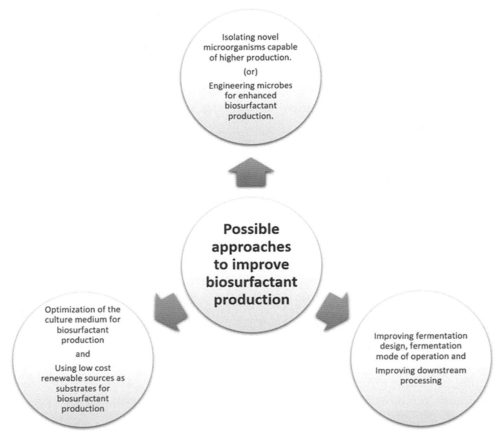

Figure 5.3 Possible approaches to improve biosurfactant production.

Some of the possible strategies that can be used to enhance biosurfactant production include the following.

5.4.1 New or engineered microbial strains

As discussed earlier, biosurfactant production depends heavily on process economics, which to a large extent, is governed by the biosurfactant yield. An important strategy to improve biosurfactant production is to focus on improving the biosurfactant yields. This can be achieved by two methods: searching for new isolates capable of producing biosurfactants at high yields or enhancing the yield of biosurfactants of already reported biosurfactant-producing microbes.

Although many reports on the isolation of biosurfactant-producing microorganisms are available in the literature, the biosurfactant yields reported by the microbes are usually very low. More efforts are required to isolate new microorganisms that can produce biosurfactants with higher yields. One of the ways to achieve this is to isolate new strains from hydrocarbon- or lipid-contaminated environments using the enrichment culture technique. Roger Merchant mentions three steps critical in claiming biosurfactant production from new strains (Marchant, 2019): detailed identification and characterization of the producer microorganism, purification and detailed characterization of the produced biosurfactant, and accurate determination of the biosurfactant yield. The producer microorganism should be nonpathogenic and should have a broad substrate range. The biosurfactant produced by the strain should exhibit better physicochemical properties, such as low CMC, high surface and interfacial tension reduction, stability under extreme conditions, and high biodegradability and should be produced in high yields. Most studies estimate the biosurfactant yield gravimetrically by taking the dry weight of the biosurfactant after its extraction and purification. The method has been criticized for its inaccurate estimation of biosurfactant yields (Marchant, 2019). Yields based on GCMS and LCMS may provide a better and more accurate estimate of biosurfactant yields.

Another method for improving the biosurfactant yield is to engineer a strain to produce biosurfactants with higher yields. This can be achieved using metabolic and cellular engineering techniques to divert the metabolic flux toward biosurfactant production. Lee et al. achieved improved biosurfactant yields by upregulating genes encoding regulatory proteins involved in biosurfactant production (Lee et al., 2005). A threefold increase in biosurfactant production was obtained by Mulligan et al. by mutating ArgC4 and HisA1residues in *Bacillus subtilis* ATCC 21332 (Mulligan et al., 1989a). Another successful improvement in the biosurfactant yield has been reported by performing the MFE-2 gene deletion in *Candida bombicola*, resulting in the blockage of the β-oxidation pathway and diversion of the metabolic flux toward biosurfactant production (Dogan et al., 2006; Koch et al., 1988; Van Bogaert et al., 2009). Another engineering approach is cloning the genes involved in biosurfactant production in a heterologous host under the control of a strong promotor has also been reported to increase biosurfactant production. Cloning of rhl genes in *Pseudomonas putida* and cloning of genes encoding for surfactin synthase of *Bacillus licheniformis* into *E. coli* are some success stories for improving the biosurfactant yields using this method (Anburajan et al., 2015; Cha et al., 2008; Wittgens et al., 2017). Although several groups have cloned genes in heterologous hosts, a large increase in yields desired for the industrial production of biosurfactant has not been obtained (Marchant, 2019). In the future, engineering of cells to produce a range of normal congeners or new-to-nature biosurfactants will be performed. The success of this approach has previously been reported by Roelants et al. in *Starmerella bombicola* (Roelants et al., 2019). However, better control of biosurfactant production and metabolic improvements needs a detailed understanding of the cellular pathways and mechanisms involved in biosurfactant production.

5.4.2 Optimizing the composition of culture medium

As discussed earlier, the culture medium is a critical factor having a profound effect on the growth of the microorganism and affects biosurfactant production. The individual effect of each component of the medium on bacterial growth and biosurfactant production has been described earlier. The components of the medium also interact with each other, which affects biosurfactant production. Therefore, an optimum culture medium not only considers the individual components required for the growth and production of the biosurfactant but also optimizes the interactions among the components to increase the overall biosurfactant production. An efficiently designed culture medium supplies all the components nutrients at the optimum concentration required for the microorganism's growth and biosurfactant production. Therefore, optimizing the culture medium for enhancing the biosurfactant production can help increase the yield of the biosurfactants. Statistical optimization of the culture medium has been reported to enhance biosurfactant production. Various methods used for statistical optimization of medium composition include response surface methodology, Taguchi and Plackett-Burman methods, and methods based on artificial intelligence such as Artificial Neural Network combined with Genetic Algorithm (ANN-GA) (Dos Santos et al., 2016; Eswari et al., 2016; Hassan et al., 2016). Zargar et al. reported a 9.6-fold increase in saponin production by *Bacillus* sp. after statistical optimization of the culture medium (Zargar et al., 2022a). Similarly, in another study, statistical optimization of the culture medium resulted in 10.6-fold increase in biosurfactant production by *Bacillus pumilus* 2A (Marchut-Mikołajczyk et al., 2021). Similar increases in biosurfactant production after optimization of culture media have been reported by various groups suggesting the importance of statistical optimizations in increasing the overall yields of biosurfactants (Almeida et al., 2017; Moshtagh et al., 2018; Yaraguppi et al., 2020).

5.4.3 Alternative substrates as raw materials

The cost of raw materials is a significant factor that increases the overall cost of biosurfactant production. Raw materials account for up to 30% of the total production cost for biosurfactants (Henkel et al., 2014). One of the attractive alternatives to reduce the cost is to use inexpensive renewable raw materials for biosurfactant production (Khan et al., 2014). Agricultural and industrial byproducts, as well as waste materials, can be used as substrates for inexpensive production of biosurfactants. Kosaric proposed using industrial or municipal wastewater high in organic pollutants to reduce pollution while producing beneficial products (Kosaric & Sukan, 2014). Khan et al. have summarized some of the cheap raw materials that have been previously used for the production of biosurfactants. These include oils derived from plants, waste cooking oil, starch waste, lactic whey, and distillery wastes (Khan et al., 2014). Various oils derived from plants, such as rapeseed, jatropha, babassu, castor, sunflower, mesua, soybean, ramtil, and jojoba, have previously been used for the production of rhamnolipids, sophorolipidss, and mannosylerythritol lipids and other biosurfactants. The use of these oils has the potential to reduce the overall cost of biosurfactant production. The earliest reports for biosurfactant production from cheap plant wastes dates back to 1993 when Mercade et al. demonstrated the production of rhamnolipid biosurfactant from *Pseudomonas aeruginosa* 47T2 using olive oil mill effluent as a carbon source (Mercade et al., 1993). In the same year, Kitamoto et al. studied the interfacial properties of biosurfactant produced by *C. antarctica* using soyabean oil as a carbon source (Kitamoto et al., 1993). Since then, a wide variety of cheap carbon sources have been tried for production of biosurfactants. Currently, a major impediment in the utilization of cheap raw materials for the production of biosurfactants is finding the waste material with the right balance of nutrients to allow higher growth and biosurfactant accumulation. The use of mixed wastes to optimize the nutrient composition for high-scale production of biosurfactants should be further studied. Successful utilization of such waste mixtures may ensure futuristic scope for large-scale economic production of biosurfactants. An example of such a strategy is the study conducted by Luna et al., where the group used two industrial wastes, namely, corn steep liquor and ground nut oil refinery residue, as a substrate for biosurfactant production (Luna et al., 2012). The group obtained a yield of 9 g/L, and the biosurfactant showed surface tension reduction to 25 mN/m.

5.4.4 Efficient fermentation design

The development of an optimized fermentation process is a fundamental step toward commercialization of any biotechnological products. The reactor vessel configuration and the mode of fermentation operation determine the overall productivity of the process. Both solid-state and submerged fermentations have been used for biosurfactants production. Most studies have focused on submerged fermentations, but a common problem faced in most cases has been excessive foaming. Accumulation of the foam in the headspace of the production vessel has been a problem of major concern because the accumulated foam traps a considerable fraction of the biomass and which reduces its amount in the bulk of fermentation broth and leads to inconsistencies in the fermentation run (Winterburn & Martin, 2012). Besides, excess accumulation of foam inside the vessel makes the system prone to contamination. It is important to concentrate on the bioprocess advancements that aim at reducing foaming in the submerged biosurfactant production processes. As addition of antifoam can alter the characteristics of the biosurfactant, their addition to control foaming during biosurfactant production is not a viable option. Therefore, the mechanical architecture of the bioreactor should be changed to improve foam disruption. Various mechanical foam breakers should be explored to help address this issue without causing shear damage to the cells. Recently, an interesting strategy developed for managing foam has been reported by Winterburn and Martin, which focuses on foam stripping and broth recirculation via inlet and outlet ports on the headspace of the bioreactor vessel to mitigate the problem caused by foaming (Winterburn & Martin, 2012). Similarly, foam fractionation for the enrichment of surface-active compounds is another strategy that has been reported by Chen et al. to gain control over foaming during biosurfactant production (Chen et al., 2006a,b). In this technique, gas is sparged through the surfactant solution to create a new gas—liquid interface, where biosurfactants adhere and produce a rising foam with a reducing liquid component. Similar process developments to address foaming during biosurfactant production are the need of the hour.

Solid-state fermentation is another evolving strategy for the production of biosurfactants that eliminates the problem caused by foam (Camilios-Neto et al., 2011; Das & Mukherjee, 2007). Recent developments in this strategy have shown the good potential of this method in biosurfactant production. Slivinki et al. reported surfactin production from *Bacillus pumilus* using solid-state fermentation (Slivinski et al., 2012). The team employed a column bioreactor with forced aeration and a medium composed of sugarcane bagasse and okara. This strategy also enabled the easy separation of biomass and extraction of produced surfactin. Biosurfactant production has also been performed using soybean flour and rice straw as a substrate (Zhu et al., 2012). More research should be conducted on the potential of solid-state fermentation for large-scale production of biosurfactants, and different substrates should be screened to achieve better yields through this method.

The mode of bioreactor operation affects the biosurfactant yield and productivity. Most studies have reported biosurfactant production in batch modes (Davis et al., 2001; Müller et al., 2010; Yeh et al., 2006; Zargar et al., 2022a). It is a well-established fact that fed-batch mode of operation is more advantageous than batch fermentation because by regulating feed rate, substrates concentrations, and growth rate can be controlled. Fed-batch mode of operation can prove to be more effective in producing biosurfactants that are non-growth-associated. Fed-batch operation can be performed to maintain the cell viability during the stationary phase to promote the continuous accumulation of the biosurfactant. Davilla et al. reported sophorolipid production using fed-batch mode of operation (Davila et al., 1992). The group successfully achieved 320 g/L of biosurfactant with a yield of 65% relative to the carbon source. The study clearly shows the potential of fed-batch mode of operation for enhancing the production of biosurfactants. More such studies must be performed to enhance the production of non-growth-associated biosurfactants. Continuous mode of operation has also been reported for producing surfactin and rhamnolipids (Chen et al., 2006b; Chen et al., 2021). Continuous mode of operation reduces the vessel's downtime and increases productivity. However, the use of continuous systems for the production of biosurfactants is limited by intensive foaming, which traps the cells, affects the steady state, and causes poor downstream processing for continuous extraction of the biosurfactants. As discussed earlier, novel vessel designs to minimize foaming and new techniques that enable continuous product recovery should be developed to enable the continuous production of biosurfactants. Continuous form stripping followed by liquid recycling is a promising method enabling the continuous production of biosurfactants.

Other strategies that can improve fermentation and enhance biosurfactant production can be cell immobilization, addition of specific solid support carriers or inducers, use of metal nanoparticles, or use of biphasic reactors (Abouseoud et al., 2008b; Dehghannoudeh et al., 2019; Dos Santos et al., 2016; Heyd et al., 2011; Kiran et al., 2014; Liu et al., 2013; Sahebnazar et al., 2018; Srivastava et al., 2021; Yeh et al., 2005). Coproduction of biosurfactants with other commercially significant chemicals has also been suggested to help improve the overall process economics (Amin et al., 2013; Colla et al., 2010; Hmidet et al., 2019; Kavuthodi et al., 2015; Raheb & Hajipour, 2011).

5.4.5 Improving downstream processing

The recovery of the biosurfactant from the culture broth is the most important step in producing biosurfactants as it determines the overall cost of the biosurfactant produced. The cost of downstream processing alone accounts for nearly 60% of overall production costs and is a key limiting factor that lowers commercial demand for biosurfactants (Satpute et al., 2010). The steps involved in downstream processing depend on the required purity of the biosurfactant, which in turn depends on the end application of the biosurfactant (Thavasi & Banat, 2019a). Conventional methods for extraction of biosurfactants include solvent extraction using chloroform-methanol, butanol, dichloromethane-methanol, ethyl acetate, and hexane. The solvents utilized for biosurfactant extraction are toxic, and their toxicity limits their use for large-scale downstream processing. Other methods used for the extraction of biosurfactants are acid precipitation, precipitation using ammonium sulfate, foam precipitation, adsorption and crystallization centrifugation, and foam fractionation (Thavasi & Banat, 2019a). Precipitation using ammonium sulfate does not require any high-end infrastructure and can be used as an efficient method for extraction of crude biosurfactants required in the petroleum or detergent industry. Acid precipitation, followed by neutralization and lyophilization, has been reported to extract the biosurfactant produced by *Bacillus coagulans* and *Pseudomonas aeruginosa* EBN-8 (Cameotra & Singh, 1990). These methods work well in the extraction of crude biosurfactants for applications in which the purity of the biosurfactant obtained can be low. Extraction of biosurfactants with high purity involves other techniques such as membrane ultrafiltration, ion exchange chromatography, adsorption-desorption, and high-pressure liquid chromatography (Bertrand et al., 2018). Although these methods result in highly pure biosurfactants, a major drawback is the low yield, as only a small amount of crude extract can be treated. Aqueous two-phase systems (ATPS) is a novel technique that has recently gained interest for downstream processing in other biomolecules. This technique has been successfully used for the downstream processing of enzymes and other micro- and macromolecules. Studies on biosurfactant extraction using ATPS should be performed to determine the efficiency of this method for the extraction of biosurfactants.

Statistical optimization of the extraction conditions has been reported to improve biosurfactant recovery (Dos Santos et al., 2016; Eswari et al., 2016; Hassan et al., 2016). The yield of biosurfactant due to optimization increased from 9.49 to 13.76 mg/L. Such a small increase in yields makes a huge difference during large-scale production of any biological compound. Therefore, more studies on the optimum design of the steps involved in downstream processing and optimization of the recovery conditions must be performed. These studies should be targeted at maximizing the yield of biosurfactants, improving the purity, use of nontoxic chemicals, and being cost-friendly.

5.5 Conclusions and future outlook

Biosurfactant production, like that of any other biotechnological product, is influenced by several factors that affect both production efficiency and market demand. The chapter attempted to discuss the factors and their effect on biosurfactant production. These factors affect biosurfactant production at various steps of the production process. The effect of the genetic make-up of the producer microorganism, culture medium components, physicochemical factors, and process conditions was discussed in detail. A brief discussion on factors that limit the large-scale production and commercialization of biosurfactants was provided. Economics of downstream processing and the cost of the raw materials were found to be the major factors that determine the overall cost of the final biosurfactant. A major advancement in reducing their cost is needed to make biosurfactants economical to improve their market demand. Finally, some strategies were discussed that could improve the production process. New techniques and engineering designs should be implemented to address the key problems faced in biosurfactant production.

References

Abbasi, H., Sharafi, H., Alidost, L., Bodagh, A., Zahiri, H. S., & Noghabi, K. A. (2013). Response surface optimization of biosurfactant produced by *Pseudomonas aeruginosa* MA01 isolated from spoiled apples. *Preparative Biochemistry and Biotechnology, 43*, 398–414.

Abdel-Mawgoud, A. M., Aboulwafa, M. M., & Hassouna, N. A.-H. (2008). Characterization of surfactin produced by *Bacillus subtilis* isolate BS5. *Applied Biochemistry and Biotechnology, 150*, 289–303.

Abouseoud, M., Maachi, R., Amrane, A., Boudergua, S., & Nabi, A. (2008a). Evaluation of different carbon and nitrogen sources in production of biosurfactant by *Pseudomonas fluorescens*. *Desalination, 223*, 143–151.

Abouseoud, M., Yataghene, A., Amrane, A., & Maachi, R. (2008b). Biosurfactant production by free and alginate entrapped cells of *Pseudomonas fluorescens*. *Journal of Industrial Microbiology and Biotechnology, 35*, 1303–1308.

Abushady, H., Bashandy, A., Aziz, N., & Ibrahim, H. (2005). Molecular characterization of *Bacillus subtilis* surfactin producing strain and the factors affecting its production. *International Journal of Agriculture and Biology, 3*, 337–344.

Almansoory, A. F., Hasan, H. A., Idris, M., Abdullah, S. R. S., & Anuar, N. (2017). Biosurfactant production by the hydrocarbon-degrading bacteria (HDB) *Serratia marcescens*: Optimization using central composite design (CCD). *Journal of industrial and engineering chemistry*, 47, 272−280.

Almeida, D. G., Soares da Silva, Rd. C. F., Luna, J. M., Rufino, R. D., Santos, V. A., & Sarubbo, L. A. (2017). Response surface methodology for optimizing the production of biosurfactant by *Candida tropicalis* on industrial waste substrates. *Frontiers in Microbiology*, 8, 157.

Amin, G., Bazaid, S., & Abd El-Halim, M. (2013). A Two-stage immobilized cell bioreactor with *Bacillus subtilis* and *Rhodococcus erythropolis* for the simultaneous production of biosurfactant and biodesulfurization of model oil. *Petroleum scIence and Technology*, 31, 2250−2257.

Anburajan, L., Meena, B., Raghavan, R. V., Shridhar, D., Joseph, T. C., Vinithkumar, N. V., Dharani, G., Dheenan, P. S., & Kirubagaran, R. (2015). Heterologous expression, purification, and phylogenetic analysis of oil-degrading biosurfactant biosynthesis genes from the marine sponge-associated *Bacillus licheniformis* NIOT-06. *Bioprocess and Biosystems Engineering*, 38, 1009−1018.

Banat, I. M., De Rienzo, M. A. D., & Quinn, G. A. (2014). Microbial biofilms: Biosurfactants as antibiofilm agents. *Applied Microbiology and Biotechnology*, 98, 9915−9929.

Banat, I. M., Franzetti, A., Gandolfi, I., Bestetti, G., Martinotti, M. G., Fracchia, L., Smyth, T. J., & Marchant, R. (2010). Microbial biosurfactants production, applications and future potential. *Applied Microbiology and Biotechnology*, 87, 427−444.

Bednarski, W., Adamczak, M., Tomasik, J., & Płaszczyk, M. (2004). Application of oil refinery waste in the biosynthesis of glycolipids by yeast. *Bioresource Technology*, 95, 15−18.

Bertrand, B., Martínez-Morales, F., Rosas-Galván, N. S., Morales-Guzmán, D., & Trejo-Hernández, M. R. (2018). Statistical design, a powerful tool for optimizing biosurfactant production: A review. *Colloids and Interfaces*, 2, 36.

Beuker, J., Syldatk, C., & Hausmann, R. (2014). *Bioreactors for the production of biosurfactants. Biosurfactants: Production and utilization; processes, technologies, and economics* (pp. 117−128). Routledge.

Cai, Q., Zhang, B., Chen, B., Zhu, Z., Lin, W., & Cao, T. (2014). Screening of biosurfactant producers from petroleum hydrocarbon contaminated sources in cold marine environments. *Marine Pollution Bulletin*, 86, 402−410.

Cameotra, S. S., & Singh, H. D. (1990). Purification and characterization of alkane solubilizing factor produced by Pseudomonas PG-1. *Journal of Fermentation and Bioengineering*, 69, 341−344.

Camilios-Neto, D., Bugay, C., de Santana-Filho, A. P., Joslin, T., de Souza, L. M., Sassaki, G. L., Mitchell, D. A., & Krieger, N. (2011). Production of rhamnolipids in solid-state cultivation using a mixture of sugarcane bagasse and corn bran supplemented with glycerol and soybean oil. *Applied Microbiology and Biotechnology*, 89, 1395−1403.

Cha, M., Lee, N., Kim, M., Kim, M., & Lee, S. (2008). Heterologous production of *Pseudomonas aeruginosa* EMS1 biosurfactant in *Pseudomonas putida*. *Bioresource Technology*, 99, 2192−2199.

Chen, C., Li, D., Li, R., Shen, F., Xiao, G., & Zhou, J. (2021). Enhanced biosurfactant production in a continuous fermentation coupled with in situ foam separation. *Chemical Engineering and Processing-Process Intensification*, 159, 108206.

Chen, C. Y., Baker, S. C., & Darton, R. C. (2006a). Batch production of biosurfactant with foam fractionation. *Journal of Chemical Technology & Biotechnology: International Research in Process, Environmental & Clean Technology*, 81, 1923−1931.

Chen, C. Y., Baker, S. C., & Darton, R. C. (2006b). Continuous production of biosurfactant with foam fractionation. *Journal of Chemical Technology & Biotechnology: International Research in Process, Environmental & Clean Technology*, 81, 1915−1922.

Colla, L. M., Rizzardi, J., Pinto, M. H., Reinehr, C. O., Bertolin, T. E., & Costa, J. A. V. (2010). Simultaneous production of lipases and biosurfactants by submerged and solid-state bioprocesses. *Bioresource Technology*, 101, 8308−8314.

Coronel-León, J., de Grau, G., Grau-Campistany, A., Farfan, M., Rabanal, F., Manresa, A., & Marqués, A. M. (2015). Biosurfactant production by AL 1.1, a *Bacillus licheniformis* strain isolated from Antarctica: Production, chemical characterization and properties. *Annals of Microbiology*, 65, 2065−2078.

Couto, C. Rd. A., Alvarez, V. M., Marques, J. M., Jurelevicius, Dd. A., & Seldin, L. (2015). Exploiting the aerobic endospore-forming bacterial diversity in saline and hypersaline environments for biosurfactant production. *BMC Microbiology*, 15, 1−17.

Darvishi, P., Ayatollahi, S., Mowla, D., & Niazi, A. (2011). Biosurfactant production under extreme environmental conditions by an efficient microbial consortium, ERCPPI-2. *Colloids and Surfaces B: Biointerfaces, 84,* 292–300.

Daryasafar, A., Jamialahmadi, M., Moghaddam, M. B., & Moslemi, B. (2016). Using biosurfactant producing bacteria isolated from an Iranian oil field for application in microbial enhanced oil recovery. *Petroleum Science and Technology, 34,* 739–746.

Das, K., & Mukherjee, A. K. (2007). Comparison of lipopeptide biosurfactants production by *Bacillus subtilis* strains in submerged and solid state fermentation systems using a cheap carbon source: Some industrial applications of biosurfactants. *Process Biochemistry, 42,* 1191–1199.

Das, P., Mukherjee, S., & Sen, R. (2008). Genetic regulations of the biosynthesis of microbial surfactants: An overview. *Biotechnology and Genetic Engineering Reviews, 25,* 165–186.

Davila, A.-M., Marchal, R., & Vandecasteele, J.-P. (1992). Kinetics and balance of a fermentation free from product inhibition: Sophorose lipid production by *Candida bombicola*. *Applied Microbiology and Biotechnology, 38,* 6–11.

Davis, D., Lynch, H., & Varley, J. (2001). The application of foaming for the recovery of surfactin from *B. subtilis* ATCC 21332 cultures. *Enzyme and Microbial Technology, 28,* 346–354.

De Almeida, D. G., Soares Da Silva, Rd. C. F., Luna, J. M., Rufino, R. D., Santos, V. A., Banat, I. M., & Sarubbo, L. A. (2016). Biosurfactants: Promising molecules for petroleum biotechnology advances. *Frontiers in Microbiology, 7,* 1718.

Dehghannoudeh, G., Kiani, K., Moshafi, M. H., Dehghannoudeh, N., Rajaee, M., Salarpour, S., & Ohadi, M. (2019). Optimizing the immobilization of biosurfactant-producing *Pseudomonas aeruginosa* in alginate beads. *Journal of Pharmacy & Pharmacognosy Research, 7,* 413–420.

Desai, J., & Desai, A. J. (1993). Production of biosurfactants. *Biosurfactants: Production, Properties, Applications, 48,* 65–97.

Desai, J. D., & Banat, I. M. (1997). Microbial production of surfactants and their commercial potential. *Microbiology and Molecular Biology Reviews, 61,* 47–64.

Dhanya, M. (2021). *Biosurfactant-enhanced bioremediation of petroleum hydrocarbons: Potential issues, challenges, and future prospects. Bioremediation for environmental sustainability* (pp. 215–250). Elsevier.

Diniz Rufino, R., Moura de Luna, J., de Campos Takaki, G. M., & Asfora Sarubbo, L. (2014). Characterization and properties of the biosurfactant produced by *Candida lipolytica* UCP 0988. *Electronic Journal of Biotechnology, 17,* 6.

Dogan, I., Pagilla, K. R., Webster, D. A., & Stark, B. C. (2006). Expression of Vitreoscilla hemoglobin in Gordonia amarae enhances biosurfactant production. *Journal of Industrial Microbiology and Biotechnology, 33,* 693–700.

Donio, M. B. S., Ronica, F. A., Viji, V. T., Velmurugan, S., Jenifer, J. S. C. A., Michaelbabu, M., Dhar, P., & Citarasu, T. (2013). Halomonas sp. BS4, A biosurfactant producing halophilic bacterium isolated from solar salt works in India and their biomedical importance. *SpringerPlus, 2,* 1–10.

Dos Santos, A. S., Pereira, N., Jr, & Freire, D. M. (2016). Strategies for improved rhamnolipid production by *Pseudomonas aeruginosa* PA1. *PeerJ, 4,* e2078.

El-Sheshtawy, H., Aiad, I., Osman, M., Abo-ELnasr, A., & Kobisy, A. (2015). Production of biosurfactant from *Bacillus licheniformis* for microbial enhanced oil recovery and inhibition the growth of sulfate reducing bacteria. *Egyptian Journal of Petroleum, 24,* 155–162.

Eswari, J. S., Anand, M., & Venkateswarlu, C. (2016). Optimum culture medium composition for lipopeptide production by *Bacillus subtilis* using response surface model-based ant colony optimization. *Sadhana, 41,* 55–65.

Finnerty, W. R. (1994). Biosurfactants in environmental biotechnology. *Current Opinion in Biotechnology, 5,* 291–295.

Fontes, G. C., Fonseca Amaral, P. F., Nele, M., & Zarur Coelho, M. A. (2010). Factorial design to optimize biosurfactant production by *Yarrowia lipolytica*. *Journal of Biomedicine and Biotechnology, 2010,* 821306.

Fracchia, L., Ceresa, C., & Banat, I. M. (2019). Biosurfactants in cosmetic, biomedical and pharmaceutical industry. In I. M. Banat, & R. Thavasi (Eds.), *Microbial biosurfactants and their environmental and industrial applications* (pp. 258–288). CRC Press.

Gayathiri, E., Prakash, P., Karmegam, N., Varjani, S., Awasthi, M. K., & Ravindran, B. (2022). Biosurfactants: Potential and eco-friendly material for sustainable agriculture and environmental safety—a review. *Agronomy, 12*, 662.

Geetha, S., Banat, I. M., & Joshi, S. J. (2018). Biosurfactants: Production and potential applications in microbial enhanced oil recovery (MEOR). *Biocatalysis and Agricultural Biotechnology, 14*, 23−32.

George, S., & Jayachandran, K. (2013). Production and characterization of rhamnolipid biosurfactant from waste frying coconut oil using a novel *Pseudomonas aeruginosa* D. *Journal of Applied Microbiology, 114*, 373−383.

Georgiou, G., Lin, S.-C., & Sharma, M. M. (1992). Surface−active compounds from microorganisms. *Biotechnology, 10*, 60−65.

Ghribi, D., & Ellouze-Chaabouni, S. (2011). Enhancement of *Bacillus subtilis* lipopeptide biosurfactants production through optimization of medium composition and adequate control of aeration. *Biotechnology Research International, 2011*, 653654.

Gomaa, E. Z. (2013). Antimicrobial activity of a biosurfactant produced by *Bacillus licheniformis* strain M104 grown on whey. *Brazilian Archives of Biology and Technology, 56*, 259−268.

Gudiña, E. J., Fernandes, E. C., Rodrigues, A. I., Teixeira, J. A., & Rodrigues, L. R. (2015). Biosurfactant production by *Bacillus subtilis* using corn steep liquor as culture medium. *Frontiers in Microbiology, 6*, 59.

Gudiña, E. J., Rodrigues, A. I., de Freitas, V., Azevedo, Z., Teixeira, J. A., & Rodrigues, L. R. (2016). Valorization of agro-industrial wastes towards the production of rhamnolipids. *Bioresource Technology, 212*, 144−150.

Guerra-Santos, L., Käppeli, O., & Fiechter, A. (1984). *Pseudomonas aeruginosa* biosurfactant production in continuous culture with glucose as carbon source. *Applied and Environmental Microbiology, 48*, 301−305.

Hassan, M., Essam, T., Yassin, A. S., & Salama, A. (2016). Optimization of rhamnolipid production by biodegrading bacterial isolates using Plackett−Burman design. *International Journal of Biological Macromolecules, 82*, 573−579.

Henkel, M., Müller, M. M., Kügler, J. H., Lovaglio, R. B., Contiero, J., Syldatk, C., & Hausmann, R. (2012). Rhamnolipids as biosurfactants from renewable resources: Concepts for next-generation rhamnolipid production. *Process Biochemistry, 47*, 1207−1219.

Henkel, M., Sydatk, C., & Hausmann, R. (2014). *The prospects for the production of Rhamnolipids on renewable resources. Biosurfactants: Production and utilization-processes technologies, and economics* (1st edn., pp. 83−99). Florida: CRC Press.

Heyd, M., Franzreb, M., & Berensmeier, S. (2011). Continuous rhamnolipid production with integrated product removal by foam fractionation and magnetic separation of immobilized *Pseudomonas aeruginosa*. *Biotechnology Progress, 27*, 706−716.

Hmidet, N., Jemil, N., & Nasri, M. (2019). Simultaneous production of alkaline amylase and biosurfactant by Bacillus methylotrophicus DCS1: Application as detergent additive. *Biodegradation, 30*, 247−258.

Hu, X., Wang, C., & Wang, P. (2015). Optimization and characterization of biosurfactant production from marine Vibrio sp. strain 3B-2. *Frontiers in Microbiology, 6*, 976.

Jain, R. M., Mody, K., Joshi, N., Mishra, A., & Jha, B. (2013). Effect of unconventional carbon sources on biosurfactant production and its application in bioremediation. *International Journal of Biological Macromolecules, 62*, 52−58.

Jimoh, A. A., & Lin, J. (2019a). Biosurfactant: A new frontier for greener technology and environmental sustainability. *Ecotoxicology and Environmental Safety, 184*, 109607.

Jimoh, A. A., & Lin, J. (2019b). Enhancement of Paenibacillus sp. D9 lipopeptide biosurfactant production through the optimization of medium composition and its application for biodegradation of hydrophobic pollutants. *Applied Biochemistry and Biotechnology, 187*, 724−743.

Joshi, J., Sanket, S., Yadav, S., Nerurkar, A., & Desai, A. J. (2007). Statistical optimization of medium components for the production of biosurfactant by Bacillus licheniformis K51. *Journal of Microbiology and Biotechnology, 17*, 313−319.

Ju, L.-K., Dashtbozorg, S., Vongpanish, N. (2016). *Wound dressings with enhanced gas permeation and other beneficial properties*. Google Patents: US9468700B2.

Kashif, A., Rehman, R., Fuwad, A., Shahid, M. K., Dayarathne, H., Jamal, A., Aftab, M. N., Mainali, B., & Choi, Y. (2022). Current advances in the classification, production, properties and applications of microbial biosurfactants—A critical review. *Advances in Colloid and Interface Science, 306*, 102718.

Kavuthodi, B., Thomas, S. K., & Sebastian, D. (2015). Co-production of pectinase and biosurfactant by the newly isolated strain *Bacillus subtilis* BKDS1. *British Microbiology Research Journal, 10*, 1−12.

Kebbouche-Gana, S., Gana, M., Khemili, S., Fazouane-Naimi, F., Bouanane, N., Penninckx, M., & Hacene, H. (2009). Isolation and characterization of halophilic Archaea able to produce biosurfactants. *Journal of Industrial Microbiology and Biotechnology, 36*, 727−738.

Khan, M. S. A., Singh, B., & Cameotra, S. S. (2014). Biological applications of biosurfactants and strategies to potentiate commercial production. *Biosurfactants, 159*, 269.

Kim, J.-T., Kang, S. G., Woo, J.-H., Lee, J.-H., Jeong, B. C., & Kim, S.-J. (2007). Screening and its potential application of lipolytic activity from a marine environment: Characterization of a novel esterase from *Yarrowia lipolytica* CL180. *Applied Microbiology and Biotechnology, 74*, 820−828.

Kiran, G. S., Nishanth, L. A., Priyadharshini, S., Anitha, K., & Selvin, J. (2014). Effect of Fe nanoparticle on growth and glycolipid biosurfactant production under solid state culture by marine Nocardiopsissp. MSA13A. *BMC Biotechnology, 14*, 1−10.

Kitamoto, D., Yanagishita, H., Shinbo, T., Nakane, T., Kamisawa, C., & Nakahara, T. (1993). Surface active properties and antimicrobial activities of mannosylerythritol lipids as biosurfactants produced by *Candida antarctica*. *Journal of Biotechnology, 29*, 91−96.

Koch, A. K., Reiser, J., Käppeli, O., & Fiechter, A. (1988). Genetic construction of lactose-utilizing strains of *Pseudomonas aeruginosa* and their application in biosurfactant production. *Biotechnology, 6*, 1335−1339.

Konishi, M., Nishi, S., Fukuoka, T., Kitamoto, D., Watsuji, T.-o, Nagano, Y., Yabuki, A., Nakagawa, S., Hatada, Y., & Horiuchi, J.-i (2014). Deep-sea Rhodococcus sp. BS-15, lacking the phytopathogenic fas genes, produces a novel glucotriose lipid biosurfactant. *Marine Biotechnology, 16*, 484−493.

Kosaric, N., & Sukan, F. V. (2010). *Biosurfactants: Production, properties and applications*. CRC Press.

Kosaric, N., & Sukan, F. V. (2014). *Biosurfactants: Production and utilization—processes, technologies, and economics*. CRC press.

Kronemberger, F. Ad, Anna, L. M. M. S., Fernandes, A. C. L. B., Menezes, R. Rd, Borges, C. P., & Freire, D. M. G. (2007). Oxygen-controlled biosurfactant production in a bench scale bioreactor. *Biotechnology for fuels and chemicals* (pp. 401−413). Springer.

Le Roes-Hill, M., Durrell, K. A., & Kügler, J. H. (2019). *Biosurfactants from actinobacteria: State of the art and future perspectives. Microbial biosurfactants and their environmental and industrial applications* (pp. 174−208). CRC Press.

Lee, C.-L., Hsieh, M.-T., & Fang, M.-D. (2005). Aliphatic and polycyclic aromatic hydrocarbons in sediments of Kaohsiung Harbour and adjacent coast, Taiwan. *Environmental Monitoring and Assessment, 100*, 217−234.

Liu, J., Vipulanandan, C., Cooper, T. F., & Vipulanandan, G. (2013). Effects of Fe nanoparticles on bacterial growth and biosurfactant production. *Journal of Nanoparticle Research, 15*, 1−13.

Llamas, I., Amjres, H., Mata, J. A., Quesada, E., & Béjar, V. (2012). The potential biotechnological applications of the exopolysaccharide produced by the halophilic bacterium Halomonas almeriensis. *Molecules, 17*, 7103−7120.

Luna, J., Rufino, R., Campos, G., & Sarubbo, L. (2012). Properties of the biosurfactant produced by *Candida sphaerica* cultivated in low-cost substrates. *Chemical Engineering Journal, 27*, 67−72.

Makkar, R., & Cameotra, S. S. (2002). Effects of various nutritional supplements on biosurfactant production by a strain of *Bacillus subtilis* at 45 C. *Journal of Surfactants and Detergents, 5*, 11−17.

Makkar, R. S., Cameotra, S. S., & Banat, I. M. (2011). Advances in utilization of renewable substrates for biosurfactant production. *AMB Express, 1*, 1−19.

Maqsood, M. I., & Jamal, A. (2011). Factors affecting the rhamnolipid biosurfactant production. *Pakistan Journal of Biotechnology, 8*, 1−5.

Marchant, R. (2019). The future of microbial biosurfactants and their applications. In I. M. Banat, & R. Thavasi (Eds.), *Microbial biosurfactants and their environmental and industrial applications* (pp. 364−370). CRC Press,.

Marchant, R., & Banat, I. M. (2012). Microbial biosurfactants: Challenges and opportunities for future exploitation. *Trends in Biotechnology, 30*, 558−565.

Marchut-Mikołajczyk, O., Drożdżyński, P., Polewczyk, A., Smułek, W., & Antczak, T. (2021). Biosurfactant from endophytic *Bacillus pumilus* 2A: Physicochemical characterization, production and optimization and potential for plant growth promotion. *Microbial Cell Factories*, *20*, 1–11.

Mercade, M., Manresa, M., Robert, M., Espuny, M., De Andres, C., & Guinea, J. (1993). Olive oil mill effluent (OOME). New substrate for biosurfactant production. *Bioresource Technology*, *43*, 1–6.

Mnif, I., Ellouze-Chaabouni, S., & Ghribi, D. (2013). Optimization of inocula conditions for enhanced biosurfactant production by *Bacillus subtilis* SPB1, in submerged culture, using Box–Behnken design. *Probiotics and Antimicrobial Proteins*, *5*, 92–98.

Moshtagh, B., Hawboldt, K., & Zhang, B. (2018). Optimization of biosurfactant production by *Bacillus subtilis* N3-1P using the brewery waste as the carbon source. *Environmental Technology*, *40*, 3371–3380.

Müller, M. M., Hörmann, B., Syldatk, C., & Hausmann, R. (2010). *Pseudomonas aeruginosa* PAO1 as a model for rhamnolipid production in bioreactor systems. *Applied Microbiology and Biotechnology*, *87*, 167–174.

Mulligan, C. N., Chow, T. Y.-K., & Gibbs, B. F. (1989a). Enhanced biosurfactant production by a mutant *Bacillus subtilis* strain. *Applied Microbiology and Biotechnology*, *31*, 486–489.

Mulligan, C. N., Mahmourides, G., & Gibbs, B. F. (1989b). The influence of phosphate metabolism on biosurfactant production by *Pseudomonas aeruginosa*. *Journal of Biotechnology*, *12*, 199–209.

Najafi, A., Rahimpour, M., Jahanmiri, A., Roostaazad, R., Arabian, D., & Ghobadi, Z. (2010). Enhancing biosurfactant production from an indigenous strain of *Bacillus mycoides* by optimizing the growth conditions using a response surface methodology. *Chemical Engineering Journal*, *163*, 188–194.

Nalini, S., & Parthasarathi, R. (2018). Optimization of rhamnolipid biosurfactant production from Serratia rubidaea SNAU02 under solid-state fermentation and its biocontrol efficacy against Fusarium wilt of eggplant. *Annals of Agrarian Science*, *16*, 108–115.

Nercessian, D., Di Meglio, L., De Castro, R., & Paggi, R. (2015). Exploring the multiple biotechnological potential of halophilic microorganisms isolated from two Argentinean salterns. *Extremophiles*, *19*, 1133–1143.

Neu, T. R. (1996). Significance of bacterial surface-active compounds in interaction of bacteria with interfaces. *Microbiological Reviews*, *60*, 151–166.

Nitschke, M., & Pastore, G. M. (2006). Production and properties of a surfactant obtained from *Bacillus subtilis* grown on cassava wastewater. *Bioresource Technology*, *97*, 336–341.

Onlamool, T., Saimmai, A., Meeboon, N., & Maneerat, S. (2020). Enhancement of glycolipid production by Stenotrophomonas acidaminiphila TW3 cultivated in low cost substrate. *Biocatalysis and Agricultural Biotechnology*, *26*, 101628.

Pacwa-Płociniczak, M., Płaza, G. A., Piotrowska-Seget, Z., & Cameotra, S. S. (2011). Environmental applications of biosurfactants: Recent advances. *International Journal of Molecular Sciences*, *12*, 633–654.

Pal, M. P., Vaidya, B. K., Desai, K. M., Joshi, R. M., Nene, S. N., & Kulkarni, B. D. (2009). Media optimization for biosurfactant production by *Rhodococcus erythropolis* MTCC 2794: Artificial intelligence versus a statistical approach. *Journal of Industrial Microbiology and Biotechnology*, *36*, 747–756.

Perfumo, A., Banat, I. M., & Marchant, R. (2018). Going green and cold: Biosurfactants from low-temperature environments to biotechnology applications. *Trends in Biotechnology*, *36*, 277–289.

Petrides, D. (2000). *Bioprocess design and economics. Bioseparations science and engineering* (pp. 1–83). Oxford University Press.

Praharyawan, S., Susilaningsih, D., & Syamsu, K. (2013). Statistical screening of medium components by Plackett-Burman experimental design for biosurfactant production by Indonesian indigenous Bacillus sp. DSW17. *Asian Journal of Microbiology, Biotechnology & Environmental Sciences*, *15*, 805–813.

Raheb, J., & Hajipour, M. (2011). The stable rhamnolipid biosurfactant production in genetically engineered pseudomonas strain reduced energy consumption in biodesulfurization. *Energy Sources, Part A: Recovery, Utilization, and Environmental Effects*, *33*, 2113–2121.

Ramírez, I. M., Tsaousi, K., Rudden, M., Marchant, R., Alameda, E. J., Román, M. G., & Banat, I. M. (2015). Rhamnolipid and surfactin production from olive oil mill waste as sole carbon source. *Bioresource Technology*, *198*, 231–236.

Randhawa, K. K. S. (2014). Biosurfactants produced by genetically manipulated microorganisms. *Biosurfactants: Production and Utilization—Processes, Technologies, and Economics*, *159*, 49.

Rikalović, M. G., Vrvić, M. M., & Karadžić, I. M. (2015). Rhamnolipid biosurfactant from *Pseudomonas aeruginosa*: From discovery to application in contemporary technology. *Journal of the Serbian Chemical Society*, *80*, 279–304.

Robert, M., Mercade, M., Bosch, M., Parra, J., Espuny, M., Manresa, M., & Guinea, J. (1989). Effect of the carbon source on biosurfactant production by *Pseudomonas aeruginosa* 44T1. *Biotechnology Letters*, *11*, 871–874.

Rodrigues, L., Banat, I. M., Teixeira, J., & Oliveira, R. (2006). Biosurfactants: Potential applications in medicine. *Journal of Antimicrobial Chemotherapy*, *57*, 609–618.

Roelants, S., Van Renterghem, L., Maes, K., Everaert, B., Redant, E., Vanlerberghe, B., Demaeseneire, S. L., & Soetaert, W. (2019). Microbial biosurfactants: From lab to market. In I. M. Banat, & R. Thavasi (Eds.), *Microbial biosurfactants and their environmental and industrial applications* (1, pp. 340–362). CRC Press,.

Ron, E. Z., & Rosenberg, E. (2014). Enhanced bioremediation of oil spills in the sea. *Current Opinion in Biotechnology*, *27*, 191–194.

Sahebnazar, Z., Mowla, D., & Karimi, G. (2018). Enhancement of *Pseudomonas aeruginosa* growth and rhamnolipid production using iron-silica nanoparticles in low-cost medium. *Journal of Nanostructures*, *8*, 1–10.

Santos, D. K. F., Rufino, R. D., Luna, J. M., Santos, V. A., & Sarubbo, L. A. (2016). Biosurfactants: Multifunctional biomolecules of the 21st century. *International Journal of Molecular Sciences*, *17*, 401.

Sarafin, Y., Donio, M. B. S., Velmurugan, S., Michaelbabu, M., & Citarasu, T. (2014). Kocuria marina BS-15 a biosurfactant producing halophilic bacteria isolated from solar salt works in India. *Saudi Journal of Biological Sciences*, *21*, 511–519.

Satpute, S. K., Banpurkar, A. G., Dhakephalkar, P. K., Banat, I. M., & Chopade, B. A. (2010). Methods for investigating biosurfactants and bioemulsifiers: A review. *Critical Reviews in Biotechnology*, *30*, 127–144.

Schultz, J., & Rosado, A. S. (2020). Extreme environments: A source of biosurfactants for biotechnological applications. *Extremophiles*, *24*, 189–206.

Sen, R., & Swaminathan, T. (1997). Application of response-surface methodology to evaluate the optimum environmental conditions for the enhanced production of surfactin. *Applied Microbiology and Biotechnology*, *47*, 358–363.

Sharma, J., Kapley, A., Sundar, D., & Srivastava, P. (2022). Characterization of a potent biosurfactant produced from Franconibacter sp. IITDAS19 and its application in enhanced oil recovery. *Colloids and Surfaces B: Biointerfaces*, *214*, 112453.

Sharma, J., Sundar, D., & Srivastava, P. (2021). Biosurfactants: Potential agents for controlling cellular communication, motility and antagonism. *Frontiers in Molecular Biosciences*, *8*, 727070.

Silva, S., Farias, C., Rufino, R., Luna, J., & Sarubbo, L. (2010). Glycerol as substrate for the production of biosurfactant by *Pseudomonas aeruginosa* UCP0992. *Colloids and Surfaces B: Biointerfaces*, *79*, 174–183.

Slivinski, C. T., Mallmann, E., de Araújo, J. M., Mitchell, D. A., & Krieger, N. (2012). Production of surfactin by *Bacillus pumilus* UFPEDA 448 in solid-state fermentation using a medium based on okara with sugarcane bagasse as a bulking agent. *Process Biochemistry*, *47*, 1848–1855.

Smyth, T., Perfumo, A., Marchant, R., & Banat, I. (2010). *Isolation and analysis of low molecular weight microbial glycolipids. Handbook of hydrocarbon and lipid microbiology* (pp. 3705–3723). Springer.

Srivastava, S., Mondal, M. K., & Agrawal, S. B. (2021). *Biosurfactants for heavy metal remediation and bioeconomics. Biosurfactants for a sustainable future: Production and applications in the environment and biomedicine* (pp. 79–98). Wiley.

Syldatk, C., Lang, S., Wagner, F., Wray, V., & Witte, L. (1985). Chemical and physical characterization of four interfacial-active rhamnolipids from Pseudomonas spec. DSM 2874 grown on n-alkanes. *Zeitschrift für Naturforschung C*, *40*, 51–60.

Tabatabaei, A., Nouhi, A. A., Sajadian, V., & Mazaheri, A. M. (2005). Isolation of biosurfactant producing bacteria from oil reservoirs. *Iranian Journal of Environmental Health Science and Engineering*, *2*, 6–12.

Thavasi, R., & Banat, I. M. (2019a). *Downstream processing of microbial biosurfactants. Microbial biosurfactants and their environmental and industrial applications* (pp. 16–27). CRC Press.

Thavasi, R., & Banat, I. M. (2019b). *Introduction to microbial biosurfactants. Microbial biosurfactants and their environmental and industrial applications* (pp. 1–15). CRC Press.

Toda, H., & Itoh, N. (2012). Isolation and characterization of styrene metabolism genes from styrene-assimilating soil bacteria Rhodococcus sp. ST-5 and ST-10. *Journal of Bioscience and Bioengineering*, *113*, 12–19.

Van Bogaert, I. N., Sabirova, J., Develter, D., Soetaert, W., & Vandamme, E. J. (2009). Knocking out the MFE-2 gene of *Candida bombicola* leads to improved medium-chain sophorolipid production. *FEMS Yeast Research, 9*, 610−617.

Varjani, S. J., & Upasani, V. N. (2016). Carbon spectrum utilization by an indigenous strain of *Pseudomonas aeruginosa* NCIM 5514: Production, characterization and surface active properties of biosurfactant. *Bioresource Technology, 221*, 510−516.

Varjani, S. J., & Upasani, V. N. (2017). Critical review on biosurfactant analysis, purification and characterization using rhamnolipid as a model biosurfactant. *Bioresource Technology, 232*, 389−397.

Vasileva-Tonkova, E., & Gesheva, V. (2004). Potential for biodegradation of hydrocarbons by microorganisms isolated from Antarctic soils. *Zeitschrift für Naturforschung C, 59*, 140−145.

Vecino, X., Rodríguez-López, L., Gudiña, E. J., Cruz, J., Moldes, A., & Rodrigues, L. (2017). Vineyard pruning waste as an alternative carbon source to produce novel biosurfactants by *Lactobacillus paracasei*. *Journal of Industrial and Engineering Chemistry, 55*, 40−49.

Vieira, I. M. M., Santos, B. L. P., Ruzene, D. S., & Silva, D. P. (2021). An overview of current research and developments in biosurfactants. *Journal of Industrial and Engineering Chemistry, 100*, 1−18.

Vigneshwaran, C., Sivasubramanian, V., Vasantharaj, K., Krishnanand, N., & Jerold, M. (2018). Potential of *Brevibacillus* sp. AVN 13 isolated from crude oil contaminated soil for biosurfactant production and its optimization studies. *Journal of Environmental Chemical Engineering, 6*, 4347−4356.

Vijayakumar, S., & Saravanan, V. (2015). Biosurfactants-types, sources and applications. *Research Journal of Microbiology, 10*, 181.

Vollú, R. E., Jurelevicius, D., Ramos, L. R., Peixoto, R. S., Rosado, A. S., & Seldin, L. (2014). Aerobic endospore-forming bacteria isolated from Antarctic soils as producers of bioactive compounds of industrial interest. *Polar Biology, 37*, 1121−1131.

Wei, Y.-H., Lai, C.-C., & Chang, J.-S. (2007). Using Taguchi experimental design methods to optimize trace element composition for enhanced surfactin production by *Bacillus subtilis* ATCC 21332. *Process Biochemistry, 42*, 40−45.

Wicke, C., Hüners, M., Wray, V., Nimtz, M., Bilitewski, U., & Lang, S. (2000). Production and structure elucidation of glycoglycerolipids from a marine sponge-associated Microbacterium species. *Journal of Natural Products, 63*, 621−626.

Wicken, A., & Knox, K. (1980). Bacterial cell surface amphiphiles. *Biochimica et Biophysica Acta (BBA)-Biomembranes, 604*, 1−26.

Winterburn, J., & Martin, P. (2012). Foam mitigation and exploitation in biosurfactant production. *Biotechnology Letters, 34*, 187−195.

Wittgens, A., Kovacic, F., Müller, M. M., Gerlitzki, M., Santiago-Schübel, B., Hofmann, D., Tiso, T., Blank, L. M., Henkel, M., & Hausmann, R. (2017). Novel insights into biosynthesis and uptake of rhamnolipids and their precursors. *Applied Microbiology and Biotechnology, 101*, 2865−2878.

Yaraguppi, D. A., Bagewadi, Z. K., Muddapur, U. M., & Mulla, S. I. (2020). Response surface methodology-based optimization of biosurfactant production from isolated *Bacillus aryabhattai* strain ZDY2. *Journal of Petroleum Exploration and Production Technology, 10*, 2483−2499.

Yeh, M.-S., Wei, Y.-H., & Chang, J.-S. (2006). Bioreactor design for enhanced carrier-assisted surfactin production with *Bacillus subtilis*. *Process Biochemistry, 41*, 1799−1805.

Yeh, M. S., Wei, Y. H., & Chang, J. S. (2005). Enhanced production of surfactin from *Bacillus subtilis* by addition of solid carriers. *Biotechnology progress, 21*, 1329−1334.

Zargar, A. N., Kumar, M., & Srivastava, P. (2023). Biosurfactants: Challenges and future outlooks. In *Advancements in Biosurfactants Research*, (pp. 551−576). Cham: Springer International Publishing.

Zargar, A. N., Lymperatou, A., Skiadas, I., Kumar, M., & Srivastava, P. (2022a). Structural and functional characterization of a novel biosurfactant from Bacillus sp. IITD106. *Journal of Hazardous Materials, 423*, 127201.

Zargar, A. N., Mishra, S., Kumar, M., & Srivastava, P. (2022b). Isolation and chemical characterization of the biosurfactant produced by Gordonia sp. IITR100. *PLoS One, 17*, e0264202.

Zarinviarsagh, M., Ebrahimipour, G., & Sadeghi, H. (2017). Lipase and biosurfactant from *Ochrobactrum intermedium* strain MZV101 isolated by washing powder for detergent application. *Lipids in Health and Disease, 16*, 1−13.

Zhu, Z., Zhang, G., Luo, Y., Ran, W., & Shen, Q. (2012). Production of lipopeptides by Bacillus amyloliquefaciens XZ-173 in solid state fermentation using soybean flour and rice straw as the substrate. *Bioresource Technology, 112*, 254–260.

Zinjarde, S. S., Kale, B. V., Vishwasrao, P. V., & Kumar, A. R. (2008). Morphogenetic behavior of tropical marine yeast *Yarrowia lipolytica* in response to hydrophobic substrates. *Journal of Microbiology and Biotechnology, 18*, 1522–1528.

CHAPTER 6

Crude oil storage tank clean-up using biosurfactants

Mohammad Mobin[1], Kanika Cial[1], Ruby Aslam[2] and Mosarrat Parveen[1]
[1]Corrosion Research Laboratory, Department of Applied Chemistry, Faculty of Engineering and Technology, Aligarh Muslim University, Aligarh, Uttar Pradesh, India
[2]School of Civil Engineering and Architecture, Chongqing University of Science and Technology, Chongqing, P.R. China

6.1 Introduction

Biosurfactants (BioS) are amphiphilic compounds having varying applications in cosmetics, the oil and gas industry, household items, corrosion inhibitors, emulsifiers, wetting agents, etc. (Nikolova & Gutierrez, 2021; Płaza et al., 2020). The genera *Rhodococcus, Thiobacillus, Leuconostoc, Citrobacter, Candida, Corynebacterium, Penicillium, Ustilago, Aspergillus, Saccharomyces, Enterobacter, and Lactobacillus* are capable of producing a variety of BioS, ranging from low-molecular-weight to high-molecular-weight BioS (Ambaye et al., 2021). The many forms of BioS include glycolipids (mannosylerythritol, rhamnolipids, sophorolipids, xylolipid, cellobiose lipids, and trehalose lipids), lipopeptides (subtilisin, vixcosin, serrawetin, surfactin, polymyxin, and iturin), and polysaccharide—protein complexes. Glycolipids and lipopeptides are most commonly synthesized as low-molecular-weight surface-active substances.

The other category, known as bioemulsifiers, is frequently utilized substitutivity with BioS to represent surface-active biomolecules. Although surface-active, bioemulsifiers do not really reduce surface tension; instead, they produce stable emulsions between water mixtures and hydrocarbons (liquids). Often, BioS are referred to as bioemulsifiers. As diverse mixtures of protein, heteropolysaccharides, lipopolysaccharides, and lipoproteins, bioemulsifiers have large molecular weights. They are also known as exopolysaccharides or high-molecular-weight biopolymers. As the most commercially valuable biotechnological products, biosurfactants pose a challenge to conventional surfactants in various industrial contexts (Jimoh & Lin, 2019). Fig. 6.1 shows various properties of biosurfactants.

6.2 Biosurfactants production strategy

A wide variety of microorganisms are responsible for the production of BioSs. These bacteria attach themselves intracellularly or extracellularly during the majority of their development (Banat et al., 2010). Most biosynthetic activity in microorganisms occurs either during the exponential growth phase or during the stationary growth phase,

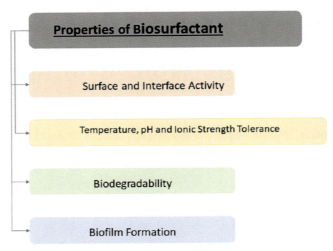

Figure 6.1 Properties of biosurfactants.

depending on the presence of nutrient-limiting circumstances in the production medium. The production of BioSs may occur spontaneously, or be stimulated by the introduction of various substances, and by adjusting the pH, temperature, inoculum size, aeration, stress, and agitation speed. In general, it has been observed that the output of biosulfuric acid is regulated by components such as carbon, nitrogen, iron, sulfur, phosphorus, and manganese (Santos et al., 2016). To get a higher BioS production yield, it is necessary to optimize various element ratios, such as carbon to nitrogen, carbon to phosphorus, carbon to iron, or carbon to magnesium, and to do so at different speeds.

6.3 Factors affecting biosurfactants production

Fig. 6.2 shows the various factors that affect the production of biosurfactants. These include carbon (glycerol, oil, glucose, mannitol), nitrogen (ammonium salts, urea, peptone), fermentation duration, ion concentration, pH, temperature, and oxygen levels. The quantity and kind of BioSs rely mostly on the producing microorganism. To produce BioSs, microbes use carbon as their source, and the most commonly used are glycerol and glucose, both being renewable and widely used in industrial feedstock.

6.4 Production methods of biosurfactants

There are currently two methods for the production of BioS: (1) submerged fermentation (SmF), which is also known as liquid fermentation and (2) solid-state fermentation (SSF). SSF refers to a microbiological activity that occurs mostly on the surface of

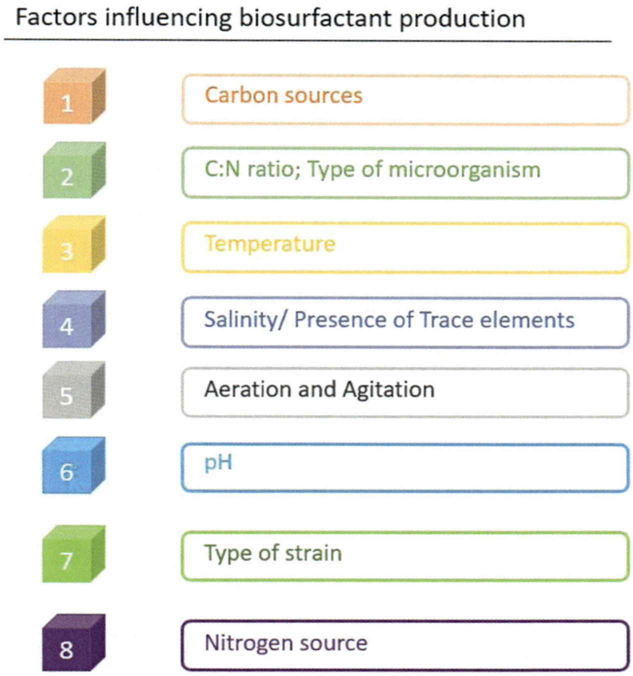

Factors influencing biosurfactant production

1. Carbon sources
2. C:N ratio; Type of microorganism
3. Temperature
4. Salinity/ Presence of Trace elements
5. Aeration and Agitation
6. pH
7. Type of strain
8. Nitrogen source

Figure 6.2 Factors influencing the production of biosurfactants.

solid substances that can absorb or hold water, with or without soluble nutrition (Lizardi-Jiménez & Hernández-Martínez, 2017). While SmF is the most well-known approach in the scientific literature and patents, SSF is still in the very early stages of development and occupies a very limited area. Both methods may use the same producing microorganisms, but the outcomes can be quite different owing to the substantial changes in circumstances required for each kind of culture regimen. Furthermore, compared with liquid fermentation, SSF is known to often result in lower overall costs for a specific bioprocess (Sadh et al., 2018). The tiny amount of water that is used in SSF has a significant effect on the process's overall profitability, mostly as a result of the smaller bioreactor capacity, decreased downstream processing, and lower sterilizing costs. In addition, a significant number of SSF procedures are centered on using low-cost agroindustrial waste as a culture medium (Venil et al., 2017). However, it is not restricted to SSF as studies that use SmF for BioS production also include the valorization of agro-industrial wastes.

6.5 Classification of biosurfactants

Biosurfactants are amphiphilic molecules, meaning that they are generated by two different moieties: hydrophilic and lipophilic. The hydrophilic component of BioS comprises carbohydrate chains, amino acid sequences, protein sequences, phosphates, carboxylic acid sequences, and alcohol motif sequences. Chains of carbon atoms, similar to those found in fatty acids, make up the lipophilic fragments. Both molecular components are put together by the coupling of several biological functions. These functionalities include ethers (C—O—C), amides (N—C=O), and esters (O—C=O). BioS are commonly categorized as glycolipids, lipopolysaccharides, lipopeptides, phospholipids, and fatty acids Fig. 6.3 according to the nature of each component, each category has particular physicochemical properties and physiological significance (De Almeida et al., 2016). Due to the ease with which they may be created compared with other forms of BioS, such as lipoproteins, glycolipids have the most potential to be manufactured on a massive scale among all types of BioS. It is anticipated that BioS manufactured at better yields would reduce the manufacturing costs (Dhanarajan & Sen, 2014). The condensation of fatty acids (also known as lipids) with carbohydrates results in the formation of glycolipids. Their names are derived from the identities of the carbohydrates that make up their bodies. Therefore, glycolipids that include sophorose are referred to as sophorolipids; glycolipids that contain rhamnose are known as rhamnolipids; glycolipids that contain trehalose are referred to as trehalose lipids; and so on. Of the many varieties of glycolipids, sophorolipids and rhamnolipids are among the most extensively researched species. Several different structures may be discovered for sophorolipids; the most common of which are open and cyclic configurations (Delbeke et al., 2016). Open sophorolipids are characterized by the presence of the chemical functionality of carboxylic acid (COOH) at the terminal point of the lipophilic chain. The condensation that takes place between the fatty acid and one of the hydroxyl motifs of the sophorose results in the formation of

Biosurfactants

Low Molecular Weight

Class of biosurfactants	Type of biosurfactant	Sources of biosurfactant:
Glycolipids	Rhamnolipids	P. aeruginosa, T. aquaticus and Burkholderia sp
	Sophorolipids	T. apicola, Candida kuol and Torulopsis bombicola
	Trehalolipids	Rhodococcus erythropolis, Gordonia sp. and Nocardia sp.
	Mannosylerythritol lipids	Ustilago scitaminea, Candida antartica and Pseudozyma sp.
Lipopeptides	Surfactin	Bacillus subtilis, Bacillus mojavensis and Bacillus licheniformis
	Iturin	Bacillus subtilis and Bacillus amyloliquefaciens
	Lichenysin	Bacillus licheniformis
	Viscosin	Pseudomonas fluorescens
	Serrawettin	Serratia marcescens
	Arthrofactin	Arthrobacter sp.
	Polymyxin	Bacillus polymyxa
Fatty acids and phospholipids	Corynomycolic acid	C. Diphtheriae and Corynebacterium lepus
	Spiculisporic acid	Talaromyces trachyspermus
	Phosphatidylethanolamine	Acinetobacter sp.

High Molecular Weight

Class of biosurfactant	Type of biosurfactant	Sources of biosurfactant:
Polymeric	Emulsan	Acinetobacter calcoaceticus RAG-1
	Liposan	Candida lipolytica
	Biodispersan	Acinetobacter calcoaceticus A2
	Lipomanan	Candida tropicalis
	Mannoproteins	Acinetobacter sp.
	Alasan	Acinetobacter radioresistens KA53
Particulate biosurfactants	Vesicles	P. marginilis, and Serratia marcescens
	Whole cells	Cyanobacteria

Figure 6.3 Classification of biosurfactants.

cyclic forms. These forms an ester functionality. Cyclic esters are also known as lactones. Similarly, acidic and lactonized sophorolipids also exist. Other variations at the molecular level include (1) the presence or absence of acetyl groups on the carbohydrate moiety, (2) the length of the alkyl chain, (3) the number of unsaturation in the fatty chain, (4) the position of the hydroxyl group in the fatty chain, and (5) the position of the hydroxyl group of the carbohydrate that is etherified with the fatty alcohol.

6.6 Oil storage tanks in the industry

6.6.1 Crude oil

Crude oil held in the tanks is in a liquid form and has several distinct features such as a dark black color, a foul smell due to bisulfite, high viscosity, high flammability, and a density lower than that of water Fig. 6.4. Crude oil is composed of intricate chains of heavy hydrocarbons, and it may also contain lighter hydrocarbons and other components but in much lesser proportions.

The following primary hydrocarbons may be found in crude oil: paraffins (saturated hydrocarbons alkanes, e.g., octane), olefins (unsaturated hydrocarbons alkenes, e.g., pentene), naphthenes or cycloparaffins (saturated hydrocarbons cycloalkanes, e.g., cyclohexane) and aromatic hydrocarbons, such as arenes (including benzene). According to the amount of sulfur it contains, it may be described as either sweet or sour: sweet crude oil has less than 0.5% sulfur content and sour crude oil has sulfur >0.5%.

The process of extracting crude oil from the earth involves several processing phases determined by the characteristics of the oil itself before being stored in tanks. After extraction, the most important and common step is removing of water and gases using three phase separators, followed by the desulfurization process. Desulfurization is because crude oil must be delivered refineries with a low sulfur content. Therefore, it

Figure 6.4 Crude oil tank bottom sludge. *From Chrysalidis, A. & Kyzas, G.Z. (2020). Applied cleaning methods of oil residues from industrial tanks. Processes, 8(5). https://doi.org/10.3390/PR8050569.*

is transported to refineries via tankers, where it is put through fractional distillation to separate its constituent parts and make the final products.

6.6.2 Storage tanks

A container or reservoir that temporarily contains oil while being processed into other oil products or before being utilized or consumed is known as an oil storage tank. This kind of tank may also be referred to as an oil reservoir. Tanks used for oil storage may be categorized as surface or above ground, semi-subterranean, subterranean, or underground, depending on their depth and proximity to the earth.

There are many different types of crude oil tanks, including bolted steel tanks (API 12B), welded steel tanks (API 12F-BS 2654), flat-sided tanks (API 620), and field-welded tanks (API 12D-API 650). In most cases, it is made from carbon steel, which can be painted on the inside and/or the outside, galvanized, or coated with a polymeric material to protect it from corrosion. The kind of tank, its size, and the material use change according to the storage conditions of the oil (pressure and temperature), the characteristics of the crude (composition and toxicity), and the quantity of storage space needed. The shell, bottom, roof (which may be fixed or floating), nozzles, pipes, instruments, cathodic protection system, and steel structures that are for the service of the personnel are its primary components.

6.6.3 Crude oil storage tank clean-up: a historical perspective

Tanks in the industry are constructed to last for several years. To accomplish this goal, a tank that stores or processes petroleum liquids must be examined, maintained, and repaired regularly so that it can continue to function safely and effectively. When a crude oil storage tank has been in operation for many years, the accumulation of heavier hydrocarbons eventually leads to the formation of numerous residues within the tank.

These residues along with water and solid particles already present in the crude oil composition sink to the bottom of the tank and lead to various issues, including degradation in the quality of the product being stored, a reduction in the capacity of the tank, and even obstruction of the suction lines. In most cases, this substance is referred to as "petroleum sludge." It forms a slurry gel that coats the bottom, shell, and other sections within the tank. Because this coating prevents interference with the metal sections of the tank, the tank has to be thoroughly cleaned before any of the aforementioned duties, such as inspection and maintenance. When the cleaning process is complete, the dirt must be removed from every tank surface, including the bottom, shell, ceiling, studs, tubing, and numerous attachments. The process of cleaning the tank, as well as all other operations to be carried out within the tank, demands that the tank be devoid of gas in the first place. This is necessary for the safe entrance of employees inside the tank area as well as the safety of the premises. Manual and automated cleaning techniques are the primary types of commercially accessible cleaning procedures. The owner's needs, which are often concerned with aspects such as safety, cost, execution time, and environmental consciousness, serve as the primary criteria for making the technique decision (Chrysalidis & Kyzas, 2020).

6.6.3.1 Manual method

Cleaning of oil tank requires many different techniques, one of which is the manual use of specialized labor assigned for this work. This method involves staff entering tanks and using pressurized water, buckets, fire hoses, vacuum hoses, and pumps to remove as much sludge as possible from inside the tank (http://www.ansonindustry.com/how-much-do-you-know-about-oil-tank-cleaning-procedure.html).

While this appears to be a simple and affordable option for tank cleaning, it has been widely criticized in recent years for its numerous drawbacks. Some of these are listed below:

1. The effectiveness of manual cleaning is very low, and staff that works in tight spaces can work effectively for short periods of time using heavy protective gear and a breathing apparatus.
2. Plenty of valuable oil is lost along with sludge, resulting in large volumes of waste that later require specialized disposal.
3. Safety is an even bigger issue, as the Health and Safety Executive (HSE) concern applies while the cleaning is being performed.
4. There is significant use of equipment and manpower.
5. Cost impact might be multiple times that of the mechanical cleaning cost.
6. Utilize high-pressure water jetting.

All these result in a small workforce willing to do the job. Overall, the manual cleaning method seems to have very few benefits compared to its many downsides.

6.6.3.2 Automated cleaning

In recent years, the sector has been subjected to more stringent rules due to the detrimental impacts that manual tank cleaning has caused. These regulations concern working in tight spaces with hazardous substances and preserving the environment. Tanks are now automatically cleaned Fig. 6.5 (Chrysalidis & Kyzas, 2020) because there was a demand for a sanitation technique that was less hazardous to people, environment, and infrastructure, which led to the development of this practice. The business of auto-cleaning began to take off in the middle of the 20th century, with the primary objective of removing the need for humans to perform work in tight areas containing potentially hazardous materials (Chrysalidis & Kyzas, 2020). The process of cleaning a tank with an automated system includes (1) installation, (2) tank padding (blanketing), (3) cleaning and sludge extraction, and (4) removal of the equipment.

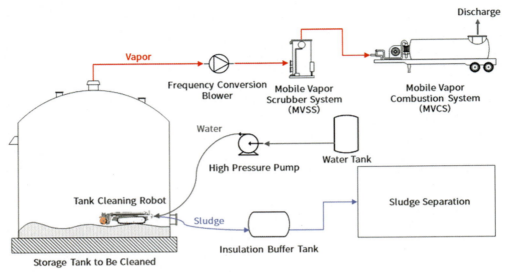

Figure 6.5 Non-man entry tank cleaning flowchart.

A downside to this method is that the installation process takes 7 days, and it cannot be done without a crane, as many Nelements need to be installed on the tank roof. Also, human personnel are required to enter the tank at some point in the process.

6.7 Use of biosurfactants in cleaning of crude oil storage tanks

During the process of extraction, transportation, the cleaning of crude oil storage tanks, and the treatment of waste from crude oil production, BioS may be used. The lowering of surface and interfacial tension, as well as variations in wettability, emulsification, deemulsification, adsorption, solubilization, and viscosity reduction, are the primary processes at work in BioS-assisted enhanced oil recovery. Biosurfactants are amphipathic molecules containing hydrophilic and hydrophobic moieties, found mainly on the cell surface or excreted to culture medium by a broad range of microorganisms (http://www.haaenclean.com/services/downstream/tankclean?lang = en). These products increase the aqueous dispersion of poorly soluble compounds by many orders of magnitude and change the affinity between microbial cells and hydrocarbons by increasing cell surface hydrophobicity (Karanth et al., 1999).

6.7.1 Mechanism of action

6.7.1.1 Biosurfactants in tank clean-up: Literature Review

Matsui et al. (2012) examined the efficacy of a novel BioS, JE1058BS, in the treatment of oil tank bottom sludge. This BioS was generated by the actinomycete Gordonia sp. The BioS maintained its stability for at least 3 weeks longer than a chemical or plant-derived surfactant, and its dispersion activity of 10 g/L was superior to that attained with either.

Diab and Din (2013) have reported that the supernatant from *P. aeruginosa* SH 29 can be used to clean oil-contaminated vessels. Within 15 minutes of addition, 97% oil was extracted from the bottom and walls of the containers and then allowed to float on the supernatant as its own separate phase. The authors demonstrated that the BioS found in the sterile supernatant of *P. aeruginosa* SH 29 may be used directly for the purpose of cleaning oil storage tanks and other vessels used for the transportation and storage of crude oil. Silva et al. (2014a,b) reported that a clearance rate of 80% was observed when an isolated BioS from *Pseudomonas cepacia* CCT6659 was examined for the possibility of cleaning beaker walls contaminated with a coating of oil. This removal rate of oil shows that this BioS may be used in the cleaning of storage tanks. Rocha et al. (2014) have isolated five strains of *B. subtilis*, out of which several *B. subtilis* isolates were able to degrade heavier n-alkanes in various paraffinic combinations, regardless of whether the circumstances were aerobic or anaerobic. Furthermore, several isolates (namely, 191 and 309) showed both the capacity for BioS synthesis and hydrocarbon degradation synergistically.

Another study by Saeki et al. (2009) reported how sludge from the bottom of a tank could be treated by employing a BioS called JE1058BS, derived from Gordonia spp. The treatment produced an outstanding outcome that was effective for 21 days. The efficacy values for 0.5%, 1%, 2%, and 3% JE1058BS/crude oil ratios were 33.4%, 70.7%, 79.5%, and 81.4%, respectively.

A strain of *P. aeruginosa* DQ8 has the potential to develop on diesel oil and crude oil as the only source of carbon and energy, and it is capable of effectively degrading both of these oils. *Pseudomonas aeruginosa* DQ8 could break down two of the most important components of crude oil, namely, n-alkanes and PAHs. As a result, *P. aeruginosa* strain DQ8 may have a significant impact not only on the bioremediation of oil-contaminated soils but also on the biotreatment of oil wastewater (Saeki et al., 2009).

Abadhsapouri et al. (Zhang et al., 2011) have studied the synergistic effect of silica nanoparticles and biosurfactants for oil tank clean-up where it has been reported that efficiency has been increased synergistically. In another study, BioS produced by stains of *Pseudomonas aeruginosa* removed 49%—54% of crude oil at room temperature, 52%—57% at 70°C and 58%—62% at 90°C (Abadshapouri et al., 2021).

Another BioS produced from *Pseudomonas aeruginosa* (RS29) (Bordoloi & Konwar, 2008) has been investigated, and it was observed in the laboratory that this strain was effective in removing hydrocarbon from refinery sludge by applying whole bacterial culture and culture supernatant to varying concentrations of sand—sludge mixture. The investigation was conducted with the intention of determining its efficacy. After 20 days, there was a record of the removal of hydrocarbon. The generated BioS was analyzed to obtain information on its stability and composition. The complete bacterial culture and culture supernatant was able to extract up to 85.3% and 55.4.5% of hydrocarbon from refinery sludge, respectively. Also, the stability of BioS was reported up to 121°C.

Cameotra et al. (Saikia et al., 2012) explored the effect of microbial consortium in which two species *Pseudomonas aeruginosa* and *Rhodococcus* were used as consortium degrading the crude oil sludge to 90% in 6 weeks.

Further, the use of the BioSs that were produced by five different bacterial isolates from the Bacillus sp. LBBMA 111A culture collection proved to be extremely effective for the treatment of oily sludge. Oily sludge hydrocarbons were emulsified by the BioSs during mixing, which was readily observed from mixing onset, and this led to a recovery of 95% of the oil that was contained in the oily sludge produced by a fuel oil storage tank Fig. 6.6 (Cameotra & Singh, 2008). The amount of residual oil that was found in the waste materials that were left over after the BioS treatment was minimal. Treatment of this kind conducted on an industrial scale would reduce the cost of waste disposal and mitigate the danger of polluting the environment with oil contained in the leftover solids of discarded oily sludge.

6.8 Challenges and future outlook

To compete with petrochemical surfactants, biosurfactants should be superior in three main areas: cost, functionality, and manufacturing capacity in relation to the applications they are designed for. The majority of research efforts have been focused on the creation of high-yielding strains, the implementation of bioreactors, the optimization of production procedures, and the procurement of inexpensive substrates. The selection of a low-cost substrate is essential to the economics of the process, given that the substrate

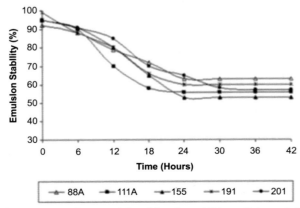

Figure 6.6 Water/oily sludge emulsion stabilized by the addition of the cell-free supernatant (isolate Dietzia maris sp. LBBMA 191).

may account for up to 50% of the total cost of production. Fortunately, biosurfactants can be manufactured using renewable resources. Undoubtedly, the petroleum sector represents the most lucrative market for biosurfactants. However, due to the vast volumes of biosurfactants required for the remediation of polluted water and soil, it appears that the use of biosurfactants can only be economically viable when they are prepared in a crude form. The food, cosmetic, and pharmaceutical sectors all require lower amounts of biosurfactants but can bear the higher costs associated with their downstream use.

6.9 Conclusions

The biosurfactant business, which exploits the biodegradation and development advantages of the pharmaceutical, cosmetic, petroleum, and food industries for renewable energy substrates, is a very lucrative and competitive sector of the economy. As the market for biosurfactants is still developing, they are employed in several specialized applications. Technical restrictions, particularly their prices and drawbacks in terms of mixing technology, are the sector's stumbling blocks. Virulent strain screening and the development of process technology will aid in reducing production costs as the understanding of biosurfactant-generating strains has to be expanded to encompass morphology, genetics, and biochemistry. Environmentally friendly surfactants are being produced by farmers from various natural and renewable sources as they rapidly gain popularity as a product. It may be possible to lower manufacturing costs and make biosurfactants economically feasible and competitive with synthetic surfactants using agro-industrial waste of both animal and plant origin. Bio-based surfactants are designed to be utilized in the treatment of heavy metals, contaminated soils and water, skin problems, improved oil restoration, food preservation, and plant disease eradication. The use of biosurfactants to treat oily sludges is an economically and environmentally viable technology, considering the small volume of microbial culture required for the treatment, as it may decrease the viscosity of sludge and oil deposits through the formation of an oil-in-water emulsion that facilitates the pumping of waste.

Acknowledgments

The authors gratefully acknowledge the Council of Scientific and Industrial Research (CSIR), New Delhi, India, for the funding under the research project [File number: 22(0832)/20/EMR-II].

References

Abadshapouri, A., Amani, H., Hajimohammadi, R., & Soltani, H. (2021). Heavy oil storage tanks clean-up using biosurfactants and investigation of the synergistic effect with silica nanoparticles. *Tenside Surfactants Detergents*, *58*(4).

Ambaye, T. G., Vaccari, M., Prasad, S., & Rtimi, S. (2021). Preparation, characterization and application of biosurfactant in various industries: A critical review on progress, challenges and perspectives. *Environmental Technology & Innovation, 24*, 102090.

Banat, I. M., Franzetti, A., Gandolfi, I., Bestetti, G., Martinotti, M. G., Fracchia, L., et al. (2010). Microbial biosurfactants production, applications and future potential. *Applied Microbiology and Biotechnology, 87*, 427–444. Available from https://doi.org/10.1007/s00253-010-2589-0.

Bordoloi, N. K., & Konwar, B. K. (2008). Microbial surfactant-enhanced mineral oil recovery under laboratory conditions. *Colloids and Surfaces B, 63*, 73–82.

Cameotra, S. S., & Singh, P. (2008). Bioremediation of oil sludge using crude biosurfactants. *International Biodeterioration & Biodegradation, 62*, 274–280.

Chrysalidis, A., & Kyzas, G. Z. (2020). Applied cleaning methods of oil residues from industrial tanks. *Processes, 8*, 569. Available from https://doi.org/10.3390/pr8050569.

De Almeida, D. G., Soares Da Silva, R. D. C. F., Luna, J. M., Rufino, R. D., Santos, V. A., Banat, I. M., & Sarubbo, L. A. (2016). Biosurfactants: Promising molecules for petroleum biotechnology advances. *Frontiers in Microbiology, 7*, 1718. Available from https://doi.org/10.3389/fmicb.2016.01718.

Delbeke, E. I. P., Everaert, J., Uitterhaegen, E., Verweire, S., Verlee, A., Talou, T., Soetaert, W., Van Bogaert, I. N. A., & Stevens, C. V. (2016). Petroselinic acid purification and its use for the fermentation of new sophorolipids. *AMB Express, 6*(Article number: 28).

Dhanarajan, G., & Sen, R. (2014). Cost analysis of biosurfactant production from a scientist's perspective. In *Biosurfactants*.

Diab, A., & Din, G. E. (2013). Application of the biosurfactants produced by Bacillus spp. (SH 20 and SH 26) and P. aeruginosa SH 29 isolated from the rhizosphere soil of an Egyptian salt marsh plant for the cleaning of oil – Contaminated vessels and enhancing the biodegradation. *African Journal of Environmental Science and Technology, 7*(7).

Jimoh, A. A., & Lin, J. (2019). Biosurfactant: A new frontier for greener technology and environmental sustainability. *Ecotoxicology and Environmental Safety, 184*, 109607. Available from https://doi.org/10.1016/j.ecoenv.2019.109607.

Karanth, N. G. K., Deo, P. G., & Veenanadig, N. K. (1999). Microbial production of biosurfactant and their importance. *Special Section: Fermentation – Science and Technology, 77*, 116–126.

Lizardi-Jiménez, M. A., & Hernández-Martínez, R. (2017). Solid state fermentation (SSF): Diversity of applications to valorize waste and biomass. *3 Biotech, 7*(1), 44.

Matsui, T., Namihira, T., Mitsuta, T., & Saeki, H. (2012). Removal of oil tank bottom sludge by novel biosurfactant, JE1058BS. *Journal of the Japan Petroleum Institute, 55*(2), 138–141.

Nikolova, C., & Gutierrez, T. (2021). Biosurfactants and their applications in the oil and gas industry: Current state of knowledge and future perspectives. *Frontiers in Bioengineering and Biotechnology, 9*, 626639.

Płaza, G. A., Płaza, G. A., & Achal, V. (2020). Biosurfactants: Eco-friendly and innovative biocides against biocorrosion. *International Journal of Molecular Sciences, 21*(6), 2152.

Rocha e Silva, N. M. P., Rufino, R. D., Luna, J. M., Santos, V. A., & Sarubbo, L. A. (2014). Screening of *Pseudomonas* species for biosurfactant production using low-cost substrates. *Biocatalysis and Agricultural Biotechnology, 3*(2), 132–139.

Sadh, P. K., Duhan, S., & Duhan, J. S. (2018). Agro-industrial wastes and their utilization using solid state fermentation: A review. *Bioresources and Bioprocessing, 5*(1).

Saeki, H., Sasaki, M., Komatsu, K., Miura, A., & Matsuda, H. (2009). Oil spill remediation by using the remediation agent JE1058BS that contains a biosurfactant produced by Gordonia sp. strain JE-1058. *Bioresource Technology, 100*, 572–577.

Saikia, R. R., Deka, S., Deka, M., & Banat, I. M. (2012). Isolation of biosurfactant-producing *Pseudomonas aeruginosa* RS29 from oil-contaminated soil and evaluation of different nitrogen sources in biosurfactant production. *Annals of Microbiology, 62*, 753–763.

Santos, D. K. F., Rufino, R. D., Luna, J. M., Santos, V. A., & Sarubbo, L. A. (2016). Biosurfactants: Multifunctional biomolecules of the 21st century. *International Journal of Molecular Sciences, 17*, 401. Available from https://doi.org/10.3390/ijms17030401.

Silva, E. J., Silva, N. M. P. R. E., Rufino, R. D., Luna, J. M., Silva, R. O., & Sarubbo, L. A. (2014a). Characterization of a biosurfactant produced by *Pseudomonas cepacia* CCT6659 in the presence of industrial wastes and its application in the biodegradation of hydrophobic compounds in soil. *Colloids and Surfaces. B, Biointerfaces, 117,* 36−41.

Silva, N. M. P. R., Rufino, R. D., Luna, J. M., Santos, V. A., & Sarubbo, L. A. (2014b). Screening of Pseudomonas species for biosurfactant production using low-cost substrates. *Biocatalysis and Agricultural Biotechnology, 3*(2), 132−139.

Venil, C. K., Yusof, N. Z. B., & Ahmad, W. A. (2017). Solid state fermentation utilizing agro-industrial waste for microbial pigment production. In A. Dhanarajan (Ed.), *Sustainable agriculture towards food security.* Singapore: Springer. Available from https://doi.org/10.1007/978-981-10-6647-4_20.

Zhang, Z., Hou, Z., Yang, C., Ma, C., Tao, F., & Xu, P. (2011). Degradation of n-alkanes and polycyclic aromatic hydrocarbons in petroleum by a newly isolated Pseudomonas aeruginosa DQ8. *Bioresource Technology, 102,* 4111−4116.

CHAPTER 7

Pollution mitigation utilizing biosurfactants

Asif Jamal[1], Muhammad Ishtiaq Ali[1], Aetsam Bin Masood[1], Maryam Khan Wazir[1], Ahsan Ullah[1] and Ramla Rehman[2]
[1]Faculty of Biological Sciences, Department of Microbiology, Quaid-i-Azam University, Islamabad, Pakistan
[2]Institute of Industrial Biotechnology, Government College University, Lahore, Punjab, Pakistan

7.1 Introduction

The chemically driven economic cycle leads to the generation of huge amounts of highly contaminated liquid and solid wastes which are being added into our ecosystem without proper treatment (Jouhara et al., 2018). On the other hand, rapidly growing populations and industrial developments are associated with significant ecological damage and abrupt climate changes (Nguyen et al., 2021). It has been estimated that these anthropogenic activities have continued to increase the carbon footprint of the ecosystem (Terhaar et al., 2021). About 55.3 Gigatons of CO_2 is released annually into the atmosphere by human activities, raising serious concerns for the sustainability of biogeochemical cycles and future food production (Caragnano et al., 2020). Soil and groundwater pollution is increasing with every passing day, especially in technologically less developed countries. The heavy use of pesticides and uncontrolled release of domestic and industrial wastewater pollutes the freshwater and groundwater reserves (Hassaan & El Nemr, 2020). In this situation, there is a dire need to develop innovative water treatment technologies to mitigate organic pollution (Santucci et al., 2018). Biosurfactants are one of the important biotechnological products that can effectively remove pollutants from the ecosystem (Sajadi Bami et al., 2022). Biosurfactants are surface-active compounds produced by hydrocarbon-degrading microorganisms with the ability to alter the surface and interfacial tension of the complex insoluble substrates (Bezerra et al., 2018). They are secreted by various soil microorganisms with astonishing chemical diversity. Their release into the contaminated soil enhances the bioavailability and transportation of hydrophobic contaminants and facilitates their disappearance from the affected environment (Kaczorek et al., 2018). To mobilize the pollutant from the bulk phase, biosurfactants entrap organic molecules in their micellar core and facilitate their uptake by the microbial cells, thereby causing their mineralization using an oxygenase system (Jahan et al., 2020). Therefore, biosurfactants are considered one of the most important microbial metabolites that can help mitigate environmental pollution (Jadeja et al., 2019). Besides extensive bioremediation trials and field-scale applications, the actual potential of these wonder molecules yet remains fully capitalized (Johnson et al., 2021).

7.2 Sources of pollution and their impact on the ecosystem

Organic contaminants are being added continuously into the environment from various anthropogenic sources, leading to the imbalance of biogeochemical cycles and perturbations in the agricultural properties of the soils (Rojas & Horcajada, 2020). In addition, these chemicals have continued to increase the carbon footprint of the ecosystem. Synthetic and man-made chemicals are causing disturbances in the food chain, leading to compromised species diversity and harmony in the ecosystem (Yongming et al., 2018). It has been reported that chemical pollution causes melting of polar ice caps, resulting in floods and the elimination of fertile soils (Barua et al., 2022). The chemical pesticides are one of the most important environmental pollutants which are being added into the agriculture fields in order to increase the crop yield. However, these chemicals pose serious ecological toxicity and are associated with various health related issues in plant, animals and humans (Shah et al., 2021). The complex hydrocarbons particularly polycyclic aromatic hydrocarbons (PAHs) and polychlorinated biphenyls (PCBs), infiltrate deeper soil layers and cause soil and water contamination (Gorovtsov et al., 2018). The concentration of PAHs varies spatio-temporally depending on the intensity of prevailing environmental conditions and seasonal fluctuations (Rocha et al., 2021). Besides various measures, the contamination of the soil and water bodies continued to increase with an order of magnitude (Yazdi et al., 2021). Generally, hydrocarbons are difficult to degrade due to their complex structures and hydrophobic nature making them persistant for biodegradation (Al-Hawash et al., 2018). Another problem associated with the complex hydrocarbons like PAHs is their adsorption and strong binding to the soil particles, reducing their mobilization in the soil system (Herz-Thyhsen et al., 2021). This leads to mass transfer limitation and decreased transport of the contaminants from the bulk to the aqueous phase (Balseiro-Romero et al., 2018). Due to complex chemistry and limited bioavailability, PAHs load in the agricultural soils and ground water has been increasing to an alarming level. There are several reports on the cellular and ecological toxicity of polyaromatic hydrocarbons. It has been reported that PAHs can cause cancer, gene mutations, immune disorders and brain damage (Nabi et al., 2019). There are several sources from where complex and toxic hydrocarbons are being added into the environment such as municipal waste water, oil spills, use of pesticides and agrochemicals, industrial effluents and untreated solid waste (Ferronato & Torretta, 2019; Du & Jing, 2018). Some of the important sources of the PAHs are presented in the following text.

7.2.1 Industrial manufacturing

Various industries are the continued source of PAHs in the environment, including steel, aluminum, cement, synthetic rubber and tires, dyes, oil refineries, coal gasification, and power generation units (Smułek et al., 2020).

7.2.2 Automotive and transportation

Automotive and vehicle exhausts are another dominant source of PAH pollution. Jets and aircraft, exhaust from ships, heavy and light vehicles, and trains release tons of PAHs into the environment (Arias et al., 2022).

7.2.3 Domestic waste

PAHs are also generated by domestic activities, such as coal burning for cooking, kerosene use, and burning and dumping of domestic solid waste (Li et al., 2019).

7.2.4 Agriculture formulations/chemicals

In many agricultural-based countries, open burning of woody biomass is the major contributor of PAHs in the soil and air because of incomplete and sub-optimized combustion process. In addition, applying agri-chemicals in huge quantities, such as pesticides, fungicides, and herbicides, is considered an important add on of PAHs pollution in the ecosystem (Klimkowicz-Pawlas et al., 2019).

7.3 Natural role of biosurfactants

Microbial cells can feed on complex hydrocarbon substrates and use them as a carbon source to satisfy their cell growth (Sangwan et al., 2022). Many soil-dwelling microorganisms are known to produce biosurfactants with exceptional chemical diversity (Phulpoto et al., 2021). Rhamnolipids (RLs) are one of the most studied glycolipids biosurfactants produced by different bacterial strains, including *Pseudomonas aeruginosa, Burkholderia* and *B. pseudomallei, P. chlororaphis* (Toribio et al., 2010). RLs are very effective in reducing the surface and interfacial tensions between the two phases and can improve the biodegradation of PAHs (Zhao et al., 2020). Another important class of biosurfactants is lipopeptides (LPs), which are produced predominantly by *Bacillus* species (Fira et al., 2018). LP biosurfactants are produced following thiotemplate pathway with the involvement of a multimodular enzyme system known as a nonribosomal peptide synthetase system (NRPS). They possess surfactant-like properties, making them multifunctional products for the mitigation of environmental pollution (Drakontis & Amin, 2020). In the contaminated environment, the release of biosurfactants gives a competitive advantage to the producers. It has been demonstrated that the extracellular release of RLs, LPs, and other biosurfactants helps bacterial colonization, cell-to-cell communication, substrate accession, solubilization of hydrophobic pollutants, and restoration of the soil habitat (Rawat et al., 2020). Therefore, it can be suggested that biosurfactant production is an adaptive strategy of the cells to survive in the hostile chemical environment (Płaza & Achal, 2020). It has been postulated that biosurfactant release increases the cell surface hydrophobicity, thereby increasing the transport of the hydrophobic materials across the cell membrane by a passive diffusion mechanism (Naughton et al., 2019). In addition, biosurfactants self-assemble to form microemulsions, thus increasing the solubility and bioavailability of the pollutants (Karlapudi et al., 2018). Fig. 7.1 explains the putative mechanism of the biosurfactants-mediated bioremediation of the complex hydrophobic contaminants.

7.4 Polycyclic aromatic hydrocarbons

Polycyclic aromatic hydrocarbons (PAHs) are the chemical compounds containing two fused aromatic rings with significant environmental toxicity (Sahoo et al., 2020). They are produced as a result of their application in wide-ranging industrial processes, including plastics, resins, paints, rubbers, and synthetic insecticides precursors (de Almeida et al., 2018). In addition, fossil fuel is one of the potent sources of PAHs in the environment (Nam et al., 2021). Polychlorinated aromatic hydrocarbons or

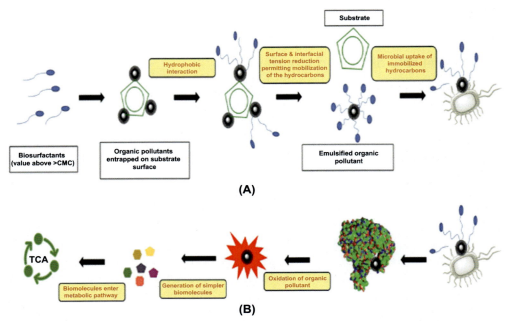

Figure 7.1 Biosurfactants-mediated pollutant removal (A) Increased solubilization and reduction of surface tension (B) Degradation of pollutant in bacterial cells (Bhadra et al., 2022).

chlorinated PAHs are currently the most widely present ecological toxicants (Yang et al., 2019). Chemically, these molecules are lipophilic, fat-soluble and strongly hydrophobic in nature (Jing et al., 2021). They enter into the environment through multiple routes predominantly by accidental spills, waste disposal, transportation, and utilization in various industrial sectors (Wang, Peng, et al., 2019; Wang, Yu, et al., 2019). From these sources, PAHs are transferred to humans through direct inhalation of the contaminated air or ingestion of contaminated food (Magalhães et al., 2022). Because of their complex and recalcitrant structure, they pose significant challenges for cleanup and removal from the ecosystem (Wassenaar & Verbruggen, 2021). As mentioned previously, PAHs are responsible for various physiological disorders in humans/animals, including endocrine disruption, cardiotoxicity, and neurotoxicity, for being carcinogenic and mutagenic (Pirsaheb et al., 2015). Microorganisms that have the metabolic traits to mineralize the PAHs are reported by many researchers in both laboratory and field experiments (Zang et al., 2021). It has been suggested that hydrocarbon-contaminated sites harbor excellent diversity of PAHs degrading microbes (Rabodonirina et al., 2019). These microorganisms use contaminants as the sole source of carbon and energy by producing surface-active compounds/biosurfactants (Premnath et al., 2021). It is important to note that biosurfactant production is an important feature of hydrocarbon-degrading microorganisms enabling these to solubilize and transport complex aromatics in both soil and aqueous systems (Yesankar et al., 2023). Fig. 7.2 depicts various metabolic routes of PAHs biodegradation.

7.5 Biosurfactants-mediated biodegradation of polycyclic aromatic hydrocarbons

Many microorganisms are natural degraders of PAHs and chlorinated polycyclic aromatic hydrocarbons (Cl-PAHs) (Rabodonirina et al., 2019). Owing to their positive charge, PAHs tend to sorb onto the soil and adhere strongly with soil particles by hydrophobic interactions (Mazarji et al., 2022). Their poor solubility and bioavailability to the microbial cells are the limiting factors for their bioremediation from the contaminated soil. In order to address mass transfer limitation, many microorgansisms produce surface active agents to enhance aqueous solubility and mobility of the hydrophobic contaminants in the soil matrix (Sun et al., 2019). It has been suggested that *P. aeruginosa* secretes rhamnolipids biosurfactants (RLs) to solubilize and mobilize PAHs (Pourfadakari et al., 2019). The extracellular release of surfactants by the microbial cells causes reduction in the surface and interfacial tension thereby improving solubility and transport of hydrophobic PAHs across the cell membrane. Once these chemicals enter into the interior of the cells, different enzymes catalyze their degradation into simple, nontoxic compounds (Yesankar et al., 2023). The biodegradation of the PAHs is complex and multistep process in which surfactants play a central role. It

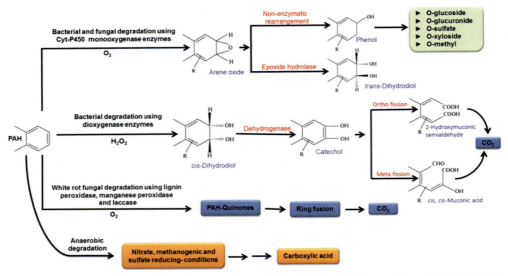

Figure 7.2 Biochemical routes for the biodegradation of PAHs (Shahsavari et al., 2019). *PAHs*, Polycyclic aromatic hydrocarbons.

has been well documented that biosurfactants form micelles when they are produced above a concentration technically known as critical micelle concentration (CMC) (Jahan et al., 2020; Yang et al., 2022). RLs have CMC values ranging from 10 to 200 mg/L (Mokhtari et al., 2019), and sophorolipids congeners showed 40–250 mg/L CMC values (Ashby et al., 2008). In contrast, surfactin, a lipopeptide biosurfactant, has a CMC of 13–25 mg/L (Deleu et al., 1999). These values are quite lower than the synthetic surfactants, giving an obvious advantage to the biosurfactants, in addition to their environmental compatibility (Gouda et al., 2020).

Biosurfactants are unique microbial amphiphiles that can form micelles of different shapes, sizes and geometrical arrangements using their innate self-assembly properties (Baccile et al., 2019). It has been suggested that RL forms spherical micelles of an average size of 80–90 nm (Mendes et al., 2015). Other biosurfactants such as sophorolipids, trehalose lipid, surfactin and fengycin can also generate micelles of different sizes and geometries (Zhu et al., 2021). Generally, a micelle is a circular arrangement of the surfactants monomers in which hydrophilic heads are located outwards and hydrophobic tails toward the core. The micelle core traps hydrocarbons by hydrophobic interactions, and the hydrophilic region on the outside fuses with the microbial membrane (Christopher et al., 2021). In this way the entrapped hydrocarbon can easily be transported into the microbial cells and presented to the enzymatic systems for biodegradation (Channashettar et al., 2022; Imron et al., 2020). It has been reported that the biodegradation of PAHs is carried out by the collective action of surface tension reduction, alteration in interfacial tension, formation of micelle and microemulsion, solubility improvement and rapid membrane transport of the pollutant with the help of biosurfactants (Bzdek et al., 2020). Surface tension of different biosurfactants is quite variable; for example, rhamnolipids reduce the surface tension (ST) of water from 28 to 30 mN/m (Gong et al., 2020), while sophorolipids can lower it between 35–36 mN/m (Kumari et al., 2021). In addition, trehalose lipids exhibit ST value around 34 mN/m (Janek et al., 2018). Owing to aforementioned remarkable features, biosurfactants have been applied to various hydrocarbon degradation laboratory and field experiments (Suganthi et al., 2018). For example, Yu et al. (2014) showed that efficiency of PAH degradation was improved to 97% with the application of rhamnolipids. In another study, Dave et al. (2014) reported the efficiency of *Achromobacter xylosoxidans* in PAH degradation better than the conventional surface active agents. They evaluated the biodegradation of 11 different PAHs compounds and suggested that lipopeptides production by the *A. xylosoxidans* was able to enhance the degradation of low-molecular-weight PAHs up to 2.8 folds. In contrast, high-molecular-weight PAHs showed a relatively higher biodegradation efficiency of 7.59-folds. Swaathy et al. (2014) reported the ability of *Bacillus licheniformis* in anthracene biodegradation with the help of biosurfactants. They found 95% anthracene degradation under the action of biosurfactant. The biodegradation of the anthracene was linked to the chemical structure of the biosurfactants. Bezza and Chirwa (2016) showed the effectiveness of PAHs degrading consortia consisting of *Pseudomonas* and *Bacillus* sp. The degradation rate with lipopeptide was 85.5%, while 57% without biosurfactant, demonstrating significance of biosurfactants in the degradation of hydrocarbon contaminants.

In soil, production of biosurfactants facilitates microbial colonization and establishment of the biofilm (Shen et al., 2020). The development of the biofilm in the roots not only helps nutrients exchange between soil and plants but also promotes biodegradation of hydrophobic contaminants and PAHs. It has been well documented that rhizospheric bacteria produce surface active agents and facilitate biofilm establishment through quorum sensing networks (Preda & Săndulescu, 2019). This intricate communication is mediated by the production of N-acyl homoserine lactones in the gram-negative cells, particularly in *P. aeruginosa* (Abe et al., 2020; Ostapska et al., 2018; Shin et al., 2019). The aggregation of the mixed microbial communities at the interfaces is an important ecological event which greatly helps solubilization and degradation of poly aromatic hydrocarbons and pesticides (Kaczorek et al., 2018). It has been reported that microorgansims use complex hydrocarbons as carbon source (Tang et al., 2018). The biodegradation of the PAHs is a multiple step process involving a range of intracellular enzymes particularly mono and dioxygenases (Cheng et al., 2022). The biodegradation pathways of various PAHs have been reported in the literature describing the role of these enzymes in the breakdown of complex hydrocarbons (Da Silva et al., 2020; Mukherjee et al., 2019). It is important to understand that the action of biosurfactants is mandatory for improving bioavailability, solubility, mobilization and transport of the pollutants across the microbial membranes. After the entry of the hydrocarbons into the cells, microorgansims employ various mechanisms to degrade these toxic materials. In the following text, the cellular degradation of PAHs will be described on the basis of prevailing knowledge.

Microorganisms can degrade organic pollutants under two physiological conditions, that is, aerobic and anaerobic (Yang et al., 2021). In the former, microbes employ dioxygenases to convert the ringed aromatic hydrocarbons into dihydrodiols and oxygen as the terminal electron acceptor (Cheng et al., 2022). The degradation process is carried out through oxygenic cleavage with ortho- and meta-cleavage pathways (Zhuk et al., 2022). Microbes degrade ringed aromatics containing five carbon atoms easily, and these catabolic processes are further continued to convert dihydrodiols into catechol and protocatechuates (Lv et al., 2022). The ringed aromatics are degraded by microbes under anaerobic conditions, which break the rings and convert the complex hydrocarbons into open chained compounds (Nzila, 2018). These compounds are then converted to carbon dioxide and water in the terminal biodegradation steps. The bioremediation efficiency depends upon the microbial community and position of certain atoms in the constituent molecules (Lászlová et al., 2018).

Anaerobic degradation of PAHs uses nitrates and sulfates as electron acceptors (Zhang et al., 2021). Carbonic acid side chains and cyclohexane are the products of anaerobic respiration in *Marinobacter* sp. (Zan et al., 2022). PAH degradation leads to better soil health and crop yield, leading to a sustainable environment (Cao et al., 2022). The details of microbes capable of degrading hydrocarbons are enlisted in Table 7.1. The efficiency of biodegradation of the biosurfactants-producing strains is mentioned along with the nature of substrates. Fig. 7.3 represents the pathway of PAH degradation under the influence of microbes and enzymes they produce.

7.6 Polychlorinated biphenyls (PCBs)

Polychlorinated biphenyls (PCBs) are derivatives of biphenyls, having hydrogens replaced with chlorines in their chemical structure (Reddy et al., 2019). The use of these compounds has been banned because of their ecological toxicity and carcinogenic nature (Cetin et al., 2018). Because of their complex and recalcitrant structure and greater persistence, they have the ability to alter soil structure similarly as PAHs and CI-PAHs (Erkul & Eker Şanli, 2020). PCBs can adsorb to soil particles and persist in the environment for a longer time compared with other hydrocarbons (Adeyinka & Moodley, 2022). They are difficult to remove from the environment and act as important indicators of soil pollution (Rivera-Pérez et al., 2022). PCBs are insoluble in water but are readily soluble in non-polar solvents, making them inaccessible in aqueous medium (Park et al., 2019). Consequently, these chemicals are accumulated on the surfaces owing to their poor solubility and greater hydrophobicity (Ahmad et al., 2019). These factors are associated with their restricted accessibility to microbial

Table 7.1 Biosurfactant mediated biodegradation of complex hydrocarbons and PAHs.

S. No.	Strain	Biosurfactant type	Hydrocarbon degraded	Degradation efficiency (%)	Application of bacterial strain/s	References
1	*Bacillus methylotrophicus*	Rhamnolipid	Petroleum Hydrocarbons	80.24	Soil fertility enhancement, bioremediation, enhancement of plant growth, increase in plant immunity	Alao and Adebayo (2022), Goveas et al. (2022), Nikolova et al. (2021), Rong et al. (2021), Ahmad et al. (2021)
2	*Klebsiella* sp.		Phenanthrene	43.7	Bioremediation, breakdown of recalcitrant compounds, enhancement of plant yield, soil fertility	
3	Consortia (*Enterococcus* (98.6%), *Vagococcus* (1.0%), *Sphingomonas* (0.3%) and *Proteus* (0.1%))	Sophorolipid	Petroleum hydrocarbons	57.7	Soil texture enhancement, remediation studies, heavy metal elimination, plant growth	Mahmud et al. (2022), Bidyarani et al. (2021), Gupta and Pathak (2020), Feng et al. (2021), Kang et al. (2010)
4	*Candida bombicola*		Crude Oil	80	Soil fertility, plant immunity, soil reclamation studies, improvement of crop yield	
5	*Rhodococcus erythropolis*	Trehalose lipid	Hexadecane Phenanthrene	40 65	Complex hydrocarbon breakdown, plant growth and immunity enhancement, Increase in crop yield, soil texture improvement	Thi et al. (2022), Siddiqui et al. (2021), Vaishnavi et al. (2021), Chang et al. (2004)

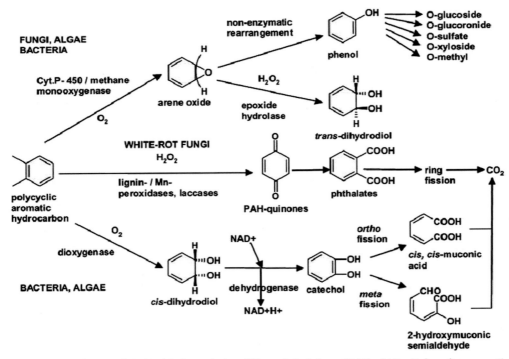

Figure 7.3 Pathway of PAHs biodegradation (Obayori & Salam, 2010). *PAHs*, Polycyclic aromatic hydrocarbons.

cells and, thus, biodegradation (Kothuru et al., 2020). The solubility of PCBs decreases with the increase in chlorination of the hydrocarbon structure and lower soil temperature (Elangovan et al., 2019). As these molecules show less tendency to biodegradation, they are known to induce significant phytotoxicity and pose a serious challenge for sustainable food production (Yazdi et al., 2021). Various scientific studies have shown that the concentration of PCBs has been increasing in marine ecosystem resulting severe diseases in marine animals and plants. Therefore, consumption of PCB contaminated sea food is also posing serious health issues (Bersuder et al., 2020). The general structure of a typical PCB is depicted in Fig. 7.4.

7.7 Biosurfactants-mediated degradation of polychlorinated biphenyls

Biosurfactants can be a very effective molecular tool for the biodegradation of the PCBs owing to their excellent surface and interfacial maneuvering properties (Lászlová et al., 2018). Biosurfactants can influence their reduced aqueous solubility, thus enhancing the bioavailability of PCBs (Pathiraja et al., 2019). When biosurfactants

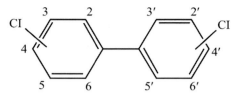

Figure 7.4 Structure of PCB (Faroon & Ruiz, 2016). *PCBs*, Polychlorinated biphenyls.

are released, they tend to display at the cell surfaces causing alteration in the membrane hydrophobicity pattern, membrane permeability, and physiological activities of the cells (Oliva et al., 2020). These changes are a function of surfactant concentration, which should be lower than the lytic concertation with the least negative effect on the architecture of the biological membrane (Hegazy et al., 2022). After reaching a critical concentration, biosurfactants begin to self-assemble in the form of micelles (Abruzzo et al., 2021). These micelles then entrap the contaminants into their core creating nano-emulsion and facilitating surfactants-encased hydrocarbon transport to the microbial cells (Al-Sakkaf & Onaizi, 2022). Various researchers have explained the complexity of pollutant–membrane interactions. Sandhu et al. (2022) showed the efficiency of RLs produced by *Brucella anthropi* sp. in PCBs degradation with 66.15% removal efficiency. In the same study, *P. aeruginosa* showed 62.06%, *P. plecoglossicida* 57.02% and *Priesta megaterium* 56.55% removal of PCBs. *Lysinibaciilus* sp. has the potential to produce lipopeptides biosurfactants and improve PCB biodegradation. Pathiraja et al., 2019 evaluated the potential of *Lysinibaciilus* sp. in PCB degradation and achieved chlorine removal up to 9.16 ± 0.8 mg/L, while the PCB solubility was noted to be 14.7 ± 0.93 mg/L. In another study conducted by Chen et al. (2015), reported the PCBs degradation by *Pseudomonas* sp. capable of producing rhamnolipids type of biosurfactants. Under optimum pH of 6, PCBs degradation was observed up to 69.6% with a dechlorination efficiency of 92.6%. Steliga et al., 2020 reported 84.5% PCB degradation using *Rhodococcus* sp. under labratory conditions. These studies suggested a supportive role of biosurfactants in the biodegradation of PCBs.

Basically, the biodegradation of PCBs is carried out in two phases: (1) biosurfactants-mediated transport of PCBs into the cells and (2) enzymatic degradation of complex aromatic rings in the cell cytoplasm (Pathiraja et al., 2019). Both processes are linked in a very precise and intricate way because intracellular degradation of the PCBs is directly associated with their transmembrane transport (Lászlová et al., 2018). Within the cell, PCBs are at the disposal of various enzymes, such as dehalogenases, monooxygenases, and dioxygenases (Oyewusi et al., 2020). The intracellular biodegradation process is quite complicated and involves the expression of certain genes related to the PCB degradative pathway. The most common genes related to PCB degradation are *bph*A, *bph*B, and *bph*C (Ren et al., 2022). It is documented that *bph*A gene is responsible for biphenyl ring deoxygenation, *bph*B for dehydrogenation reaction, and *bph*C in the deoxygenation of ring and cleavage reactions (Cao et al., 2021). The products of aromatic ring degradation are phenyl catechol and dyhidrodiols (Elangovan et al., 2019). Bacteria capable of degrading PCBs by aerobic oxidation utilize *bph*A gene to make simpler products such as chlorobenzoic acids, while the reductive dehalogenation of PCBs results in the formation of simpler chlorinated congeners (Ines et al., 2021). The biodegradation process starts with the dehalogenation of PCBs to form biphenyls that act as starting substrates for microbes. They are converted initially into biphenyl-2,3-dihydrodiol by the action of biphenyl-2,3-dioxygenase, followed by their conversion into 2,3-dihydroxybiphenyl with the help of biphenyl-2,3-dihydrodiol-2,3-dehydrogenase. The enzyme 2,3-dihydroxybiphenyl-1,2-dioxygenase then converts 2,3-dihydroxybiphenyl into 2-hydroxy-6-oxo-6-phenylhexa-2,4-dienoic acid. After this reaction, 2-hydroxy-6-oxo-6-phenylhexa-2,4-dienoic acid is converted into two products, i.e., 2-hydroxypenta-2,4-dienoic acid and benzoic acid by the action of 2-hydroxy-6-oxo-6-phenylhexa-2,4-dienoate hydrolase. In the next reaction, 4-hydroxy-2-oxovaleric acid is formed by 2-hydroxypenta-2,4-dienoate hydratase, which is further transformed into acetaldehyde and pyruvic acid by 4-hydroxy-2-oxovalerate aldolase. At the end, acetyl-CoA is formed by acetaldehyde dehydrogenase. The products formed by the degradation of PCBs are considered as nontoxic for being TCA cycle intermediates (Seeger et al., 1997). Some microbes can perform dechlorination reactions naturally with the help of biosurfactants and co-metabolism strategies (Yan et al., 2021). However, there are several gray areas of understanding regarding the biodegradation of these compounds, demanding further scientific attending. The exploitation of more efficient biosurfactants degraders, surfactants behavior, and the role of field conditions could provide better insight into PCBs bioremediation (Debnath et al., 2021). A typical pathway of degradation of PCBs after dechlorination reaction is shown in Fig. 7.5.

7.8 Organopesticides

Population explosion has resulted in an increased demand for agricultural products. To meet such needs, extensive use of organic pesticides has become common (Reeves et al., 2019). There are two main types of organopesticides: organophosphate and organochlorine (Semu et al., 2019). These organic pesticides are hydrophobic in nature and pose significant challenge for their biodegradation. According to the reliable scientific data, organopesticides (OPs) are present in significantly higher concentrations in edible and nonedible parts of plants and increase the risk of diabetes mellitus type 2 in humans (Lee et al., 2011). Malathion, parathion, diazinon, fenthion, dichlorvos, chlorpyrifos, and ethion are the major

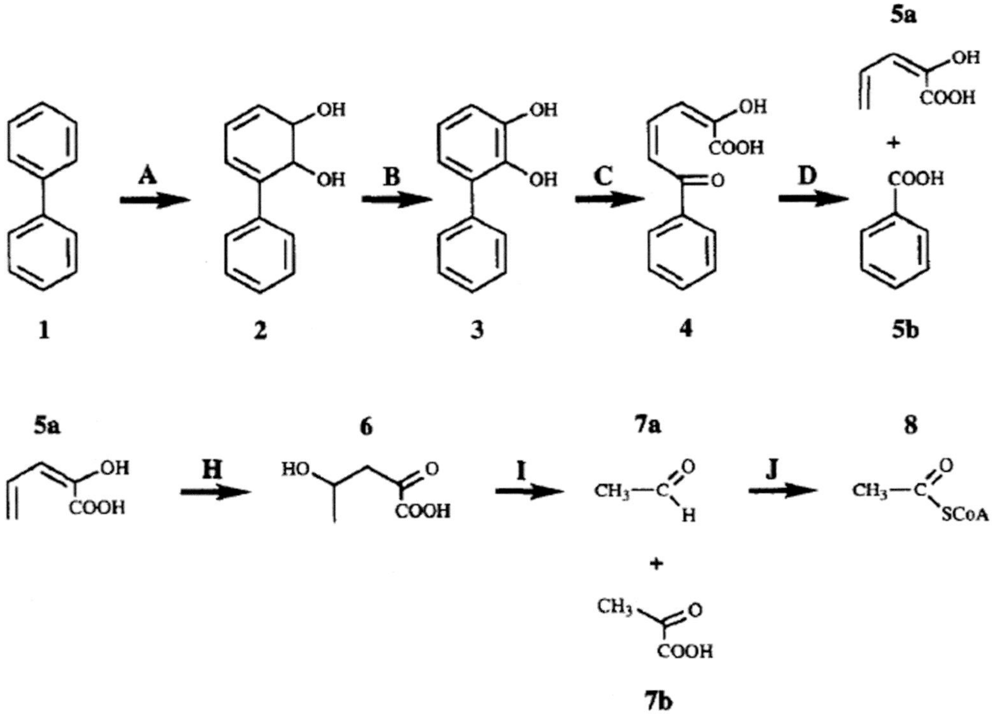

Figure 7.5 Degradation pathway of PCBs (Seeger et al., 1997). *PCBs*, Polychlorinated biphenyls.

organophosphate fertilizers used in the agro-industry and thus found in the environment (Heydari et al., 2021).

7.9 Surfactants enhanced degradation of organopesticides

Biosurfactants are considered important components of modern bioremediation techniques for reclamation of pesticide-contaminated soils (Lamilla et al., 2021). Owing to the hydrophobic nature of pesticides, they exhibit low solubility in an aqueous medium and limited biodegradation by the soil microorganisms (Wang, Peng, et al., 2019; Wang, Yu, et al., 2019). In order to address these challenges, microorgansims use their unique cellular properties of biosurfactant production. Biosurfactants play their role following three major steps: (1) reduction in surface tension of organic contaminants, (2) enhanced mobilization via micelle formation above cmc values, and (3) enzymatic breakdown after cellular intake. Scientific studies on the biodegradation of some commercially available pesticides, including DDT (dichlorodiphenyltrichloroethane), Chlorpyrifos, Quinalphos, and PHE, are explained below in order to provide stimulatory role of biosurfactants in pollution mitigation and ecosytem restoration.

7.9.1 Dichlorodiphenyltrichloroethane

Agricultural applications of DDT are diverse, but its use ends with diverse side effects too. The deterioration of soil texture, agriculture properties and fertility have been reviewed extensively in addition to their negative impact on plants and animals health. DDT removal has been reported in various scientific studies with the help of biosurfactants. In a study, Bhatt et al. (2021) showed the effectiveness of RLs in DDT degradation up to 64%. Purnomo et al. (2017) suggested that a higher degradation rate of 86% was achieved for DDT with the help of RLs. In various other studies application of mixed microbial consortium was suggested in order to get higher biodegradation efficiency. The consortium of *F. pinicola* and *P. aeruginosa* was applied in a shake flask study to investigate the combined efficacy of the aforementioned strains. It was indicated that these strains caused 68% DDT degradation, which was directly associated with their biosurfactant production abilities. In addition, microorganisms capable of producing lipopeptide biosurfactants, *B. subtilis* and *F. pinicola*, resulted in 86% degradation of DDT with their combined effect (Sariwati & Purnomo, 2018; Sariwati et al., 2017). Basically, biosurfactants help transport these materials across the microbial membranes by various previously described mechanisms for their degradation within the cells. These mechanisms are operated by the production of biosurfactants, which increase the pesticides' bioavailability to the degraders and stimulate biodegradation and bioremediation (Ahmad et al., 2018). The glycolipids are taking lead in the biodegradation of pesticides because of their higher production rates, better surface and interfacial properties and creation of mixed micelle system to encase the wide variety of contaminants in their micelle core. The increased availability of the pesticides in aqueous media has been reported for *Burkholderia cenocepacia* which caused significant improvement in DDT biodegradation as compared to the nonbiosurfactants producers (Almatawah, 2017). Interestingly, about 40% degradation rate of fluorinated pesticide was recorded by *Streptomyces* species with the co-inoculation of *Pseudomonas aeruginosa* producing RL molecules (Alexandrino et al., 2022). These studies are indicative of the supportive role of biosurfactants in pesticide biodegradation by making the substrate available and ready for the cell. A similar mechanism is responsible for chlorinated pesticides such as DDT (Wang et al., 2017). Besides extensive studies, the exact mechanisms behind the biosurfactants-mediated pesticide degradation and interaction of these chemicals with biological membranes remain elucidative.

7.9.2 Chlorpyrifos

Chlorpyrifos is a widely applied as pesticide for improving the productivity of the agriculture products, however, its application results deterioration of soil quality and properties.

Various bacterial and fungal species have been studied with respect to the chlorpyrifos degradation under different experimental conditions. Singh et al. (2016), reported biosurfactant mediated biodegradation by *P. aeruginosa* which was capable of producing rhamnolipids and enhanced degradation of chlorpyrifos up to 82%. The higher biodegradation of the chlorpyrifos (98%) was achieved under the influence of biosurfactants (Singh et al., 2009). The potential of *Pseudomonas marginalis* capable of producing glycopeptides and glycolipids producing *P. rhodesiae* was evaluated for chlorpyrifos degradation. These surfactants reduced surface tension of the aqueous medium up to 34.47 and 37.44 mN/m, respectively, and led to 51% chlorpyrifos biodegradation (Lamilla et al., 2021). For the evaluation of the degradation potential of a consortium, Kumar et al. (2022) showed the efficiency of fungal and bacterial species in the complete degradation of chlorpyrifos, suggested that a mixture of these species was able to remove almost all pesticides from the medium. On a similar note, Kavitha et al. (2016) investigated glycolipids biosurfactants for the degradation of chlorpyrifos and suggested incremental biodegradation rates of 65% with the production of biosurfactants. These studies suggested that soil contaminated with chlorpyrifos can be rehabilitated by applying biosurfactants-producing microorganisms.

7.9.3 Quinalphos

Quinalphos is an organophosphorus pesticide that is difficult to degrade due to its lesser solubility and complex structure. It has been declared as moderately hazardous by the WHO, but its presence in the soil is associated with ecological toxicity, plant and animal diseases, and altered soil properties. Several studies have been conducted for screening and isolation of biosurfactant-producing microbes that can degrade quinalphos. In a study by Nair et al. (2015), RL-producing *Pseudomonas aeruginosa* was evaluated for its potential to degrade quinalphos. A degradation rate of 94% was achieved with phosphorothioic acid and 2-hydroxy quinoxaline as the degradation products. In another study, lipopeptides-producing *Bacillus thuringiensis* was evaluated for quinalphos degradation. Gangireddygari et al. (2017) found that 94.74% degradation was achieved for the aforementioned pesticide at 35°C–37°C and 6.5–7.5 pH. Surfactin produced from *Bacillus subtilis* has also been studied with respect to quinalphos degradation with promising results of 92.8 %. On the basis of these studies, it is generally believed that biosurfactants and biosurfactants producing microorganisms are the important biotechnological tools for the bioremediation of the pesticide contaminated soil and water. However, the application of biosurfactants producing microorganisms or use of purified biosurfactants needs further optimization considering different environmental and soil conditions.

7.9.4 Phenanthrene

Phenanthrene (PHE) is a PAH with three benzene rings. It is widely applied in pesticide formulation, dyes, and plastics. Several microbes have the ability to utilize it as a

Table 7.2 Types of organochlorinated pesticides.

S. No	Classes	Examples	Uses	References
1	Chlorinated derivatives of ethane	Dichlorodiphenyltrichloroethane DDT, methoxychlor	Used against flies, cockroaches, mosquitoes, agricultural crops, livestock, home garden, pets, poultry, etc.	Chen et al. (2015), Fuentes et al. (2014), Mulliken, Zambone, Rolph, & Env't (2004)
2	Chlorinated derivatives of benzene and cyclohexane	Lindane	Insecticide used for fruits, grains, plants, seed treatment, head lice, etc.	Humphreys et al. (2008), Walker, Vallero, Lewis, and Technology (1999)
3	Cyclodienes	Endosulfan, Aldrin, Dieldrin, Endrin	Food crops such as tea, corn, wheat, hay, barley, vegetables, rye, oats, grains, citrus crops, tobacco, cotton, and carpet and textile manufacturing	Jia et al. (2010), Jorgenson (2001)
4	ChlorinatedCamphenes	Toxaphene and Chlordecone	Fruits, banana fields, cotton, soybean	Multigner et al. (2016), Multigner et al. (2010), Wallace (2014)

carbon source and degrade it to simpler products. The biodegradation of PHE can be facilitated by the production of biosurfactants in order to enhance solubilization and mobilization of these pesticides. A study conducted by Chang et al. (2004) demonstrated the effectiveness of *Rhodococcus erythropolis* in phenanthrene degradation. This strain produces trehalose lipids, which enhance PHE degradation up to 30 folds. Ahmad et al. (2021), demonstrated the role of *Klebsiella* sp. in PHE degradation via the production of mono-RLs. The strain was able to emulsify insoluble substrate with a recorded emulsification index of >40% and reduced the surface tension of the medium up to 30%. Other soil microorgansims such as *Sphingomonas* sp. also have potential to degrade complex structures of PHE and have the metabolic traits to produce glycolipid biosurfactants. Xiao-Hong et al. (2010) demonstrated the effectiveness of RLs biosurfactants and achieved 99.5% degradation of PHE. These studies show the effectiveness of biosurfactants aided biodegradation of PHE. *Bacillus subtilis* has also been studied for the biosynthesis of lipopeptide biosurfactants and their role in the biodegradation of PHE. Ni'Fatimah and Sumarsih (2017) evaluated the potential of *Bacillus subtilis* in PHE degradation and achieved 24.8% removal.

Some important studies on hydrocarbon degradation are summarized in Table 7.1, along with degradation efficiencies, applications, and types of biosurfactants used in the bioremediation process.

7.10 Plant-based surfactants and their role in bioremediation

Saponins are structurally diverse natural non-ionic organic compounds derived from different plants. They are composed of a hydrophobic region composed of sapogenin and a

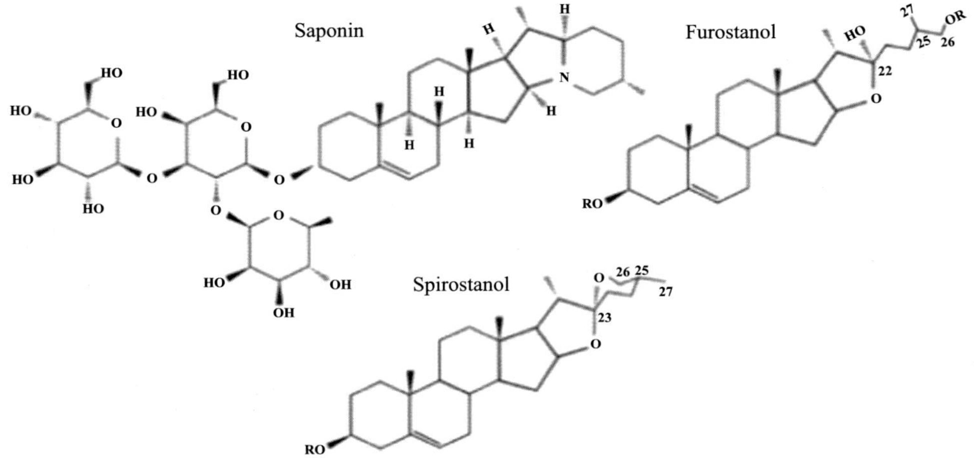

Figure 7.6 Structure of saponin molecules (Ku-Vera et al., 2020).

Table 7.3 Biosurfactant patents.

S. No	Type of biosurfactant/producing microbes	Patent holder	Title of the patent	Publication date	Publication number
1	Hydrocarbon-degrading	Denis Altman, Robin L. Brigmon, Sandra Story, Christopher J. Berry	Surfactant biocatalyst for remediation of recalcitrant organics and heavy metals	28 June 2005	PI 0519962–0 A2
2	Sophorolipid	Doncel GF Gross RA, Shah V	Spermicidal and virucidal properties of various forms of sophorolipids	29 September 2005	WO 2005089522 A2
3	*C. lipolytica*, *C. tropicalis*, *C. albicans*, *C. rugosa*, *C. torulopsis*	Awada M, Awada S, Spendlove R,	Microbial biosurfactants as agents for controlling pests	1 December 2005	US 2005266036 A1
4	*Pseudomonas aeruginosa*	Roberto Rodrigues De Souza, Silvanito Alves Barbosa,	Biosurfactant production for the development of biodegradable detergent	16 May 2011	PI 1102592–1 A2

hydrophilic one consisting of triterpenoidal or steroidal nature (Tucker et al., 2020). Fig. 7.6 shows a typical structure of saponins. These unique compounds are considered as a part of the immune response of plants due to their antimicrobial activity (Zaynab et al., 2021). These can regulate nutrient transport, transmembrane activity, and signaling in plants (Zhou et al., 2018). Saponins can degrade pollutants due to their ability to trap hydrophobic contaminants within the micelle core and emulsify them. Saponins show surfactants like properties and can reduce the surface and interfacial tensions effectively. With these properties saponins are able to enhance the solubility of contaminants and make them biologically available for the microbes and thus facilitate biodegradation (Hoang et al., 2022; Oyewusi et al., 2020). Various studies have been conducted to evaluate the bioremediation potential of plant derived surfactants including saponins. Among the contaminants, fluorene, acenaphthene, naphthalene, and phenanthrene degradation has been observed by the application *Quillaja saponaria*-derived saponins. Davin et al. (2018) showed that a 4 g/L concentration of saponins was effective in the removal of naphthalene (294 ± 124 g/L), fluorene (354 ± 78 g/L), Acenaphtene (864 ± 121 g/L), and Phenanthrene (459 ± 152 g/L). Among other plant species, *Sapindus mukorossi* has been studied in order to investigate their role as natural surfactants in the process of hydrocarbon degradation. Being a natural surface active agent, saponin can improve the cell surface hydrophobicity of the bacteria and facilitate bioremediation of the pollutants. Smułek et al. (2016) showed the effectiveness of diesel biodegradation with saponin (100 mg/L). A higher degradation rate for *Sphingomonas* sp. (41%) and *Pseudomonas alcaligenes* (56%) was achieved by the application of plant driven saponin. In addition to the excellent surface active properties, biodegradability, and environmental compatibility, saponins are considered better because of easy extraction process as compared to the microbial products (Le et al., 2018). In this connection, the potential of saponins needs to be investigated both at lab and field scales under different soil conditions to maximize its commercial use (Di Gioia & Petropoulos, 2019). The application of biosurfactants is increasing for the bioremediation of hydrocarbons contaminated soil. Table 7.3 shows list of commercial biosurfactants based products.

Some important details of biosurfactants patent are enlisted in the Table 7.3 below.

7.11 Conclusion and future directions

The gap in demand and supply chain is evident considering the ongoing population growth, urbanization and industrialization trends. It has been estimated that food resources will become scarce by 2050 because of increase in population, environmental pollution, climate changes, soil degradation, and biodiversity losses. The application of man made and toxic chemicals for improving agriculture output is leading to serious ecological and health challenges. The chemical pollution is continued to increase owing to the deliberate release of organics from various anthropogenic sources into the freshwater and marine ecosystem. There is a dire need to adopt environmental friendly and sustainable options to solve these pertinent issues. Biosurfactants provide a sustainable solution as an ecofriendly alternative to the toxic agro-chemicals. They have antimicrobial properties and can be very effective in controlling environmental pollution. Microorganisms and plants are the natural sources of the biosurfactants. There are compelling scientific evidences that biosurfactants are very effective in improving soil quality, properties and reduce the incidence of plant infections. In addition, biosurfactants play significant role in the biodegradation of complex and hydrophobic pollutants including PAHs, organopesticides and related toxic chemicals. Besides extensive research, the field potential of the biosurfactants yet remains elucidative. The future work for exploring new biosurfactants from microbial and plant resources, with better chemical and surface active properties and understanding their behavior under different environmental and soil conditions could enhance biosurfactants applications and market potential.

References

Abe, K., Nomura, N., & Suzuki, S. (2020). Biofilms: Hot spots of horizontal gene transfer (HGT) in aquatic environments, with a focus on a new HGT mechanism. *FEMS Microbiology Ecology, 96*(5), fiaa031.

Abruzzo, A., Parolin, C., Corazza, E., Giordani, B., di Cagno, M. P., Cerchiara, T., & Luppi, B. (2021). Influence of lactobacillus biosurfactants on skin permeation of hydrocortisone. *Pharmaceutics, 13*(6), 820.

Adeyinka, G. C., & Moodley, B. (2022). Effect of sorbate stereochemistry on the sorption/desorption of polychlorinated biphenyls between natural soil and an aqueous solution. *Soil and Sediment Contamination: An International Journal, 32*, 1–16.

Ahmad, I., Weng, J., Stromberg, A. J., Hilt, J. Z., & Dziubla, T. D. (2019). Fluorescence based detection of polychlorinated biphenyls (PCBs) in water using hydrophobic interactions. *Analyst, 144*(2), 677–684.

Ahmad, Z., Imran, M., Qadeer, S., Hussain, S., Kausar, R., Dawson, L., & Khalid, A. (2018). Biosurfactants for sustainable soil management. *Advances in Agronomy, 150*, 81–130.

Ahmad, Z., Zhang, X., Imran, M., Zhong, H., Andleeb, S., Zulekha, R., & Coulon, F. (2021). Production, functional stability, and effect of rhamnolipid biosurfactant from Klebsiella sp. on phenanthrene degradation in various medium systems. *Ecotoxicology and Environmental Safety, 207*, 111514.

Alao, M. B., & Adebayo, E. A. (2022). Current advances in microbial bioremediation of surface and ground water contaminated by hydrocarbon. *Development in wastewater treatment research and processes* (pp. 89–116). Elsevier.

Alexandrino, D. A., Mucha, A. P., Almeida, C. M. R., & Carvalho, M. F. (2022). Atlas of the microbial degradation of fluorinated pesticides. *Critical Reviews in Biotechnology, 42*(7), 991–1009.

Al-Hawash, A. B., Dragh, M. A., Li, S., Alhujaily, A., Abbood, H. A., Zhang, X., & Ma, F. (2018). Principles of microbial degradation of petroleum hydrocarbons in the environment. *The Egyptian Journal of Aquatic Research, 44*(2), 71–76.

Almatawah, Q. (2017). An indigenous biosurfactant producing Burkholderia cepacia with high emulsification potential towards crude oil. *Journal of Environmental and Analytical Toxicology, 7*(528), 2161-0525.

Al-Sakkaf, M. K., & Onaizi, S. A. (2022). Rheology, characteristics, stability, and pH-responsiveness of biosurfactant-stabilized crude oil/water nanoemulsions. *Fuel, 307*, 121845.

Arias, S., Molina, F., Palacio, R., López, D., & Agudelo, J. R. (2022). Assessment of carbonyl and PAH emissions in an automotive diesel engine fueled with butanol and renewable diesel fuel blends. *Fuel, 316*, 123290.

Ashby, R. D., Solaiman, D. K., & Foglia, T. A. (2008). Property control of sophorolipids: Influence of fatty acid substrate and blending. *Biotechnology Letters*, *30*(6), 1093−1100.

Baccile, N., Delbeke, E. I., Brennich, M., Seyrig, C., Everaert, J., Roelants, S. L., & Stevens, C. V. (2019). Asymmetrical, symmetrical, divalent, and Y-shaped (bola) amphiphiles: The relationship between the molecular structure and self-assembly in amino derivatives of sophorolipid biosurfactants. *The Journal of Physical Chemistry. B*, *123*(17), 3841−3858.

Balseiro-Romero, M., Monterroso, C., & Casares, J. J. (2018). Environmental fate of petroleum hydrocarbons in soil: Review of multiphase transport, mass transfer, and natural attenuation processes. *Pedosphere*, *28*(6), 833−847.

Barua, R., Bardhan, N., & Banerjee, D. (2022). Impact of the polar ice caps melting on ecosystems and climates. In Handbook of Research on Water Sciences and Society, (pp. 722−735). IGI Global.

Bersuder, P., Smith, A. J., Hynes, C., Warford, L., Barber, J. L., Losada, S., & Lyons, B. P. (2020). Baseline survey of marine sediments collected from the Kingdom of Bahrain: PAHs, PCBs, organochlorine pesticides, perfluoroalkyl substances, dioxins, brominated flame retardants and metal contamination. *Marine Pollution Bulletin*, *161*, 111734.

Bezerra, K. G. O., Rufino, R. D., Luna, J. M., & Sarubbo, L. A. (2018). Saponins and microbial biosurfactants: Potential raw materials for the formulation of cosmetics. *Biotechnology Progress*, *34*(6), 1482−1493.

Bezza, F. A., & Chirwa, E. M. N. (2016). Biosurfactant-enhanced bioremediation of aged polycyclic aromatic hydrocarbons (PAHs) in creosote contaminated soil. *Chemosphere*, *144*, 635−644.

Bhadra, S., Chettri, D., & Kumar Verma, A. (2022). Biosurfactants: Secondary metabolites involved in the process of bioremediation and biofilm removal. *Applied Biochemistry and Biotechnology*, 1−27.

Bhatt, P., Verma, A., Gangola, S., Bhandari, G., & Chen, S. (2021). Microbial glycoconjugates in organic pollutant bioremediation: Recent advances and applications. *Microbial Cell Factories*, *20*(1), 1−18.

Bidyarani, N., Jaiswal, J., Shinde, P., & Kumar, U. (2021). Recent developments and future prospects of fungal sophorolipids. *Progress in mycology* (pp. 573−591). Springer.

Bzdek, B. R., Reid, J. P., Malila, J., & Prisle, N. L. (2020). The surface tension of surfactant-containing, finite volume droplets. *Proceedings of the National Academy of Sciences*, *117*(15), 8335−8343.

Cao, S., Davis, A., & Kjellerup, B. V. (2021). Presence of bacteria capable of PCB biotransformation in stormwater bioretention cells. *FEMS Microbiology Ecology*, *97*(12), fiab159.

Cao, X., Cui, X., Xie, M., Zhao, R., Xu, L., Ni, S., & Cui, Z. (2022). Amendments and bioaugmentation enhanced phytoremediation and micro-ecology for PAHs and heavy metals co-contaminated soils. *Journal of Hazardous Materials*, *426*, 128096.

Caragnano, A., Mariani, M., Pizzutilo, F., & Zito, M. (2020). Is it worth reducing GHG emissions? Exploring the effect on the cost of debt financing. *Journal of Environmental Management*, *270*, 110860.

Cetin, B., Yurdakul, S., Gungormus, E., Ozturk, F., & Sofuoglu, S. C. (2018). Source apportionment and carcinogenic risk assessment of passive air sampler-derived PAHs and PCBs in a heavily industrialized region. *Science of the Total Environment*, *633*, 30−41.

Chang, J. S., Radosevich, M., Jin, Y., & Cha, D. K. (2004). Enhancement of phenanthrene solubilization and biodegradation by trehalose lipid biosurfactants. *Environmental Toxicology and Chemistry: An International Journal*, *23*(12), 2816−2822.

Channashettar, V., Srivastava, S., Lal, B., Singh, A., & Rathore, D. (2022). Bioremediation of petroleum hydrocarbons (phc) using biosurfactants. *Microbial surfactants* (pp. 226−240). CRC Press, el biodegradation by bacteria isolates: A review. Journal of Cleaner Production, 251, 119716.

Chen, F., Hao, S., Qu, J., Ma, J., & Zhang, S. (2015). Enhanced biodegradation of polychlorinated biphenyls by defined bacteria-yeast consortium. *Annals of Microbiology*, *65*(4), 1847−1854.

Cheng, M., Chen, D., Parales, R. E., & Jiang, J. (2022). Oxygenases as powerful weapons in the microbial degradation of pesticides. *Annual Review of Microbiology*, *76*.

Christopher, J. M., Sridharan, R., Somasundaram, S., & Ganesan, S. (2021). Bioremediation of aromatic hydrocarbons contaminated soil from industrial site using surface modified amino acid enhanced biosurfactant. *Environmental Pollution*, *289*, 117917.

Da Silva, S., Gonçalves, I., Gomes de Almeida, F. C., Padilha da Rocha e Silva, N. M., Casazza, A. A., Converti, A., & Asfora Sarubbo, L. J. E. (2020). Soil bioremediation: Overview of technologies and trends. *Energies*, *13*(18), 4664.

Dave, B. P., Ghevariya, C. M., Bhatt, J. K., Dudhagara, D. R., & Rajpara, R. K. (2014). Enhanced biodegradation of total polycyclic aromatic hydrocarbons (TPAHs) by marine halotolerant Achromobacter xylosoxidans using Triton X-100 and β-cyclodextrin—A microcosm approach. *Marine Pollution Bulletin*, *79*(1–2), 123–129.

Davin, M., Starren, A., Deleu, M., Lognay, G., Colinet, G., & Fauconnier, M. L. (2018). Could saponins be used to enhance bioremediation of polycyclic aromatic hydrocarbons in aged-contaminated soils? *Chemosphere*, *194*, 414–421.

de Almeida, M., do Nascimento, de Oliveira Mafalda, P., Jr., Patire, V.F., & de Albergaria-Barbosa. (2018). Distribution and sources of polycyclic aromatic hydrocarbons (PAHs) in surface sediments of a Tropical Bay influenced by anthropogenic activities: 137 (pp. 399–407). os Santos Bay, BA, Brazil: Todos.

Debnath, M., Chauhan, N., Sharma, P., & Tomar, I. (2021). Potential of nano biosurfactants as an eco-friendly green technology for bioremediation. *Handbook of nanomaterials for wastewater treatment* (pp. 1039–1055). Elsevier.

Deleu, M., Razafindralambo, H., Popineau, Y., Jacques, P., Thonart, P., & Paquot, M. (1999). Interfacial and emulsifying properties of lipopeptides from Bacillus subtilis. *Colloids and Surfaces A: Physicochemical and Engineering Aspects*, *152*(1–2), 3–10.

Di Gioia, F., & Petropoulos, S. A. (2019). Phytoestrogens, phytosteroids and saponins in vegetables: Biosynthesis, functions, health effects and practical applications. *Advances in food and nutrition research* (90, pp. 351–421). Academic Press.

Drakontis, C. E., & Amin, S. (2020). Biosurfactants: Formulations, properties, and applications. *Current Opinion in Colloid & Interface Science*, *48*, 77–90.

Du, J., & Jing, C. (2018). Anthropogenic PAHs in lake sediments: A literature review (2002–2018). *Environmental Science: Processes & Impacts*, *20*(12), 1649–1666.

Elangovan, S., Pandian, S. B. S., Geetha, S. J., & Joshi, S. J. (2019). Polychlorinated biphenyls (PCBs): Environmental fate, challenges and bioremediation. *Microbial metabolism of xenobiotic compounds* (pp. 165–188). Singapore: Springer.

Erkul, S., & Eker Şanli, G. İ. Z. E. M. (2020). Determination of Polychlorinated Biphenyl (PCB) concentrations of olive groves in spring season. *Journal of the Faculty of Engineering and Architecture of Gazi University*, *35*(2).

Faroon, O., & Ruiz, P. (2016). Polychlorinated biphenyls: New evidence from the last decade. *Toxicology and Industrial Health*, *32*(11), 1825–1847.

Feng, L., Jiang, X., Huang, Y., Wen, D., Fu, T., & Fu, R. (2021). Petroleum hydrocarbon-contaminated soil bioremediation assisted by isolated bacterial consortium and sophorolipid. *Environmental Pollution*, *273*, 116476.

Ferronato, N., & Torretta, V. (2019). Waste mismanagement in developing countries: A review of global issues. *International Journal of Environmental Research and Public Health*, *16*(6), 1060.

Fira, D., Dimkić, I., Berić, T., Lozo, J., & Stanković, S. (2018). Biological control of plant pathogens by Bacillus species. *Journal of Biotechnology*, *285*, 44–55.

Fuentes, M. S., Alvarez, A., Sáez, J. M., Benimeli, C. S., & Amoroso, M. J. (2014). Methoxychlor bioremediation by defined consortium of environmental Streptomyces strains. *International Journal of Environmental Science and Technology*, *11*, 1147–1156.

Gangireddygari, V. S. R., Kalva, P. K., Ntushelo, K., Bangeppagari, M., Djami Tchatchou, A., & Bontha, R. R. (2017). Influence of environmental factors on biodegradation of quinalphos by Bacillus thuringiensis. *Environmental Sciences Europe*, *29*(1), 1–10.

Gong, Z., He, Q., Che, C., Liu, J., & Yang, G. (2020). Optimization and scale-up of the production of rhamnolipid by *Pseudomonas aeruginosa* in solid-state fermentation using high-density polyurethane foam as an inert support. *Bioprocess and Biosystems Engineering*, *43*(3), 385–392.

Gorovtsov, A. V., Sazykin, I. S., Sazykina, M. A. J. E. S., & Research, P. (2018). The influence of heavy metals, polyaromatic hydrocarbons, and polychlorinated biphenyls pollution on the development of antibiotic resistance in soils. *Environmental Science and Pollution Research*, *25*(10), 9238–9292.

Gouda, R. K., Pathak, M., & Khan, M. K. (2020). A biosurfactant as prospective additive for pool boiling heat transfer enhancement. *International Journal of Heat and Mass Transfer*, *150*, 119292.

Goveas, L. C., Selvaraj, R., Vinayagam, R., Alsaiari, A. A., Alharthi, N. S., & Sajankila, S. P. (2022). Nitrogen dependence of rhamnolipid mediated degradation of petroleum crude oil by indigenous Pseudomonas sp. WD23 in seawater. *Chemosphere, 304*, 135235.

Gupta, S., & Pathak, B. (2020). Mycoremediation of polycyclic aromatic hydrocarbons. *Abatement of environmental pollutants* (pp. 127–149). Elsevier.

Hassaan, M. A., & El Nemr, A. (2020). Pesticides pollution: Classifications, human health impact, extraction and treatment techniques. *The Egyptian Journal of Aquatic Research, 46*(3), 207–220.

Hegazy, G. E., Abu-Serie, M. M., Abou-Elela, G. M., Ghozlan, H., Sabry, S. A., Soliman, N. A., & Abdel-Fattah, Y. R. (2022). Bioprocess development for biosurfactant production by Natrialba sp. M6 with effective direct virucidal and anti-replicative potential against HCV and HSV. *Scientific Reports, 12*(1), 1–19.

Herz-Thyhsen, R. J., Miller, Q. R., Rother, G., Kaszuba, J. P., Ashley, T. C., & Littrell, K. C. (2021). Nanoscale Interfacial Smoothing and Dissolution during unconventional reservoir stimulation: Implications for hydrocarbon mobilization and transport. *ACS Applied Materials & Interfaces, 13*(13), 15811–15819.

Heydari, M., Jafari, M. T., Saraji, M., Soltani, R., & Dinari, M. (2021). Covalent triazine-based framework-grafted functionalized fibrous silica sphere as a solid-phase microextraction coating for simultaneous determination of fenthion and chlorpyrifos by ion mobility spectrometry. *Microchimica Acta, 188*(1), 1–11.

Hoang, S. A., Lamb, D., Sarkar, B., Seshadri, B., Lam, S. S., Vinu, A., & Bolan, N. S. (2022). Plant-derived saponin enhances biodegradation of petroleum hydrocarbons in the rhizosphere of native wild plants. *Environmental Pollution, 313*, 120152.

Humphreys, E. H., Janssen, S., Heil, A., Hiatt, P., Solomon, G., & Miller, M. D. (2008). Outcomes of the California ban on pharmaceutical lindane: Clinical and ecologic impacts. *Environmental Health Perspectives, 116*(3), 297–302.

Imron, M. F., Kurniawan, S. B., Ismail, N. I., & Abdullah, S. R. S. (2020). Future challenges in diesel biodegradation by bacteria isolates: A review. *Journal of Cleaner Production, 251*, 119716.

Ines, P., Vlasta, D., Sanja, F., Ana, B. K., Dubravka, H., Fabrice, M. L., & Nikolina, U. K. (2021). Unraveling metabolic flexibility of rhodococci in PCB transformation. *Chemosphere, 282*, 130975.

Jadeja, N. B., Moharir, P., & Kapley, A. (2019). Genome sequencing and analysis of strains Bacillus sp. AKBS9 and Acinetobacter sp. AKBS16 for biosurfactant production and bioremediation. *Applied Biochemistry and Biotechnology, 187*(2), 518–530.

Jahan, R., Bodratti, A. M., Tsianou, M., & Alexandridis, P. (2020). Biosurfactants, natural alternatives to synthetic surfactants: Physicochemical properties and applications. *Advances in Colloid and Interface Science, 275*, 102061.

Janek, T., Krasowska, A., Czyżnikowska, Ż., & èukaszewicz, M. (2018). Trehalose lipid biosurfactant reduces adhesion of microbial pathogens to polystyrene and silicone surfaces: An experimental and computational approach. *Frontiers in Microbiology, 9*, 2441.

Jia, H., Liu, L., Sun, Y., Sun, B., Wang, D., Su, Y., Kannan, K., & Li, Y. F. (2010). Monitoring and modeling endosulfan in Chinese surface soil. *Environmental Science & Technology, 44*(24), 9279–9284. Available from https://doi.org/10.1021/es102791n.

Jing, Q., Chen, L., Zhao, Q., Zhou, P., Li, Y., Wang, H., & Wang, X. (2021). Effervescence-assisted dual microextraction of PAHs in edible oils using lighter-than-water phosphonium-based ionic liquids and switchable hydrophilic/hydrophobic fatty acids. *Analytical and Bioanalytical Chemistry, 413*(7), 1983–1997.

Johnson, P., Trybala, A., Starov, V., & Pinfield, V. J. (2021). Effect of synthetic surfactants on the environment and the potential for substitution by biosurfactants. *Advances in Colloid and Interface Science, 288*, 102340.

Jorgenson, J. L. (2001). Aldrin and dieldrin: A review of research on their production, environmental deposition and fate, bioaccumulation, toxicology, and epidemiology in the United States. *Environmental Health Perspectives, 109*(Suppl 1), 113–139. Available from https://doi.org/10.1289/ehp.01109s1113.

Jouhara, H., Khordehgah, N., Almahmoud, S., Delpech, B., Chauhan, A., & Tassou, S. A. (2018). Waste heat recovery technologies and applications. *Thermal Science and Engineering Progress, 6*, 268–289.

Kaczorek, E., Pacholak, A., Zdarta, A., & Smułek, W. (2018). The impact of biosurfactants on microbial cell properties leading to hydrocarbon bioavailability increase. *Colloids and Interfaces, 2*(3), 35.

Kang, S. W., Kim, Y. B., Shin, J. D., & Kim, E. K. (2010). Enhanced biodegradation of hydrocarbons in soil by microbial biosurfactant, sophorolipid. *Applied Biochemistry and Biotechnology*, *160*(3), 780–790.

Karlapudi, A. P., Venkateswarulu, T. C., Tammineedi, J., Kanumuri, L., Ravuru, B. K., Ramu Dirisala, V., & Kodali, V. P. (2018). Role of biosurfactants in bioremediation of oil pollution – A review. *Petroleum*, *4*(3), 241–249.

Kavitha, D., Sureshkumar, M., & Senthilkumar, B. (2016). Screening of pesticide degrading and biosurfactant producing bacteria from chlorpyrifos contaminated soil. *International Journal of Pharmacy and Biological Sciences*, *7*(3), 525–532.

Klimkowicz-Pawlas, A., Maliszewska-Kordybach, B., & Smreczak, B. (2019). Triad-based screening risk assessment of the agricultural area exposed to the long-term PAHs contamination. *Environmental Geochemistry and Health*, *41*(3), 1369–1385.

Kothuru, A., Singh, A. P., Varaprasad, B. K. S. V. L., & Goel, S. (2020). Plasma TREATMENT and copper metallization for reliable plated-through-holes in microwave PCBS for space electronic packaging. *IEEE Transactions on Components, Packaging and Manufacturing Technology*, *10*(11), 1921–1928.

Kumar, G., Lal, S., Soni, S. K., Maurya, S. K., Shukla, P. K., Chaudhary, P., & Garg, N. (2022). Mechanism and kinetics of chlorpyrifos co-metabolism by using environment restoring microbes isolated from rhizosphere of horticultural crops under subtropics. *Frontiers in Microbiology*, *13*, 2796.

Kumari, A., Kumari, S., Prasad, G. S., & Pinnaka, A. K. (2021). Production of sophorolipid biosurfactant by insect derived novel yeast metschnikowia Churdharensis fa, sp. nov., and its antifungal activity against plant and human pathogens. *Frontiers in Microbiology*, *12*, 678668.

Ku-Vera, J. C., Jiménez-Ocampo, R., Valencia-Salazar, S. S., Montoya-Flores, M. D., Molina-Botero, I. C., Arango, J., & Solorio-Sánchez, F. J. (2020). Role of secondary plant metabolites on enteric methane mitigation in ruminants. *Frontiers in Veterinary Science*, *7*, 584.

Lamilla, C., Schalchli, H., Briceño, G., Leiva, B., Donoso-Piñol, P., Barrientos, L., & Diez, M. C. (2021). A pesticide biopurification system: A source of biosurfactant-producing bacteria with environmental biotechnology applications. *Agronomy*, *11*(4), 624.

Lászlová, K., Dudášová, H., Olejníková, P., Horváthová, G., Velická, Z., Horváthová, H., & Dercová, K. (2018). The application of biosurfactants in bioremediation of the aged sediment contaminated with polychlorinated biphenyls. *Water, Air, & Soil Pollution*, *229*(7), 1–18.

Le, Q. U., Lay, H. L., Wu, M. C., Nguyen, T. H. H., & Nguyen, D. L. (2018). Phytoconstituents and biological activities of Panax vietnamensis (Vietnamese Ginseng): A precious ginseng and call for further research-a systematic review. *Natural Product Communications*, *13*(10), 1934578X1801301036.

Lee, D. H., Steffes, M. W., Sjödin, A., Jones, R. S., Needham, L. L., & Jacobs, D. R., Jr. (2011). Low dose organochlorine pesticides and polychlorinated biphenyls predict obesity, dyslipidemia, and insulin resistance among people free of diabetes. *PloS One*, *6*(1)e15977.

Li, Z., Wang, Y., Guo, S., Li, Z., Xing, Y., Liu, G., & Yan, Y. (2019). PM2. 5 associated PAHs and inorganic elements from combustion of biomass, cable wrapping, domestic waste, and garbage for power generation. *Aerosol and Air Quality Research*, *19*(11), 2502–2517.

Lv, L., Sun, L., Yuan, C., Han, Y., & Huang, Z. (2022). The combined enhancement of RL, nZVI and AQDS on the microbial anaerobic-aerobic degradation of PAHs in soil. *Chemosphere*, *307*, 135609.

Magalhães, K. M., Carreira, R. S., Rosa Filho, J. S., Rocha, P. P., Santana, F. M., & Yogui, G. T. (2022). Polycyclic aromatic hydrocarbons (PAHs) in fishery resources affected by the 2019 oil spill in Brazil: Short-term environmental health and seafood safety. *Marine Pollution Bulletin*, *175*, 113334.

Mahmud, T., Sabo, I. A., Lambu, Z. N., Danlami, D., & Shehu, A. A. (2022). Hydrocarbon degradation potentials of fungi: A review. *Journal of Environmental Bioremediation and Toxicology*, *5*(1), 50–56.

Mazarji, M., Minkina, T., Sushkova, S., Mandzhieva, S., Bayero, M. T., Fedorenko, A., & Soldatov, A. (2022). Metal-organic frameworks (MIL-101) decorated biochar as a highly efficient bio-based composite for immobilization of polycyclic aromatic hydrocarbons and copper in real contaminated soil. *Journal of Environmental Chemical Engineering*, *10*, 108821.

Mendes, A. N., Filgueiras, L. A., Pinto, J. C., & Nele, M. (2015). Physicochemical properties of rhamnolipid biosurfactant from *Pseudomonas aeruginosa* PA1 to applications in microemulsions. *Journal of Biomaterials and Nanobiotechnology*, *6*(01), 64.

Mishra, S., Lin, Z., Pang, S., Zhang, Y., Bhatt, P., & Chen, S. (2021). Biosurfactant is a powerful tool for the bioremediation of heavy metals from contaminated soils. *Journal of Hazardous Materials*, *418*, 126253.

Mokhtari, M., Hajizadeh, Y., Ebrahimi, A. A., Shahi, M. A., Jafari, N., & Abdolahnejad, A. (2019). Enhanced biodegradation of n-hexane from the air stream using rhamnolipid in a biofilter packed with a mixture of compost, scoria, sugar beet pulp and poplar tree skin. *Atmospheric Pollution Research*, *10*(1), 115–122.

Mukherjee, A., Singh, K., & Das, S. (2019). Biodegradation and detoxification of chlorinated pesticide endosulfan by soil microbes. *Uses, toxicological profile and regulation* (pp. 247–274). Nova Science Publishers.

Mulliken, D. L., Zambone, J. D., & Rolph, C. G. (2004). DDT: A persistent lifesaver. *Nat. Resources & Env't.*, *19*, 3.

Multigner, L., Kadhel, P., Rouget, F., Blanchet, P., & Cordier, S. (2016). Chlordecone exposure and adverse effects in French West Indies populations. *Environmental Science and Pollution Research*, *23*(1), 3–8.

Multigner, L., Ndong, J. R., Giusti, A., Romana, M., Delacroix-Maillard, H., Cordier, S., ... Blanchet, P. (2010). Chlordecone exposure and risk of prostate cancer. *Journal of Clinical Oncology*, *28*(21), 3457–3462.

Nabi, G., Ali, M., Khan, S., & Kumar, S. (2019). The crisis of water shortage and pollution in Pakistan: Risk to public health, biodiversity, and ecosystem. *Environmental Science and Pollution Research*, *26*, 10443–10445.

Nair, A. M., Rebello, S., Rishad, K. S., Asok, A. K., & Jisha, M. S. (2015). Biosurfactant facilitated biodegradation of quinalphos at high concentrations by *Pseudomonas aeruginosa* Q10. *Soil and Sediment Contamination: An International Journal*, *24*(5), 542–553.

Nam, K. J., Li, Q., Heo, S. K., Tariq, S., Loy-Benitez, J., Woo, T. Y., & Yoo, C. K. (2021). Inter-regional multimedia fate analysis of PAHs and potential risk assessment by integrating deep learning and climate change scenarios. *Journal of Hazardous Materials*, *411*, 125149.

Naughton, P. J., Marchant, R., Naughton, V., & Banat, I. M. (2019). Microbial biosurfactants: Current trends and applications in agricultural and biomedical industries. *Journal of Applied Microbiology*, *127*(1), 12–28.

Nguyen, T. H., Sahin, O., & Howes, M. (2021). Climate change adaptation influences and barriers impacting the Asian agricultural industry. *Sustainability*, *13*(13), 7346.

Ni'Fatimah., & Sumarsih, S. (2017). Biodegradation of naphthalene and phenanthren by *Bacillus subtilis* 3KP. *AIP Conference Proceedings*, *1854*(1), 020026, AIP Publishing LLC.

Nikolova, C., Ijaz, U., Magill, C., Kleindienst, S., Joye, S. B., & Gutierrez, T. (2021). Rhamnolipid, a naturally produced oil dispersant, may improve oil spill remediation.

Nzila, A. (2018). Biodegradation of high-molecular-weight polycyclic aromatic hydrocarbons under anaerobic conditions: Overview of studies, proposed pathways and future perspectives. *Environmental Pollution*, *239*, 788–802.

Obayori, O. S., & Salam, L. B. (2010). Degradation of polycyclic aromatic hydrocarbons: Role of plasmids. *Scientific Research and Essays*, *5*(25), 4093–4106.

Oliva, A., Teruel, J. A., Aranda, F. J., & Ortiz, A. (2020). Effect of a dirhamnolipid biosurfactant on the structure and phase behaviour of dimyristoylphosphatidylserine model membranes. *Colloids and Surfaces B: Biointerfaces*, *185*, 110576.

Ostapska, H., Howell, P. L., & Sheppard, D. C. (2018). Deacetylated microbial biofilm exopolysaccharides: It pays to be positive. *PLoS Pathogens*, *14*(12), e1007411.

Oyewusi, H. A., Wahab, R. A., & Huyop, F. (2020). Dehalogenase-producing halophiles and their potential role in bioremediation. *Marine Pollution Bulletin*, *160*, 111603.

Park, H. S., Han, Y. S., & Park, J. H. (2019). Massive recycling of waste mobile phones: Pyrolysis, physical treatment, and pyrometallurgical processing of insoluble residue. *ACS Sustainable Chemistry & Engineering*, *7*(16), 14119–14125.

Pathiraja, G., Egodawatta, P., Goonetilleke, A., & Te'o, V. S. J. (2019). Solubilization and degradation of polychlorinated biphenyls (PCBs) by naturally occurring facultative anaerobic bacteria. *Science of the Total Environment*, *651*, 2197–2207.

Phulpoto, I. A., Hu, B., Wang, Y., Ndayisenga, F., Li, J., & Yu, Z. (2021). Effect of natural microbiome and culturable biosurfactants-producing bacterial consortia of freshwater lake on petroleum-hydrocarbon degradation. *Science of the Total Environment*, *751*, 141720.

Pirsaheb, M., Limoee, M., Namdari, F., & Khamutian, R. (2015). Organochlorine pesticides residue in breast milk: A systematic review. *Medical Journal of the Islamic Republic of Iran*, *29*, 228.

Płaza, G., & Achal, V. (2020). Biosurfactants: Eco-friendly and innovative biocides against biocorrosion. *International Journal of Molecular Sciences*, *21*(6), 2152.

Pourfadakari, S., Ahmadi, M., Jaafarzadeh, N., Takdastan, A., Ghafari, S., & Jorfi, S. (2019). Remediation of PAHs contaminated soil using a sequence of soil washing with biosurfactant produced by Pseudomonas aeruginosa strain PF2 and electrokinetic oxidation of desorbed solution, effect of electrode modification with Fe3O4 nanoparticles. *Journal of Hazardous Materials*, *379*, 120839.

Preda, V. G., & Săndulescu, O. (2019). Communication is the key: Biofilms, quorum sensing, formation and prevention. *Discoveries*, *7*(3).

Premnath, N., Mohanrasu, K., Rao, R. G. R., Dinesh, G. H., Prakash, G. S., Pugazhendhi, A., & Arun, A. (2021). Effect of C/N substrates for enhanced extracellular polymeric substances (EPS) production and Poly Cyclic Aromatic Hydrocarbons (PAHs) degradation. *Environmental Pollution*, *275*, 116035.

Purnomo, A. S., Ashari, K., & Hermansyah, F. T. (2017). Evaluation of the synergistic effect of mixed cultures of white-rot fungus Pleurotus ostreatus and biosurfactant-producing bacteria on DDT biodegradation. *Journal of Microbiology and Biotechnology*, *27*(7), 1306−1315.

Rabodonirina, S., Rasolomampianina, R., Krier, F., Drider, D., Merhaby, D., Net, S., & Ouddane, B. (2019). Degradation of fluorene and phenanthrene in PAHs-contaminated soil using Pseudomonas and Bacillus strains isolated from oil spill sites. *Journal of Environmental Management*, *232*, 1−7.

Rawat, G., Dhasmana, A., & Kumar, V. (2020). Biosurfactants: The next generation biomolecules for diverse applications. *Environmental Sustainability*, *3*(4), 353−369.

Reddy, A. V. B., Moniruzzaman, M., & Aminabhavi, T. M. (2019). Polychlorinated biphenyls (PCBs) in the environment: Recent updates on sampling, pretreatment, cleanup technologies and their analysis. *Chemical Engineering Journal*, *358*, 1186−1207.

Reeves, W. R., McGuire, M. K., Stokes, M., & Vicini, J. L. (2019). Assessing the safety of pesticides in food: How current regulations protect human health. *Advances in Nutrition*, *10*(1), 80−88.

Ren, H., Ding, Y., Hao, X., Hao, J., Liu, J., & Wang, Y. (2022). Enhanced rhizoremediation of polychlorinated biphenyls by resuscitation-promoting factor stimulation linked to plant growth promotion and response of functional microbial populations. *Chemosphere*, *309*, 136519.

Rivera-Pérez, A., Romero-González, R., & Garrido Frenich, A. (2022). Persistent organic pollutants (PCBs and PCDD/Fs), PAHs, and plasticizers in spices, herbs, and tea − A review of chromatographic methods from the last decade (2010−2020). *Critical Reviews in Food Science and Nutrition*, *62*(19), 5224−5244.

Rocha, M. J., Ribeiro, A. B., Campos, D., & Rocha, E. (2021). Temporal-spatial survey of PAHs and PCBs in the Atlantic Iberian northwest coastline, and evaluation of their sources and risks for both humans and aquatic organisms. *Chemosphere*, *279*, 130506.

Rojas, S., & Horcajada, P. (2020). Metal−organic frameworks for the removal of emerging organic contaminants in water. *Chemical Reviews*, *120*(16), 8378−8415.

Rong, L., Zheng, X., Oba, B. T., Shen, C., Wang, X., Wang, H., ... Sun, L. (2021). Activating soil microbial community using bacillus and rhamnolipid to remediate TPH contaminated soil. *Chemosphere*, *275*, 130062.

Sahoo, B. M., Ravi Kumar, B. V., Banik, B. K., & Borah, P. (2020). Polyaromatic hydrocarbons (PAHs): Structures, synthesis and their biological profile. *Current Organic Synthesis*, *17*(8), 625−640.

Sajadi Bami, M., Raeisi Estabragh, M. A., Ohadi, M., Banat, I. M., & Dehghannoudeh, G. (2022). Biosurfactants aided bioremediation mechanisms: A mini-review. *Soil and Sediment Contamination: An International Journal*, *31*(7), 801−817.

Sandhu, M., Paul, A. T., Proćków, J., & Jha, P. N. (2022). PCB-77 biodegradation potential of biosurfactant producing bacterial isolates recovered from contaminated soil. *Frontiers in Microbiology*, *13*, 3467.

Sangwan, S., Kaur, H., Sharma, P., Sindhu, M., & Wati, L. (2022). Routing microbial biosurfactants to agriculture for revitalization of soil and plant growth. *New and future developments in microbial biotechnology and bioengineering* (pp. 313–338). Elsevier.

Santucci, L., Carol, E., & Tanjal, C. (2018). Industrial waste as a source of surface and groundwater pollution for more than half a century in a sector of the Río de la Plata coastal plain (Argentina). *Chemosphere, 206*, 727–735.

Sariwati, A., & Purnomo, A. S. (2018). The effect of *Pseudomonas aeruginosa* addition on 1, 1, 1-trichloro-2, 2-bis (4-chlorophenyl) ethane (DDT) biodegradation by brown-rot fungus fomitopsis pinicola. *Indonesian Journal of Chemistry, 18*(1), 75–81.

Sariwati, A., Purnomo, A. S., & Kamei, I. (2017). Abilities of co-cultures of brown-rot fungus fomitopsis pinicola and *Bacillus subtilis* on biodegradation of DDT. *Current Microbiology, 74*(9), 1068–1075.

Seeger, M., Timmis, K. N., & Hofer, B. (1997). Bacterial pathways for the degradation of polychlorinated biphenyls. *Marine Chemistry, 58*(3–4), 327–333.

Semu, E., Tindwa, H., & Singh, B. R. (2019). Heavy metals and organopesticides: Ecotoxicology, health effects and mitigation options with emphasis on Sub-Saharan Africa. *Journal of Toxicology Current Research, 3*(010).

Shah, K. K., Tripathi, S., Tiwari, I., Shrestha, J., Modi, B., Paudel, N., & Das, B. D. (2021). Role of soil microbes in sustainable crop production and soil health: A review. *Agricultural Science & Technology (1313–8820), 13*(2).

Shahsavari, E., Schwarz, A., Aburto-Medina, A., & Ball, A. S. (2019). Biological degradation of polycyclic aromatic compounds (PAHs) in soil: A current perspective. *Current Pollution Reports, 5*(3), 84–92.

Shen, Y., Zou, Y., Chen, X., Li, P., Rao, Y., Yang, X., & Hu, H. (2020). Antibacterial self-assembled nanodrugs composed of berberine derivatives and rhamnolipids against *Helicobacter pylori*. *Journal of Controlled Release, 328*, 575–586.

Shin, D., Gorgulla, C., Boursier, M. E., Rexrode, N., Brown, E. C., Arthanari, H., & Nagarajan, R. (2019). N-acyl homoserine lactone analog modulators of the Pseudomonas aeruginosa RhlI quorum sensing signal synthase. *ACS Chemical Biology, 14*(10), 2305–2314.

Siddiqui, Z., Anas, M., Khatoon, K., & Malik, A. (2021). Biosurfactant-producing bacteria as potent scavengers of petroleum hydrocarbons. *Microbiomes and the global climate change* (pp. 321–348). Singapore: Springer.

Singh, P. B., Sharma, S., Saini, H. S., & Chadha, B. S. (2009). Biosurfactant production by Pseudomonas sp. and its role in aqueous phase partitioning and biodegradation of chlorpyrifos. *Letters in Applied Microbiology, 49*(3), 378–383.

Singh, P., Saini, H. S., & Raj, M. (2016). Rhamnolipid mediated enhanced degradation of chlorpyrifos by bacterial consortium in soil-water system. *Ecotoxicology and Environmental Safety, 134*, 156–162.

Smułek, W., Sydow, M., Zabielska-Matejuk, J., & Kaczorek, E. (2020). Bacteria involved in biodegradation of creosote PAH–A case study of long-term contaminated industrial area. *Ecotoxicology and Environmental Safety, 187*, 109843.

Smułek, W., Zdarta, A., èuczak, M., Krawczyk, P., Jesionowski, T., & Kaczorek, E. (2016). Sapindus saponins' impact on hydrocarbon biodegradation by bacteria strains after short-and long-term contact with pollutant. *Colloids and Surfaces B: Biointerfaces, 142*, 207–213.

Steliga, T., Wojtowicz, K., Kapusta, P., & Brzeszcz, J. (2020). Assessment of biodegradation efficiency of polychlorinated biphenyls (PCBs) and petroleum hydrocarbons (TPH) in soil using three individual bacterial strains and their mixed culture. *Molecules (Basel, Switzerland), 25*(3), 709.

Suganthi, S. H., Murshid, S., Sriram, S., & Ramani, K. (2018). Enhanced biodegradation of hydrocarbons in petroleum tank bottom oil sludge and characterization of biocatalysts and biosurfactants. *Journal of Environmental Management, 220*, 87–95.

Sun, S., Wang, Y., Zang, T., Wei, J., Wu, H., Wei, C., & Li, F. (2019). A biosurfactant-producing *Pseudomonas aeruginosa* S5 isolated from coking wastewater and its application for bioremediation of polycyclic aromatic hydrocarbons. *Bioresource Technology, 281*, 421–428.

Swaathy, S., Kavitha, V., Pravin, A. S., Mandal, A. B., & Gnanamani, A. (2014). Microbial surfactant mediated degradation of anthracene in aqueous phase by marine Bacillus licheniformis MTCC 5514. *Biotechnology Reports, 4*, 161–170.

Tang, W., Ji, H., & Hou, X. (2018). Research progress of microbial degradation of organophosphorus pesticides. *Progress in Applied Microbiology, 1*(1).

Terhaar, J., Frölicher, T. L., & Joos, F. (2021). Southern Ocean anthropogenic carbon sink constrained by sea surface salinity. *Science Advances, 7*(18), eabd5964.

Thi, Mo, L., Irina, P., Natalia, S., Irina, N., Lenar, A., Andrey, F., & Olga, P. (2022). Hydrocarbons biodegradation by rhodococcus: Assimilation of hexadecane in different aggregate states. *Microorganisms, 10*(8), 1594.

Toribio, J., Escalante, A. E., & Soberón-Chávez, G. (2010). Rhamnolipids: Production in bacteria other than Pseudomonas aeruginosa. *European Journal of Lipid Science and Technology, 112*(10), 1082–1087.

Tucker, I. M., Burley, A., Petkova, R. E., Hosking, S. L., Thomas, R. K., Penfold, J., & Welbourn, R. (2020). Surfactant/biosurfactant mixing: Adsorption of saponin/nonionic surfactant mixtures at the air-water interface. *Journal of Colloid and Interface Science, 574*, 385–392.

Vaishnavi, J., Parthipan, P., Arul Prakash, A., Sathishkumar, K., & Rajasekar, A. (2021). Biosurfactant-assisted bioremediation of crude oil/petroleum hydrocarbon contaminated soil. *Handbook of Assisted and Amendment: Enhanced Sustainable Remediation Technology* (pp. 313–329). Wiley.

Wang, B., Wang, Q., Liu, W., Liu, X., Hou, J., Teng, Y., & Christie, P. (2017). Biosurfactant-producing microorganism Pseudomonas sp. SB assists the phytoremediation of DDT-contaminated soil by two grass species. *Chemosphere, 182*, 137–142.

Wang, L., Yu, G., Li, J., Feng, Y., Peng, Y., Zhao, X., & Zhang, Q. (2019). Stretchable hydrophobic modified alginate double-network nanocomposite hydrogels for sustained release of water-insoluble pesticides. *Journal of Cleaner Production, 226*, 122–132.

Walker, K., Vallero, D. A., & Lewis, R. G. (1999). Factors influencing the distribution of lindane and other hexachlorocyclohexanes in the environment. *Environmental Science & Technology, 33*(24), 4373–4378.

Wallace, D. R. (2014). Toxaphene. In *Encyclopedia of Toxicology*, (Third Edition, pp. 606–609). Elsevier. Available from https://doi.org/10.1016/B978-0-12-386454-3.00202-5.

Wang, Q., Peng, F., Chen, Y., Jin, L., Lin, J., Zhao, X., & Li, J. Y. (2019). Heavy metals and PAHs in an open fishing area of the East China Sea: Multimedia distribution, source diagnosis, and dietary risk assessment. *Environmental Science and Pollution Research, 26*(21), 21140–21150.

Wassenaar, P. N., & Verbruggen, E. M. (2021). Persistence, bioaccumulation and toxicity-assessment of petroleum UVCBs: A case study on alkylated three-ring PAHs. *Chemosphere, 276*, 130113.

Xiao-Hong, P. E. I., Xin-Hua, Z. H. A. N., Shi-Mei, W. A. N. G., Yu-Suo, L. I. N., & Li-Xiang, Z. H. O. U. (2010). Effects of a biosurfactant and a synthetic surfactant on phenanthrene degradation by a Sphingomonas strain. *Pedosphere, 20*(6), 771–779.

Yan, N., An, M., Chu, J., Cao, L., Zhu, G., Wu, W., & Rittmann, B. E. (2021). More rapid dechlorination of 2, 4-dichlorophenol using acclimated bacteria. *Bioresource Technology, 326*, 124738.

Yang, J., Qadeer, A., Liu, M., Zhu, J. M., Huang, Y. P., Du, W. N., & Wei, X. Y. (2019). Occurrence, source, and partition of PAHs, PCBs, and OCPs in the multiphase system of an urban lake, Shanghai. *Applied Geochemistry, 106*, 17–25.

Yang, K., Zhao, Y., Ji, M., Li, Z., Zhai, S., Zhou, X., & Liang, B. (2021). Challenges and opportunities for the biodegradation of chlorophenols: Aerobic, anaerobic and bioelectrochemical processes. *Water Research, 193*, 116862.

Yang, Y., Zhang, J., Liu, Y., Wu, L., Li, Q., Xu, M., & Yang, X. (2022). pH-dependent micellar properties of edible biosurfactant steviol glycosides and their oil-water interfacial interactions with soy proteins. *Food Hydrocolloids, 126*, 107476.

Yazdi, F., Shoeibi, S., Yazdi, M. H., & Eidi, A. (2021). Effect of prevalent polychlorinated biphenyls (PCBs) food contaminant on the MCF7, LNCap and MDA-MB-231 cell lines viability and PON1 gene expression level: Proposed model of binding. *DARU. Journal of Pharmaceutical Sciences, 29*(1), 159–170.

Yesankar, P. J., Pal, M., Patil, A., & Qureshi, A. (2023). Microbial exopolymeric substances and biosurfactants as 'bioavailability enhancers' for polycyclic aromatic hydrocarbons biodegradation. *International Journal of Environmental Science and Technology, 20*, 5823–5844.

Yongming, L. U. O., Qian, Z. H. O. U., Haibo, Z. H. A. N. G., Xiangliang, P. A. N., Chen, T. U., Lianzhen, L. I., & Jie, Y. A. N. G. (2018). Pay attention to research on microplastic pollution in soil for prevention of ecological and food chain risks. *Bulletin of Chinese Academy of Sciences (Chinese Version)*, *33*(10), 1021–1030.

Yu, H., Huang, G. H., Xiao, H., Wang, L., & Chen, W. (2014). Combined effects of DOM and biosurfactant enhanced biodegradation of polycylic armotic hydrocarbons (PAHs) in soil–water systems. *Environmental Science and Pollution Research*, *21*(17), 10536–10549.

Zan, S., Wang, J., Fan, J., Jin, Y., Li, Z., & Du, M. (2022). Cyclohexanecarboxylic acid degradation with simultaneous nitrate removal by Marinobacter sp. SJ18. *Environmental Science and Pollution Research*, *30*(12), 34296–34305.

Zang, T., Wu, H., Yan, B., Zhang, Y., & Wei, C. (2021). Enhancement of PAHs biodegradation in biosurfactant/phenol system by increasing the bioavailability of PAHs. *Chemosphere*, *266*, 128941.

Zaynab, M., Sharif, Y., Abbas, S., Afzal, M. Z., Qasim, M., Khalofah, A., & Li, S. (2021). Saponin toxicity as key player in plant defense against pathogens. *Toxicon*, *193*, 21–27.

Zhang, Z., Sun, J., Guo, H., Wang, C., Fang, T., Rogers, M. J., & Wang, H. (2021). Anaerobic biodegradation of phenanthrene by a newly isolated nitrate-dependent Achromobacter denitrificans strain PheN1 and exploration of the biotransformation processes by metabolite and genome analyses. *Environmental Microbiology*, *23*(2), 908–923.

Zhao, F., Han, S., & Zhang, Y. (2020). Comparative studies on the structural composition, surface/interface activity and application potential of rhamnolipids produced by Pseudomonas aeruginosa using hydrophobic or hydrophilic substrates. *Bioresource Technology*, *295*, 122269.

Zhou, P., Xie, W., Luo, Y., Lu, S., Dai, Z., Wang, R., & Sun, X. (2018). Protective effects of total saponins of Aralia elata (Miq.) on endothelial cell injury induced by TNF-α via modulation of the PI3K/Akt and NF-κB signalling pathways. *International Journal of Molecular Sciences*, *20*(1), 36.

Zhu, Z., Zhang, B., Cai, Q., Cao, Y., Ling, J., Lee, K., & Chen, B. (2021). A critical review on the environmental application of lipopeptide micelles. *Bioresource Technology*, *339*, 125602.

Zhuk, T. S., Babkina, V. V., & Zorn, H. (2022). Aerobic C−C bond cleavage catalyzed by whole-cell cultures of the white-rot fungus dichomitus albidofuscus. *ChemCatChem*, *14*(3), e202101408.

CHAPTER 8

Strategic biosurfactant-advocated bioremediation technologies for the removal of petroleum derivatives and other hydrophobic emerging contaminants

Swathi Krishnan Venkatesan, Raja Rajeswari Devi Mandava, Venkat Ramanan Srinivasan, Megha Prasad and Ramani Kandasamy
Industrial and Environmental Sustainability Laboratory, Department of Biotechnology, School of Bioengineering, SRM Institute of Science and Technology, Kattankulathur, Chengalpattu, Tamil Nadu, India

8.1 Introduction

Today's fast-developing economies largely rely on petroleum and its derivatives to fulfill the ever-increasing demand for energy. As a result, these petroleum derivatives are released into the environment through petrochemical industries and refineries, along with several natural and anthropogenic sources from processes such as exploration, transportation, refining, leakages from oil pipelines, and storage tanks (Adipah, 2018; Gawdzik & Gawdzik, 2011). They mainly comprise saturated, unsaturated, aromatic, alicyclic, naphthalene, polyaromatic compounds, and cycloalkanes, making petroleum hydrocarbons highly hydrophobic and recalcitrant (Islam, 2015). Furthermore, they exhibit toxic, carcinogenic, and mutagenic effects on organisms, including humans, and have serious ecological consequences, making them highly concerning contaminants (Deivakumari et al., 2020). Though effective, conventional technologies, such as landfilling, incineration, gasification, and pyrolysis, are limited due to their high-energy requirements, high process costs, and possible secondary pollution (Patowary et al., 2017). Bioremediation, an environmental-friendly and cost-effective substitute, is extensively used to mitigate petroleum-contaminated sites aided by microbes and plants. Biosurfactant-based bioremediation is gaining traction due to its sustainablility, innocuity, cost-effectiveness, and higher functioning even under severe conditions, in addition to the fact that it can be directly produced from sustainable resources aided by microbial fermentation (Barin et al., 2014). Biosurfactants are highly specific, surface-active, bio-amphiphilic molecules that can enhance the pollutants' bioavailability for microbial utilization by augmenting the surface area and pseudosolubility of the substrates, thus overcoming the major shortcomings of conventional technologies (Karlapudi et al., 2018). Due to their ability to function as emulsifying, dispersing, solubilizing, and mobilizing agents, they are increasingly being utilized in the form of either purified biomolecules or biosurfactant-producing microbes for the remediation of contaminated soil and waterbodies, oils spills, storage tank oil residues, and they play a significant role in enhancing oil recovery with microbes (Rizvi et al., 2021). However, the major stumbling block for the commercialization of biosurfactants is their low yields and high production costs. Therefore, modern strategies and technologies, such as nanotechnology, genetic engineering, omics approaches, and bioinformatics, can be implemented for improved yields. However, knowledge about the mechanisms of biosurfactant production and their mode of action on organic/hydrophobic substrates is very limited.

This chapter discusses the prevalence of HCPDs in the environment and their harmful effects and how the unique properties of biosurfactants are utilized for enhanced bioremediation strategies of petroleum and its various derivatives, followed by their production economics and bottlenecks with an added note on the nanotechnological and omics approaches for the enhanced production and application of biosurfactants.

8.2 Environmental prevalence of hydrophobic contaminants/petroleum derivatives

Petroleum derivatives are a class of anthropogenic compounds, including alkanes, aromatic hydrocarbons, resins, asphaltenes, and other organic matter. They are ubiquitous and commonly detected in air, water, and soil and are proven to be highly hazardous, teratogenic, and mutagenic even at low concentrations. Thus, petroleum derivatives have been classified as priority pollutants by the United States Environmental Protection Agency (USEPA) (Wang et al., 2018). Accidental oil spills, petroleum extraction processes, storage, and transportation are the major reasons for HCPD exposure to the environment Fig. 8.1. In petroleum industries, petroleum is subjected to various treatment processes, such as isomerization, thermal cracking, catalytic treatment, desalting, exploration, and distillation, to produce its highly competent derivatives (Gawdzik & Gawdzik, 2011). The amount of water required in all these processes is so huge that the wastewater generated almost adds up to 0.4−1.6 times the quantity of the actual product synthesized in the industry (Salehi et al., 2021). The discharged wastewater from these petroleum refineries and petrochemical plants contains large amounts of toxic organic compounds such as benzene, toluene, phenolics, xylenes, and a mixture of PHCs. The total global emission of PHCs was found to be >500 Gg per year (Gg/y), with the highest by China (114 Gg/y), followed by India (90 Gg/y), and the USA (32 Gg/y) (Zhang, Liu, et al., 2020). HCPDs found in the

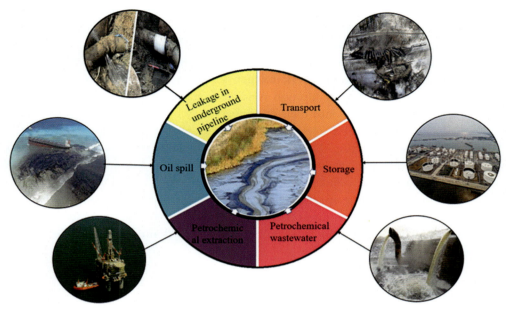

Figure 8.1 Hydrophobic pollutants environmental prevalence of hydrophobic pollutants/petroleum derivatives.

atmosphere (such as polyaromatic hydrocarbons (PAHs) from automobile exhausts) are deposited on the earth's surface by dry and wet depositions and observed to settle on the surface of lakes, streams, and oceans and dispersed by currents and ineradicably accumulate in the sediments (Abdel-Shafy & Mansour, 2016). In tropical climates, these sediments act as sink and reemission sources. The accumulated toxic effects of HCPDs subsequently lead to the poisoning of sediments, affecting the structure of benthic communities and thereby disrupting the ecosystem's balance (Charles et al., 2012). In the case of marine environments, the distribution of HCPDs depends on factors such as the availability, rate of release, sections of trapped or escaped chemicals, and degradation pace of processes (Aghadadashi et al., 2019). They have adverse effects on exposure to the aquatic ecosystem, impacting the biodiversity of its phytoplankton, invertebrate communities, fishes, seabirds, and marine mammals. The presence of these hydrophobic wastes leads to the reduced dissolved oxygen level in the water, leading to increased mortality in the aquatic species. These wastes are persistent and highly soluble in water and therefore pose the threat of infiltrating the food chain starting from the marine fauna, subsequently biomagnifying at consecutive trophic levels and also penetrating the groundwater and eventually into the soil ecosystem, affecting not only the biological balance of the soil but also its productivity (Kumar et al., 2022). Soil is considered a major sink for HCPDs, particularly for PAHs (approximately 50%–90%) in the environment (Ma et al., 2022). The presence of HCPDs changes the organic composition of the soil, unbalancing the organic carbon-to-nitrogen and carbon-to-phosphorus ratio and affecting the development of biological life. They disturb the soil's colloidal properties, such as ion exchange and pH, leading to soil acidification. Furthermore, HCPD induces the formation of emulsified layers on the soil surfaces, interrupting the supply of atmospheric oxygen to the fauna and flora of the polluted area. They penetrate as vapors through respiratory tracts, into their alimentary canal through food and water, and into their skin where they build up in the adipose tissues, and subsequently get released into the bloodstream and get precipitated in the body fluids, thus adversely affecting the nervous and reproductive systems. Their toxicity is attributed to the HCPDs' ability to disintegrate enzyme functioning by binding to nonpolar constituents of cellular membranes. For instance, diesel fuel disrupts the nitrogen cycle, this limiting species richness, distribution, and phylogenetic diversity. The immediate result of this disruption was reduced species diversity and functional genes about nitrification, leading to a community heavily dominated by species such as *Pseudomonas* (Xu et al., 2018).

8.3 Toxic impacts of petroleum hydrocarbons

HCPD present in contaminated environments ultimately reaches humans through three main pathways: dermal contact, oral ingestion, and inhalation. The metabolic degradation of these lipophilic hydrocarbon components in the living systems produces several neuro-, hepato-, nephrotoxic alcohols. Their highly lipophilic nature enables them to cross the blood—brain barrier, causing various neuropsychiatric effects, including disorders such as dementia, motor, and cognitive dysfunction, reduced impulsive control, and hallucination (Rajendran et al., 2022). Hydrocarbon-derived epoxides destabilize and distort the structure of nucleic acids and proteins and adversely affect cell mitosis and, thereby, mutating the genetic material (Gospodarek et al., 2021). On exposure to the mixture of PHCs, volatile organic compounds (VOCs), polyaromatic hydrocarbons (PAHs), and other petroleum compounds, humans have serious physical and mental health repercussions, notably kidney and liver damage, asthma, inflammation, dizziness, headache, irritation, cardiovascular disorders, and other nervous, immune, reproductive, and developmental deformities (Singh et al., 2020). Moreover, they are highly carcinogenic, especially to the lung, skin, and liver. Appallingly, studies have revealed that polyaromatic PHCs present in the environment cause up to $\sim 80\%$ of all cancers that occur in animals, proving to be the biggest group of carcinogens (Ambade et al., 2021). Prolonged human exposure to benzene and its derivatives, which lead to bone marrow mutilation, consecutively cause aplastic anemia and induce leukemia and lymphoma, and in most cases, pancytopenia (i.e., reduced production of blood cells of all types). PHCs, furthermore, are known to have a narcotizing action, exerting a paralyzing effect on the central nervous system. Extended exposure to these pollutants was also proved to contribute to the development of neurodegenerative diseases, such as Alzheimer's and Parkinson's disease, and autism spectrum disorders (Rajendran et al., 2022).

8.4 Conventional treatment technologies for the mitigation of hydrophobic contaminants/petroleum derivatives

The threatening effects of HCPD raise alarm for an urgent need for an effective treatment strategy to eradicate these lethal contaminants from the environment. Many conventional physicochemical techniques such as adsorption, photocatalytic degradation, landfilling, incineration, gasification, pyrolysis, photocatalytic combustion, ultrasonic treatment, supercritical water oxidation (SCWO) technology, microwave radiation technology, and the use of sorbents and chemical surfactants are practiced. However, their application is limited by their energy intensiveness, high expenditures, tedious operations, the need for specialized equipment and skilled labor, and most importantly, secondary pollution caused by the conversion of the pollutants from one phase to another instead of its complete removal (Rabani et al., 2020; Swathi et al., 2020). Consequently, there is a dire need for a robust, ecological, and sustainable technology to mitigate PHCs.

Over decades, bioremediation technology has utilized indigenous or specialized microbes, that is, bacteria, fungi, algae, and yeast, for degrading petroleum compounds present in the environment. These hydrocarbonoclastic microorganisms use HCPD as a carbon and energy source for their growth by secreting catalytic biomolecules, such as alkane oxidizing enzymes, as a survival mechanism in extreme environments (Khalid et al., 2021). Bioremediation has been widely and successfully applied for the treatment of industrial effluents, chemical spills, leakages, and several other environmental applications. Despite being effective, the major restraint of the bioremediation process is the longer treatment duration due to the poor availability of hydrophobic substrates, which reduces their microbial uptake, subsequently resulting in the slackened metabolic degradation of HCPD (Dell' Anno et al., 2021). This can be addressed by the application of surface-active biomolecules such as biosurfactants, which are microbially produced amphiphilic molecules with unique surface-active properties that facilitate the reduction of hydrophobicity of PHCs, thus making them bioavailable for the microbes to degrade (Kaczorek et al., 2018). Hence, biosurfactant-aided bioremediation could be a highly valuable option for the effective treatment of toxic PHCs in a very short time.

8.5 Green surfactants: bio-based surfactants and biosurfactants

Earlier, synthetic surfactants have been extensively used to manage various incidences of HCPD contamination. Even though synthetic surfactants are cost-effective, they are toxic and recalcitrant in nature, which makes them unsuitable for environmental applications. Not only are the biosurfactants more eco-friendly than their chemical counterparts, but they also have a comparatively greater capacity to solubilize hydrophobic pollutants, proving their potential in the bioremediation of HCPD (Carolin et al., 2021). Extensive research concluded that the green surfactant was superior to its synthetic counterparts in terms of its ability to eliminate pollutants more sustainably. Hence, the use of green surfactants has recently attracted immense interest due to their viable approach and eco-friendly nature.

8.5.1 Classification of biosurfactants

Biosurfactants, the 21st century's multifaceted biomolecules, have broad scope in various fields, including household, pharmaceutical, food, medical, and environmental protection. Biosurfactants are being sustainably produced using renewable substrates such as agro-industrial and oily wastes with various bacterial, viral, fungal, and yeast strains. They are dynamic surface-active molecules with two different moieties—hydrophobic tail and hydrophilic head—thereby acting on hydrophilic and hydrophobic surfaces. The major constituents of biosurfactants are carbohydrates (mono or oligo), amino acids, and proteins on the hydrophilic part and lipids, fatty acids, and alkyl derivatives on the hydrophobic part. The activity of biosurfactants depends on their chemical composition, based on which biosurfactants are classified as glycopeptides, lipopeptides, phospholipids, and high-molecular-weight biosurfactants, among which glycolipids are extensively studied. The biosurfactants classified based on their molecular charge and weight are represented in Fig. 8.2.

8.5.2 Properties of biosurfactants

8.5.2.1 Critical micelle concentration

One of the significant properties defining biosurfactants is critical micelle formation. Critical micelle concentration (CMC) can be defined as the surfactant concentration in the bulk phase above which micellization occurs. As biosurfactants start accumulating on the surface, it reaches the threshold for CMC, and thus, micelle formation occurs at this point. Efficient biosurfactants have a low CMC value, indicating a minimal quantity of molecules required to efficiently decrease the surface tension Fig. 8.3. The biosurfactant mixtures are pooled at the CMC critical micelle concentration and the following observation was found. Due to hydrophobic and Van der Waals interaction, the micelles organize themselves in a way, the encapsulating hydrophobic molecules forming an emulsion with water. It was observed that the concentration of micelle formation is higher in the soil in comparison to the aqueous system (Li et al., 2014).

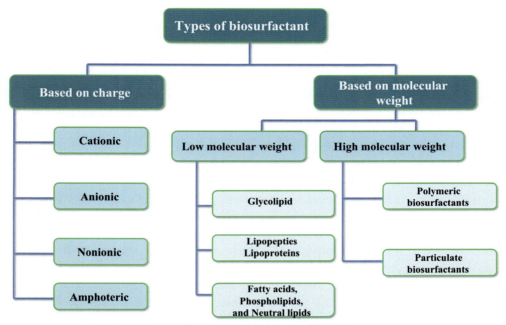

Figure 8.2 Types of biosurfactants: flowchart classification of biosurfactants.

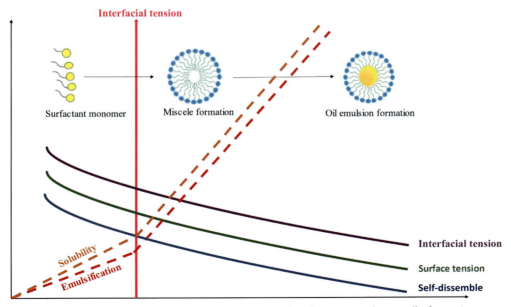

Figure 8.3 Surfactants—micelle formation mechanism of surfactants on the micelle formation.

8.5.2.2 Surface and interface activity

Adding biosurfactant molecules to a fluid reduces its interfacial tension due to their adsorption at the liquid–air or liquid–liquid interface. When the surface of the aqueous phase becomes saturated with an excess of biosurfactant, the formation of the micelle is initiated. On reaching this point, the surface tension can no longer be reduced. The classic theories of biosurfactant aggregation show that concentrations less than CMC result in biosurfactant molecules prevailing as monomers in the aqueous potion and aggregating at the liquid–liquid or air–liquid interface. These monomers assemble spherically, paving the way for the hydrophobic portion to shift toward the center, which comprises the nucleus, followed by the hydrophilic part that is turned to the sphere's surface facing the external side, forming an interface with water. Thus, hydrocarbon degradation is favored by micelle formation, reducing its surface tension and enhancing its exposure to bacteria and oxygen. When a micelle is formed, the part that makes an interface with water, that is, the hydrophilic component of the biosurfactant, is surrounded by a double, compact, electric layer called the stern layer. This process of micelle formation is represented by Stigter's approach. Certain microbes like *Bacillus subtilis* can produce surfactin, which lowers the surface tension of water and the interfacial strain between water and hexadecane to 25 and 1 mN/m, respectively. In terms of thermodynamics, surfactants can reduce the surface-free energy per unit area by adsorption, creating a new surface (closely correlated to the surface and interfacial tension). The biosurfactant interfacial behavior is driven by two forces: free energy of adsorption (ΔG_{ad}) and micellization (ΔG_{mic}) (Jahan et al., 2020). The key property of the surface and interfacial activity plays a crucial role separating the hydrophilic and hydrophobic parts for adsorption on a solid or liquid surface. This allows the biosurfactants to act as wetting and dispersing agents, which are widely used in the pharmaceutical industry for their antioxidant, antiviral, anticancer, antiaging, and antiinflammatory properties. They are also utilized in the food, soap, cosmetics, and herbicides industries (Anestopoulos et al., 2020).

8.5.2.3 Stability

The usefulness of microbial surfactants is highly dependent on their high performance or stability under various environmental circumstances, including a wide range of temperatures and pH. The biosurfactant (mono-rhamnolipids) produced by *Klebsiella* sp. KOD36 was observed to be stable under extreme conditions (60°C, pH 10, and 10% salinity), giving it the ability to degrade phenanthrene in three different environmental matrices (aqueous, soil-slurry, and soil) with high efficiency (Ahmad et al., 2021). This proves that stability plays a major role in the degradation of petroleum hydrocarbons.

8.5.2.4 Emulsification

Biosurfactants can operate as emulsifiers or de-emulsifiers, depending on the type of emulsion. An emulsion can be represented as a heterogeneous framework consisting of one immiscible fluid dispersed in a second fluid with a diameter larger than 0.1 mm of the beads. There are two forms of emulsions: oil-in-water type of emulsion and water-in-oil type of emulsion. Their stability is low and can be compensated by adding compounds like biosurfactants, which can stably preserve emulsions for years. More than 250 biosurfactants (like Emulsan) with related patents have been extended to industries mainly due to their robust emulsification capability, which is very well utilized in food formulations, cosmetic preparations, and pharmaceuticals processing and various other industrial formulations (Ribeiro et al., 2020). The bioemulsifying compounds produced by the microbe *Rhodococcus erythropolis* OSDS1 could retain the emulsification activity of up to 90% even after 168 hours (Xia et al., 2019). The emulsification capacity of the bio-emulsifier was not only limited to one but also several petroleum hydrocarbons such as diesel, mineral/crude oil and gasoline, in the decreasing order of efficacy.

8.5.2.5 Low toxicity
Biosurfactants are often regarded as nonhazardous or low-hazardous substances suitable for medicinal, curative, and nourishment applications, although there are few published studies on their toxicity. Studies show that chemically derived surfactants are 10 times more hazardous than biosurfactants like rhamnolipids. The low toxicity of the biosurfactants like sophorolipids from *Candida bombicola* makes them advantageous for environmental applications.

8.5.2.6 Self-assembly
One of the unique properties of biosurfactants is the self-assembly of the moieties in any given mixture. The hydrophobic effect and weak Van der Waals interactions in biosurfactants promote their natural tendency to self-assemble. The efficacy of the biosurfactant is measured by its capacity to reduce surface tension. The surface tension of water is 72 mN/m, which can be reduced to 30 mN/m using biosurfactants with greater efficacy. Alterations in biosurfactant concentration, temperature, pH, pressure, and salts can affect the micelle's size and form. The hydrocarbon chain structure and the peptide sequence both contribute to the micelle production of biosurfactants. Biosurfactants produce supramolecular structures with various morphologies due to the hydrogen connections between their head groups.

8.5.3 Petroleum hydrocarbons waste: a suitable carbon and energy source for microbial production of biosurfactants
The reuse of various industrial wastes is highly emphasized owing to its nutrient-enriched composition for the growth of biosurfactant-synthesizing bacteria at low costs. Carbohydrates, fats, and hydrocarbons are the three profoundly used substrates. Biosurfactant production was optimum in the substrates with high lipid or carbohydrate content. The commonly used organic and hydrophobic substrates for producing

microbial biosurfactants Fig. 8.4 include fat-rich sources such as corn steep liquor, animal fats, lard and tallows, vegetable fats, soyabean oil, chicken fat, soap stock, olive oil mill effluent (OOME), while carbohydrate-rich sources include lignocellulosic wastes and agro-industrial wastes such as wheat, rice, soy, rice, sugarcane, molasses, bagasse, sugarcane, water used to process cereals, pulses, barley bran, and corn cobs have showed significant effect (Mohanty et al., 2021). An approach of matrix integrated fermentation systems, wherein the nutrients supporting microbial growth substrates (like morsels, grains or particles of rice, beans, legumes, lentils, corn, grains, pasta, oats, oatmeal, wheat bran, wheat flour, and corn flour) was infused to obtain the solid substrate matrix, facilitated the production of varied high yield biosurfactants such as sophorolipids, rhamnolipids, trehalose lipids, iturin, surfactin, fengycin, lichenysin, and mannosylerythritol lipids. On the other hand, hydrocarbon substrates are one of the most economical and profitable substrates for the large-scale production of biosurfactants (Sivapathasekaran & Sen, 2017). The commonly used hydrocarbon substrates include hexadecane, dodecane, tetradecane, and pyrene and hydrophobic mixtures such as motor oil, diesel, crude oil, paraffin, and kerosene (Jimoh & Lin, 2019). Complex substrates such as PAHs are also greatly used for biosurfactant production. Petroleum sludge produced throughout various processes in the petrochemical industry is composed of oil, solid particles, and water, with petroleum hydrocarbons being the most toxic and abundant and rich in sulfur, nitrogen, oxygen, phosphorus, and metallic elements showing its potential to be used as

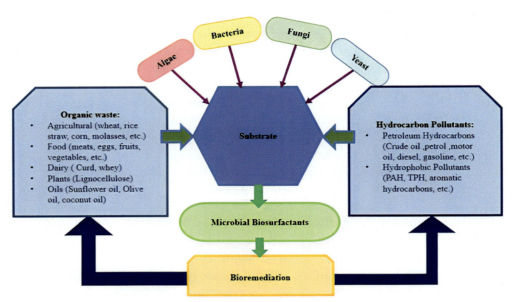

Figure 8.4 Microbial biosynthesis of biosurfactants Various organic and hydrophobic substrates utilized to produce microbial biosurfactants.

a carbon and energy substrate (Aguelmous et al., 2019). Microbes can then degrade these substrates to alleviate the physiological stress caused and utilize them as carbon and energy sources (Sah et al., 2022). In a study conducted in Assam, India, substrate PAH such as phenanthrene, fluorene, and pyrene were used with *Pseudomonas aeruginosa*, which could produce biosurfactants with a yield of 0.36–0.45 g/L capable of reducing the surface tension from 72 to 35 mN/m of pure water (Bordoloi & Konwar, 2009). In another study, the biosurfactant produced by *Aeromonas*, grown in diesel oil as a substrate, could degrade about 75% of the oil in 7 days (Nurfarahin et al., 2018). *B. subtilis* B series strains detected with mutated comK gene (encoding competence transcription factor) and srfA operon (encoding surfactin synthetase), identified from the soil samples obtained from oil wells, were subjected to the enrichment process using a minimal medium, where crude oil is the sole source of carbon, resulted in the production of biosurfactants posing 10- to 12-fold higher activity than other strains under anaerobic conditions (2020b). From previous studies, it is observed that petroleum hydrocarbon contaminants, although very complex, could act as a potent resource for the production of effective and substrate-specific biosurfactants in significant yields.

8.5.3.1 Biosynthesis of biosurfactants

Biosurfactants can be synthesized using different metabolic pathways depending on their chemical composition and the carbon substrates provided Fig. 8.5. When sugars

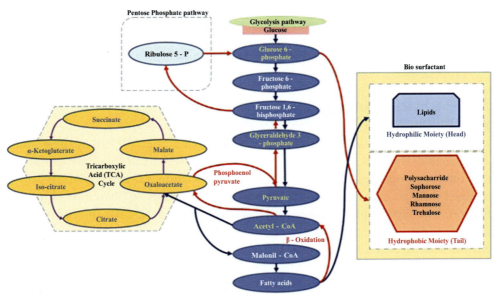

Figure 8.5 Microbial biosynthesis of biosurfactants common microbial metabolic pathway involved in the synthesis of biosurfactants.

are provided as carbon substrates, biosurfactant is synthesized via the glycolysis and lipogenesis pathway. However, when hydrocarbons act as the sole carbon source, they are used up for the synthesis of sugars and fatty acids via the gluconeogenesis and lipolysis pathways (Sah et al., 2022). The polar and nonpolar moieties of the biosurfactant are synthesized using the hydrophilic and hydrophobic segments of the hydrocarbon substrates. The biosynthesis of biosurfactant can be of any four cases: (1) independent synthesis of carbohydrate and lipid portions, (2) independent lipid synthesis, while the carbohydrate portion is synthesized based on the substrate, (3) independent carbohydrate synthesis, while the lipid portion is synthesized based on the substrate, or (4) both the portions depend on the nature of the substrate (Santos et al., 2016). Therefore, the type of hydrocarbon substrate is significant for determining the structure of biosurfactants. The experiment conducted on the substrate-specific production of manosilerythritol lipid (MEL) on microbial fermentation using *Candida antarctica* has proved that the molecular structure composed of MEL includes hydrocarbons ranging from C_{12} to C_{18} (Kitamoto et al., 2001). The metabolic pathway is slightly altered according to the type of biosurfactant. Sophorolipids (glucose molecules) are transferred via glycosyltransferases to hydroxyl acid, which is further acetylated by acetyltransferases, producing a fatty acid portion from the hydrocarbon (Joshi-Navare et al., 2013). Mannosylerythritol from *Ustilago maydis* employs the enzymes mannosyltransferase and acetyltransferase encoded by the genes *emt1*, and *mat1* and *mac1*, respectively (Hewald et al., 2006). For trehalose lipids, trehalose-6-P-UDP is synthesized from glucose-6-phosphate-via trehalose-6-phosphate synthase, which is further phosphorylated by trehalose-6-phosphate phosphatase (Rao et al., 2006). Phospholipids, generally synthesized in the cytosol under the action of various enzymes such as floppases, choline phosphotransferase, and filppases are transferred to the destined location on the cell membrane as vesicles, releasing biosurfactants on the inner leaflet. In the case of surfactin, amino acids are activated and converted to surfactin's peptide chains nonribosomally through surfactin synthase (Bickong, 2011).

8.5.4 The mechanism behind the biosurfactants on the removal of hydrophobic pollutants/petroleum derivatives

The major limitation in mitigating petroleum hydrocarbons (PAHs) is their limited bioavailability. Biosurfactants have the ability to overcome this disadvantage through their unique properties. They facilitate the degradation of PAHs, which can be explained through anyof the three phenomena: mobilization, solubilization, and emulsification. The type of phenomenon depends on the molecular mass and concentration of the biosurfactant. At concentrations below CMC, low-molecular-weight biosurfactants function through mobilization, whereas at concentrations above CMC, solubilization takes place. During mobilization, biosurfactants minimize the surface and interfacial tension between oil and the contaminated matrix (water, soil, etc.), facilitating an increased interaction with the biosurfactants, hence increasing the contact angle. This shows a reduction in capillary forces responsible for holding the oil in the matrix

(Maikudi Usman et al., 2016). Solubilization is a phenomenon in which the biosurfactant molecules come together to form micelles. The hydrophilic heads face toward the aqueous phase while the hydrophobic tails form the center of the micelle, creating a suitable environment for the hydrophobic contaminants, thus increasing the solubility of the PAHs. On the other hand, high-molecular-weight biosurfactants emulsify the PAHs, which, in simple terms, is the formation of fine oil droplets in water (O/W emulsion) (Rizvi et al., 2021). Along with the pollutants, biosurfactants simultaneously interact with the microbes. Biosurfactants are adsorbed onto the nonpolar surface of cells via van der Waals and hydrophobic forces acting between the hydrophobic tail and nonpolar components of the cell, leading to chemical and molecular alterations in the cell surface and to increased hydrophobicity. Biosurfactants neutralize the surface charges of the cells, causing a reduction in their zeta potential, resulting in reduced interaction of the cells with the matrix and, eventually, bacterial retention (Zhong et al., 2017). The mechanism involved in the uptake of hydrophobic compounds via biosurfactants and without biosurfactants in the microbial bioremediation process is shown in Fig. 8.6.

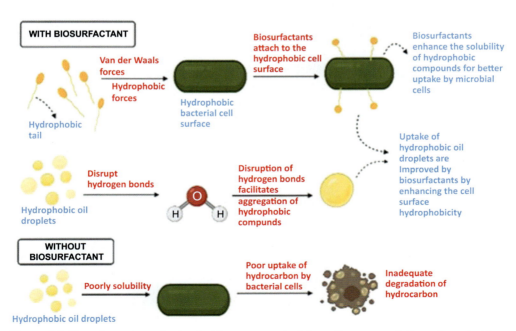

Figure 8.6 Biosurfactants on hydrophobic compounds Schematic representation of mechanisms involved in the microbial bioremediation process of hydrophobic compounds via biosurfactants and without biosurfactants.

8.5.5 Strategies involved in the application of biosurfactants on the remediation of hydrophobic pollutants/petroleum derivatives

Biosurfactants possess a plethora of biotechnological applications within the petrochemical industry itself. They are utilized in bulk operations as mobilizing agents, increasing the availability and recovery of hydrocarbons along with extraction, treatment, cleaning, and transportation processes. Due to their ability to function as emulsifying/demulsifying agents, they are utilized as anticorrosive agents, biocides for sulfate-reducing bacteria, and in processes such as fuel devising, bitumen extraction from tar sands, and various other advanced applications across the oil processing chain. Biosurfactants are considered valuable and versatile tools that can transform and modernize petroleum biotechnology (De Almeida et al., 2016). The two primary strategies involved in bioremediation are the application of (1) biosurfactant-producing microbes for bioremediation and (2) purified biosurfactants. Biosurfactant treatment involves the use of biosurfactants in their purified form, which possess the ability to dissolve oil and other hydrophobic substances. Microbial-based bioremediation, on the other hand, involves applying biosurfactant-synthesizing microbes to decompose and eliminate contaminants from the environment, including HCPD Fig. 8.7. Many studies apply biosurfactants to degrade petroleum pollutants present in the environment. Biosurfactants such

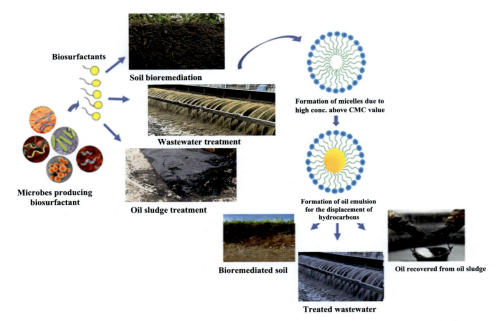

Figure 8.7 Biosurfactants and their applications in petroleum hydrocarbon-contaminated environments Applications of microbial surfactants for the remediation of HCPDs. *HCPDs*, Hydrophobic contaminant/petroleum derivatives.

as Iturin A synthesized using *Bacillus aryabhattai* showed a promising result of 74% ± 1.9% emulsification index and were able to degrade about 61.18% ± 0.85% of crude oil (Yaraguppi et al., 2022). The bacterial strain *Pseudomonas* sp. SA3 produces biosurfactants with an emulsification capacity of 43% and a surface tension reduction ability of 34.5 mN/m, indicating its effectiveness in treating petroleum-contaminated agricultural soils (Ambust et al., 2021). Lipopeptide concentrations of 0.2% and 0.6% (w/w) degraded 4-ringed and 5- and 6-ringed PAHs at 51.2%, 64.1%, 55%, and 79%, respectively (Eras-Muñoz et al., 2022). The contamination caused by hydrocarbons in the environment has prompted scientists to identify the bacterial species that can effectively remove the hydrocarbon pollutants. *Ochrobactrum intermedium* CN3 isolate and glycopeptide biosurfactant showed 70% degradation of crude oil sludge (Bezza et al., 2015). A consortium of marine bacteria consisting of *B. subtilis* AS2, *Bacillus licheniformis* AS3, and *Bacillus velezensis* AS4 under optimum temperature and pH showed significantly high efficiencies of 88%, 92%, and 97%, respectively, in the biodegradation of crude oil (Prakash et al., 2021). Both biosurfactant-based bioremediation and purified biosurfactant can effectively clean up environmental contamination. The choice of approach will depend on the specific contaminants present and the site's conditions. As the studies reported, adding biosurfactant-producing bacteria showed effective results. Microbial biosurfactants and their removal efficiency are represented in Table 8.1. However, further research must be conducted to understand and effectively evaluate the strategies and bring improvements in the combinations of surfactant type and substrate.

8.5.6 Biosurfactant-facilitated, microbial-enhanced oil/hydrophobic contaminants recovery from crude oil tank bottom oil sludge

Oil sludge is a semisolid material formed during the production, refinement, and storage of oil found at the bottom of oil storage tanks, pipelines, oil production sites, and oil-based drilling muds. It is formed when water and other impurities are mixed with oil, causing the oil to become thick and sludgy. It contains numerous contaminants like heavy metals, chemicals, and bacteria. Oil sludge can contain various PAHs, which are usually formed during the refining process, including classified carcinogens by the International Agency for Research on Cancer (IARC) and should be treated immediately. Biological methods are preferred over physicochemical methods for treating oil sludge due to their advantages, such as limited hazardous chemicals used, the requirement of mild conditions to reduce energy consumption, and the elimination of potential secondary pollution, as a result of which biosurfactants have gained traction. This is primarily due to their property of high surface/interface activity, higher biodegradability, lower toxicity, and environmental compatibility. First proposed in 1981 as an alternative to clean the oil storage sludge tanks, this treatment could recover almost 91% of hydrocarbons from the sludge, and the resulting crude oil retrieved was valuable enough to be sold (blended with fresh crude oil) and also compensate the cost for the cleaning operation due to its exceptional properties (Jayashree, 2021). Several

Table 8.1 List of microbial biosurfactants and their HCPD (hydrophobic contaminant/petroleum derivatives) removal efficiency.

Biosurfactant/Biosurfactant-producing microorganisms	HCPD	Initial concentration	Removal efficiency	References
Rhamnolipid (*Pseudomonas aeruginosa*)	Low TPH-contaminated soil (LTC) and high TPH-contaminated soil (HTC)	LTC soil containing 3000 mg TPH/kg of dry soil. HTC soil containing 9000 mg TPH/kg of dry soil	LTC–23% and HTC–63%	Lai et al. (2009)
Rhamnolipid (*P. aeruginosa* DS10–129) + Bacterial consortium *Micrococcus* sp. GS2–22 (21.7 ± 1.4 · 105 CFU/mL), *Bacillus* sp. DS6–86 (30.3 ± 0.9 · 105 CFU/mL), *Corynebacterium* sp. GS5–66 (27.4 ± 4.7 · 105 CFU/mL, *Flavobacterium* sp. DS5–73 (18.9 ± 3.6 · 105 CFU/mL), *Pseudomonas* sp. DS10–129 (32.6 ± 0.8 · 105 CFU/mL)	Oil spill and petroleum sludge	87.4% of oil and grease content in soil samples with 10% and 20% tank bottom sludge	nC8–nC11, nC12–nC21, nC22–nC31 and nC32–nC40 were degraded 100%, 83%–98%, 80%–85%, and 57%–73%, respectively in 10% tank bottom sludge and 81%–87%, 64%–83% in 20% tank bottom sludge	Rahman et al. (2003)
P. aeruginosa AHV-KH10	Diesel-contaminated sediments from Persian Gulf	1000 mg/kg of diesel hydrocarbon	70%	Pourfadakari et al. (2021)
Trehalose lipid (*Rhodococcus qingshengii* strain FF)	Polyaromatic hydrocarbon (Naphthalene)	50 mg/L Naphthalene	52.5%	Wang et al. (2019)

Sophorolipids (*Candida bombicola* ATCC 22214)	Iranian light (crude oil)	Hydrocarbon contaminants with Saturates (35% ± 2%), aromatics (28% ± 3%), and polars (37% ± 2%)	85%–97% of the total amount of hydrocarbons degradation	Kang et al. (2010)
Saponins (quillaja bark-based) and Rhamnolipids)	Crude oil-contaminated soil	54.4 mg of oil per gram of soil	80%	Urum et al. (2003)
Surfactin-like biosurfactant (*Bacillus nealsonii* S2MT)	Heavy engine oil-contaminated soil	10 and 40 mg/L	43.6% ± 0.08% and 46.7% ± 0.01%	Phulpoto et al. (2020)
Bacillus cereus BN66	Polyaromatic hydrocarbons containing crude oil	1% crude oil	C5-C9 alkanes-95%; C10-C15alkanes-100%; C16-c19alkanes -100%; benzene-84%; toluene-90%; biphenyl- 70%; Naphthalene- 95%; Phenanthrene- 58%; Anthracene- 52%	Christova et al. (2019)
Glycolipidss (*Rhodococcus erythropolis* and *Rhodococcus ruber*)	Hydrocarbons from oil	5 kg of oil shale	25% and 26% (recovery)	Haddadin et al. (2009)
Iturin A (*Bacillus aryabhattai*)	Oil in sand pack column	50 ml crude oil	61.18% ± 0.85%	Yaraguppi et al. (2022)

(*Continued*)

Table 8.1 (Continued)

Biosurfactant/Biosurfactant-producing microorganisms	HCPD	Initial concentration	Removal efficiency	References
Bacillus licheniformis	PAH (crude oil)	1 gm of crude oil	60%	El-Sheshtawy & Ahmed (2017)
Lipopeptide (*Bacillus cereus* SPL-4)	16 priority PAHs	6745.5 mg/kg	lower molecular weight (2 and 3-ring PAHs)—9.7%; 4, 5, and 6-ringed PAHs—48.9% and 44.9%, respectively	Bezza & Chirwa (2017)
Staphylococcus saprophyticus UFPEDA 800, *Serratia marcescens* UFPEDA 839, *Rhodotorula aurantiaca* UFPEDA 845d and *Candida ernobii* UFPEDA 862	Diesel oil	Diesel oil 85.9% (ww^{-1}) of carbon, 13.1% (ww^{-1}) of hydrogen, and <0.3% (ww^{-1}) nitrogen	69%	Soares da Silva et al. (2021)
Ochrobactrum intermedium CN3 isolate and glycopeptide biosurfactant	Oil sludge	20 μl of 4% (v/v) crude oil sludge	40% degradation efficiency of isolate and 70% by biosurfactant.	Bezza et al. (2015)
Pseudoxanthomonas sp. PNK-04	PAH 2-chlorobenzoic acid, 3-chlorobenzoic acid, and 1-methyl naphthalene		2-chlorobenzoic acid (75%) and 1-methyl naphthalene (60%)	Nayak et al. (2009)

Pseudomonas sp. BP10 and *Rhodococcus* sp. NJ2	PAH from crude oil	2% of crude oil	60.6% and 49.5%	Kumari et al. (2012)
Halomonas sp. (GQ169077), *P. aeruginosa* (GU72841), *Marinobactermobilis* (GQ214550), *Gaetbulibacter* sp. (FJ360684)	PAH	0.75 mL of heavy oil	74.35% and 71.79%	Nuñal et al. (2014)
Lysinibacillusbronitolerans RI18 (KF964487), *Bacillus thuringiensis* RI16 (KM111604), *Bacillusweihenstephanensis* strain RI12 (KM094930) and Consortium + SPB1 + RI7	PAH	5% (v/v) of diesel oil	31.44 ± 1.640 and 48.86 ± 1.560	Mnif et al. (2015)
Mixed culture of *Serratia marcescens* ZCF25, *Pseudomonas* sp. ZCF53	Hydrocarbons in oil sludge	0.1 g of oil sludge	(C11-C20) and long-chain (C21-C40) n-alkanes (>89%). Furthermore, reductions of 94.5 µg/L (17.5% reduction), 345.57 µg/L (64% reduction), 300.6 µg/L (55.67% reduction), and 398.76 µg/L (73.8% reduction)	Huang et al. (2020)

other previously recorded experiments also showed that biosurfactants demonstrated applications of oil storage tank cleaning, smooth flow through the pipelines, and enhanced the oil recovery from oil reservoirs. This is attributed to the potential of the biosurfactants to make oil deposits less viscous, facilitating their easy pumping. Due to their demulsification property, these molecules also contribute to breaking emulsions, thus facilitating crude oil recovery. The biosurfactant's porosity plays a crucial role in the de-oiling operation and posttreatment with biosurfactant, in which the oil sludge gets fragmented into tiny, relatively smooth particles. Several experiments have been conducted to analyze the efficiency and mechanisms associated with this process in which the biosurfactant produced by five strains of bacteria recorded a recovery of up to 95% of the oil from the total oil sludge, with only 2% in the absence of biosurfactant (Lima et al., 2011). Furthermore, SWPUEN-1 is an active interface biosurfactant that can modify the interfacial tension and wettability of particles, leading to the destruction of the oil—water interface. The contact angle between crude oil and the solid surface increases, allowing for convenient peeling of crude oil (Ren et al., 2020). Another attractive and very efficient method is the use of rhamnose tallow. It has recorded a crude oil recovery rate of 91.5% from the sludge under optimal conditions. Instrumental analysis proved the high compatibility of the lipopeptide biosurfactant in the applications such as recovering crude oil from oily sludge and the degrading organic impurities in oil, facilitating the production of cleaner oil products (Liu et al., 2018). Fewer innovative strategies were designed for the screening of the biosurfactant-producing yeasts, which showed that the application of the yeast cells such as *Starmerella bombicola, Pseudozyma aphidis, Wickerhamomyces anomalus, Pichia sydowiorum, Pichia guilliermondii,* and *Pichia lynferdii* as consortia in inactive forms could reduce the viscosity of oil by 60% and enhance the oil recovery process (2021). Despite existing technical problems in biosurfactant-enhanced oil recovery, it is an ideal economical method gaining recent traction for its ability to recover significant amounts of oil from the tank bottom sludge (Sarma et al., 2019). These research findings show that biosurfactant-assisted oil recovery is a tremendously productive method that offers reassuring performance.

8.5.7 Adjuvant for the decontamination of soil associated with petroleum derivatives

Bioremediation is an integral part of pollutant containment, removal, and cleaning of the environment. The remediation approach is mainly decided based on the site. It is specific about the nature and pollutant composition, the physicochemical and biogeological conditions of the contaminated environment, the available microbes, and the cost and time constraints. Bioremediating soil polluted with hydrocarbon depends mainly on the bioavailability of the hydrophobic contaminants. In such cases, biosurfactants function as adjuvants that help enhance the bioremediation activity of microbes and improve their growth conditions (Ossai et al., 2020). This process is referred to as biostimulation. Adding biosurfactants usually helps lower the constraints on microbial performance, improving their capacity. The degradation capacity of hydrocarbon-degrading microbial species depends on the availability and physical state of hydrocarbons, temperature, oxygen and water availability, inorganic nutrients, and pH. Biosurfactants utilize their unique mobilization, solubilization, and emulsification properties on the HCPD, resulting in the increased solubility and bioavailability of the hydrocarbons. The increase in mobilization is mainly due to the biosurfactants' ability to increase the angle of contact and reduce the capillary forces binding the soil and oil. The biosurfactants attributing to their specificity, surface area-enhancing property, emulsification activity, boost solubility (above CMC), and bioavailability of the hydrophobic contaminants facilitate the degradation and bioremediation of hydrocarbons in the soil. The diffusive length between HCPD and microbes is significantly reduced after the absorption of biosurfactants onto hydrocarbon, thereby boosting the uptake and subsequent enzyme activity of the soil microbes on hydrocarbons (Fenibo et al., 2019). The common biosurfactants utilized in bioremediation are sophorolipids, trehalolipids, rhamnolipids, lipid–peptide complexes, and carbohydrate–lipid–peptide complexes, which are either synthesized inside the cells (intracellular) or secreted outside the cells (extracellular) (Karlapudi et al., 2018). The rhamnolipids in *P. aeruginosa* were intensively characterized for their hydrocarbon-degrading property. *P. aeruginosa*, a marine bacterium isolated from oil-polluted sea, was reported to produce both mono- and di-rhamnolipid-type biosurfactants (Shin et al., 2006). These rhamnolipids were used to remediate phenanthrene-contaminated soil and were specifically observed to degrade hexadecane, octadecane, heptadecane, and nonadecane after 28 days of incubation. This particular biosurfactant has also shown the ability to degrade hydrocarbons such as 2-methylnaphthalene, tetradecane, and pristine (Karlapudi et al., 2018). Nearly 31% recovery of hexachlorobiphenyl by biosurfactant was achieved, which was three times higher than its chemical counterpart (Unaeze & Henrietta, 2020). The biosurfactant rhamnolipid has been most reported to remediate diverse kinds of hydrocarbon (Fenibo et al., 2019). Another strain, *Pseudomonas ceparia* AC 1100, produced an emulsifier with a potential activity against 2,4,5-T and chlorophenols and, therefore, works well in the bacterial degeneration of organochlorides. Hence, there is a clear correlation between the hydrocarbon that gets degraded in the oil and the type of biosurfactant (Patel et al., 2015). *Aspergillus*, a filamentous fungus, was reported to have the potential for degrading an extensive range of PAHs from the environment (Al-Hawash et al., 2019). Morever, the species of Aspergillus and Rhizopus from automobile workshops in Pudukkottai in south India displayed engine oil-degrading capacity (Thenmozhi et al., 2013). Novel bacterial strains such as *Ralstonia pickettii* srs PTA-5579, *Alcaligenes piechaudii* srs PTA-5580, and *Pseudomonas putida* biotype b srs PTA-5581 were reported to display enhanced capability on degrading various chain lengths of petroleum hydrocarbons along with the indigenous production of biosurfactants. These strains, when used as consortia, for the treatment of 4 tons of contaminated soil at a TPH concentration of 26,000 ppm, metabolized the petroleum hydrocarbons to 100 ppm at an average biodegradation rate of 62 g/day within 14 months of retention time (2011). The synergy of phytoremediation and biopiling was adapted to design an efficient remediation system called EcoPiling, especially for soils contaminated with crude oil, petroleum or diesel, heavy lubricant oils, pesticides, polychlorinated biphenyls, dioxins/furans, cyanide or polycyclic aromatic hydrocarbons, and a mixture of heavy metals. Ecopiling treatment resulted in 33% of TPH removal of the contaminated soil with a TPH of above 30000 ppm, in which the major contaminants were phenanthrene and fluoranthene and were degraded by 87% and 50%, respectively (2016).

8.5.8 Removal of petroleum derivatives from petroleum refineries and a petrochemical waste effluent stream

Petrochemical wastewater is a term that includes all oil-related industrial wastewater. It is a significant source of aquatic pollution, producing several physicochemical and biological effects on the aquatic ecosystem, resulting in the alteration of the population dynamics, destruction of the balance, and interplay within the ecosystems in the contaminated sites and severely impacting the native population. This wastewater's toxicity, degradability, and composition depend on the discharge source. Oily wastewater is discharged into the water due to the bulk release of several contaminants through oil drilling and transportation operations in metal machining, electronics industries, petroleum refineries, food processors, and palm oil mills (Zahed et al., 2022). Another source of hydrophobic waste in aquatic ecosystems is oil spills. Approximately 35 million barrels of petroleum are transported via transoceanic means while the ocean environment is unguarded in case of oil spills (Zaki et al., 2015). However, the primary contributor remains the petroleum refineries. The wastewater discharged from petrochemical industries contains a range of aromatic hydrocarbons and heavy metals, dissolved gases, oils, and other compounds such as minerals and solids. In addition, the water required in petroleum refining industries typically ranges from 30% to 50% of the crude oil processed (Ghimire & Wang, 2019). It is notable that in contrast to marine oil spills, oil in industrial wastewater is present as oil-in-water emulsions, causing several problems in the different treatment stages, such as fouling in the process equipment of treatment units, complications in biological treatment stages, and failure to meet the water discharge standards. The currently used treatment techniques include gas flotation, granular media filtration, chemical coagulation, gravity separation, cyclone separation, parallel plate coalesces, microfiltration, and ultrafiltration; however, they do not provide satisfactory treatment, are time-consuming, and have expensive initial and operating costs, making it imperative to develop an effective and economical alternative treatment method that is sustainable as well (Affandi et al., 2014). Biosurfactants fulfill this criterion. Certain groups of microbes are capable of

metabolizing hydrocarbons and other oil-related compounds and generating biosurfactants and bioemulsifiers. These biosurfactant-generating microorganisms use petroleum through two general mechanisms (Zahed et al., 2022):
1. quickening the process of complexation and solubilization of apolar compounds, increasing their bioavailability; and
2. inducing oil—water interface film deterioration by increasing the attraction between the oil—water interface and the cell surface.

The biosurfactants tend to evolve with surface-active property and increase the surface area of the water-insoluble, hydrophobic substrates, facilitating increased use of PHCs by hydrocarbonoclastic bacteria for their metabolism. The biosurfactants contribute to the hydrophobicity of the microbial cell, which enables the utilization of pollutants as carbon substrate to obtain energy, thus breaking the hydrocarbon chain and its amphiphilic nature, leading to the interaction of hydrophobic compounds with cell surfaces and to the increase in its bioavailability. The pollutants are finally transformed into CO_2, water, and minerals, resulting in efficient bioremediation (Mahjoubi et al., 2018). Contaminated aquatic environments can also be bioremediated by applying biosurfactants to remove PAHs, which can be attributed to their adsorption and solubilization capacity against PAHs. The increased solubility and bioavailability of pollutants by biosurfactants facilitated the improved biodegradation and adsorption of hydrophobic organic compounds in a contaminated marine environment (Lászlová et al., 2018).

8.5.9 Management of marine oil spills

Due to limited oil resources, oil resources are imported from oil-rich countries through waterways by ships, underwater conduits, or pipelines. Leaks or breakages caused due to corrosion, technical failures, and accidents result in the unintentional release of oils directly into the aquatic environment, leading to an oil film called oil spill, which is a colossal peril to the aquatic environment and the inhabiting species that are especially vulnerable (Patel et al., 2019). Oil spills block the sunlight and oxygen exchange from the atmosphere and get deposited on the skin and respiratory organs of the organisms, proving to be fatal (Freitas et al., 2016). Various techniques such as corralling, oil ignition, chemical surfactants, and natural sorbents are used to curb this predicament, which, although effective, are time-consuming, and therefore need more robust options. Surfactant-augmented bioremediation is gaining momentum for this purpose. Fig. 8.8, when oil molecules are encountered by biosurfactants, they undergo dispersion in the aqueous phase facilitated by biosurfactants, thereby resulting in its enhanced bioavailability for microbial utilization (Aparna et al., 2011). The biosurfactants synthesized by the yeast *Candida sphaerica* showed oil-spreading capacities up to 75% when applied to oil in seawater and observed a degradation rate of 92.6% in a hydrocarbon-contaminated soil and water (de Gusmão et al., 2010; Sobrinho et al., 2008). A novel biosurfactant, Lunasan

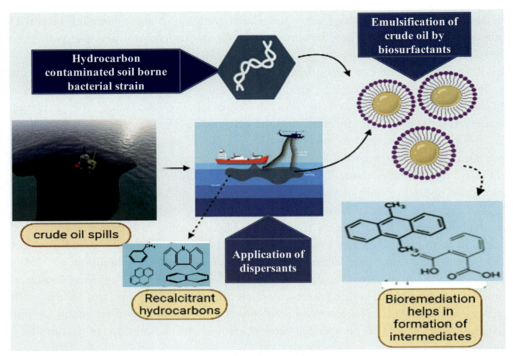

Figure 8.8 Oil spills remediation of marine oil spills using biosurfactants.

from *C. sphaerica* showed the removal of 95%, demonstrating its potential to be used in bioremediation applications (Luna et al., 2012). Another study reported a degradation of 63% in crude oil in half of the CMC of biosurfactant produced by *Pseudomonas cepacian* (Soares da Silva et al., 2021).

8.6 Omics approaches for the biosurfactants—case studies in filed level applications toward the remediation of hydrophobic pollutants/petroleum derivatives

The omics technologies are high-throughput molecular techniques that provide essential information on genetic makeup and metabolic profiling (Datta et al., 2020). Omics techniques provide an important means for the profound understanding of biosurfactants.

8.6.1 Metabolomics

As the study of the micromolecules or metabolites, metabolomics provides information about the physiological and biochemical setting of the cell (Worley & Powers, 2012). Consequently, it can be effectively utilized in the identification of various biosurfactants and their microbial strains. Untargeted metabolic profiling methods can be used to identify the synthesis and release of biosurfactants among other nonproducing strains (Gaur et al., 2022; Sirohi et al., 2021). Lipopeptide surfactin has been identified using the same technique (Floros et al., 2016). Techniques such as molecular networking and metabolomic investigation could reveal other important attributes of biosurfactants such as antimicrobial and anticancerous properties and structural relations. Moreover, manual de-replication techniques against SciFinder can be used, as in the case of glycolipid, which was then established to be the broadest molecular class (Buedenbender et al., 2021).

8.6.2 Metagenomics

Metagenomics includes techniques such as DNA isolation, genomic library construction, 16 s rRNA sequencing, and screening and interpretation techniques (Datta et al., 2020). It can simplify the analysis of the metagenomic information, metabolic pathways, and interplay of the uncultivated bacterial species, therefore aiding in the discovery and identification of new and effective biosurfactant-synthesizing microbial species (Malik et al., 2022). However, due to its inability to differentiate between expressible or nonexpressible genes, this field is commonly applied in combination with metatranscriptomics.

8.6.3 Metatranscriptomics

It deals with the analysis of the total mRNA or transcriptome of the cell and can be used to understand how environmental alterations lead to over- or underexpression of genes. Due to the instability of mRNA, the application of these techniques is limited, which is then supplemented by metaproteomics (Gaur et al., 2022; Malik et al., 2022).

8.6.4 Metaproteomics

Metaproteomics concerns cellular protein expression, which can help link and analyze the functional genes of the microbial population. It can help investigate the effect of substrate changes and detect newer functional genes and metabolic networks (Maron et al., 2007). New high-yielding variety of strains can be developed by employing metabolic engineering techniques, which can significantly improve the yield of biosurfactants (Dobler et al., 2016). Further data collected can be enhanced by adopting a multimeta-omics strategy (Malik et al., 2022). Although tremendously valuable technologies, further research must be conducted to overcome several challenges in the relatively new field.

8.7 Recent advancements in enhancing the specificity and functional properties of biosurfactants

The recent advancements in enhancing biosurfactants employed to degrade petroleum wastewater include a study in which biosurfactant-producing bacteria were inoculated in an anoxic packed-bed biofilm reactor (AnPBR) in situ (Molaei et al., 2022). The TPH and COD removal values were observed to be 99% and 96% effective, respectively. The biosurfactants can also be used with electrokinetic remediation systems called bio-electrokinetic remediation, which shows promising results in crude oil degradation from the soil. In studies conducted on marine bacteria-producing biosurfactants, identified using 16 s rDNA analysis as *B. subtilis* AS2, *Bacillus licheniformis* AS3, and *Bacillus velezensis* AS4, displayed biodegradation efficiency of crude oil as 88%, 92%, and 97% for strain AS2, AS3, and AS4, respectively, at an optimum temperature of 37°C and pH 7. As demonstrated in their result, bio-electrokinetic systems can be aided with biosurfactants to enhance the bioremediation of soil contaminated with petroleum. This is because biosurfactant in electrokinetic remediation increases the biodegradation rate of crude oil-contaminated soil by approximately 92% compared with the use of electrokinetic remediation, which alone was 60% (Prakash et al., 2021). A host cell selected from the *Pseudomonas* species (*P. putida*, *P. chlororaphis*, *P. fluorescens*, *P. alcaligenes*, *P. aeruginosa*, *P. cepacia*, *P. clemancea*, *P. collierea*, *P. luteola*, *P. stutzeri*, and *P. teessidea*) comprising rhamnosyltransferase 1 A (rhlA) and rhamnosyltransferase 1B (rhlB) gene or their orthologs were reported to achieve the yield of more than 0.18 C mol rhamnolipid/C mol substrate under the control of a heterologous promoter.

8.7.1 Targeted hydrophobicity enhancements in biosurfactants

The utilization of biosurfactants in various industries has led to an increased demand for more customized and target-specific biosurfactants. There have been reports of different types of bacteria being capable of efficiently cosynthesizing biosurfactants and value-added products such as microbial lipids, ethanol, and polyhydroxyalkanoates. Pectinases such as pectin lyase (PNL), pectate lyase (PEL), and polygalacturonase (PG) were discovered in the *B. subtilis* BKDS1 culture media-producing biosurfactants (Kavuthodi et al., 2015).

8.7.1.1 Coproduction of biosurfactants with other value-added products

Nearly 0.32–0.50 tonnes of waste, including fibers and bagasse, is produced for every ton of oil extracted from oil palms. This waste is used as fuel in processing factories that produce construction materials. Sugars such as arabinose, xylose, mannose, glucose, cellulose, hemicellulose, and galactose are present in these wastes giving them nutritional value. Using these qualities, oil palm wastes can be employed as possible additives for increasing the bioremediation of soils contaminated with oil. Sugarcane bagasse was used to manufacture rhamnolipids and ethanol by *Saccharomyces cerevisiae* and *P. aeruginosa*, respectively. As many as 86 hours of fermentation yielded 8.4 g/L of ethanol and 9.1 g/L rhamnolipids, which produced a surface tension of 35 mN/m and an emulsification index of 84% (Guzmán-López et al., 2021). Furthermore, studies on *Bacillus methylotrophicus* DCS1 strain in the coproduction of alkaline amylase and lipopeptide biosurfactant proved that 10 g/L of potato starch and 5 g/L of glutamic acid were the optimal sources of carbon and nitrogen in the synthesis of amylase and biosurfactant, respectively (Hmidet et al., 2019). The bacterial strain was grown in an incubator for 48 hours at 25°C and 150 rotations per minute. As the zymography method shows, this strain generated a novel amylase. In a single batch culture at 30°C and 28°C, the production of 36% PHA and 0.4 g/L of rhamnolipid simultaneously by *P. aeruginosa* IFO3924 on a medium of 7 g/L of palm oil contained glycerol and fatty acids Mcl-PHA was produced through a combination of fatty acid de novo synthesis and oxidation (Hori et al., 2011). In the presence of the HAA synthetase (rhlA) enzyme, the de novo fatty acid biosynthetic intermediate (R)-3-hydroxyacyl-ACP was converted into rhamnolipid precursor 3-(3-hydroxyalkanoyloxy) alkanoic acids (HAAs), which were further transformed by the activity of rhamnosyl transferase I (rhlB) into mono-rhamnolipid and di-rhamnolipid by rhamnosyl transferaseII (rhlC) with all chain reactions ending in the synthesis of rhamnolipids.

8.7.1.2 Inducers: amino acids and fatty acid

Surfactin is one of the most commercially used surfactants. Various studies have been conducted on surfactin biosynthesis via attachment of N-terminal fatty acid pathway. Surfactin is synthesized by large multifunctional NRPS, which add an amino acid to surfactin using three modules: *SrfAA*, *SrfAB*, and *SrfAC*, forming a linearly aligned array of a seven-module structure in which each module adds one amino acid with one module per residue. The production of surfactin involves three distinct steps: (1) the biosynthesis of branched-chain amino acids (L-isoleucine, L-valine, and L-leucine), (2) the branched-chain fatty acids biosynthesis, (3) precursors such as CoA-activated 3-hydroxy fatty acids, synthesis catalyzed via NRPS. Studies conducted on *Bacillus velezensis* BS-37 showed that the production of BS-37 stain was increased twofold, nearly 2 g/L when 10 mM of L-Leu was used (Zhou et al., 2019). The significance of branched-chain fatty acids and branched-chain amino acids has been demonstrated by several researches. For instance, deletion of a component responsible for converting CoA—precursors to their respective branched-chain fatty acids led to a twofold increase in the production of surfactin C14 isoform with the straight fatty acids chain compared to its wild type. This conversion of CoA—precursors was done by a component of dehydrogenase complex, encoded by a gene called *lpdV*. There is a report that in genetically engineered bacteria *B. subtilis*, when the promoter of myc operon is replaced by a constitutive promoter, mycosubtilin which is the fatty acid chain comprising of 17 carbon atoms and having glutamine in place of asparagine in position 3 of its peptide cycle, facilitating the engineered bacteria to produce a larger quantity of mycosubtilins instead of surfactin when compared to the wild type *B. subtilis* strain (Coutte et al., 2017).

8.7.2 Metabolic engineering approaches

The use of metabolic engineering to create high-yielding strains has tremendous potential to produce numerous customized and modified strains. According to the literature, the potential output of biosurfactant (0.47 gbiosurfactant/gsubstrate) may be significantly increased by genomic tailoring and the modification of metabolic pathways through metabolic engineering (Zhang, Huo, et al., 2020). This demonstrated that the alternative surfactin route boosted the cofactor, ATP balance, and biosurfactant production. There have been report that the surfactin synthesis was achieved in *B. amyloliquefaciens* LL3 containing srfA operon by the removal of 4.18% of junked genomic regions. Ultimately, a new genome-reduced strain GR167 was constructed, which showed improved growth, capacity of expression of heterologous protein, efficiency in transformation, with improved intracellular reducing power and surfactin production. The GR167ID strain was subjected to further metabolic engineering by removing the biosynthetic gene clusters for iturin and fengycin from the GR167 genome. After the exchange of promoters, PRsuc and PRtpxi from LL3 strain into GR167ID strain, GR167IDS and GR167ID were obtained; of which, the best mutant strain was found to be GR167IDS with enhanced transcription by 678-fold in the srfA operon compared to strain GR167ID with a surfactant production improved by 10.4-fold (11.35 mg/L) compared to the GR167 strain. The Sfp protein (4-phosphopantetheyl transferase) in *B. subtilis* transfers the 4'-phosphopantetheinyl moiety of coenzyme A to a serine residue, contributing to surfactin production. However, the Sfp gene has a termination codon in the sequence in *B. subtilis* 168, leading to the inactivation of Sfp protein. To restore surfactin production in 168, the whole Sfp gene was inserted into the genome of 168 at the same location from *B. amyloliquefaciens* MT45, producing a recombinant strain, designated as 168S1, that produced 0.4 g/L surfactin, and was raised by 3.3-fold after 3.8% of the genome was removed. This included genes that were involved in the biofilm formation and polyketide synthase pathways (Wu et al., 2019).

8.8 Bottlenecks in the real-time application of biosurfactants on the removal of hydrophobic pollutants/petroleum derivatives

The high cost of production of biosurfactants is a major factor that hinders its market flow despite the excellent potential to replace synthetic petroleum surfactants and extraordinary environmental applications (Mishra et al., 2021; Uddin et al., 2021). The final cost of production of biosurfactants added up to 12 times that of chemical biosurfactants. Despite the development of several strategies to counter the hurdles encountered in the commercial-scale production of biosurfactants, there are challenges such as the need for pretreatment, industrial-scale production, effective downstream process, and availability of raw materials. The major contributing factors are the very low yield of biosurfactants, subsequent issues in extraction and purification processes, and the cost of the substrate (Johnson et al., 2021). There is an urgent need to optimize the biosurfactant production process to increase its yield and subsequently reduce its cost. Hence, there is an urgent requirement to identify strains with high production capabilities at reduced costs. More studies should focus on sustainability and biosurfactants life cycle assessment, including production, consumption, and their applications

(Kubicki et al., 2019). Performance and cost-effectiveness must be improved (Soares da Silva et al., 2021). Toxicity-related issues and potential applications in wastewater treatments with a greener and more sustainable approach must be explored (Rasheed et al., 2020). Better utilization of the latest multidisciplinary technologies, that is, Omics approaches and nanotechnology, is needed for improved production. There is an immense knowledge gap in understanding the production mechanisms and their interactions with cells and other compounds, especially their mode of action in organic/hydrophobic substrates, which is more importantly required for fully understanding their bioremediation capacity, which can be done with the help of the bioinformatics technology. Better treatment strategies must be developed to achieve greater degradation of these pollutants (Sánchez, 2022).

8.8.1 Factors affecting the remediation of hydrophobic pollutants using biosurfactants

Biosurfactant activity is prominently influenced by factors such as temperature, pH, salinity, and concentration of biosurfactants. Biosurfactants are viable at general temperature settings without significantly affecting biosurfactant stability and activity. This signifies the applicability of biosurfactants for bioremediation of petroleum-contaminated waters in extremely cold places such as the north Yellow Sea. For pH, the performance and emulsification capacity of the biosurfactant seemed to show no significant changes for different values. The emulsification activity of the biosurfactant was found to be maximum at a pH of 7, while it showed the lowest activity at a pH of 2, indicating that the optimal pH for biosurfactants is near neutral conditions (Guo et al., 2022). In the case of salinity, the surfactant is stable under various salt concentrations showing its applicability in various marine environments such as salt marshes and estuaries. However, at high salt concentrations, lower biosurfactant activity was observed, which is attributed to the altered shape and sizes of the micelles due to the presence of electrolytes in the salt (De França et al., 2015). The stability of biosurfactants has contributed to a profound emulsifying activity in a wide range of environments, an advantage that can be used in MEOR applications. Another important factor that bioremediation depends on is the use of biosurfactant—degrader combination (Cameotra & Bollag, 2003). Biosurfactants are reported to have improved activity when the correct degrader is used in combination with it. Factors such as soil chemistry, micellar size, ion exchange capacity, electrolyte content, nature, concentration of pollutant, and aeration state play an essential role in determining the efficiency of contaminant removal (Shah et al., 2016).

8.8.2 Economic feasibility

The global market for biosurfactants has been on the rise, and the market revenue has increased from 1.8 billion US dollars to 2.6 billion USD from 2016 to 2023 (Singh et al., 2019). This rise in the market share depicts its growing demand, indicating its capability to replace its chemical counterparts. The current price of a 10 mg vial of 98% pure biosurfactant ranges around $153 compared with that of chemical surfactants, which costs about US$1 a pound. The global goal for the commercialization of biosurfactants can be said to be satisfied if the biosurfactants cost about $3−4 per pound (Makkar et al., 2011). Biosurfactants compete with chemically synthesized surfactants in three aspects: production capacity, cost, and functionality (Makkar & Cameotra, 2002). Although there are significant improvements in the production processes and subsequent technologies, it is yet to be completely economized in terms of industrial-scale production. The limited concentration, low yields, high cost of substrates, and low purity, in addition to other growth and upscale and downstreaming problems, make the use of biosurfactants a costly process (Syldatk & Hausmann, 2010). The cost of the raw materials itself alone accounts for about 10%−30% of the total cost. Applications such as oil recovery require huge amounts of biosurfactants, which, on average, becomes extremely expensive. Another process contributing to the expenses is the refining process, especially if the biosurfactant requires advanced techniques such as centrifugation, crystallization, foam fractionation, and adsorption. The treatment and disposal costs of the enormous number of waste generated also contribute to the expenses (Mohanty et al., 2021). Suitable strains, either isolated or transformed by recombinant strains, can overproduce biosurfactants, which can significantly contribute to reducing the cost (Kosaric et al., 1984). The process feedstock or substrate is especially significant. Alternative and waste substrates with required nutritional balance are assessed and adapted as an efficient measure to reduce production costs. Agro-based industrial wastes and other organic wastes have been extensively used as low-cost nutrition-rich substrates contributing to the economical production of biosurfactants. Also, organic matter-rich industrial and municipal wastewaters can be used to produce biosurfactants with the added advantage of their simultaneous treatment (Adesra et al., 2021). Finally, the processes adapted must try to achieve lower operating costs. This can be done by optimizing the culture condition, which can significantly increase the yield of biosurfactants. Their throughput value can be enhanced along with antimicrobial and pharmacological properties. Further optimization of the conditions related to culture medium and growth conditions can result in major improvement in the yield (Mutalik et al., 2008). For instance, an air/liquid membrane contractor was used in which the cells were immobilized on the membrane contactor, which favored the continuous production of mycosubtilin with reduced formation of foam, in turn resulting in increased biosurfactant production yield at low costs (2018a). Also, the use of experimental design strategies such as response surface methodology (RSM) and factorial design is suggested for easy optimization of the process parameters.

8.9 Nanotechnology—engineering of biosurfactants for the removal of hydrophobic pollutants/petroleum derivatives—challenges and future perspectives

Nanotechnology is an emergent sphere in the scientific domain that has revolutionized all sectors of life, including environmental remediation. Many nanoparticles such as titanium dioxide and iron-based NPs, dendrimers, silica and carbon nanomaterials, graphene-based NPs, nanotubes, polymers, micelles, and nanomembranes are being applied for the treatment of various environmental contaminants. Limited biosurfactant yield is curbing its increased utilization in the mitigation of contaminants. Nanotechnological advances can therefore be significantly applied to improve biosurfactant production and the relevant bioremediation strategies Fig. 8.9. Employing limited concentrations of metal nanoparticles helps increase biosurfactant production. Iron plays a crucial role in various microbial metabolic pathways and hence biosurfactant production (Sah et al., 2022). It was observed that 1 mg/L of iron NPs increased the biomass and biosurfactant synthesis in *Serratia* sp. by 57% and 63%, respectively (Liu et al., 2013). This can be due to increased cell proliferation in the presence of nanoparticles and subsequent cell lysis, which releases more biosurfactant (Sah et al., 2022).

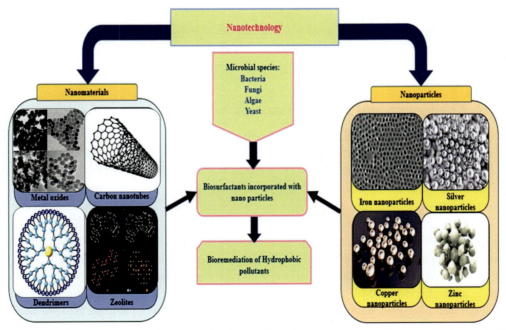

Figure 8.9 Advancements of nanotechnology: biosurfactants engineering nanotechnological approaches for the application of biosurfactants in the removal of HCPDs. *HCPDs*, Hydrophobic contaminant/petroleum derivatives.

In addition, nanoparticles can be associated with biosurfactants to induce affinity to the hydrocarbon intended to be treated by altering the hydrophobic chain of the precursor. The biosurfactant mobility can also be altered by controlling the charge density in the modified NPs (Kumari & Singh, 2016). Recent innovations in the development of novel biphasic biosurfactants biocatalyst conjugated magnetic nanometal oxide nanoparticles applied for the removal of petroleum hydrocarbons from tank bottom petroleum oil sludge (TBOS) showed significant results in the treatment process, that is, 87% ± 2% TPH was removed from the sludge within 48 hours of retention time, in which the surface tension reduction was achieved up to 38 ± 2 mN/m. Also, the process highlighted the remarkable reusability of the carrier material used in the treatment (2020a). Studies on the Submicron fumed silica particles, which are mesoporous nanomaterials of sizes ranging from 200 to 800 nm, when integrated into a surfactant system (especially glycolipid or lipopeptide), showed promising results in the prospect of oil recovery from a subterranean reservoir (2017). In another innovative approach, a nano-ferric ionosphere scaffold loaded with lipoprotein biosurfactant was employed for the targeted removal of recalcitrants and complex poly-aromatics associated with the landfill leachate via an integrative approach of using hetero-activated Fenton catalytic oxidation combined with bio-sequestration systems, which resulted in complete removal of refractory organics with 48–72 hours (2022a). These studies epitomize the multifunctional properties and applicability of biosurfactants under diverse environmental conditions. However, profound research is much needed to harness the significance of nanotechnology for the expansion of biosurfactants toward the development of sustainable environmental pollution remediation technologies.

8.10 Challenges and future prospects

Biosurfactants have successfully been employed for the purpose; however, their large-scale applicability is still a challenging task owing to their high costs. More research is needed to study the mechanisms associated with the molecule, for further understanding its possible applications. Further studies are needed to investigate the process and parameters that determine the ability of other microorganisms to produce biosurfactants. Also, more attention must be given to identifying effective treatment methods for low- to middle-income countries considering sustainability in terms of environment and economy. Furthermore, the use of molecular engineering techniques must be encouraged to facilitate the production of engineered microorganisms with function-specific genes incorporated in their genome. These techniques, therefore, enable the generation of a tailored microbial community to facilitate the production of the biosurfactant molecule for enhanced treatment efficiencies. To introduce efficient organisms, more emphasis should be given toward the use of molecular techniques in concert with fields such as proteomics, genomics, and nanotechnology. As biological treatment methods are useful for decreasing organic pollutants, but they are not as effective in removing heavy metals and inorganic chemical compounds, an integrated solution would provide optimal treatment efficiencies. Studies on the combined biological and physico/chemical processes targeting a wide range of pollutants, including organics, inorganics, and heavy metals, show promising perspective. Also, various powerful mathematical models can be adopted to predict and aid in acquiring data for higher prospects of achieving greater efficiency in treating petroleum derivatives. While innovative treatment methods are being investigated for wastewater treatment, challenges exist because of the lack of economic evaluation. Thus, new research must report the cost of remediation methods and challenges with possible pilot/full-scale implementation. Although not yet used on an industrial scale, biosurfactants have tremendous potential for use in MEOR and dispersed compositions to clean the oil-contaminated waters. Advanced research on cost-affordability and the development of biosurfactants can make wastewater treatment technology greener and more economical.

8.11 Conclusion

HCPDs are concerning pollutants released into the environment via various activities of petrochemical and petroleum refining industries and pose a humongous impact on human life. Biosurfactants have been reported to be promising solutions for extensive bioremediation applications of HCPD-contaminated sites, owing to their unique surface-active properties and degradation capacity. They can facilitate the bioavailability of petroleum hydrocarbons for microbial metabolism and operate as mobilizing, solubilizing, or bioemulsifying agents, governed by their molecular weight and critical concentrations. However, the commercialization of biosurfactants is hindered by their low yields and high production costs, needing further investigation from an economic standpoint. Modern technological tools such as nanotechnology and omics technology can be implemented and studied to advance the prospects of bioremediation and pollution abatement.

References

Abdel-Shafy, H. I., & Mansour, M. S. M. (2016). A review on polycyclic aromatic hydrocarbons: Source, environmental impact, effect on human health and remediation. *Egyptian Journal of Petroleum*, *25*(1), 107−123. Available from https://doi.org/10.1016/j.ejpe.2015.03.011.

Adesra, A., Srivastava, V. K., & Varjani, S. (2021). Valorization of dairy wastes: Integrative approaches for value added products. *Indian Journal of Microbiology*, *61*(3), 270−278. Available from https://doi.org/10.1007/s12088-021-00943-5.

Adipah, S. (2018). Introduction of petroleum hydrocarbons contaminants and its human effects. *Journal of Environmental Science and Public Health*, *03*(01). Available from https://doi.org/10.26502/jesph.96120043.

Affandi, I. E., Suratman, N. H., Abdullah, S., Ahmad, W. A., & Zakaria, Z. A. (2014). Degradation of oil and grease from high-strength industrial effluents using locally isolated aerobic biosurfactant-producing bacteria. *International Biodeterioration and Biodegradation, 95*, 33−40. Available from https://doi.org/10.1016/j.ibiod.2014.04.009.

Aghadadashi, V., Mehdinia, A., Riyahi Bakhtiari, A., Mohammadi, J., & Moradi, M. (2019). Source, spatial distribution, and toxicity potential of polycyclic aromatic hydrocarbons in sediments from Iran's environmentally hot zones, the Persian Gulf. *Ecotoxicology and Environmental Safety, 173*, 514−525. Available from https://doi.org/10.1016/j.ecoenv.2019.02.029.

Aguelmous, A., El Fels, L., Souabi, S., Zamama, M., & Hafidi, M. (2019). The fate of total petroleum hydrocarbons during oily sludge composting: A critical review. *Reviews in Environmental Science and Biotechnology, 18*(3), 473−493. Available from https://doi.org/10.1007/s11157-019-09509-w.

Ahmad, Z., Zhang, X., Imran, M., Zhong, H., Andleeb, S., Zulekha, R., Liu, G., Ahmad, I., & Coulon, F. (2021). Production, functional stability, and effect of rhamnolipid biosurfactant from Klebsiella sp. on phenanthrene degradation in various medium systems. *Ecotoxicology and Environmental Safety, 207*, 111514. Available from https://doi.org/10.1016/j.ecoenv.2020.111514.

Al-Hawash, A. B., Zhang, X., & Ma, F. (2019). Removal and biodegradation of different petroleum hydrocarbons using the filamentous fungus Aspergillus sp. RFC-1. *Microbiologyopen, 8*(1), e00619. Available from https://doi.org/10.1002/mbo3.619.

Ambade, B., Sethi, S. S., Kumar, A., Sankar, T. K., & Kurwadkar, S. (2021). Health risk assessment, composition, and distribution of Polycyclic Aromatic Hydrocarbons (PAHs) in Drinking Water of Southern Jharkhand, East India. *Archives of Environmental Contamination and Toxicology, 80*(1), 120−133. Available from https://doi.org/10.1007/s00244-020-00779-y.

Ambust, S., Das, A. J., & Kumar, R. (2021). Bioremediation of petroleum contaminated soil through biosurfactant and Pseudomonas sp. SA3 amended design treatments. *Current Research in Microbial Sciences, 2*, 100031. Available from https://doi.org/10.1016/j.crmicr.2021.100031.

Anestopoulos, I., Kiousi, D. E., Klavaris, A., Galanis, A., Salek, K., Euston, S. R., Pappa, A., & Panayiotidis, M. I. (2020). Surface active agents and their health-promoting properties: Molecules of multifunctional significance. *Pharmaceutics, 12*(7), 1−35. Available from https://doi.org/10.3390/pharmaceutics12070688.

Aparna, A., Srinikethan, G., Smitha, H. (2011). Effect of addition of biosurfactant produced by *Pseudomonas* sps. on biodegradation of crude oil Smitha Hegde. In: *2nd International Conference on Environmental Science and Technology IPCBEE* (6, pp. 71−75).

Barin, R., Talebi, M., Biria, D., & Beheshti, M. (2014). Fast bioremediation of petroleum-contaminated soils by a consortium of biosurfactant/bioemulsifier producing bacteria. *International Journal of Environmental Science and Technology, 11*(6), 1701−1710. Available from https://doi.org/10.1007/s13762-014-0593-0.

Bezza, F. A., & Chirwa, E. M. N. (2017). The role of lipopeptide biosurfactant on microbial remediation of aged polycyclic aromatic hydrocarbons (PAHs)-contaminated soil. *Chemical Engineering Journal, 309*, 563−576. Available from https://doi.org/10.1016/j.cej.2016.10.055.

Bezza, F. A., Beukes, M., & Chirwa, E. M. N. (2015). Application of biosurfactant produced by *Ochrobactrum intermedium* CN3 for enhancing petroleum sludge bioremediation. *Process Biochemistry, 50*(11), 1911−1922. Available from https://doi.org/10.1016/j.procbio.2015.07.002.

Bickong, H. A. E. (2011). Role of surfactin from *Bacillus subtilis* in protection against antimicrobial peptides produced by *Bacillus species*. Doctoral dissertation, Stellenbosch: University of Stellenbosch.

Bordoloi, N. K., & Konwar, B. K. (2009). Bacterial biosurfactant in enhancing solubility and metabolism of petroleum hydrocarbons. *Journal of Hazardous Materials, 170*(1), 495−505. Available from https://doi.org/10.1016/j.jhazmat.2009.04.136.

Buedenbender, L., Kumar, A., Blümel, M., Kempken, F., & Tasdemir, D. (2021). Genomics- and metabolomics-based investigation of the deep-sea sediment-derived yeast, rhodotorula mucilaginosa 50-3-19/20B. *Marine Drugs, 19*(1), 14. Available from https://doi.org/10.3390/md19010014.

Cameotra, S. S., & Bollag, J. M. (2003). Biosurfactant-enhanced bioremediation of polycyclic aromatic hydrocarbons. *Critical Reviews in Environmental Science and Technology, 33*(2), 111−126. Available from https://doi.org/10.1080/10643380390814505.

Carolin, C. F., Kumar, P. S., & Ngueagni, P. T. (2021). A review on new aspects of lipopeptide biosurfactant: Types, production, properties and its application in the bioremediation process. *Journal of Hazardous Materials, 407*, 124827. Available from https://doi.org/10.1016/j.jhazmat.2020.124827.

Charles, F., Nozais, C., Pruski, A. M., Bourgeois, S., Méjanelle, L., Vétion, G., Rivière, B., & Coston-Guarini, J. (2012). Ecodynamics of PAHs at a peri-urban site of the French Mediterranean Sea. *Environmental Pollution, 171*, 256−264. Available from https://doi.org/10.1016/j.envpol.2012.07.034.

Christova, N., Kabaivanova, L., Nacheva, L., Petrov, P., & Stoineva, I. (2019). Biodegradation of crude oil hydrocarbons by a newly isolated biosurfactant producing strain. *Biotechnology and Biotechnological Equipment, 33*(1), 863−872. Available from https://doi.org/10.1080/13102818.2019.1625725.

Coutte, F., Jacques, P., Lecouturier, D., Guez, J.-S., Dhulster, P., Leclere, V., Bechet, M. (2017). *Bacillus sp. biosurfactants, composition including same, method for obtaining same, and use thereof*. Available from https://patents.google.com/patent/US9688725B2/en.

Datta, S., Rajnish, K. N., Samuel, M. S., Pugazlendhi, A., & Selvarajan, E. (2020). Metagenomic applications in microbial diversity, bioremediation, pollution monitoring, enzyme and drug discovery. A review. *Environmental Chemistry Letters, 18*(4), 1229−1241. Available from https://doi.org/10.1007/s10311-020-01010-z.

De Almeida, D. G., Soares Da Silva, R. d C. F., Luna, J. M., Rufino, R. D., Santos, V. A., Banat, I. M., & Sarubbo, L. A. (2016). Biosurfactants: Promising molecules for petroleum biotechnology advances. *Frontiers in Microbiology, 7*, 1718. Available from https://doi.org/10.3389/fmicb.2016.01718.

De França, Í. W. L., Lima, A. P., Lemos, J. A. M., Lemos, C. G. F., Melo, V. M. M., De Sant'ana, H. B., & Gonçalves, L. R. B. (2015). Production of a biosurfactant by *Bacillus subtilis* ICA56 aiming bioremediation of impacted soils. *Catalysis Today, 255*, 10−15. Available from https://doi.org/10.1016/j.cattod.2015.01.046.

de Gusmão, C. A. B., Rufino, R. D., & Sarubbo, L. A. (2010). Laboratory production and characterization of a new biosurfactant from *Candida glabrata* UCP1002 cultivated in vegetable fat waste applied to the removal of hydrophobic contaminant. *World Journal of Microbiology and Biotechnology, 26*(9), 1683−1692. Available from https://doi.org/10.1007/s11274-010-0346-2.

Deivakumari, M., Sanjivkumar, M., Suganya, A. M., Prabakaran, J. R., Palavesam, A., & Immanuel, G. (2020). Studies on reclamation of crude oil polluted soil by biosurfactant producing *Pseudomonas aeruginosa* (DKB1). *Biocatalysis and Agricultural Biotechnology, 29*, 101773. Available from https://doi.org/10.1016/j.bcab.2020.101773.

Dell' Anno, F., Rastelli, E., Sansone, C., Brunet, C., Ianora, A., & Dell' Anno, A. (2021). Bacteria, fungi and microalgae for the bioremediation of marine sediments contaminated by petroleum hydrocarbons in the omics era. *Microorganisms, 9*(8), 1695. Available from https://doi.org/10.3390/microorganisms9081695.

Dobler, L., Vilela, L. F., Almeida, R. V., & Neves, B. C. (2016). Rhamnolipids in perspective: Gene regulatory pathways, metabolic engineering, production and technological forecasting. *New Biotechnology, 33*(1), 123−135. Available from https://doi.org/10.1016/j.nbt.2015.09.005.

El-Sheshtawy, H. S., & Ahmed, W. (2017). Bioremediation of crude oil by *Bacillus licheniformis* in the presence of different concentration nanoparticles and produced biosurfactant. *International Journal of Environmental Science and Technology, 14*(8), 1603−1614. Available from https://doi.org/10.1007/s13762-016-1190-1.

Eras-Muñoz, E., Farré, A., Sánchez, A., Font, X., & Gea, T. (2022). Microbial biosurfactants: A review of recent environmental applications. *Bioengineered, 13*(5), 12365−12391. Available from https://doi.org/10.1080/21655979.2022.2074621.

Fenibo, E. O., Ijoma, G. N., Selvarajan, R., & Chikere, C. B. (2019). Microbial surfactants: The next generation multifunctional biomolecules for applications in the petroleum industry and its associated environmental remediation. *Microorganisms, 7*(11), 581. Available from https://doi.org/10.3390/microorganisms7110581.

Floros, D. J., Jensen, P. R., Dorrestein, P. C., & Koyama, N. (2016). A metabolomics guided exploration of marine natural product chemical space. *Metabolomics. Official Journal of the Metabolomic Society, 12*(9), 5. Available from https://doi.org/10.1007/s11306-016-1087-5.

Freitas, B. G., Brito, J. G. M., Brasileiro, P. P. F., Rufino, R. D., Luna, J. M., Santos, V. A., & Sarubbo, L. A. (2016). Formulation of a commercial biosurfactant for application as a dispersant of petroleum and by-products spilled in oceans. *Frontiers in Microbiology, 7*, 646. Available from https://doi.org/10.3389/fmicb.2016.01646.

Gaur, V. K., Sharma, P., Gupta, S., Varjani, S., Srivastava, J. K., Wong, J. W. C., & Ngo, H. H. (2022). Opportunities and challenges in omics approaches for biosurfactant production and feasibility of site remediation: Strategies and advancements. *Environmental Technology and Innovation*, *25*, 102132. Available from https://doi.org/10.1016/j.eti.2021.102132.

Gawdzik, B., & Gawdzik, J. (2011). Impact of pollution with oil derivatives on the natural environment and methods of their removal. *Ecological Chemistry and Engineering S*, *18*(3), 345–357.

Ghimire, N., & Wang, S. (2019). *Biological treatment of petrochemical wastewater*. IntechOpen. Available from https://doi.org/10.5772/intechopen.79655.

Gospodarek, J., Rusin, M., Barczyk, G., & Nadgórska-Socha, A. (2021). The effect of petroleum-derived substances and their bioremediation on soil enzymatic activity and soil invertebrates. *Agronomy*, *11*(1), 80. Available from https://doi.org/10.3390/agronomy11010080.

Guo, P., Xu, W., Tang, S., Cao, B., Wei, D., Zhang, M., Lin, J., & Li, W. (2022). Isolation and characterization of a biosurfactant producing strain Planococcus sp. XW-1 from the cold marine environment. *International Journal of Environmental Research and Public Health*, *19*(2), 782. Available from https://doi.org/10.3390/ijerph19020782.

Guzmán-López, O., Cuevas-Díaz, M., del, C., Martínez Toledo, A., Contreras-Morales, M. E., Ruiz-Reyes, C. I., Ortega Martínez, A., & del, C. (2021). Fenton-biostimulation sequential treatment of a petroleum-contaminated soil amended with oil palm bagasse (Elaeis guineensis). *Chemistry and Ecology*, *37*(6), 573–588. Available from https://doi.org/10.1080/02757540.2021.1909003.

Haddadin, M. S. Y., Abou Arqoub, A. A., Abu Reesh, I., & Haddadin, J. (2009). Kinetics of hydrocarbon extraction from oil shale using biosurfactant producing bacteria. *Energy Conversion and Management*, *50*(4), 983–990. Available from https://doi.org/10.1016/j.enconman.2008.12.015.

Hewald, S., Linne, U., Scherer, M., Marahiel, M. A., Kämper, J., & Bölker, M. (2006). Identification of a gene cluster for biosynthesis of mannosylerythritol lipids in the basidiomycetous fungus Ustilago maydis. *Applied and Environmental Microbiology*, *72*(8), 5469–5477. Available from https://doi.org/10.1128/AEM.00506-06.

Hmidet, N., Jemil, N., & Nasri, M. (2019). Simultaneous production of alkaline amylase and biosurfactant by *Bacillus methylotrophicus* DCS1: Application as detergent additive. *Biodegradation*, *30*(4), 247–258. Available from https://doi.org/10.1007/s10532-018-9847-8.

Hori, K., Ichinohe, R., Unno, H., & Marsudi, S. (2011). Simultaneous syntheses of polyhydroxyalkanoates and rhamnolipids by *Pseudomonas aeruginosa* IFO3924 at various temperatures and from various fatty acids. *Biochemical Engineering Journal*, *53*(2), 196–202. Available from https://doi.org/10.1016/j.bej.2010.10.011.

Huang, Y., Zhou, H., Zheng, G., Li, Y., Xie, Q., You, S., & Zhang, C. (2020). Isolation and characterization of biosurfactant-producing *Serratia marcescens* ZCF25 from oil sludge and application to bioremediation. *Environmental Science and Pollution Research*, *27*(22), 27762–27772. Available from https://doi.org/10.1007/s11356-020-09006-6.

Islam, B. (2015). Petroleum sludge, its treatment and disposal: A review. *International Journal of Chemical Sciences*, *13*(4), 1584–1602. Available from http://www.sadgurupublications.com/ContentPaper/2015/4_2091_13(4)2015_IJCS.pdf.

Jahan, R., Bodratti, A. M., Tsianou, M., & Alexandridis, P. (2020). Biosurfactants, natural alternatives to synthetic surfactants: Physicochemical properties and applications. *Advances in Colloid Interface Science*, *275*, 102061. Available from https://doi:10.1016/j.cis.2019.102061.

Jayashree, R. (2021). Chapter-3 Biosurfactants and its application to recover petroproducts in petroleum industry. *In Research Trends in Environmental Science*, *8*, 29–50. Available from https://doi.org/10.13140/RG.2.2.27269.47847.

Jimoh, A. A., & Lin, J. (2019). Biosurfactant: A new frontier for greener technology and environmental sustainability. *Ecotoxicology and Environmental Safety*, *184*, 607. Available from https://doi.org/10.1016/j.ecoenv.2019.109607.

Johnson, P., Trybala, A., Starov, V., & Pinfield, V. J. (2021). Effect of synthetic surfactants on the environment and the potential for substitution by biosurfactants. *Advances in Colloid and Interface Science*, *288*, 102340. Available from https://doi.org/10.1016/j.cis.2020.102340.

Joshi-Navare, K., Khanvilkar, P., & Prabhune, A. (2013). Jatropha oil derived sophorolipids: Production and characterization as laundry detergent additive. *Biochemistry Research International*. Available from https://doi.org/10.1155/2013/169797.

Kaczorek, E., Pacholak, A., Zdarta, A., & Smułek, W. (2018). The impact of biosurfactants on microbial cell properties leading to hydrocarbon bioavailability increase. *Colloids and Interfaces*, *2*(3), 35. Available from https://doi.org/10.3390/colloids2030035.

Kang, S. W., Kim, Y. B., Shin, J. D., & Kim, E. K. (2010). Enhanced biodegradation of hydrocarbons in soil by microbial biosurfactant, sophorolipid. *Applied Biochemistry and Biotechnology*, *160*(3), 780–790. Available from https://doi.org/10.1007/s12010-009-8580-5.

Kavuthodi, B., Thomas, S., & Sebastian, D. (2015). Co-production of pectinase and biosurfactant by the newly isolated strain *Bacillus subtilis* BKDS1. *British Microbiology Research Journal*, *10*(2), 1–12. Available from https://doi.org/10.9734/bmrj/2015/19627.

Karlapudi, A. P., Venkateswarulu, T. C., Tammineedi, J., Kanumuri, L., Ravuru, B. K., Dirisala, V. r., & Kodali, V. P. (2018). Role of biosurfactants in bioremediation of oil pollution-a review. *Petroleum*, *4*(3), 241–249. Available from https://doi.org/10.1016/j.petlm.2018.03.007.

Khalid, F. E., Lim, Z. S., Sabri, S., Gomez-Fuentes, C., Zulkharnain, A., & Ahmad, S. A. (2021). Bioremediation of diesel contaminated marine water by bacteria: A review and bibliometric analysis. *Journal of Marine Science and Engineering*, *9*(2), 1–19. Available from https://doi.org/10.3390/jmse9020155.

Kitamoto, D., Ikegami, T., Suzuki, G. T., Sasaki, A., Takeyama, Y. I., Idemoto, Y., Koura, N., & Yanagishita, H. (2001). Microbial conversion of n-alkanes into glycolipid biosurfactants, mannosylerythritol lipids, by Pseudozyma (*Candida antarctica*). *Biotechnology Letters*, *23*(20), 1709–1714. Available from https://doi.org/10.1023/A:1012464717259.

Kosaric, N., Cairns, W. L., Gray, N. C. C., Stechey, D., & Wood, J. (1984). The role of nitrogen in multiorganism strategies for biosurfactant production. *Journal of the American Oil Chemists' Society*, *61*(11), 1735–1743. Available from https://doi.org/10.1007/BF02582138.

Kubicki, S., Bollinger, A., Katzke, N., Jaeger, K. E., Loeschcke, A., & Thies, S. (2019). Marine biosurfactants: Biosynthesis, structural diversity and biotechnological applications. *Marine Drugs*, *17*(7). Available from https://doi.org/10.3390/md17070408.

Kumar, B., Verma, V. K., & Kumar, S. (2022). Source apportionment and risk of polycyclic aromatic hydrocarbons in Indian sediments: A review. *Arabian Journal of Geosciences*, *15*(6). Available from https://doi.org/10.1007/s12517-022-09771-3.

Kumari, B., & Singh, D. P. (2016). A review on multifaceted application of nanoparticles in the field of bioremediation of petroleum hydrocarbons. *Ecological Engineering*, *97*, 98–105. Available from https://doi.org/10.1016/j.ecoleng.2016.08.006.

Kumari, B., Singh, S. N., & Singh, D. P. (2012). Characterization of two biosurfactant producing strains in crude oil degradation. *Process Biochemistry*, *47*(12), 2463–2471. Available from https://doi.org/10.1016/j.procbio.2012.10.010.

Lai, C. C., Huang, Y. C., Wei, Y. H., & Chang, J. S. (2009). Biosurfactant-enhanced removal of total petroleum hydrocarbons from contaminated soil. *Journal of Hazardous Materials*, *167*(1–3), 609–614. Available from https://doi.org/10.1016/j.jhazmat.2009.01.017.

Lászlová, K., Dudášová, H., Olejníková, P., Horváthová, G., Velická, Z., Horváthová, H., & Dercová, K. (2018). The application of biosurfactants in bioremediation of the aged sediment contaminated with polychlorinated biphenyls. *Water, Air, and Soil Pollution*, *229*(7). Available from https://doi.org/10.1007/s11270-018-3872-4.

Li, H., Chen, J., & Jiang, L. (2014). Elevated critical micelle concentration in soil–water system and its implication on PAH removal and surfactant selecting. *Environmental Earth Sciences*, *71*(9), 3991–3998. Available from https://doi.org/10.1007/s12665-013-2783-3.

Lima, M. S. T., Fonseca, A. F. A., Leão, B. A., & Mounteer, A. (2011). Oil recovery from fuel oil storage tank sludge using biosurfactants. *Journal of Bioremediation and Biodegradation*, *02*(04). Available from https://doi.org/10.4172/2155-6199.1000125.

Liu, C., Zhang, Y., Sun, S., Huang, L., Yu, L., Liu, X., Lai, R., Luo, Y., Zhang, Z., & Zhang, Z. (2018). Oil recovery from tank bottom sludge using rhamnolipids. *Journal of Petroleum Science and Engineering*, *170*, 14–20. Available from https://doi.org/10.1016/j.petrol.2018.06.031.

Liu, J., Vipulanandan, C., Cooper, T. F., & Vipulanandan, G. (2013). Effects of Fe nanoparticles on bacterial growth and biosurfactant production. *Journal of Nanoparticle Research*, *15*(1). Available from https://doi.org/10.1007/s11051-012-1405-4.

Luna, J. M., Rufino, R. D., Campos-Takaki, G. M., & Sarubbo, L. A. (2012). Properties of the biosurfactant produced by *Candida sphaerica* cultivated in low-cost substrates. In Chemical Engineering Transactions. *Italian Association of Chemical Engineering — AIDIC*. Available from https://doi.org/10.3303/CET1227012.

Ma, W., Hu, J., Li, J., Li, J., Wang, P., & Okoli, C. P. (2022). Distribution, source, and health risk assessment of polycyclic aromatic hydrocarbons in the soils from a typical petroleum refinery area in south China. *Environmental Monitoring and Assessment*, *194*(10). Available from https://doi.org/10.1007/s10661-022-10281-8.

Mahjoubi, M., Cappello, S., Souissi, Y., Jaouani, A., & Cherif, A. (2018). Microbial bioremediation of petroleum hydrocarbon—contaminated marine environments. IntechOpen. Available from https://doi.org/10.5772/intechopen.72207.

Maikudi Usman, M., Dadrasnia, A., Tzin Lim, K., Fahim Mahmud, A., & Ismail, S. (2016). Application of biosurfactants in environmental biotechnology; remediation of oil and heavy metal. *AIMS Bioengineering*, *3*(3), 289—304. Available from https://doi.org/10.3934/bioeng.2016.3.289.

Makkar, R. S., Cameotra, S. S., & Banat, I. M. (2011). Advances in utilization of renewable substrates for biosurfactant production. *AMB Express*, *1*(1), 1—19. Available from https://doi.org/10.1186/2191-0855-1-5.

Makkar, R., & Cameotra, S. (2002). An update on the use of unconventional substrates for biosurfactant production and their new applications. *Applied Microbiology and Biotechnology*, *58*(4), 428—434. Available from https://doi.org/10.1007/s00253-001-0924-1.

Malik, G., Arora, R., Chaturvedi, R., & Paul, M. S. (2022). Implementation of genetic engineering and novel omics approaches to enhance bioremediation: A focused review. *Bulletin of Environmental Contamination and Toxicology*, *108*(3), 443—450. Available from https://doi.org/10.1007/s00128-021-03218-3.

Maron, P. A., Ranjard, L., Mougel, C., & Lemanceau, P. (2007). Metaproteomics: A new approach for studying functional microbial ecology. *Microbial Ecology*, *53*(Issue 3), 486—493. Available from https://doi.org/10.1007/s00248-006-9196-8.

Mishra, S., Lin, Z., Pang, S., Zhang, Y., Bhatt, P., & Chen, S. (2021). Biosurfactant is a powerful tool for the bioremediation of heavy metals from contaminated soils. *Journal of Hazardous Materials*, *418*, 126253. Available from https://doi.org/10.1016/j.jhazmat.2021.126253.

Mnif, I., Mnif, S., Sahnoun, R., Maktouf, S., Ayedi, Y., Ellouze-Chaabouni, S., & Ghribi, D. (2015). Biodegradation of diesel oil by a novel microbial consortium: Comparison between co-inoculation with biosurfactant-producing strain and exogenously added biosurfactants. *Environmental Science and Pollution Research*, *22*(19), 14852—14861. Available from https://doi.org/10.1007/s11356-015-4488-5.

Mohanty, S. S., Koul, Y., Varjani, S., Pandey, A., Ngo, H. H., Chang, J. S., Wong, J. W. C., & Bui, X. T. (2021). A critical review on various feedstocks as sustainable substrates for biosurfactants production: A way towards cleaner production. *Microbial Cell Factories*, *20*(1). Available from https://doi.org/10.1186/s12934-021-01613-3.

Molaei, S., Moussavi, G., Talebbeydokhti, N., & Shekoohiyan, S. (2022). Biodegradation of the petroleum hydrocarbons using an anoxic packed-bed biofilm reactor with in-situ biosurfactant-producing bacteria. *Journal of Hazardous Materials*, *421*, 126699. Available from https://doi.org/10.1016/j.jhazmat.2021.126699.

Mutalik, S. R., Vaidya, B. K., Joshi, R. M., Desai, K. M., & Nene, S. N. (2008). Use of response surface optimization for the production of biosurfactant from Rhodococcus spp. MTCC 2574. *Bioresource Technology*, *99*(16), 7875—7880. Available from https://doi.org/10.1016/j.biortech.2008.02.027.

Nayak, A. S., Vijaykumar, M. H., & Karegoudar, T. B. (2009). Characterization of biosurfactant produced by Pseudoxanthomonas sp. PNK-04 and its application in bioremediation. *International Biodeterioration and Biodegradation*, *63*(1), 73—79. Available from https://doi.org/10.1016/j.ibiod.2008.07.003.

Nuñal, S. N., Leon, S.-D., Bacolod, S. M. S., Koyama, E., Uno, J., Hidaka, S., Yoshikawa, T., M., & Maeda, H. (2014). Bioremediation of heavily oil-polluted seawater by a bacterial consortium immobilized in cocopeat and rice hull powder. *Biocontrol Science*, *19*(1), 11—22. Available from https://doi.org/10.4265/bio.19.11.

Nurfarahin, A. H., Mohamed, M. S., & Phang, L. Y. (2018). Culture medium development for microbial-derived surfactants production—an overview. *Molecules (Basel, Switzerland)*, *23*(5). Available from https://doi.org/10.3390/molecules23051049.

Ossai, I. C., Ahmed, A., Hassan, A., & Hamid, F. S. (2020). Remediation of soil and water contaminated with petroleum hydrocarbon: A review. *Environmental Technology and Innovation*, *17*, 100526. Available from https://doi.org/10.1016/j.eti.2019.100526.

Patel, J., Borgohain, S., Kumar, M., Rangarajan, V., Somasundaran, P., & Sen, R. (2015). Recent developments in microbial enhanced oil recovery. *Renewable and Sustainable Energy Reviews*, *52*, 1539–1558. Available from https://doi.org/10.1016/j.rser.2015.07.135.

Patel, S., Homaei, A., Patil, S., & Daverey, A. (2019). Microbial biosurfactants for oil spill remediation: Pitfalls and potentials. *Applied Microbiology and Biotechnology*, *103*(1), 27–37. Available from https://doi.org/10.1007/s00253-018-9434-2.

Patoway, K., Patoway, R., Kalita, M. C., & Deka, S. (2017). Characterization of biosurfactant produced during degradation of hydrocarbons using crude oil as sole source of carbon. *Frontiers in Microbiology*, *8*, 279. Available from https://doi.org/10.3389/fmicb.2017.00279.

Phulpoto, I. A., Yu, Z., Hu, B., Wang, Y., Ndayisenga, F., Li, J., Liang, H., & Qazi, M. A. (2020). Production and characterization of surfactin-like biosurfactant produced by novel strain Bacillus nealsonii S2MT and it's potential for oil contaminated soil remediation. *Microbial Cell Factories*, *19*(1). Available from https://doi.org/10.1186/s12934-020-01402-4.

Pourfadakari, S., Ghafari, S., Takdastan, A., & Jorfi, S. (2021). A salt resistant biosurfactant produced by moderately halotolerant *Pseudomonas aeruginosa* (AHV-KH10) and its application for bioremediation of diesel-contaminated sediment in saline environment. *Biodegradation*, *32*(3), 327–341. Available from https://doi.org/10.1007/s10532-021-09941-2.

Prakash, A. A., Prabhu, N. S., Rajasekar, A., Parthipan, P., AlSalhi, M. S., Devanesan, S., & Govarthanan, M. (2021). Bio-electrokinetic remediation of crude oil contaminated soil enhanced by bacterial biosurfactant. *Journal of Hazardous Materials*, *405*, 4061. Available from https://doi.org/10.1016/j.jhazmat.2020.124061.

Rabani, M. S., Habib, A., & Gupta, M. K. (2020). *Polycyclic aromatic hydrocarbons: Toxic effects and their bioremediation strategies* (pp. 65–105). Springer Science and Business Media LLC. Available from https://doi.org/10.1007/978-3-030-48690-7_4.

Rahman, K. S. M., Rahman, T. J., Kourkoutas, Y., Petsas, I., Marchant, R., & Banat, I. M. (2003). Enhanced bioremediation of n-alkane in petroleum sludge using bacterial consortium amended with rhamnolipid and micronutrients. *Bioresource Technology*, *90*(2), 159–168. Available from https://doi.org/10.1016/S0960-8524(03)00114-7.

Rajendran, R., Ragavan, R. P., Al-Sehemi, A. G., Uddin, M. S., Aleya, L., & Mathew, B. (2022). Current understandings and perspectives of petroleum hydrocarbons in Alzheimer's disease and Parkinson's disease: A global concern. *Environmental Science and Pollution Research*, *29*(8), 10928–10949. Available from https://doi.org/10.1007/s11356-021-17931-3.

Rao, K. N., Kumaran, D., Seetharaman, J., Bonanno, J. B., Burley, S. K., & Swaminathan, S. (2006). Crystal structure of trehalose-6-phosphate phosphatase-related protein: Biochemical and biological implications. *Protein Science*, *15*(7), 1735–1744. Available from https://doi.org/10.1110/ps.062096606.

Rasheed, T., Shafi, S., Bilal, M., Hussain, T., Sher, F., & Rizwan, K. (2020). Surfactants-based remediation as an effective approach for removal of environmental pollutants—A review. *Journal of Molecular Liquids*, *318*, 113960. Available from https://doi.org/10.1016/j.molliq.2020.113960.

Ren, H., Zhou, S., Wang, B., Peng, L., & Li, X. (2020). Treatment mechanism of sludge containing highly viscous heavy oil using biosurfactant. *Colloids and Surfaces A: Physicochemical and Engineering Aspects*, *585*, 124117. Available from https://doi.org/10.1016/j.colsurfa.2019.124117.

Ribeiro, B. G., Guerra, J. M. C., & Sarubbo, L. A. (2020). Biosurfactants: Production and application prospects in the food industry. *Biotechnology Progress*, *36*(5). Available from https://doi.org/10.1002/btpr.3030.

Rizvi, H., Verma, J. S., & Ashish. (2021). *Biosurfactants for oil pollution remediation* (pp. 197–212). Springer Science and Business Media LLC. Available from https://doi.org/10.1007/978-981-15-6607-3_9.

Sah, D., Rai, J. P. N., Ghosh, A., & Chakraborty, M. (2022). A review on biosurfactant producing bacteria for remediation of petroleum contaminated soils. *3 Biotech*, *12*(9). Available from https://doi.org/10.1007/s13205-022-03277-1.

Salehi, S., Abdollahi, K., Panahi, R., Rahmanian, N., Shakeri, M., & Mokhtarani, B. (2021). Applications of biocatalysts for sustainable oxidation of phenolic pollutants: A review. *Sustainability (Switzerland)*, *13*(15). Available from https://doi.org/10.3390/su13158620.

Sánchez, C. (2022). A review of the role of biosurfactants in the biodegradation of hydrophobic organopollutants: Production, mode of action, biosynthesis and applications. *World Journal of Microbiology and Biotechnology*, *38*(11). Available from https://doi.org/10.1007/s11274-022-03401-6.

Santos, D. K. F., Rufino, R. D., Luna, J. M., Santos, V. A., & Sarubbo, L. A. (2016). Biosurfactants: Multifunctional biomolecules of the 21st century. *International Journal of Molecular Sciences*, *17*(3). Available from https://doi.org/10.3390/ijms17030401.

Sarma, H., Bustamante, K. L. T., & Prasad, M. N. V. (2019). Biosurfactants for oil recovery from refinery sludge: Magnetic nanoparticles assisted purification. *Industrial and municipal sludge: Emerging concerns and scope for resource recovery* (pp. 107−132). Elsevier. Available from https://doi.org/10.1016/B978-0-12-815907-1.00006-4.

Shah, A., Shahzad, S., Munir, A., Nadagouda, M. N., Khan, G. S., Shams, D. F., Dionysiou, D. D., & Rana, U. A. (2016). Micelles as soil and water decontamination agents. *Chemical Reviews*, *116*(10), 6042−6074. Available from https://doi.org/10.1021/acs.chemrev.6b00132.

Shin, K. H., Kim, K. W., & Ahn, Y. (2006). Use of biosurfactant to remediate phenanthrene-contaminated soil by the combined solubilization-biodegradation process. *Journal of Hazardous Materials*, *137*(3), 1831−1837. Available from https://doi.org/10.1016/j.jhazmat.2006.05.025.

Singh, H., Bhardwaj, N., Arya, S. K., & Khatri, M. (2020). Environmental impacts of oil spills and their remediation by magnetic nanomaterials. *Environmental Nanotechnology, Monitoring & Management*, *14*, 100305. Available from https://doi.org/10.1016/j.enmm.2020.100305.

Singh, P., Patil, Y., & Rale, V. (2019). Biosurfactant production: Emerging trends and promising strategies. *Journal of Applied Microbiology*, *126*(1), 2−13. Available from https://doi.org/10.1111/jam.14057.

Sirohi, R., Joun, J., Choi, H. I., Gaur, V. K., & Sim, S. J. (2021). Algal glycobiotechnology: Omics approaches for strain improvement. *Microbial Cell Factories*, *20*(1). Available from https://doi.org/10.1186/s12934-021-01656-6.

Sivapathasekaran, C., & Sen, R. (2017). Origin, properties, production and purification of microbial surfactants as molecules with immense commercial potential. *Tenside, Surfactants, Detergents*, *54*(2), 92−104. Available from https://doi.org/10.3139/113.110482.

Soares da Silva, R. d. C. F., Luna, J. M., Rufino, R. D., & Sarubbo, L. A. (2021). Ecotoxicity of the formulated biosurfactant from *Pseudomonas cepacia* CCT 6659 and application in the bioremediation of terrestrial and aquatic environments impacted by oil spills. *Process Safety and Environmental Protection*, *154*, 338−347. Available from https://doi.org/10.1016/j.psep.2021.08.038.

Sobrinho, H. B. S., Rufino, R. D., Luna, J. M., Salgueiro, A. A., Campos-Takaki, G. M., Leite, L. F. C., & Sarubbo, L. A. (2008). Utilization of two agroindustrial by-products for the production of a surfactant by *Candida sphaerica* UCP0995. *Process Biochemistry*, *43*(9), 912−917. Available from https://doi.org/10.1016/j.procbio.2008.04.013.

Swathi, K. V., Muneeswari, R., Ramani, K., & Sekaran, G. (2020). Biodegradation of petroleum refining industry oil sludge by microbial-assisted biocarrier matrix: Process optimization using response surface methodology. *Biodegradation*, *31*(4−6), 385−405. Available from https://doi.org/10.1007/s10532-020-09916-9.

Syldatk, C., & Hausmann, R. (2010). Microbial biosurfactants. *European Journal of Lipid Science and Technology*, *112*(6), 615−616. Available from https://doi.org/10.1002/ejlt.201000294.

Thenmozhi, R., Arumugam, K., Nagasathya, A., & Thajuddin, N. (2013). Studies on Mycoremediation of used engine oil contaminated soil samples. *Advances in Applied Science Research*, *4*(2), 110−118.

Uddin, M., Swathi, K. V., Anil, A., Boopathy, R., Ramani, K., & Sekaran, G. (2021). Biosequestration of lignin in municipal landfill leachate by tailored cationic lipoprotein biosurfactant through *Bacillus tropicus* valorized tannery solid waste. *Journal of Environmental Management*, *300*, 113755. Available from https://doi.org/10.1016/j.jenvman.2021.113755.

Unaeze., & Henrietta, C. (2020). Application of biosurfactants in environmental remediation. *IOSR Journal of Environmental Science*, *14*(10), 30−42. Available from https://doi.org/10.9790/2402-1410043042.

Urum, K., Pekdemir, T., & Gopur, M. (2003). Optimum conditions for washing of crude oil-contaminated soil with biosurfactant solutions. *Process Safety and Environmental Protection: Transactions of the Institution of Chemical Engineers, Part B*, *81*(3), 203−209. Available from https://doi.org/10.1205/095758203765639906.

Wang, D., Ma, J., Li, H., & Zhang, X. (2018). Concentration and potential ecological risk of PAHs in different layers of soil in the petroleum-contaminated areas of the loess plateau. China. *International Journal of Environmental Research and Public Health*, *15*(8), 1785. Available from https://doi.org/10.3390/ijerph15081785.

Wang, Y., Nie, M., Diwu, Z., Lei, Y., Li, H., & Bai, X. (2019). Characterization of trehalose lipids produced by a unique environmental isolate bacterium Rhodococcus qingshengii strain FF. *Journal of Applied Microbiology*, *127*(5), 1442−1453. Available from https://doi.org/10.1111/jam.14390.

Worley, B., & Powers, R. (2012). Multivariate analysis in metabolomics. *Current Metabolomics*, *1*(1), 92−107. Available from https://doi.org/10.2174/2213235X11301010092.

Wu, Y., Xu, M., Xue, J., Shi, K., & Gu, M. (2019). Characterization and enhanced degradation potentials of biosurfactant-producing bacteria isolated from a marine environment. *ACS Omega*, *4*(1), 1645−1651. Available from https://doi.org/10.1021/acsomega.8b02653.

Xia, M., Fu, D., Chakraborty, R., Singh, R. P., & Terry, N. (2019). Enhanced crude oil depletion by constructed bacterial consortium comprising bioemulsifier producer and petroleum hydrocarbon degraders. *Bioresource Technology*, *282*, 456−463. Available from https://doi.org/10.1016/j.biortech.2019.01.131.

Xu, X., Liu, W., Tian, S., Wang, W., Qi, Q., Jiang, P., Gao, X., Li, F., Li, H., & Yu, H. (2018). Petroleum hydrocarbon-degrading bacteria for the remediation of oil pollution under aerobic conditions: A perspective analysis. *Frontiers in Microbiology*, *9*, 2885. Available from https://doi.org/10.3389/fmicb.2018.02885.

Yaraguppi, D. A., Bagewadi, Z. K., Mahanta, N., Singh, S. P., Yunus Khan, T. M., Deshpande, S. H., Soratur, C., Das, S., & Saikia, D. (2022). Gene expression and characterization of iturin a lipopeptide biosurfactant from *Bacillus aryabhattai* for enhanced oil recovery. *Gels*, *8*(7). Available from https://doi.org/10.3390/gels8070403.

Zahed, M. A., Matinvafa, M. A., Azari, A., & Mohajeri, L. (2022). Biosurfactant, a green and effective solution for bioremediation of petroleum hydrocarbons in the aquatic environment. *Discover Water*, *2*(1). Available from https://doi.org/10.1007/s43832-022-00013-x.

Zaki, M. S., Authman, M. M., & Abbas, H. H. (2015). Bioremediation of petroleum contaminants in aquatic environments (Review Article). *Life Science Journal*, *12*(5), 109−121.

Zhang, F., Huo, K., Song, X., Quan, Y., Wang, S., Zhang, Z., Gao, W., & Yang, C. (2020). Engineering of a genome-reduced strain Bacillus amyloliquefaciens for enhancing surfactin production. *Microbial Cell Factories*, *19*(1). Available from https://doi.org/10.1186/s12934-020-01485-z.

Zhang, T., Liu, Y., Zhong, S., & Zhang, L. (2020). AOPs-based remediation of petroleum hydrocarbons-contaminated soils: Efficiency, influencing factors and environmental impacts. *Chemosphere*, *246*, 125726. Available from https://doi.org/10.1016/j.chemosphere.2019.125726.

Zhong, H., Liu, G., Jiang, Y., Yang, J., Liu, Y., Yang, X., Liu, Z., & Zeng, G. (2017). Transport of bacteria in porous media and its enhancement by surfactants for bioaugmentation: A review. *Biotechnology Advances*, *35*(4), 490−504. Available from https://doi.org/10.1016/j.biotechadv.2017.03.009.

Zhou, D., Hu, F., Lin, J., Wang, W., & Li, S. (2019). Genome and transcriptome analysis of *Bacillus velezensis* BS-37, an efficient surfactin producer from glycerol, in response to d-/l-leucine. *MicrobiologyOpen*, *8*(8). Available from https://doi.org/10.1002/mbo3.794.

CHAPTER 9

Removal of hydrophobic contaminant/petroleum derivate utilizing biosurfactants

Chiamaka Linda Mgbechidinma and Chunfang Zhang
Institute of Marine Biology and Pharmacology, Ocean College, Zhejiang University, Zhoushan, Zhejiang, P.R. China

9.1 Introduction

Environmentally, biosurfactants have received much attention in recent years due to their broad spectrum of properties, consisting of their lower toxicity, higher biodegradability, and stability at extreme temperatures, pH, and salinity (Wu et al., 2022). It is well known that synthetic surfactants, including anionic, nonionic, cationic, and mixed surfactants, coupled with chelating agents, can be used to improve soil washing and remove hydrophobic organic compounds (HOCs) from polluted soils and groundwater (Zhang et al., 2022). Some synthetic surfactants have been shown to increase nonpolar chemical concentrations in the aqueous phase during HOC removal. However, residual synthetic surfactants in soils and groundwater pose a risk of toxicity to the environment and human health (Pandey et al., 2022). Biosurfactants are thus an improved technique for environmental cleansing instead of synthetic surfactants.

According to reports in several journals, most previous studies have focused on a few biosurfactants, although the neglected biosurfactants may have more promising qualities that are environmentally compatible. When evaluating the techno-economic significance, the possible production of biosurfactants from renewable feedstock has revealed probable sustainability in terms of production and application (Dierickx et al., 2022; Mgbechidinma, Akan et al., 2022). However, the major limitation in the use of biosurfactants is at the microproduction scale (Silva et al., 2022), as most biosynthesis practices are at the laboratory stage. The discovery of new biosurfactants with novel fermentation and recovery procedures combined with the use of low-cost raw materials (such as agro-industry wastes as carbon sources) will increase the availability of more affordable biosurfactants that can be used for remediation processes (Mgbechidinma, Zhang, et al., 2022). In producing low-cost biosurfactants, Subsanguan et al. (2022) recently established that a hydrophilic—lipophilic balance could be achieved by formulating a synergistic glycolipid: lipopeptide mixture from

Weissella cibaria PN3 and *Bacillus subtilis* GY19. However, employing the cell immobilization technique in a stirred tank bio-fermenter ensures sustainable production.

Interestingly, it has been observed that bioavailability or mass transfer limitation of hydrophobic contaminant/petroleum derivate (HC/PD) in terrestrial soils and aquatic sediments leads to poor microbial accessibility and thus reduced biodegradation efficiency in contaminated ecosystems (Guo et al., 2022; Kariyawasam et al., 2022). The penetration of HC/PD into the soil/sediment matrix is a highly complex process, including physical, chemical, and biological components, and as HC/PD are highly hydrophobic materials, their mobility is limited (Bolan et al., 2023; Carolin et al., 2021). Hence, HC/PDs are water-insoluble as their components bind to soil/sediment particles, reducing their bioavailability and restricting their mass transfer rate for biodegradation. In the mathematical study to elucidate hydrocarbon bioremediation, Martins et al. (2022) emphasized that HC/PD in soils can take several physical forms, including being dissolved in pore water, adsorbed onto soil particles, absorbed into soil particles, or existing as a separate phase that can be liquid or solid. Therefore, a fundamental mechanism for increasing the bioavailability of HC/PD pollutants is by transporting them to the aqueous bulk phase. In short, the use of biosurfactants to accelerate the desorption and solubilization of petroleum hydrocarbons promotes their assimilation by microorganisms, and this represents one of the most efficient strategies to increase HC/PD bioavailability (or the solubility) in aqueous phase (Cecotti et al., 2018; Wang et al., 2021; Zhang et al., 2022).

Biosurfactants are soluble in water at low concentrations and produce micelles in solution as concentrations increase, forming a critical micelle concentration (CMC). Micelle formation gives rise to biosurfactants solubilization of HC/PD in soil–water systems (Zhou et al., 2021), improving techniques such as bioremediation (Carolin et al., 2021), phytoremediation (Sonowal et al., 2022), bio-electrokinetic remediation (Prakash et al., 2021), microbial electrochemical treatment (Ambaye et al., 2022), and other remediation processes. As such, biosurfactants can break down HC/PD that threaten environmental sustainability and clean water provision through pollution (Goveas et al., 2022; Li et al., 2021; Rizzo et al., 2021). Given the detrimental effects of HC/PD, there is a need to remove these pollutants from the environment using biosurfactants. Therefore, this chapter seeks to reveal relevant literature on the sources, occurrence, fate, health, and ecological implications of HC/PD in marine and terrestrial ecosystems to demonstrate their prevalence in the environment. Key factors influencing the accumulation and transport of HC/PD within ecosystems will be discussed. In addition, the eco-sustainable biosurfactant-based remediation approach as a contributor to achieve some selected United Nations Sustainable Development Goals (UN-SDGs) consistent with possible future outlooks will be enumerated. This knowledge is essential for assessing the environmental risk of HC/PD and developing conservation and management strategies to achieve environmental sustainability.

9.2 Hydrophobic contaminant/petroleum derivate: sources, occurrence, fate, and implications

9.2.1 Sources and occurrence of hydrophobic contaminant/petroleum derivate in aquatic and terrestrial environments

Several diverse ecosystems in the aquatic and terrestrial environments are increasingly threatened by multiple anthropogenic disturbances that result in chemical pollution of sediments and soils (Ambaye et al., 2022). Although hydrophobic pollution is rarely reported in terrestrial ecosystems compared to aquatic environments, the latter also significantly helps maintain planetary health and function by supporting nutrient biogeochemical cycling and biodiversity (Ambaye et al., 2022; Dell'Anno et al., 2018; Mapelli et al., 2017; Montas et al., 2022). HC/PDs enter these ecosystems through atmospheric deposition, oceanic transport, and waste disposal and through anthropogenic activities (seabed mining, fishing, ship traffic, accidental spillage, industrial, and mining activities) (Negreiros et al., 2022; Pandey et al., 2022). Global and regional conventions have been created to govern and protect several ecosystems from HC/PD pollution (Table 9.1); however, studies have reported the prevalence of varying HC/PD in diverse ecosystems. HC/PDs such as polycyclic aromatic hydrocarbons (PAHs), polychlorinated biphenyls (PCBs), dichlorodiphenyltrichloroethane (DDT), organochlorine pesticides, polybrominated diphenyl ethers, polycyclic aromatic hydrocarbons, polyfluoroalkyl substances, pesticides, plastics, and their additives have been detected in surface soils, landfill, vegetable greenhouse soils, industrial areas, agricultural soils, bay sediments, lake sediment cores, sediments near wetland plants, seawater, deep sea, and other extreme environments (Akan et al., 2021; Kumar et al., 2021).

According to Sanganyado et al. (2021), before the ban and production restrictions by the Stockholm Convention, about 2.79 million metric tons of DDT and 1.51 million metric tons of PCBs were produced globally. Furthermore, between 1970 and 2002, the chemical derivatives of perfluorooctane sulfonate accounted for about 96,000 tons of consumption, while perfluoroalkyl carboxylic acids (PFCAs) production reached 21,400 tons within 64 years. With China leading in production, a whopping sum of 30,000 tons of hexabromocyclododecanes was produced globally before being banned. Due to the extensive application of these hydrocarbons in consumer and industrial products, HC/PDs have high potential for long-distance transport. Thus, there are not only detected in rare ecosystems but linked to several lethal diseases in animals and humans, while studies on their effect on plants are not well elucidated.

9.2.2 Fate of hydrophobic contaminant/petroleum derivate in the environment

HC/PDs are continually discharged into the marine and terrestrial environment, where they are usually degraded or accumulated. It is recognized that when crude oil,

Table 9.1 Some conventions and regulations for the environmental prevention of HC/PD (hydrophobic contaminant/petroleum derivate) pollution.

Geographical domain	Conventions/regulation	Measures and policies
Global/International	International Convention for the Prevention of Pollution from Ships (MARPOL)	Prevention of pollution of the marine environment by ships from operational or accidental causes. The regulatory annexes include Prevention of pollution by oil (enforced on October 2, 1983). Control of pollution by noxious liquid substances in bulk (enforced on October 2, 1983). Prevention of pollution by harmful substances carried by sea in packaged form (enforced on July 1, 1992). Prevention of pollution by sewage from ships (enforced on September 27, 2003). Preventing pollution by garbage from ships (enforced on December 31, 1988). Prevention of air pollution from ships (enforced on May 19, 2005)
	United Nations Convention on the Law of the Sea	Section 5 emphasizes on the international rules and national legislation to prevent, reduce, and control pollution of the marine environment. Thus, contingent plans were created and enforced to prevent pollution from land-based sources, seabed activities, activities in the area, by dumping, from vessels, and from or through the atmosphere
	Convention on the Prevention of Marine Pollution by Dumping of Wastes and Other Matter (also called the "London Convention" or "LC 72")	One of the first global conventions to protect the marine environment from human activities and has been in force since 1975. Currently, 87 states are parties to this Convention. The reverse list approach adopted implies that all dumping is prohibited unless explicitly permitted. Incineration of wastes at sea is prohibited. Export of wastes for the purpose of dumping or incineration at sea is prohibited

Regional	Convention for the Protection of the Marine Environment in the North-East Atlantic (OSPAR Convention)	Enables cooperation between 15 governments and the EU to prevent and eliminate pollution by protecting the maritime area against the adverse effects of human activities. The focus is to safeguard human health and conserve marine ecosystems, and, when practicable, restore marine areas which have been adversely affected. The participating countries include Belgium, Germany, the Netherlands, Sweden, Denmark, Iceland, Norway, Switzerland, Finland, Ireland, Portugal, the United Kingdom, France, Luxembourg, Spain, and the European Union
	Convention on the Protection of the Marine Environment in the Baltic Sea Area (HELCOM) https://www.eea.europa.eu/	Regulated hazardous substances include hexachlorobenzene (HCB), benzo[a]pyrene, dichlorodiphenyltrichloroethane (DDT), and polychlorinated biphenyl (PCB)
	Convention for the Protection of Marine Environment and the Coastal Region of the Mediterranean https://www.eea.europa.eu/	The Convention for the Protection of the Mediterranean Sea Against Pollution (the Barcelona Convention) was adopted on February 16, 1976. The Convention provides guidelines for securing the sustainable use of water for humans and different sectors such as health, agriculture, energy production, transport, and nature
	Convention for the protection of the Black Sea https://www.eea.europa.eu/	The Convention is also referred to as "Bucharest Convention" and was signed in Bucharest in April 1992. Bucharest Convention is ratified by all six legislative assemblies of the Black Sea countries at the beginning of 1994. Significantly, the Convention is the basic legal framework for regional cooperation to protect the coastal and marine environment in Russia, Turkey, Ukraine, Georgia, Bulgaria, and Romania

petroleum products, and other HC/PD are released into the environment, these pollutants enter the surface/groundwater and soil/sediments by various means. Generally, the fate of HC/PD in the environments varies.

HC/PD in soils undergoes migration, adsorption, and degradation through physical, chemical, and biological processes. HC/PD migration can reach the deep soil horizon by relative free flow at the top layer and under the action of capillary force by planar diffusion motion and gravity impact along the soil depth in the presence of moisture (Kariyawasam et al., 2022). HC/PD do not entirely persist on the surface; thus, some particles penetrate deep into the soil and, after a long time, are influenced by hydraulic, gravity, diffusion, and mixing factors while gradually forming a more stable state (Jamal, 2022; Kariyawasam et al., 2022). Soils with deeper water tables are associated with lesser surface HC/PD accumulation and hydrodynamic drive. The expansion of HC/PD is more significant transversely in the capillary zone near the groundwater level than along the groundwater level and transversely along the direction of groundwater flow (Bolan et al., 2023; Kariyawasam et al., 2022). HC/PD adsorption depends on the physical and chemical properties, including hydrophobicity, viscosity, intermolecular force, and electrostatic attraction. However, the interference of other factors such as organic matter in the soil promotes HC/PD dissolution and distribution. HC/PD migration and adsorption in the soil are very weak; as such, these pollutants are mostly absorbed and concentrated in the surface soil undergoing self-purification by volatilization into the atmosphere. Other natural transformative pathways include self-oxidation and degradation through slow processes such as photolysis, mechanical degradation, and biodegradation. Biodegradation mediated by microbes is the most widely accepted endpoint for HC/PD in the environment. Some of these pollutants serve as nutrient and energy sources for microbes, while the rest are oxidized or reduced to simple organic or inorganic substances, such as methane, carbon dioxide, and water (Zhou et al., 2020, 2021). At high pollution intensity with large number of small hydrocarbon molecules, HC/PD tends to migrate to groundwater aquifers. The fate of HC/PD in liquid/aqueous media includes weathering, spreading, drifting, evaporation, oxidation, dissolution, photolysis, biodegradation, and formation of both oil-in-water and water-in-oil emulsions. Several factors influence biosurfactant-enhanced remediation, as illustrated in Fig. 9.1.

9.2.3 Ecological and health implications of hydrophobic contaminant/petroleum derivate

Monitoring and characterizing anthropogenic disturbances and their effect on key species is one of the essential means for a robust assessment of the health and ecological implications of HC/PD in the environment. However, the increasing population, coupled with rapid industrialization and urbanization, remains the major contributor to HC/PD emissions in the environment. In the aquatic environment, the presence of

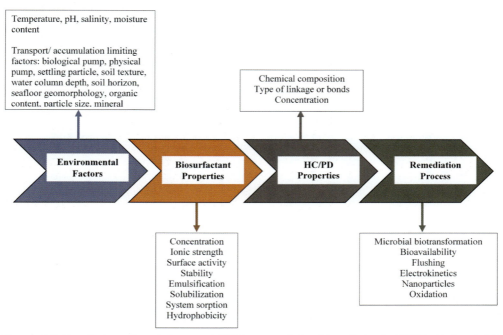

Figure 9.1 Key factors that influence HC/PD removal by biosurfactant. *HC/PD*, Hydrophobic contaminant/petroleum derivate.

HC/PD disrupts the primary food producers, thereby interfering with the transfer of nutrients to the higher organisms, which affects the food chain and ecosystem balance. While the negative economic consequences are seen from the perspectives of tourism industry, fishery, aquaculture, and agricultural deterioration, the social impacts are mainly on the health sector (Fig. 9.2).

So far, some regions have higher hydrophobic contamination in their soils than the national standard threshold, despite the intervention of several conventions at the global and regional levels, which could be attributed to the significant effect of past and recent oil spill incidences and the depletion of HC/PD (Montas et al., 2022; Negreiros et al., 2022), such as the Exxon Valdez oil spill in 1989 (37,000 tonnes), Nigeria in 1998 (5456 tones), United States in 1976 (24,961), Trinidad and Tobago in 1979 (287,000 tones), Kuwait in 1991 (7,557,935 tones), Brazil in 2000 (31,491 tones), Deepwater Horizon oil spill in 2010 (4,900 Mbbl), Mediterranean coast of Israel, and Lebanon in 2021 (hundreds of tons of tar on the beaches along a 160-km (99 mi) stretch), and Peru in 2022 (6000 barrels in the sea). Thus, the aquatic and terrestrial environments are targeted by HC/PD through migration, detention, and deposition, while plants, animals, and humans cope through several mechanisms for survival.

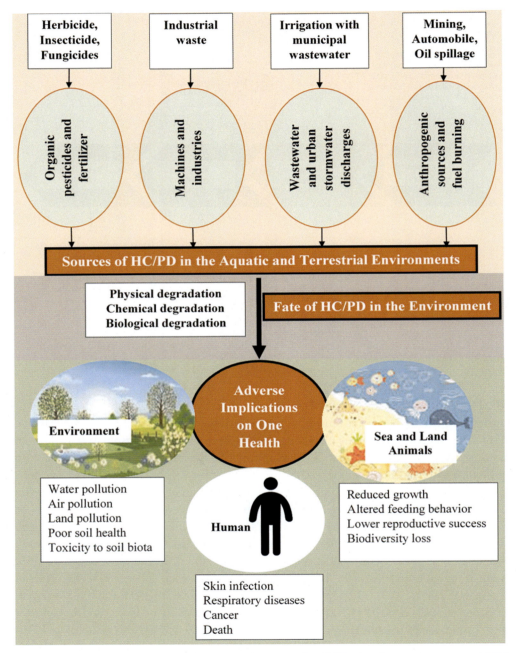

Figure 9.2 Sources, fate, and adverse implication of HC/PD in the environment. *HC/PD*, Hydrophobic contaminant/petroleum derivate.

The health implication of HC/PD in humans and animals simulates a myriad of toxicity pathways in different species, some of which include DNA damage, carcinogenicity, impaired growth, oil ingestion, respiratory dysfunction, improper immune function, morphological abnormalities, accumulation of contaminants in tissues, poor reproductive rate, cardiac dysfunction, and mass mortality of eggs and larvae (Dai et al., 2022; Goveas et al., 2022; Negreiros et al., 2022). Although some plants (primary producers) can convert HC/PD components (carbon, hydrogen, oxygen, nitrogen, and other substances) into nutrients for growth/energy, the stress-inducing signaling pathways have shown HC/PD as a significant pollutant whereby biosurfactants can ensure sustainable agriculture (Gayathiri et al., 2022; Ram et al., 2019; Sangwan et al., 2022).

9.3 Biosurfactants for hydrophobic contaminant/petroleum derivate removal

Reports on the removal of HC/PD using biosurfactants are becoming increasingly attractive, emphasizing the mechanism of action, which depends on the chemical composition of the surfactants. Biosurfactants are classified mainly according to their origin and chemical composition, with a hydrophilic moiety associated with an anion or cation of peptide or amino acid, di- or polysaccharide and a hydrophobic moiety of saturated or unsaturated fatty acids form the basic structure. The two major biosurfactant classifications with details on their types, characteristics, examples, and class of microbial producers have been previously reported (Mgbechidinma, Akan, et al., 2022). Exploring these biomolecules benefits humanity and sustainable environment by integrating green economy concepts (Mgbechidinma, Akan, et al., 2022). A more comprehensive description of the biosurfactant types and their application in HC/PD removal is revealed in this study.

9.3.1 Low-molecular-mass biosurfactants

9.3.1.1 Glycolipids

Glycolipids are the most widely isolated and reported group of biosurfactants, consisting of carbohydrates and chains of aliphatic hydroxyl groups. Examples of glycolipids mainly used for HC/PD removal are rhamnolipids, sophorolipids, and trehalose lipids.

Rhamnolipids have rhamnose as their unique chemical composition and were first isolated from *Pseudomonas aeruginosa*. Rhamnose forms the hydrophilic head, whereas the tail consists of a varying length of 3-(hydroxy-alkanoyloxy) alkanoic acid (HAA) fatty acid (Dobler et al., 2020; Phulpoto et al., 2022; Sharma et al., 2021). Besides the synthesis of mono-rhamnolipids by *P. aeruginosa* rhlA and rhlB genes, di-rhamnolipids production is known to occur in the presence of the rhlC gene encoding rhamnosyltransferase 2. So far, thin-layered chromatography and mass spectrometry are preferred for rhamno-biosurfactant detection. Goveas et al. (2022) showed the dependency of rhamnolipid at low CMC concentration on nitrogen availability during petroleum crude oil remediation in seawater. In addition to the crude petroleum oil serving as the sole carbon source, the degradation efficiency observed was 81.8% ± 0.67% in 28 days on optimization by central composite design. Rhamnolipid alteration at higher concentrations in petroleum hydrocarbon-contaminated soils increases microbial electrochemical remediation through surface tension reduction, anodic electron transfer increase, and improved microbial density (Ambaye et al., 2022).

Sophorolipids comprise two beta-1,2-linked glucose units and are commonly synthesized by yeast strains. The hydroxyl group is often acylated and has a lipid component linked to its reducing end through a glycosidic connection. A binary mixed biosurfactant system consisting of rhamnolipid/sophorolipid has a strong synergistic effect as oily sludge detergents during thermal washing (Bao et al., 2022). Techno-economically, the process has a high level of efficiency and a low biosurfactant consumption rate. Trehalose lipids, also called cord factors, are produced by different microbes that reflect morphological and structural differences. Commonly reported disaccharide trehalose linked at C6 and C6 to oxalic acid is mainly associated with *Mycobacterium*, *Nocardia*, and *Corynebacterium* spp. In agriculture, trehalose lipids function as an effective dispersant in detecting and removing organophosphorus pesticides, a type of HC/PD in cabbage (Hassan et al., 2021). The advantages observed in using trehalose lipid microdroplets were its rapidity for microextraction using gas chromatography flame ionization detection (GC-FID) and its inexpensive/eco-friendly nature. Similarly, the cross-comparative evaluation of different marine biosurfactants revealed that trehalose lipids have higher oil affinity, biodegradation rate, and lower toxicity (Cai et al., 2021).

9.3.1.2 Lipopeptides

Lipopeptides are one of the best eco-friendly techniques in HC/PD remediation by overcoming the downside of their chemical counterparts. Different lipopeptides have varying structures and isoforms, which contain two main regions, acyl tail(s) and a short linear oligopeptide sequence having different D and L amide acids (Carolin et al., 2021; Sharma et al., 2021; Sonowal et al., 2022). Hence, the hydrophobic tail includes a hydrocarbon chain, and the hydrophilic head includes a peptide sequence with cationic/anionic residues and possibly nonproteinaceous amino acids. Examples of lipopeptides mainly used for HC/PD removal are surfactin, subtilisin, serrawettin, viscosin, and lichenysin.

Surfactin is a common cyclic lipopeptide known to have a unique chiral sequence and effectively reduce surface tension. Structurally, surfactin consists of cyclic heptapeptide connected by β-hydroxyl fatty acid with 13−15 carbon atoms and a series of seven amino acids such as L-asparagine, two L-leucine, glutamic acid, L-valine, and two D-leucines connected by lactone linkage (Carolin et al., 2021). For efficient oil recovery from waste crude oil, surfactin is an effective pH-switchable bio-demulsifier in a challenging oil separation operation (Yang et al., 2020). Surfactin can significantly enhance the removal rate of alkylaromatic (1.5%—87.2%) and saturated hydrocarbons during crude oil bioremediation, showing alkyldibenzothiophenes and triaromatic steroids as powerful degradation indicators (Wang et al., 2020). An LC-MS metabolite assay shows that South African hot spring *Bacillus* spp. produces subtilisin with high bioremediation potential toward PAHs in hazardous polluted water (Jardine, 2022).

Serratia marcescens isolated from petroleum-contaminated soil produce serrawettin W1 capable of degrading hydrocarbons with the concomitant production of biosurfactants at 20°C—30°C (Zhang et al., 2021). Serrawettin W1 reduced surface tension (30 mN/m), increased cell surface hydrophobicity (CSH), and stabilized surface/emulsion activity at 20°C—100°C temperatures, 2—10 pH and 0—50 g/L NaCl concentrations while enhancing the bioavailability of hydrocarbon pollutants for degradation. A cyclic lipopeptide, viscosin is produced by soil and marine bacteria possessing viscosinamides, pseudodesmins, massetolides, and pseudophomins (Carolin et al., 2021; Mgbechidinma, Akan, et al., 2022; Mishra et al., 2021), mostly *Pseudomonas fluorescens* (Gammaproteobacteria). A biosurfactant consists of nine amino acids joined with the β-hydroxydecanoyl C10-C12 fatty acid and can stimulate alkane mineralization with emulsification indices on hexadecane of 20%—31% at 90 mg/L, comparable to synthetic surfactant Tween 80 (Bak et al., 2015). Notably, *B. licheniformis* produces lichenysin, a similar structural biomolecule to surfactin, except for replacing glutamic acid with glutaminyl residue. The biosurfactant has excellent environmental stability at different pH, temperature, and salinity ranges, but the functional properties change in line with the amino acids present. While lichenysin has various applications for HC/PD removal in the petroleum industry, bioremediation, and the food industry, its production can be significantly improved by repressing the transcriptional factor CodY under an amino acid-rich condition (Zhu et al., 2017). However, numerous unanswered questions exist in the molecular investigation of biosurfactant production and application.

9.3.2 High-molecular-mass biosurfactants
9.3.2.1 Polymeric and particulate surfactants
Among the high-molecular-mass biosurfactants, the best studied for HC/PD removal are emulsan, liposan, mannoprotein, and other polysacc—protein complexes. Emulsan consisting of heteropolysaccharide backbone with repeating trisaccharide of N-acetyl-D-galactosamine, N-acetylgalactosamine uronic acid, and unidentified N-acetyl amino sugar is commonly produced by *Acinetobacter* spp. (Mujumdar et al., 2019). As the name implies, emulsan effectively serves as emulsifiers even at low concentrations. Emulsan, liposan, and alasan are the best examples of commercially used bioemulsifers

for microbial-enhanced oil recovery and biodegradation of toxic HC/PD compounds. A detailed description of the characterization of emulsan and its application in the removal of HC/PD has been reported by (Mujumdar et al., 2019). Fig. 9.3 shows the major environmental sectors that utilize biosurfactants.

9.4 Removal mechanisms of hydrophobic contaminant/petroleum derivate in the presence of biosurfactants

The exposure of microorganisms to HC/PD is significant for degradation, although it is well known that the hydrophobicity of hydrocarbons limits their contact with microbes. Biosurfactants also improve the CSH by sorbing on the bacterial surface and inducing the release of lipopolysaccharides (LPS) from the cell. These two phenomena work synergistically to promote the sorption of pseudosoluble or encapsulated emulsified HC/PDs by cells (Mgbechidinma, Zhang et al., 2022; Sharma et al., 2021). Generally, solubilization, emulsification, and mobilization are the three main mechanisms of HC/PD

Figure 9.3 Biosurfactant application in the environmental sectors.

removal, which depend on the biosurfactant's molecular weight. Furthermore, these mechanisms are affected by a complex interplay of features such as the microbial community and environmental conditions (temperature, pH, and NaCl concentration). Fig. 9.4 shows the mechanisms of HC/PD removal by biosurfactants.

9.4.1 Hydrophobic contaminant/petroleum derivate solubilization

The addition of biosurfactants increases the HC/PD solubilization in a liquid medium, resulting in the formation of micelles and the segregation of the HC/PD hydrocarbons from the hydrophobic domain. Solubilization increases with increasing biosurfactant concentration and can be quantified by solubility enhancement factor, molar solubilization ratio, and micelle—water partition coefficient (Song et al., 2016). NK biosurfactant produced by *Rhodococcus erythropolis* HX-2 shows high solubilization efficiency for petroleum and PAHs (Hu et al., 2020). Coupled with the surface tension-reducing potential of NK (28.89 mN/m) and the CMC value (100 mg/L), the biosurfactant can serve as a substitute for chemically synthesized surfactants. Moreover, a binary rhamnolipid—sophorolipid mixture exhibited greater solubilization than individual glycolipids at increasing temperatures and decreased with increasing salinity (Song et al., 2016).

Rhamnolipid, derived from agro-industrial waste valorization by *Planomicrobium okeanokoites*, has good potential in solubilizing petroleum hydrocarbons (279.17 mg/L removals of crude oil, as compared to 90.92 mg/L by control) from cotton matrix, with higher efficiency (41 mg/L fold) in the aqueous phase (Gaur et al., 2022). Mono-rhamnolipids are more effective solubilizers of PAHs than di-rhamnolipids; however, the slow release of hydrocarbon from mono-rhamnolipid micelles makes di-rhamnolipids a preferred option for biodegradation (Phulpoto et al., 2022). Rhamnolipids can be used in fieldwork microbial-enhanced oil recovery and environmental bioremediation to increase the bioavailability of HC/PD. Commonly reported in laboratory studies, HC/PD solubilization assay is conducted with an increasing concentration of biosurfactant (0.05—2 g/L) and constant HC/PD concentration. After incubation for 72 h at 25°C with shaking (200 r/min) in the dark, the residual HC/PD hydrocarbons can be filtered, concentrated by Na_2SO_4, and transferred into n-hexane for gas chromatography (GC) analysis (Fegade, 2021; Khondee et al., 2022; Pandey et al., 2022; Zhou et al., 2020). Solubilization is expressed as a percentage and calculated according to the following formula:

$$\text{Solubilization (\%)} = \frac{Sa - Sc}{Si} \times 100$$

where Sa is the concentration of HC/PD in the n-hexane extracts of assay samples; Sc is the concentration of HC/PD in control experiments having zero biosurfactants, and Si is the initial concentration of HC/PD before the solubilization test.

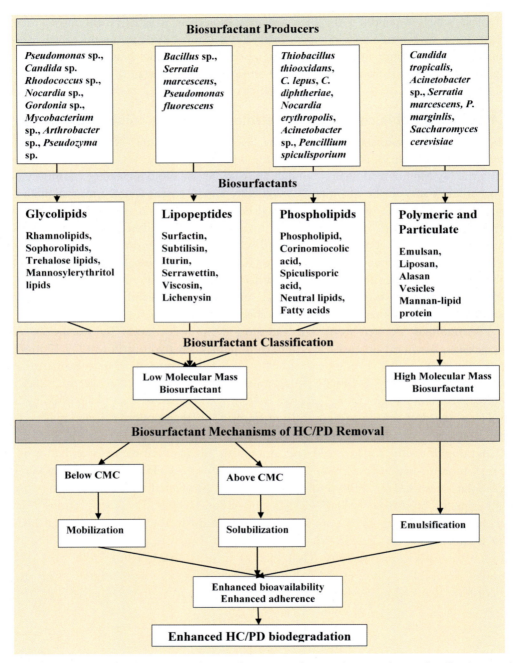

Figure 9.4 Systematic illustration of biosurfactant producers, their products, classification, and mechanisms of HC/PD removal. *HC/PD*, Hydrophobic contaminant/petroleum derivate.

9.4.2 Hydrophobic contaminant/petroleum derivate emulsification

Biosurfactants emulsify high-concentration hydrophobic solvents into aqueous media to improve aqueous solubility. The efficiency depends on the ability of the biosurfactant/HC/PD mixture to form a stable middle-phase microemulsion. Solubilization of hydrocarbons plays an important role in facilitating the formation of microemulsions, which helps in HC/PD removal. Multifunctional di-rhamnolipid from *Pseudomonas* sp. S1WB has a maximum emulsification index of 60.15% ± 0.07%, with high stability under different environmental conditions, i.e., temperature, pH, and various NaCl concentrations for 30 days of incubation (Phulpoto et al., 2022). Although polymeric and particulate biosurfactants are known to produce stable emulsions below CMC, low-molecular-mass biosurfactants produce fairly stable emulsions above CMC. Several studies have confirmed that emulsification is an important parameter of highly potent biosurfactants for removing immiscible HC/PD (Cai et al., 2021; Li et al., 2021; Muneeswari et al., 2022; Zhao et al., 2021). The ability of emulsifying biosurfactants is key to (1) enhancing the complexity and solubility of nonpolar substrates, thereby promoting HC/PD bioavailability, and (2) improving affinity between cell surfaces and oil–water interfaces through metabolism, promoting deformation of the HC/PD–water interface film. The emulsification index of biosurfactant is determined by formulating an equal ratio of biosurfactant and HC/PD in a tube, followed by homogenization and resting for 24 hours. The process leads to the formation of micro- and macroemulsions, which collapse to create a uniform, stable emulsion (Khondee et al., 2022; Zhou et al., 2020). The emulsification index is calculated as a percentage of the formula stated as follows:

$$\text{Emulsification index } (E1_{24}\%) = \frac{\text{Height of emulsion}}{\text{Total height of liquid columm}} \times 100$$

9.4.3 Hydrophobic contaminant/petroleum derivate mobilization

Mobilization is a significant property of biosurfactants for removing HC/PD in soil and sediment, owing to the high hydrophobicity, low solubility, immobilization, or sorption of the hydrocarbons. Immobilization is the dormant process in which the HC/PD hydrocarbons are partitioned from the gaseous and aqueous phases onto the solid phase, resulting in limited pollutant dissolution, which in turn impacts their fate and behavior in the environment (Bolan et al., 2023). Mobilization can be achieved using biosurfactants to foster desorption, allowing the flow and bioavailability of HC/PD in a porous media. In addition to the structure, composition, and various internal processes within the interfacial layers of an HC/PD-embedded matrix, mobilization depends on two principles: dispersion and displacement.

The dispersion efficiency of HC/PD hydrocarbons into the aqueous phase depends on the biosurfactant concentration and interfacial tension. Chen et al. (2022) illustrated an oiled sand cleanup technique using chitosan/rhamnolipid complex dispersion to achieve a high efficiency of HC/PD removal from oil spills. However, the operation was negatively impacted by increasing salinity. In addition, displacement (also known as oil spreading) is a function of the hydrocarbon kinetics through the absorbent or porous medium at reduced surface/interfacial tension caused by disabling capillary forces in the system (Fdez-Sanromán et al., 2021; Goveas et al., 2022; Pandey et al., 2022; Zhou et al., 2020). Displacement occurs between oil and aqueous solutions, commonly determined by the diameter of the halo formed when a biosurfactant is dropped into a petri-dish containing water and floating hydrocarbon (Dobler et al., 2020). HC/PD mobilization is significantly enhanced at below CMC biosurfactant concentrations, creating molecular aggregates.

9.5 Impact of biosurfactant-mediated hydrophobic contaminant/petroleum derivate removal on microbial community

Biosurfactant application for HC/PD removal is a double-edged sword, as it encourages pollutant solubilization but concurrently affects the dynamics of endogenic microbial communities that play an important role in HC/PD biodegradation. So far, the dramatic shift in microbial communities, in terms of their diversity and population during HC/PD removal in the presence of biosurfactant, is debatable. Some studies have revealed that the increased addition of biosurfactants during HC/PD removal in marine or terrestrial ecosystems favors microbial community; however, other reports have shown that biosurfactant addition significantly influences the composition and activities of indigenous microbes by selecting those that survive and function under those conditions (Kariyawasam et al., 2022; Li et al., 2018; Mishra et al., 2021; Muneeswari et al., 2022; Phulpoto et al., 2022; Zhao et al., 2021). However, the knowledge of microbial communities during HC/PD removal is important for determining the potential for bioremediation, transformation, or persistence of pollutants.

Microorganisms, whether bacteria, yeast, or filamentous fungi, comprise a significant segment of the biosphere, and their involvement in several biogeochemical cycles cannot be neglected. HC/PD-contaminated sites are a rich source of biosurfactant producers, which are in direct or indirect interaction with other organisms (Mgbechidinma, Zhang, et al., 2022). In a study on microbial communities using phospholipid fatty acid (PLFA)-based assay after biosurfactant amendment during HC/PD removal, Li et al. (2018) reported an increased microbial metabolism of indigenous soil microbes to assimilate hydrocarbons, with Gram-negative bacteria are the major hydrocarbon degraders. The microbial mechanisms during petroleum hydrocarbon remediation showed three transformation patterns on treatment by natural attenuation, rhamnolipids enhancement, and Tween 80. Notably, rhamnolipids enhancement increased total soil biomass than natural attenuation but less than that observed with Tween 80 after 17 days of treatment (Li et al., 2018).

Interestingly, Wang et al. (2021) revealed a changing microbial community in PAH-contaminated soil after treatment with combined rhamnolipid and agricultural wastes. Microbial diversity increased among the fungi community, although the addition of rhamnolipid significantly impacted the abundance of the bacteria and fungi communities, especially among the genus *Sphingomonas*, *Altererythrobacter*, *Lysobacter*, and *Humicola*. Following microbial enumeration and molecular pyrosequencing of the 16 S rRNA gene, the dynamic of the bacteria community changes with the addition of surfactant at different CMC doses during aged PAH bioremediation (Cecotti et al., 2018), following the trend: surface tension reduction, PAH sorption, PAH elimination, and selected enrichment of PAH degrading microbial community. Similarly, Zhou et al. (2020) confirmed that indigenous biosurfactant-producing *Acinetobacter* sp. Y2 enhances the bioremediation of hydraulic fracturing flow back by increasing bacteria growth, microbial activity, and gene expression. The taxonomic analysis among treatments revealed apparent shifts in the bacteria community structure with a relative abundance from Bacteroidetes (44.95%), Spirochaetes (27.02%), and Proteobacteria (17.82%) to Bacteroidetes (34.74%), Spirochaetes (4.01%), and Proteobacteria (75.68%) at the phyla level.

9.6 Eco-sustainable biosurfactant-based hydrophobic contaminant/petroleum derivate removal approach toward achieving some selected United Nations Sustainable Development Goals

The removal of HC/PD is directly and indirectly linked to some Sustainable Development Goals (SDGs). The goal of achieving sustainable life below water, life on land, and clean water/sanitation through environmental sustainability is complex. However, the waste-to-biosurfactant initiative has exposed a solution pipeline of integrating green economy concepts by incorporating bio-economy, low carbon economy, and circular economy (Mgbechidinma, Akan, et al., 2022). At the forefront is the valorization of aqua wastes, agro-food, and agro-industrial wastes by microbial fermentation to substitute the use of conventional substrate in biosurfactant production (Silva et al., 2022). This is for the primary purpose of reducing production costs; however, the perspective strongly aligns with some waste management schemes, which closely connect with the sustainability agenda of the United Nations (UN). Among the 17 SDGs proposed by the UN, SDGs-2, 3, 6, 12, 13, 14, and 15 directly or indirectly support biosurfactant-mediated removal of HC/PD from the environment. Despite the potential of biosurfactant application in addressing some SDGs, focused research in this field is limited. Therefore, recent biosurfactant-mediated HC/PD remediation and environmental reclamation studies with prospects toward contributing to the UN-SDGs are reviewed as follows.

The SDGs directly connected to biosurfactant-mediated removal of HC/PD are SDGs 6, 14, and 15. SDG-6 aims to ensure the availability and sustainable management of water and sanitation for all. Research on wastewater treatments has progressively highlighted the benefit of biosurfactant incorporation, especially in the removal of HC/PD. Some techniques available for wastewater purification include adsorption, solvent extraction, photodegradation, electrocoagulation, and ion exchange, in which scientists and intellectuals have emphasized extensive research (Fegade, 2021). The drawback of toxicity, solubility, thermal stability, and inefficiency for seawater desalination has not been fully addressed; thus, recent studies are incorporating biosurfactants. This new frontier is to improve the treatment of effluent waste, polluted wastewater, and sewage sludge, hence promoting SDG-3, which focuses on good health and well-being. Some of the properties of biosurfactants widely explored in this field are biodegradability, low toxicity, waste valorization as raw materials, physical stability, and surface/interfacial activity (Mgbechidinma, Zhang, et al., 2022). Apparently, biosurfactants can function as an effective collector (97% oil-removal rate) in a water—oil separation system during oily water treatment by an induced presaturation process (Bandeira dos Santos et al., 2021). Native Bilge water bacteria *Marinobacter* sp. BIC3M3, *Alcanivorax* sp. BIC1A5, and *Halomonas* sp. BIC1H44 produce biosurfactants with a total yield of 3.148, 2.922, and 2.596 g/L, respectively, which can remove hydrocarbons during wastewater treatment (Rizzo et al., 2021).

SDG-14 aims to conserve and sustain the use of oceans, seas, and marine resources for sustainable development. Wastes from water bodies, especially fish by-catch, marine organisms at ocean shorelines due to hypoxia, and overfishing, are a rich source of nutrients containing protein, fat, and lipid (Mgbechidinma et al., 2023). These nutrients can serve as substrates for biosurfactant production, reducing costs and promoting sustainable consumption and production (Mgbechidinma, Zhang, et al., 2022), which is the sole objective of SDG-12. This practice allows the flow of the integrated green economy, with the substrate as a biowaste, the process of reducing the waste contribution to the carbon footprint, and the biosurfactant product having applicability in HC/PD removal. Biosurfactant produced from 50-L bioreactor using *Bacillus cereus* yielded 4.7 g/L and has a 32 mN/m surface tension that can be successfully applied as a seawater bioremediation agent since the survival rate of fish *Poecilia vivipara* can be higher than 90% (Durval et al., 2020). In contrast, some scientific investigations have highlighted the impact of lipopeptide SPH6 produced by the bacterium *Pseudomonas* sp. H6 on marine organisms. Lipopeptide SPH6 has an antiparasitic effect on green algae, cyanobacteria, crustaceans, and zebrafish at 170, 20, 27, and 80 mg/L, respectively (Korbut et al., 2022). Hence, biosurfactant application for HC/PD removal in the marine environment should be conducted under a regulated policy to ensure that marine organisms are not endangered. Apart from it application in HC/PD marine remediation, biosurfactants formulated from *Pseudomonas cepacia* CCT 6659 can serve as a dispersing agent for the bioremediation of oil spills in terrestrial environments, with 76.55% removal in soil and 84.50% in sea stones (Soares da Silva et al., 2021).

SDG-15 is focused on addressing life on land by ensuring the protection, restoration, and promotion of sustainable terrestrial ecosystem use. Similar to aquatic environments, studies have revealed several improvements in land restoration through biosurfactant application. These translate to promoting land use for agricultural purposes, thus fulfilling the objectives of SDG-2 centered on zero hunger to achieve food and nutrition security. HC/PD persistence in terrestrial ecosystems has streamlined the possibility of achieving these goals by consequently eliciting modifications in soil microbial population and processes, reducing soil fertility, inducing stress-related metabolic pathways (oxidative damage, plant signaling, and biosynthetic pathways), and stimulating abnormal genetic behavior and reproductive responses (Dai et al., 2022; Goveas et al., 2022; Negreiros et al., 2022). However, recent studies have proven the effectiveness of biosurfactants in combatting these challenges by reversing land degradation and halting loss of diversity. Sangwan et al. (2022) emphasized the role of microbial biosurfactants in the revitalization of agriculture by serving as a plant growth stimulant, seed germination enhancer, soil nutrient mobilizer, and biocontrol agents. These functions ultimately enhance productivity, improve soil health, and reduce environmental pollution to achieve sustainable agriculture. Sustainability in agriculture productivity and improvements in the physical, chemical, and biological properties of soil can also be greatly achieved if valorized products of the crop residues are used appropriately. Remarkably, screening of biosurfactant producers for environmental use should be more holistic. For instance, the alkaliphilic bacterium *Brevibacterium casei* NK8 is capable of fermenting alkaline/hydrothermally pretreated lignocellulosic wastes (corn husk, coconut oil cake, and defatted rice bran) to zwitterionic biosurfactants (1.14–1.32 g/L yield) with stability at high pH, salinity, and temperature (Khondee et al., 2022), promoting the waste-to-wealth initiative. Using lignocellulose waste as biosurfactant substrates instead of burning the waste is a direct attempt at reducing greenhouse gas (GHG) emissions, revealing new frontiers and possible crosstalks between biosurfactant production and climate actions. Hence, this is a subtle approach to combat climate change and its impacts with respect to SDG-13.

9.7 Conclusion and possible future outlooks

Hydrophobic contaminants and petroleum derivatives (HC/PDs) through natural and anthropogenic sources are a serious environmental concern because of their long-term persistency, hazardous nature, and toxicity to the ecosystem and human health. Although synthetic surfactants can be used to remove HC/PDs in contaminated ecosystems, biosurfactants are a safer alternative with highly beneficial properties for HC/PD solubilization–mobilization–emulsification and their sequent removal through soil washing, bioremediation, and phytoremediation processes. Due to the increasing awareness of environmental sustainability, coupled with the known toxicity and HC/PD leaching potential of synthetic surfactants, more research is needed to ensure optimal biosurfactant production for application at the most suitable techno-economic level. Moreover, there is limited knowledge on the biochemical interactions, fate, transport, contaminant mobilization, ecotoxicity, and consequences of spent biosurfactants in aquatic and terrestrial ecosystems despite the numerous published articles on the environmental application of biosurfactants.

References

Akan, O. D., Udofia, G. E., Okeke, E. S., Mgbechidinma, C. L., Okoye, C. O., Zoclanclounon, Y. A. B., Atakpa, E. O., & Adebanjo, O. O. (2021). Plastic waste: Status, degradation and microbial management options for Africa. *Journal of Environmental Management*, *292*, 112758. Available from https://doi.org/10.1016/j.jenvman.2021.112758.

Ambaye, T. G., Formicola, F., Sbaffoni, S., Franzetti, A., & Vaccari, M. (2022). Insights into rhamnolipid amendment towards enhancing microbial electrochemical treatment of petroleum hydrocarbon contaminated soil. *Chemosphere*, *307*, 136126. Available from https://doi.org/10.1016/j.chemosphere.2022.136126.

Bak, F., Bonnichsen, L., Jørgensen, N. O. G., Nicolaisen, M. H., & Nybroe, O. (2015). The biosurfactant viscosin transiently stimulates n-hexadecane mineralization by a bacterial consortium. *Applied Microbiology and Biotechnology*, *99*(3), 1475–1483. Available from https://doi.org/10.1007/s00253-014-6054-3.

Bandeira dos Santos, L., de Cássia Freire Soares da Silva, R., Pinto Ferreira Brasileiro, P., Dias Baldo, R., Sarubbo, L. A., & dos Santos, V. A. (2021). Oily water treatment in a multistage tower operated under a novel induced pre-saturation process in the presence of a biosurfactant as collector. *Biotechnology Reports*, *30*. Available from https://doi.org/10.1016/j.btre.2021.e00638.

Bao, Q., Huang, L., Xiu, J., Yi, L., Zhang, Y., & Wu, B. (2022). Study on the thermal washing of oily sludge used by rhamnolipid/sophorolipid binary mixed bio-surfactant systems. *Ecotoxicology and Environmental Safety*, *240*, 113696. Available from https://doi.org/10.1016/j.ecoenv.2022.113696.

Bolan, S., Padhye, L. P., Mulligan, C. N., Alonso, E. R., Saint-Fort, R., Jasemizad, T., Wang, C., Zhang, T., Rinklebe, J., Wang, H., Siddique, K. H. M., Kirkham, M. B., & Bolan, N. (2023). Surfactant-enhanced mobilization of persistent organic pollutants: Potential for soil and sediment remediation and unintended consequences. *Journal of Hazardous Materials*, *443*, 130189. Available from https://doi.org/10.1016/j.jhazmat.2022.130189.

Cai, Q., Zhu, Z., Chen, B., Lee, K., Nedwed, T. J., Greer, C., & Zhang, B. (2021). A cross-comparison of biosurfactants as marine oil spill dispersants: Governing factors, synergetic effects and fates. *Journal of Hazardous Materials*, *416*, 126122. Available from https://doi.org/10.1016/j.jhazmat.2021.126122.

Carolin, C. F., Kumar, P. S., & Ngueagni, P. T. (2021). A review on new aspects of lipopeptide biosurfactant: Types, production, properties and its application in the bioremediation process. *Journal of Hazardous Materials*, *407*, 124827. Available from https://doi.org/10.1016/j.jhazmat.2020.124827.

Cecotti, M., Coppotelli, B. M., Mora, V. C., Viera, M., & Morelli, I. S. (2018). Efficiency of surfactant-enhanced bioremediation of aged polycyclic aromatic hydrocarbon-contaminated soil: Link with bioavailability and the dynamics of the bacterial community. *Science of the Total Environment*, *634*, 224–234. Available from https://doi.org/10.1016/j.scitotenv.2018.03.303.

Chen, Z., An, C., Wang, Y., Zhang, B., Tian, X., & Lee, K. (2022). A green initiative for oiled sand cleanup using chitosan/rhamnolipid complex dispersion with pH-stimulus response. *Chemosphere*, *288*, 132628. Available from https://doi.org/10.1016/j.chemosphere.2021.132628.

Dai, C., Han, Y., Duan, Y., Lai, X., Fu, R., Liu, S., Leong, K. H., Tu, Y., & Zhou, L. (2022). Review on the contamination and remediation of polycyclic aromatic hydrocarbons (PAHs) in coastal soil and sediments. *Environmental Research*, *205*, 112423. Available from https://doi.org/10.1016/j.envres.2021.112423.

Dell'Anno, F., Sansone, C., Ianora, A., & Dell'Anno, A. (2018). Biosurfactant-induced remediation of contaminated marine sediments: Current knowledge and future perspectives. *Marine Environmental Research*, *137*, 196−205. Available from https://doi.org/10.1016/j.marenvres.2018.03.010.

Dierickx, S., Castelein, M., Remmery, J., De Clercq, V., Lodens, S., Baccile, N., De Maeseneire, S. L., Roelants, S. L. K. W., & Soetaert, W. K. (2022). From bumblebee to bioeconomy: Recent developments and perspectives for sophorolipid biosynthesis. *Biotechnology Advances*, *54*, 107788. Available from https://doi.org/10.1016/j.biotechadv.2021.107788.

Dobler, L., Ferraz, H. C., Araujo de Castilho, L. V., Sangenito, L. S., Pasqualino, I. P., Souza dos Santos, A. L., Neves, B. C., Oliveira, R. R., Guimarães Freire, D. M., & Almeida, R. V. (2020). Environmentally friendly rhamnolipid production for petroleum remediation. *Chemosphere*, *252*, 126349. Available from https://doi.org/10.1016/j.chemosphere.2020.126349.

Durval, I. J. B., Mendonça, A. H. R., Rocha, I. V., Luna, J. M., Rufino, R. D., Converti, A., & Sarubbo, L. A. (2020). Production, characterization, evaluation and toxicity assessment of a Bacillus cereus UCP 1615 biosurfactant for marine oil spills bioremediation. *Marine Pollution Bulletin*, *157*, 111357. Available from https://doi.org/10.1016/j.marpolbul.2020.111357.

Fdez-Sanromán, A., Pazos, M., Rosales, E., & Sanromán, M. Á. (2021). Prospects on integrated electrokinetic systems for decontamination of soil polluted with organic contaminants. *Current Opinion in Electrochemistry*, *27*, 100692. Available from https://doi.org/10.1016/j.coelec.2021.100692.

Fegade, U. (2021). Application of biosurfactant for treatment of effluent waste, polluted wastewater treatment, and sewage sludge. In *Green Sustainable Process for Chemical and Environmental Engineering and Science: Biosurfactants for the Bioremediation of Polluted Environments* (pp. 1−19). Elsevier. Available from https://doi.org/10.1016/B978-0-12-822696-4.00020-6.

Gaur, V. K., Gupta, P., Tripathi, V., Thakur, R. S., Regar, R. K., Patel, D. K., & Manickam, N. (2022). Valorization of agro-industrial waste for rhamnolipid production, its role in crude oil solubilization and resensitizing bacterial pathogens. *Environmental Technology and Innovation*, *25*, 102108. Available from https://doi.org/10.1016/j.eti.2021.102108.

Gayathiri, E., Prakash, P., Karmegam, N., Varjani, S., Awasthi, M. K., & Ravindran, B. (2022). Biosurfactants: Potential and eco-friendly material for sustainable agriculture and environmental safety—A review. *Agronomy*, *12*(3), 662. Available from https://doi.org/10.3390/agronomy12030662.

Goveas, L. C., Selvaraj, R., Vinayagam, R., Alsaiari, A. A., Alharthi, N. S., & Sajankila, S. P. (2022). Nitrogen dependence of rhamnolipid mediated degradation of petroleum crude oil by indigenous Pseudomonas sp. WD23 in seawater. *Chemosphere*, *304*, 135235. Available from https://doi.org/10.1016/j.chemosphere.2022.135235.

Guo, S., Liu, X., Wang, L., Liu, Q., Xia, C., & Tang, J. (2022). Ball-milled biochar can act as a preferable biocompatibility material to enhance phenanthrene degradation by stimulating bacterial metabolism. *Bioresource Technology*, *350*, 126901. Available from https://doi.org/10.1016/j.biortech.2022.126901.

Hassan, F. W. M., Raoov, M., Kamaruzaman, S., Mohamed, A. H., Ibrahim, W. N. W., Hanapi, N. S. M., Zain, N. N. M., Yahaya, N., & Chen, D. D. Y. (2021). A rapid and efficient dispersive trehalose biosurfactant enhanced magnetic solid phase extraction for the sensitive determination of organophosphorus pesticides in cabbage (Brassica olearaceae var. capitate) samples by GC-FID. *Journal of Food Composition and Analysis*, *102*, 104057. Available from https://doi.org/10.1016/j.jfca.2021.104057.

Hu, X., Qiao, Y., Chen, L. Q., Du, J. F., Fu, Y. Y., Wu, S., & Huang, L. (2020). Enhancement of solubilization and biodegradation of petroleum by biosurfactant from Rhodococcus erythropolis HX-2. *Geomicrobiology Journal*, *37*(2), 159−169. Available from https://doi.org/10.1080/01490451.2019.1678702.

Jamal, M. T. (2022). Enrichment of potential halophilic marinobacter consortium for mineralization of petroleum hydrocarbons and also as oil reservoir indicator in Red Sea, Saudi Arabia. *Polycyclic Aromatic Compounds*, *42*(2), 400−411. Available from https://doi.org/10.1080/10406638.2020.1735456.

Jardine, J. L. (2022). Potential bioremediation of heavy metal ions, polycyclic aromatic hydrocarbons and biofilms with South African hot spring bacteria. *Bioremediation Journal, 26*(3), 261−269. Available from https://doi.org/10.1080/10889868.2021.1964429.

Kariyawasam, T., Doran, G. S., Howitt, J. A., & Prenzler, P. D. (2022). Polycyclic aromatic hydrocarbon contamination in soils and sediments: Sustainable approaches for extraction and remediation. *Chemosphere, 291*, 132981. Available from https://doi.org/10.1016/j.chemosphere.2021.132981.

Khondee, N., Ruamyat, N., Luepromchai, E., Sikhao, K., & Hawangchu, Y. (2022). Bioconversion of lignocellulosic wastes to zwitterionic biosurfactants by an alkaliphilic bacterium: Process development and product characterization. *Biomass and Bioenergy, 165*, 106568. Available from https://doi.org/10.1016/j.biombioe.2022.106568.

Korbut, R., Skjolding, L. M., Mathiessen, H., Jaafar, R., Li, X., Jørgensen, L. v G., Kania, P. W., Wu, B., & Buchmann, K. (2022). Toxicity of the antiparasitic lipopeptide biosurfactant SPH6 to green algae, cyanobacteria, crustaceans and zebrafish. *Aquatic Toxicology, 243*, 106072. Available from https://doi.org/10.1016/j.aquatox.2021.106072.

Kumar, M., Bolan, N. S., Hoang, S. A., Sawarkar, A. D., Jasemizad, T., Gao, B., Keerthanan, S., Padhye, L. P., Singh, L., Kumar, S., Vithanage, M., Li, Y., Zhang, M., Kirkham, M. B., Vinu, A., & Rinklebe, J. (2021). Remediation of soils and sediments polluted with polycyclic aromatic hydrocarbons: To immobilize, mobilize, or degrade? *Journal of Hazardous Materials, 420*, 126534. Available from https://doi.org/10.1016/j.jhazmat.2021.126534.

Li, M., Zhou, J., Xu, F., Li, G., & Ma, T. (2021). An cost-effective production of bacterial exopolysaccharide emulsifier for oil pollution bioremediation. *International Biodeterioration & Biodegradation, 159*, 105202. Available from https://doi.org/10.1016/j.ibiod.2021.105202.

Li, X., Fan, F., Zhang, B., Zhang, K., & Chen, B. (2018). Biosurfactant enhanced soil bioremediation of petroleum hydrocarbons: Design of experiments (DOE) based system optimization and phospholipid fatty acid (PLFA) based microbial community analysis. *International Biodeterioration and Biodegradation, 132*, 216−225. Available from https://doi.org/10.1016/j.ibiod.2018.04.009.

Mapelli, F., Scoma, A., Michoud, G., Aulenta, F., Boon, N., Borin, S., Kalogerakis, N., & Daffonchio, D. (2017). Biotechnologies for marine oil spill cleanup: Indissoluble ties with microorganisms. *Trends in Biotechnology, 35*(9), 860−870. Available from https://doi.org/10.1016/j.tibtech.2017.04.003.

Martins, G., Campos, S., Ferreira, A., Castro, R., Duarte, M. S., & Cavaleiro, A. J. (2022). A mathematical model for bioremediation of hydrocarbon-contaminated soils. *Applied Sciences (Switzerland), 12*(21), 69. Available from https://doi.org/10.3390/app122111069.

Mgbechidinma, C. L., Akan, O. D., Zhang, C., Huang, M., Linus, N., Zhu, H., & Wakil, S. M. (2022). Integration of green economy concepts for sustainable biosurfactant production − A review. *Bioresource Technology, 364*, 128021. Available from https://doi.org/10.1016/j.biortech.2022.128021.

Mgbechidinma, C. L., Zhang, X., Wang, G., Atakpa, E. O., Jiang, L., & Zhang, C. (2022). Advances in biosurfactant production from marine waste and its potential application. In *The marine environment* (pp. 223−254). Informa UK Limited. Available from https://doi.org/10.1201/9781003307464-10.

Mgbechidinma, C. L., Zheng, G., Baguya, E. B., Zhou, H., Okon, S. U., & Zhang, C. (2023). Fatty acid composition and nutritional analysis of waste crude fish oil obtained by optimized milder extraction methods. *Environmental Engineering Research, 28*(2). Available from https://doi.org/10.4491/eer.2022.034, 220034−0.

Mishra, S., Lin, Z., Pang, S., Zhang, Y., Bhatt, P., & Chen, S. (2021). Biosurfactant is a powerful tool for the bioremediation of heavy metals from contaminated soils. *Journal of Hazardous Materials, 418*, 126253. Available from https://doi.org/10.1016/j.jhazmat.2021.126253.

Montas, L., Ferguson, A. C., Mena, K. D., Solo-Gabriele, H. M., & Paris, C. B. (2022). PAH depletion in weathered oil slicks estimated from modeled age-at-sea during the Deepwater Horizon oil spill. *Journal of Hazardous Materials, 440*, 129767. Available from https://doi.org/10.1016/j.jhazmat.2022.129767.

Mujumdar, S., Joshi, P., & Karve, N. (2019). Production, characterization, and applications of bioemulsifiers (BE) and biosurfactants (BS) produced by Acinetobacter spp.: A review. *Journal of Basic Microbiology, 59*(3), 277−287. Available from https://doi.org/10.1002/jobm.201800364.

Muneeswari, R., Swathi, K. V., Sekaran, G., & Ramani, K. (2022). Microbial-induced biosurfactant-mediated biocatalytic approach for the bioremediation of simulated marine oil spill. *International*

Journal of Environmental Science and Technology, 19(1), 341–354. Available from https://doi.org/10.1007/s13762-020-03086-0.

Negreiros, A. C. S. V. d, Lins, I. D., Maior, C. B. S., & Moura, M. J. d C. (2022). Oil spills characteristics, detection, and recovery methods: A systematic risk-based view. *Journal of Loss Prevention in the Process Industries*, 80, 104912. Available from https://doi.org/10.1016/j.jlp.2022.104912.

Pandey, R., Krishnamurthy, B., Singh, H. P., & Batish, D. R. (2022). Evaluation of a glycolipopepetide biosurfactant from Aeromonas hydrophila RP1 for bioremediation and enhanced oil recovery. *Journal of Cleaner Production*, 345, 131098. Available from https://doi.org/10.1016/j.jclepro.2022.131098.

Phulpoto, I. A., Yu, Z., Li, J., Ndayisenga, F., Hu, B., Qazi, M. A., & Yang, X. (2022). Evaluation of di-rhamnolipid biosurfactants production by a novel Pseudomonas sp. S1WB: Optimization, characterization and effect on petroleum-hydrocarbon degradation. *Ecotoxicology and Environmental Safety*, 242, 113892. Available from https://doi.org/10.1016/j.ecoenv.2022.113892.

Prakash, A. A., Prabhu, N. S., Rajasekar, A., Parthipan, P., AlSalhi, M. S., Devanesan, S., & Govarthanan, M. (2021). Bio-electrokinetic remediation of crude oil contaminated soil enhanced by bacterial biosurfactant. *Journal of Hazardous Materials*, 405, 124061. Available from https://doi.org/10.1016/j.jhazmat.2020.124061.

Ram, H., Kumar Sahu, A., Said, M. S., Banpurkar, A. G., Gajbhiye, J. M., & Dastager, S. G. (2019). A novel fatty alkene from marine bacteria: A thermo stable biosurfactant and its applications. *Journal of Hazardous Materials*, 380, 120868. Available from https://doi.org/10.1016/j.jhazmat.2019.120868.

Rizzo, C., Caldarone, B., De Luca, M., De Domenico, E., & Giudice, A. L. (2021). Native bilge water bacteria as biosurfactant producers and implications in hydrocarbon-enriched wastewater treatment. *Journal of Water Process Engineering*, 43, 102271. Available from https://doi.org/10.1016/j.jwpe.2021.102271.

Sanganyado, E., Chingono, K. E., Gwenzi, W., Chaukura, N., & Liu, W. (2021). Organic pollutants in deep sea: Occurrence, fate, and ecological implications. *Water Research*, 205, 117658. Available from https://doi.org/10.1016/j.watres.2021.117658.

Sangwan, S., Kaur, H., Sharma, P., Sindhu, M., & Wati, L. (2022). *Routing microbial biosurfactants to agriculture for revitalization of soil and plant growth* (pp. 313–338). Elsevier BV. Available from https://doi.org/10.1016/b978-0-323-85581-5.00015-x.

Sharma, J., Sundar, D., & Srivastava, P. (2021). Biosurfactants: Potential agents for controlling cellular communication, motility, and antagonism. *Frontiers in Molecular Biosciences*, 8, 727070. Available from https://doi.org/10.3389/fmolb.2021.727070.

Silva, G. F. d, Gautam, A., Duarte, I. C. S., Delforno, T. P., Oliveira, V. M. d, & Huson, D. H. (2022). Interactive analysis of biosurfactants in fruit-waste fermentation samples using BioSurfDB and MEGAN. *Scientific Reports*, 12(1). Available from https://doi.org/10.1038/s41598-022-11753-0.

Soares da Silva, R. d C. F., Luna, J. M., Rufino, R. D., & Sarubbo, L. A. (2021). Ecotoxicity of the formulated biosurfactant from *Pseudomonas cepacia* CCT 6659 and application in the bioremediation of terrestrial and aquatic environments impacted by oil spills. *Process Safety and Environmental Protection*, 154, 338–347. Available from https://doi.org/10.1016/j.psep.2021.08.038.

Song, D., Liang, S., Yan, L., Shang, Y., & Wang, X. (2016). Solubilization of polycyclic aromatic hydrocarbons by single and binary mixed rhamnolipid-sophorolipid biosurfactants. *Journal of Environmental Quality*, 45(4), 1405–1412. Available from https://doi.org/10.2134/jeq2015.08.0443.

Sonowal, S., Nava, A. R., Joshi, S. J., Borah, S. N., Islam, N. F., Pandit, S., Prasad, R., Sarma, H. (2022). Biosurfactant-assisted phytoremediation of potentially toxic elements in soil: Green technology for meeting the United Nations Sustainable Development Goals. In: *Pedosphere* (Vol. 32, Issue 1, pp. 198–210). Soil Science Society of China. https://doi.org/10.1016/S1002-0160(21)60067-X.

Subsanguan, T., Khondee, N., Rongsayamanont, W., & Luepromchai, E. (2022). Formulation of a glycolipid:lipopeptide mixture as biosurfactant-based dispersant and development of a low-cost glycolipid production process. *Scientific Reports*, 12(1), 3. Available from https://doi.org/10.1038/s41598-022-20795-3.

Wang, J., Bao, H., Pan, G., Zhang, H., Li, J., Li, J., Cai, J., & Wu, F. (2021). Combined application of rhamnolipid and agricultural wastes enhances PAHs degradation via increasing their bioavailability and changing microbial community in contaminated soil. *Journal of Environmental Management*, 294, 112998. Available from https://doi.org/10.1016/j.jenvman.2021.112998.

Wang, X., Cai, T., Wen, W., Ai, J., Ai, J., Zhang, Z., Zhu, L., & George, S. C. (2020). Surfactin for enhanced removal of aromatic hydrocarbons during biodegradation of crude oil. *Fuel, 267*, 117272. Available from https://doi.org/10.1016/j.fuel.2020.117272.

Wu, B., Xiu, J., Yu, L., Huang, L., Yi, L., & Ma, Y. (2022). Biosurfactant production by *Bacillus subtilis* SL and its potential for enhanced oil recovery in low permeability reservoirs. *Scientific Reports, 12*(1). Available from https://doi.org/10.1038/s41598-022-12025-7.

Yang, Z., Zu, Y., Zhu, J., Jin, M., Cui, T., & Long, X. (2020). Application of biosurfactant surfactin as a pH-switchable biodemulsifier for efficient oil recovery from waste crude oil. *Chemosphere, 240*. Available from https://doi.org/10.1016/j.chemosphere.2019.124946.

Zhang, K., Tao, W., Lin, J., Wang, W., & Li, S. (2021). Production of the biosurfactant serrawettin W1 by Serratia marcescens S-1 improves hydrocarbon degradation. *Bioprocess and Biosystems Engineering, 44*(12), 2541–2552. Available from https://doi.org/10.1007/s00449-021-02625-4.

Zhang, X., Zhang, X., Wang, S., & Zhao, S. (2022). Improved remediation of co-contaminated soils by heavy metals and PAHs with biosurfactant-enhanced soil washing. *Scientific Reports, 12*(1). Available from https://doi.org/10.1038/s41598-022-07577-7.

Zhao, F., Zhu, H., Cui, Q., Wang, B., Su, H., & Zhang, Y. (2021). Anaerobic production of surfactin by a new Bacillus subtilis isolate and the in situ emulsification and viscosity reduction effect towards enhanced oil recovery applications. *Journal of Petroleum Science and Engineering, 201*, 108508. Available from https://doi.org/10.1016/j.petrol.2021.108508.

Zhou, H., Huang, X., Liang, Y., Li, Y., Xie, Q., Zhang, C., & You, S. (2020). Enhanced bioremediation of hydraulic fracturing flowback and produced water using an indigenous biosurfactant-producing bacteria Acinetobacter sp. Y2. *Chemical Engineering Journal, 397*, 125348. Available from https://doi.org/10.1016/j.cej.2020.125348.

Zhou, H., Jiang, L., Li, K., Chen, C., Lin, X., Zhang, C., & Xie, Q. (2021). Enhanced bioremediation of diesel oil-contaminated seawater by a biochar-immobilized biosurfactant-producing bacteria Vibrio sp. LQ2 isolated from cold seep sediment. *Science of The Total Environment, 793*, 148529. Available from https://doi.org/10.1016/j.scitotenv.2021.148529.

Zhu, C., Xiao, F., Qiu, Y., Wang, Q., He, Z., & Chen, S. (2017). Lichenysin production is improved in codY null Bacillus licheniformis by addition of precursor amino acids. *Applied Microbiology and Biotechnology, 101*(16), 6375–6383. Available from https://doi.org/10.1007/s00253-017-8352-z.

CHAPTER 10

Role of biosurfactants in improving target efficiency of drugs and designing novel drug delivery systems

Ramla Rehman[1], Asif Jamal[2], Irfan Ali[3], Munira Quddus[2] and Aziz ur Rehman[2]
[1]Institute of Industrial Biotechnology, Government College University, Lahore, Punjab, Pakistan
[2]Faculty of Biological Sciences, Department of Microbiology, Quaid-i-Azam University, Islamabad, Pakistan
[3]Centre of Agricultural Biochemistry and Biotechnology (CABB), University of Agriculture Faisalabad, Faisalabad, Punjab, Pakistan

10.1 Introduction

The pharmaceutical industry is facing huge challenges in providing effective medicines for various emerging infections and other disorders (Sharma et al., 2020). In this pursuit, various innovative products have been emerging in the international market (Ratcliffe et al., 2021). Biosurfactants are innovative surface-active molecules produced by different soil-dwelling microorganisms (De Giani et al., 2021). They can solubilize insoluble materials/drugs in the aqueous medium and improve the drug target efficiency with molecular properties to interact with biological membranes (Liu et al., 2020). Owing to their unique chemical structure, diverse chemistry, and fine supramolecular architecture, they are paving their way as an alternative drug delivery system for the development of future molecular medicine (Bjerk et al., 2021). In the past few years, different types of biosurfactants have been tested for their suitability and efficacy for drug design and their application as a drug carrier system (Naughton et al., 2019). In this regard, glycolipids and lipopeptide biosurfactants are taking the lead and the subject of various interesting investigations for their biomedical applications (Zhang et al., 2017). One of the most exciting studies highlights their putative role in curtailing COVID infections (Senten et al., 2020). The sophorolipids produced by *Candida bombicola* have been used to inactivate SARS-COVID viruses in vitro (Daverey et al., 2021). Another important finding suggested that they are better candidates for improving drug-targeted efficiency of insoluble drugs, such as ciprofloxacin and telmisartan (Rofeal et al., 2021). These studies motivate scientists to further investigate the unique role of these magic molecules in developing innovative drugs and pharmaceutical products for future needs (Saikia et al., 2021). This chapter will provide a detailed understanding of the qualifications, limitations, and opportunities of biosurfactants for improving drug delivery systems, typically considering their interaction with the biological membranes.

10.2 Unique self-assembly features of biosurfactants and their suitability for drug adaptation and target improvement

Biosurfactants are promising biomolecules having immense biotechnological applications. Their superior properties and self-assembly patterns have enabled their wide spread applications in various fields. The monomers of biosurfactants can interact independently and form diverse structural aggregates (Tucker et al., 2021). These molecules can act as vehicles for transferring desired molecules to the target site with improved efficiency (Smith et al., 2020). In the pharmaceutical industry, low-molecular-weight biosurfactants are of particular interest due to their rapid release in the extracellular environment (Phale et al., 2020). Furthermore, the extraordinary surface-active properties of biosurfactants enhance their potential as safe, efficient, and reliable drug delivery vectors.

Biosurfactants exhibit different morphological forms depending on their type, the nature of the hydrophilic head, the aliphatic chain length, critical micelle concentration (CMC), and the extracellular chemical environment. For example, biosurfactants form reverse micelles when surrounded by the oil phase in contrast to the aqueous system (Fig. 10.1) (Warisnoicharoen et al., 2000). It is important to note that biosurfactants performance in a particular application depends upon surfactants concentration. CMC is the critical concentration of biosurfactants at which the monomers aggregate and self-assemble to form particular micellar structures (Table 10.1). The self-assembly of the surfactant monomers relies on various physiological factors, including temperature, pH, and salt concentration (Md, 2012). The process of self-assembly is driven by packaging

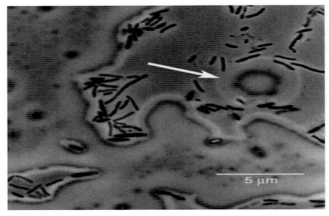

Figure 10.1 Representation of a biosurfactant micelle. *From Perfumo, A., Smyth, T. J. P., Marchant, R., & Banat, I. M. (2010). Production and roles of biosurfactants and bioemulsifiers in accessing hydrophobic substrates (pp. 1501–1512). Springer Science and Business Media LLC. https://doi.org/10.1007/978-3-540-77587-4_103.*

Table 10.1 CMC (critical micelle concentration) values of biosurfactants.

Biosurfactant	Critical micelle concentration (mg/L)	Source organism	References
Surfactin	13	Bacillus subtilis	Sen and Swaminathan (2005)
Lichenysin	22	Bacillus licheniformis	Suthar and Nerurkar (2016)
Fengycin	20.5	B. subtilis	Mnif and Ghribi (2015)
Sophorolipids	10	Candida bombicola	Kulakovskaya and Kulakovskaya (2014)
Trehalolipids	250	Mycobacterium, Corynebacterium	White et al. (2013)
Rhamnolipids	20	Pseudomonas aeruginosa	Shreve and Makula (2019)
Viscosin	13.5	Pseudomonas fluorescens	Renard et al. (2019)

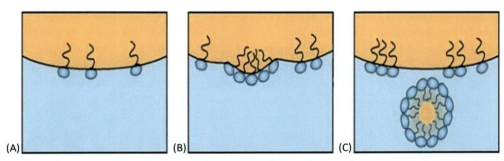

Figure 10.2 Biosurfactants monomers forming micelles at the interface. *From Perfumo, A., Smyth, T. J. P., Marchant, R., & Banat, I. M. (2010). Production and roles of biosurfactants and bioemulsifiers in accessing hydrophobic substrates (pp. 1501–1512). Springer Science and Business Media LLC. https://doi.org/10.1007/978-3-540-77587-4_103.*

principles and supramolecular forces to yield an orderly arrangement of the surfactant monomers (Fig. 10.2). The standard free energy of micellization governs the orientation and phase distribution of biosurfactant monomers to make the reaction thermodynamically feasible (Molyneux et al., 1965). Several noncovalent interactions are involved in the self-assembly of biosurfactant monomers. These include the repulsive forces between the water molecules and the hydrophobic tails and, the steric hindrance induced by the interaction of hydrophilic heads of the adjacent monomers. The combined effect of these intermolecular interactions results in the formation of a stable micellar aggregates comprising a hydrophobic core and a hydrophilic head (Baccile et al., 2021).

The combined effect of noncovalent forces play a pivotal role in determining the physicochemical properties of micelles. The tails of monomers are responsible for directing the attractive forces toward the core, while the head groups repel each other to form a characteristic circular morphology. It has been reported that degree of micellization and geometrical arrangement of surfactants molecules are determined by the nature and length of the aliphatic chains of a typical biosurfactant molecule (Baccile et al., 2012). In the case of morphology, the monomers of biosurfactants have lamellar, spherical and hexagonal symmetry that is respectively assembled into micelles, liposomes, and niosomes. The formation of different structural aggregates depends on the biosurfactants concentration. At lower concentrations, micelles are the preferred structures; however, with the increase in concentration, the formation of many other structural aggregates can be observed. These include the lamellae and cubic forms that vary according to the external environmental conditions (Kitamoto et al., 2009).

Because of aforementioned traits, biosurfactants can encapsulate different drugs efficiently in their core allowing their application as an alternative and green drug delivery tool. Studies have shown the efficacy of different biosurfactants in the solubilization and delivery of both hydrophilic and hydrophobic drugs (Nakanishi et al., 2009). The accurate process and the nature of the drug encapsulation is usually determined by the type of biosurfactants used in the drug delivery system. Slight variations in the physicochemical conditions, such as temperature, pH, and constituents, might directly affect the geometry of the surfactants aggregates and performance of the system (Chauhan et al., 2018). Liposomes are the biosurfactant aggregates containing cholesterol as their integral structural component, whereas niosomes lack the cholesterol chains. The nanostructures of liposomes and niosomes are bigger than the usual micelles. In the bilayer architecture, the outside surface and the inside core are hydrophilic, while the inter-bilayer space comprises aliphatic chains and represents the hydrophobic region. The water-soluble drugs are contained in the hydrophilic center, whereas the hydrophobic drugs can attach themselves to the hydrocarbon chains (Igarashi et al., 2006; Magalhães et al., 2021). Due to these distinctive properties, the suitability of biosurfactants is evident for their use as drug delivery agents. The other factors giving advantage to the biosurfactants over conventional systems include tolerance to extreme physicochemical conditions, i.e., pH, temperature, and salinity (Rodrigues et al., 2006). In addition, biosurfactants are biodegradable and have very low toxicity, thus considered important biotechnological products for designing ecologically and biologically safe drug delivery systems. In vivo studies by Sahnoun et al. provide experimental evidence for the low toxicity of liposomes even at high concentrations (Sahnoun et al., 2014).

Biosurfactants with their unique properties are continued to dominate conventional drug delivery systems with a rapid pace. The diverse role of biosurfactants in the target improvement for effective drug delivery has been studied in a great deal. It has been reported that direct interaction of the biosurfactants with the biological membranes and adapting appropriate molecular geometry at the interfaces could be the most important aspects of the biosurfactants based drug delivery systems. In addition, the effective interplay of surfactants and membrane molecules supports easy release of the drug to the target site with greater precision. Biosurfactants make hydrophobic interactions with the lipid membranes, especially cholesterol molecules, that distort the membrane. This disturbs the structural integrity of the plasma membrane and results in the formation of irregular patches. The perforated plasma membrane paves the way for easy and targeted release of drugs (Nazari et al., 2012). The process of drug release from micelles can also be controlled by manipulating environmental and chemical conditions (Cheow & Hadinoto, 2012). It has been suggested that conjugating the biosurfactant vesicles with different antigens improved the efficacy of the drug release at a specified target site (Sajid et al., 2020).

10.3 Solubility and emulsion formation by biosurfactants for hydrophobic drug bioavailability

The emulsion is defined as the dispersion of two immiscible liquids to develop a uniform phase. In general, an emulsion is made up of two phases that lack balance. The phase with relatively less concentration is termed the dispersed phase, whereas the more concentrated phase that usually surrounds the dispersed phase is called the continuous phase (Deore, 2017). Oil-in-water (O/W) system is the best example of the emulsion, where oil is the dispersed phase and water is the continuous phase. The process of emulsion formation is a regulated phenomenon that has been exploited for different industrial applications. In an emulsion system, the surfactant molecule plays the role of an emulsifier that lowers the surface tension between the two phases and makes them miscible. Chemically derived surfactants were used for the said purpose; however, their high toxicity and nondegradability have raised some serious environmental concerns. Experimental evidences are available that further show the inadequacy of chemical surfactants to be used for in vivo studies, particularly in drug delivery systems (Furrer et al., 2000).

Biosurfactants are considered a safe, nontoxic, and eco-friendly alternative to their chemical counterparts for stable emulsion formation (Farias et al., 2021). In a kinetically stabilized emulsion, the interfacial tension between the solvent molecules is reduced to enhance the miscibility of the two liquid layers. Biosurfactants are more effective than chemical surfactants in reducing the surface tensions at the interfaces. Some biosurfactants take assistance from a co-surfactant molecule to carry out the emulsification; however, emulsion formation without using a co-surfactant has also been reported (Worakitkanchanakul et al., 2008). Different types of micellar assemblies have been observed for biosurfactants in the formation of stable emulsions. Hydrophilic—lipophilic balance (HLB) is one of the parameters used to determine the stability of the emulsion system. Biosurfactants with less HLB values form water-in-oil microemulsions, whereas biosurfactants with high HLB values form oil-in-water emulsions (Akbari et al., 2018). Both kinds of emulsions are used for different purposes following their nature. The other parameters that define the nature and stability of emulsions are the Winsor-R ratio and critical packing parameter (CPP) (Nguyen & Sabatini, 2011). The cumulative effect of these parameters determines the self-assembly of biosurfactants, which in turn directly affects the stability of an emulsion system.

The two most important types of biosurfactant-mediated emulsions effectively used in green drug delivery systems are microemulsions and nanoemulsions. Microemulsion-based drug delivery systems (MDDS) were developed to enhance the solubility and the bioavailability of hydrophobic drugs. These microemulsions incorporate water-insoluble drugs for their effective and targeted release. Microemulsions are known for their well-structured assembly and thermodynamic stability (Boonme, 2007). Different theories have been proposed for the stability of microemulsions, including the reduction of interfacial tension and repulsion theory. According to repulsion theory, biosurfactants encapsulate only one of the phases and repel the other phase, thus stabilizing the microemulsion system in its dispersed state (Deore, 2017). The microemulsions can adapt different phases depending on the Winsor R-ratio. These include (1) the two layers with the layer of oil on the top and microemulsion at the bottom, (2) the microemulsion at the top and a layer of water at the bottom, (3) a layer of microemulsion sandwiched between the upper oil layer and lower water layer, or (4) assembling all constituents in a monophase state (Ruckenstein, 1996).

The preparation of a microemulsion formulation is a sequential process that involves the addition of oil, water, and biosurfactants in an appropriate concentration to the mixture of the hydrophobic drug (Fig. 10.3). Homogenization is carried out to obtain uniformity within the reaction mixture (Yadav et al., 2018). The resulting microemulsions and nanoemulsions are known as self-micro-emulsifying drug delivery systems (SMEDDS) and are used to enhance the bioavailability of water-insoluble drugs. SMEDDS can tolerate extreme pH and temperature alterations. Moreover, their biphasic nature helps them rapidly profuse through the barrier of biological membranes. Additional modifications include the release of the drugs under controlled conditions. The use of ionic biosurfactants is preferred as it enhances the mobility of drugs toward the biological membranes due to the attractive forces developed between the membranes and charged head groups of biosurfactants, hence maximizing the target efficiency. In vivo studies by Gharbavi et al. and others have proved the nontoxicity and successful applications of these microemulsion-based DDS (Gharbavi et al., 2019).

Nanoemulsions are amphiphasic emulsions that are used as drug delivery systems. The basic components of the nanoemulsions system are quite similar to that of microemulsions, i.e., an oil and water phase and a surfactant molecule. However, the two systems differ in terms of structural variations and mode of solubilization (Table 10.2).

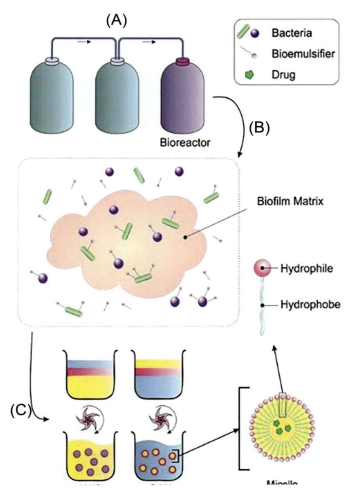

Figure 10.3 Microemulsion-based drug delivery system (MDDS). *From Alizadeh-Sani, M., Hamishehkar, H., Khezerlou, A., Azizi-Lalabadi, M., Azadi, Y., Nattagh-Eshtivani, E., Fasihi, M., Ghavami, A., Aynehchi, A., & Ehsani, A. (2018). Bioemulsifiers derived from microorganisms: Applications in the drug and food industry. Advanced Pharmaceutical Bulletin, 8(2), 191–199. https://doi.org/10.15171/apb.2018.023. 30023320.*

In a nanoemulsion, the drug to be delivered is kept inside the hydrophobic core while the rest of the constituents are dispersed in the surrounding hydrophilic phase (Shah et al., 2010). The surface of nanoemulsions is covered with some exogenous moieties that target the specific cellular sites to trigger a rapid immune response (Fig. 10.4). The role of biosurfactants in a nanoemulsion is critical as these biomolecules solubilize the oil and water phases, making them kinetically stable, therefore preventing

Table 10.2 Difference in properties of microemulsions and nanoemulsions.

Property	Microemulsion	Nanoemulsion	References
Size	10–300 nm	10–1000 nm	Fink. (2021), Jaiswal et al. (2015)
Preparation requirements	High temperature facilitates self-assembly	Other than the heat, ultrasonication, and vigorous homogenization	Rao and McClements (2011)
Requirement of biosurfactant	High concentration	Less concentration	Deore (2017)
Drug loading capacity	1%–3% (~10 mg/g)	~45 mg/mL	Patel et al. (2010), Sabale and Vora (2012), Sun et al. (2012)
pH range	7–10	4.9–6.8	Yang et al. (2013), Bernardi et al. (2011)
Conductivity test	0.046–0.136 m S/cm	0.4542–0.5523 m S/cm	Gurpreet and Singh (2018), Moghimipour et al. (2013)
Appearance	Transparent or translucent due to reduction in particle size	Cloudy or turbid in some cases due to scattered light	Zhang and McClements (2018)
Particle structural variability	The structure of particles can be spherical or may vary with a considerable radius as the surface tension is not that high	The particle structure is spherical, having a small radius. This is due to the high surface tension and pressure on the particles	McClements (2012)

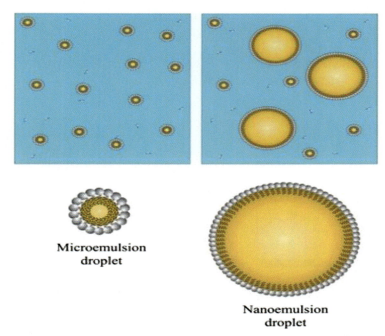

Figure 10.4 Difference between micro- and nanoemulsion droplets. *From McClements, D. J. (2012). Nanoemulsions versus microemulsions: Terminology, differences, and similarities. Soft Matter, 8(6), 1719−1729. https://doi.org/10.1039/c2sm06903b.*

flocculation or breaking down of the emulsion (Lewińska et al., 2020). Various approaches have been used in numerous studies for the preparation of a stable nanoemulsion. These include (1) ultrasonication, (2) temperature alterations in the formulation process, and (3) homogenization (Kumar et al., 2019). Analyzing the physicochemical traits of a nanoemulsion provides insight into their effective role as drug delivery systems. Analytical techniques such as dynamic light scattering and electron microscopy can be used to reveal the details of size and shape, along with the pH and temperature-induced variations in these thermodynamically stable systems (Grapentin et al., 2015). The nano-size of these emulsions increases the stability and allows the system to overcome different biological barriers present in living systems. Hydrophobic drugs, when encapsulated in a nanoemulsion system, can easily be delivered across the skin (Fig. 10.5). The surface-active properties of biosurfactants play a significant role in reducing the interfacial tension, which, along with the solubilizing nanoemulsions, facilitates the bioavailability of drugs. It was observed that nanoemulsions-mediated drugs retained their potency when taken orally. Another characteristic feature of the nanoemulsions system is the improvement in the mode of treatment of already available drugs. Many infectious microorganisms have developed

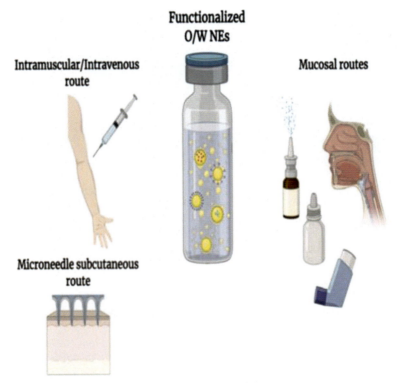

Figure 10.5 Application of nanoemulsions as drug carriers for different routes of administration. *From Tayeb, H.H., Felimban, R., Almaghrabi, S., & Hasaballah, N. (2021). Nanoemulsions: Formulation, characterization, biological fate, and potential role against COVID-19 and other viral outbreaks. Colloid and Interface Science Communications, 45, 100533.*

resistance to antibiotics by altering their membrane compositions and inhibiting the drug influx. These nanoemulsions can carry the drug across the plasma membrane and help kill multidrug-resistant pathogens. Furthermore, the development of highly efficient nanoemulsion systems has led to some recent advancements in anticancer therapy (Tayeb et al., 2021). Therefore, in a stable emulsion system, biosurfactants increase the solubility and bioavailability of hydrophobic drugs. The enhanced interactions of these biomolecules with the biological membranes pave the way for the efficacious release of drug to the target site.

10.4 Interaction of biosurfactants with bio-interfaces

Biological molecules interact in various ways within the natural systems. One such interaction is the contact between their organic/inorganic components, which results in the formation of a bio-interface (Chilkoti & Hubbell, 2005). The interaction of biosurfactants with a bio-interface depends on their nature, type, and concentration. The other determining factors include the structural composition, biochemistry, and the nature of the other biomolecules involved in forming the bio-interface (Heerklotz et al., 2004). The most common bio-interface is the membrane that acts as the point of contact. Biosurfactants interact with the major components of membranes and destabilize them. Some biosurfactants attack the ionic parts of the membrane, resulting in the formation of ion channels, increasing the membrane permeability. Other biosurfactants solubilize the hydrophobic portion of the biological membranes and increase their permeability. Loss of water is another strategy that decreases the fluidity of membranes and compromises their structural integrity. Such an adaptive strategy has been reported in the case of surfactin, a lipopeptide biosurfactant (Seydlová & Svobodová, 2008). This biosurfactant—biointerface interaction is responsible for the antimicrobial activity of biosurfactants against pathogens. The concentration plays a pivotal role in enhancing the solubility of drugs. When added in low quantity, biosurfactant monomers tend to solubilize the membrane through micelle formation. In contrast, if the biosurfactant concentration is increased, distinct domains are formed due to the increasing dissociation of the intermembrane linkages (Coronel et al., 2017). As a result, the membrane is perforated and eventually loses its conformity. This process of distorting the membranous interfacial molecules holds significant biomedical importance.

Interfacial interactions are well understood when the interplay of biosurfactant—biomolecule is considered. Lipids are the most abundant biomolecules to be found at bio-interfaces. The phenomenon of biosurfactants—lipids interaction has been divided into different stages. Initially, the biosurfactant monomers migrate and position themselves at the interface, separating the two immiscible phases. This step is followed by the state where biosurfactants and lipids micelles exist in equilibrium. Finally, when the concentration of biosurfactant monomers increases, the lipid membrane is solubilized (Aguirre-Ramírez et al., 2021). Biosurfactants increase membrane fluidity by altering its composition. This is usually done by converting the saturated lipids into unsaturated ones or by inducing certain changes in the phospholipid head groups. This results in an increase in the fluidity of biological membranes and a decrease in the system's surface tension (Ortiz et al., 2009). Some biosurfactant monomers can cross the secondary barrier and form small vesicular structures. Once inside the cell, these monomers also affect the membrane of organelles. Such activity has been observed for iturin-like biosurfactants (Jiang et al., 2020). Saponins—cholesterol interaction is aided by compounds such as glycyrrhizic acid, which results in inactive membrane segregation (Fig. 10.6). In another study, the potential of biosurfactants as a drug delivery system was explored due to their key role in the controlled and sustained transfusion of drugs across biological membranes (Liao et al., 2021).

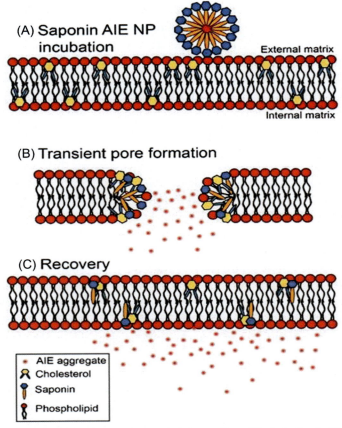

Figure 10.6 Saponin membrane interaction and delivery. *From Nicol, A., Kwok, R.T., Chen, C., Zhao, W., Chen, M., Qu, J., & Tang, B.Z. (2017). Ultrafast delivery of aggregation-induced emission nanoparticles and pure organic phosphorescent nanocrystals by saponin encapsulation. Journal of the American Chemical Society, 139(41), 14792–14799.*

Proteins and polysaccharides are also found in abundance at the bio-interfaces as an integral component of the biological membranes. Biosurfactants are preferred over chemical surfactants due to their better compatibility with proteins. The biosurfactants—protein interaction at bio-interfaces is regulated by the noncovalent binding forces. These forces are a function of the type, concentration, and ionic character of the biosurfactants involved (Li & Lee, 2019). Several other parameters also decide the fate of biosurfactants—proteins interaction. For example, the secondary structure of the protein is one of the deciding factors as to whether the denaturing effect will make its way or not. Reports show the inability of some biosurfactant molecules to alter the protein conformation (Zaragoza et al., 2012). When present in low concentration, the monomers of biosurfactants assist the folding of amino acid chains to form a stable protein confirmation. However, the higher concentration of biosurfactants drastically affects the protein structure and causes denaturation (Khan et al., 2015). Polysaccharides are also known to develop unique interactions with biosurfactants, forming complex molecular aggregates. Therefore, it can be concluded that biosurfactants play a critical role in forming bio-interfaces, highlighting their significance in various innovative therapeutic applications.

10.5 Biosurfactants as delivery carriers for DNA-/RNA-based drug vehicles

The use of DNA- and RNA-based vectors in advanced drug delivery systems (DDSs) has attracted growing attention across the globe. Previously used DDSs were less efficient in terms of target selectivity and poor water solubility of drugs. However, current developments in the field of nanobiotechnology have resolved some of the important technical issues. This approach has several advantages over conventional drug delivery systems including high stability, better drug-carrying capacity, and improved feasibility to encapsulate lipophilic or hydrophilic drugs (Gelperina et al., 2005). One of the exciting features of this technique is the conjugation of DNA or RNA molecules with the drugs. This results in the formation of a highly sophisticated nano-architecture that can treat various diseases with limited side effects. The use of small DNA fragments has been frequently reported in complex nanomedicines. The DNA-based drug nano-architecture disassembles in a controlled and precise manner for the effective drug release at the respective target (Komiyama et al., 2022). Furthermore, RNA-based therapies are playing a dominating role in manipulating gene expression and regulating the production of proteins having medical importance. This approach is typically used to make drug formulations of known target recipients. Various types of cancers, neurological disorders, and immune and infectious diseases have been investigated in the context (Paunovska et al., 2022). Biosurfactants enhance the efficacy of the RNA-/DNA-based nano-sized drug delivery systems by encapsulating the drug in the micellar core (Table 10.3). Ueno et al. (2007) indicated that in mammalian cells, mannosylerythritol lipids (MEL)-A, and their associated liposomes are effective in gene transfection (Ueno et al., 2007). Liposomes are derived from lipid—bilayer vesicles and can be used as efficient carriers for DNA transfection. Some previous reports have shown that liposomes and biosurfactants can encase the hydrophobic or hydrophilic drugs and therefore exhibit improved efficiency for target drug delivery (Inoh et al., 2004). In another study, Farhood et al. revealed, that complexes of DNA-liposomes can be used to treat metastatic melanoma with comparatively less complications and high safety to the human body (Farhood et al., 1994). Similarly, Maitani et al. (2006) reported that the aggregates of plasmid DNA (pDNA) and liposomes along with the biosurfactants MEL-A and

Table 10.3 Biosurfactant-based carriers (liposomes, niosomes) used for RNA-/DNA-based drug vehicles.

Nanoparticles	Co-delivered drug	RNA/DNA vehicle	Targeted pathology	References
Liposomal nanocarriers/PEGylated liposome	Docetaxel (DTX)	BCL2 siRNA	Lung cancer	Qu et al. (2014)
pH-sensitive carboxy methyl chitosan liposome	Sorafenib	siRNA	Liver cancer	Yao et al. (2015)
Thermo-sensitive magnetic liposome	Duxorubicin	shRNA	Gastric cancer	Peng et al. (2014)
Folic acid-modified liposome	Duxorubicin	Bmi1 siRNA	HeLa, KB, Hep3B cells	Yang et al. (2014)
Liposome	Duxorubicin	Vimentin siRNA	Hepatic cancer	Oh et al. (2016)
Modified Liposome	Duxorubicin	Vascular endothelial growth factor siRNA	Brain tumor/Glioma	Yang, Li, Wang, et al. (2014)
Lipid-based nanoparticle	Photodynamic therapy (PDT)	siRNA	Head and neck tumor	Chen et al. (2015)
Niosomes	Thymoquinone (TQ)	Akt siRNA	Breast cancer	Rajput et al. (2015)
	Duxorubicin	MAC 15 A	subcutaneous tumor	Uchegbu et al. (1996)
Liposomes (MEL-A and β-sitosterol β-D-glucoside)	—	pDNA	Hepatoblastoma	Maitani et al. (2006)

beta-sitosterol beta-D-glucoside have the potential to lower the cytotoxicity of HepG2 cell line. The beta-sitosterol beta-D-glucoside (sit-G) liposome/DNA complex was used as a potential vector in gene therapy and can be used in intravenous and intratumoral injections in future (Maitani et al., 2006).

The emergence of multidrug resistance (MDR), particularly in cancer patients, has been quite a concern for health officials (Fig. 10.7). RNA interference (RNAi) is an innovative approach that was recently introduced as a powerful tool to overcome MDR. Small interfering (siRNA) is an extension of the biochemical mechanism of RNA interference used to silence the gene expression involved in MDR. The combined use of siRNA and other chemotherapeutics has been reported to significantly

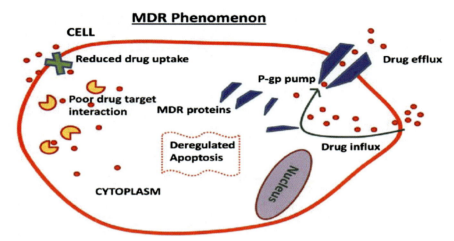

Figure 10.7 An animation illustrates the mechanisms of multidrug resistance (MDR) in cancerous cells. Increased drug efflux, MDR protein synthesis, lower drug absorption, insufficient drug-target binding, and down-regulated apoptosis are a few of the key mechanisms. *From Babu, A., Munshi, A., & Ramesh, R. (2017). Combinatorial therapeutic approaches with RNAi and anticancer drugs using nanodrug delivery systems.* Drug Development and Industrial Pharmacy, 43(9), 1391–1401.

improve the cancer treatment process. However, the co-delivery of anticancer drugs and siRNA through an effective and stable drug delivery system requires nanocarriers. Nanoparticles deliver multiple therapeutic agents simultaneously, as they can encapsulate and release drugs in a regulated manner at a specific target. Therefore, lipid molecules or liposome-based nano-vehicles were used (Babu et al., 2017). It was observed that the nano carrier-based co-delivery vectors of siRNA have the potential to simultaneously target multiple genes responsible for therapeutic inefficiency and MDR (Fig. 10.8). Overall, these nanocarrier-based co-delivery systems are fundamental for developing a sustainable approach to cancer therapy.

10.6 New biosurfactants with improved drug target efficiency

Biosurfactants are eco-friendly replacements for existing therapeutics for cancer treatment and DDS (Gangwar et al., 2012). The principal attributes of DDS are (1) drug-loading capacity with an optimal rate, improved bioavailability, and high target specificity and (2) controlled release of the drugs. Due to the amphiphilic nature of biosurfactants, they possess significant antibacterial, antiviral, antifungal, antimalarial, antiinflammatory, and anticancerous activity. Numerous studies have reported the role of biosurfactants in cancer treatment by arresting cell division, preventing metastasis, inhibiting growth, and inducing apoptosis (Wu et al., 2017). Various microbial biosurfactants are reported to stimulate cytotoxic effects against different cancerous cell lines, i.e., lung cancer, melanoma,

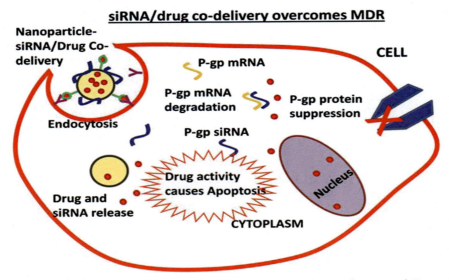

Figure 10.8 Illustrating Multidrug resistance (MDR) phenomena in cancer therapy and Nanoparticle/Drug co-delivery to deal with MDR by siRNA-mediated p-glycoprotein (p-GP) reduction and triggering cellular apoptotic processes by increasing drug accumulation. *From Babu, A., Munshi, A., & Ramesh, R. (2017). Combinatorial therapeutic approaches with RNAi and anticancer drugs using nanodrug delivery systems. Drug Development and Industrial Pharmacy, 43(9), 1391–1401.*

leukemia, breast cancer, and colon cancer, consequently limiting the progression of malignant cells. The incompatibility of anticancer drugs to penetrate tumor cells, and their harmful effects on healthy cells, has led to the exploration of the anticancer potential of biosurfactants. Nanoparticles, niosomes, core–shell type nanocapsules, and liposomes are a few examples of biosurfactants-based novel DDS (Varun et al., 2012).

10.6.1 Niosomes

Niosomes are the nonionic nanostructures derived from the bilayer lipid vesicles of biosurfactants. The amphiphilic nature of biosurfactants is vital for hydrophilic or hydrophobic drug encapsulation and safe adsorption (Ray et al., 2018). Niosomes exhibit diverse properties due to variations in their composition, tapped volume, size, surface charge, and concentration. However, their stability as a drug delivery vehicle significantly depends on the type of biosurfactants used in vesicle formation, biochemical properties of membrane-spanning lipids, the nature of the drug being encapsulated, interfacial polymerization of biosurfactants, storage temperature and detergents (Fig. 10.9) (Naughton et al., 2019). Recently, surfactin has been reported as a part of the nano-formulation of niosomes due to its surface-active and hydrophilic properties. The surfactin-derived niosomes can efficiently

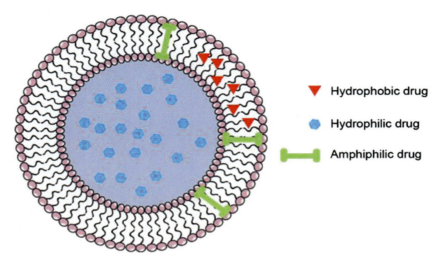

Figure 10.9 Representation of different types of drugs in niosomes. *From Saraswathi, T. S., Mothilal, M., & Jaganathan, M. K. (2019). Niosomes as an emerging formulation tool for drug delivery-a review.* International Journal of Applied Pharmaceutics, *11(2), 7−15. https://doi.org/10.22159/ijap.2019v11i2.30534.*

incorporate the drug into their core—shell structure (Wu et al., 2017). It has been reported that the niosomes acting as the carrier for doxorubicin showed double antitumor activity in comparison to the free doxorubicin drug administered for the treatment of subcutaneous tumors in mice. Similarly, the use of niosomal formulations against ovary tumors has shown significant improvement in the treatment of cancer (Uchegbu et al., 1996).

10.6.2 Nanoparticles

The synthesis of nanoparticles (NPs) has been introduced with the exciting roles of biosurfactants as their reducing, capping, or stabilizing agents (Table 10.4) (Rodrigues, 2015). Sophorolipids have been used as capping and reducing agents for silver nanoparticles. Redy et al. revealed that surfactin can also be used as a stabilizing agent in the formation of gold and silver NPs (Reddy et al., 2009). NPs are reported to be used in combination with chemotherapeutic drugs as they can be easily loaded into nanoformulations to augment cancer treatment (Wu et al., 2017). In another study, Huang et al. developed a combination of an anticancer drug named doxorubicin (DOX) with a surfactin-based nanoparticle termed DOX@SUR. Surfactin-based NPs loaded with DOX exhibited better cytotoxicity against DOX-resistant breast cancer cell line (MCF-7/ADR cells) than the free DOX. This was achieved by regulating the expression of p-glycoprotein by inhibiting its cellular efflux. Moreover, in vivo studies have also shown higher tumor suppression and lower side effects for DOX@SUR (Huang et al., 2018).

Table 10.4 Variety of biosurfactants and microemulsions as drug delivery agents.

Type of biosurfactant	Microemulsion formulations	Role of biosurfactants	Function of microemulsions	References
Surfactin	Lipopeptide, containing self-micro-emulsifying system (SMEDDS) along with Docosahexaenoic acid	Act as stabilizer	Sustained drug delivery system	He et al. (2017)
	Lipopeptide with vitamin E, containing SMEDDS	Active stabilizer	Stable drug delivery	Kural and Gürsoy (2010)
Rhamnolipid	A mixture of biosurfactants, water, or oil	Act as an emulsifier, hydrophobic/hydrophilic linker	Application in detergency and drug delivery system	Nguyen and Sabatini (2009)
	A mixture of glycolipid biosurfactant, water, n-butanol, or n-heptane	Act as stabilizer and emulsifier	Sustained drug delivery	Xie et al. (2007)
	Combination of n-heptane, water, alcohol, and glycolipid biosurfactant	Active emulsifier and stabilizer	Drug delivery	Xie et al. (2007)
Sophorolipids	A mixture of biosurfactants with a variety of oils and lecithin	Act as amphiphilic linker and stabilizer	Drug delivery, cosmetics, and detergency	Nguyen and Sabatini (2009)

Another drug Paclitaxel (PTX), which belongs to class 4 of biopharmaceutic drugs (BCS), is used in the treatment of multiple tumors. However, on administration, PTX faces hindrances in the host body because of water insolubility and membrane impermeability. Moreover, the excipients used in the formulation of PTXs for stability also obstruct the carriage and release of the drug to the target. Limitations in drug loading and the drawbacks of excipients used in this particular DDS have led to the development of a new DDS. Katiyar revealed that the use of core—shell type nanocapsules for PTX has significantly increased the drug efficacy and reduced the infusion time in the breast cancer model. It has been reported that the core—shell type nanocapsules contain drugs in their core, which is further surrounded by lipids or other polymers in the form of an envelope. The nontoxic and biodegradable biosurfactants are used to stabilize the core—shell-type nanocapsules instead of other excipients. These biosurfactant-based nano formulations or core—shell-type nanocapsules have the potential to overcome DR commonly observed in the chemo-treatment of patients with cancer (Table 10.5). These types of nano

Table 10.5 Diverse biosurfactants with anticancer activity.

Biosurfactants	Targeted cancer	Cell line	Description of activity	References
Surfactin or surfactin-like molecules	Breast cancer	MCF7	Apoptosis induction, growth inhibition	Ma et al. (2014), Lee et al. (2012), Cao et al. (2010, 2011)
	Colon-adenocarcinoma	LoVo	Inhibition of growth, Induction of apoptosis	Kim et al. (2007)
	Myelogenous leukemia	K562	Arrest cell cycle, apoptosis induction, growth inhibition	Wang et al. (2007), Cao et al. (2009)
	Hepatocellular carcinoma	BEL7402	Induce apoptosis, inhibition of growth	Cao et al. (2009)
Sophorolipids	Pancreatic cancer	HPAC	Cause necrosis	Fu et al. (2008)
	Lung cancer	A549	Induce apoptosis	Chen et al. (2006)
	Promyelocytic leukemia	HL60	Differentiation of cell, Inhibit the growth	Isoda, et al. (1997)
	Liver cancer	H7402	Induce apoptosis, Arrest cell cycle, Inhibit the growth	Chen et al. (2006)
Viscosin	Metastatic prostate cancer	PC3M	Inhibit migration	Saini et al. (2008)
Monolein	Leukemia cancer	U937	Inhibit of growth	Chiewpattanakul et al. (2010)
	Cervical cancer	HeLa	Inhibit growth	Chiewpattanakul et al. (2010)
Serratamolide	B-chronic lymphocytic leukemia	BCLL	Induce apoptosis	Escobar-Díaz et al. (2005)
Succinoyltrehalose lipids (STLs)	Basophilic leukemia	KU812	Inhibit growth	Isoda et al. (1996)
	Promyelocytic leukemia	HL60	Differentiation, inhibit the growth	Sudo et al. (2000)
Mannosylerythritol lipids (MELs)	Myelogenous leukemia	K562	Differentiation and inhibition of growth	Isoda and Nakahara (1997)

formulations act as a safe and efficient drug delivery carriers; however, a thorough insight into the overall mechanism for carrying the drug and subsequent target release is still needed (Katiyar et al., 2020).

10.7 Patents concerning the applications of biosurfactants for the pharmaceutical industry

Owing to their exceptional qualities and promising future, biosurfactants have been widely utilized in wide-ranging industrial applications. Global Industry Insight estimates that biosurfactants will account for 5.5% of the surfactants market by 2024, increasing $1.55 billion at a 5.5% CAGR from 2019 (Ahuja & Singh, 2020). Due to their unusual lyotropic liquid crystalline phase behavior, biosurfactants have seen a significant increase in the number of international patents. Paul Damien Price's patented research shows that rhamnolipids improve the bioavailability of antimicrobial drugs, particularly lactam antibiotics. When used together, lactam and rhamnolipids proved highly effective in combating infectious biofilm. Lactam antibiotics have not been widely used due to their low water solubility. The water solubility of the lactam medicines, however, has been greatly enhanced with the addition of rhamnolipids biosurfactants (Google Patents WO2017029175A1, 2016). Wang and Xuan-Rui have patented a medication delivery system that uses surfactin (Wang & Xuan-Rui, 2014). Surfactin is a drug delivery vehicle with antiaging and antimicrobial properties that boost the efficacy of other therapeutic ingredients such as polysaccharides, vitamins, and gold nanoparticles (Fig. 10.10). Because of their amphiphilic character and enhanced solubility across biological membranes, biosurfactants are the most appropriate carriers for transdermal administration of cosmetic products (Fig. 10.11).

Patented by Prabhune et al., sophorolipids were reported to improve the transport efficiency of curcumin (Fig. 10.12). Curcumin has been studied for its potential therapeutic uses as antioxidant, antiinflammatory, and anticarcinogenic properties. According to their claim, curcumin was solubilized and encapsulated in acidic sophorolipids. The entrapment of curcumin in the surfactant micelles greatly improved its solubilization and bioavailability (Kumar Singh et al., 2015).

Ishigami et al. patented liposomes containing Rhamnolipid A and Rhamnolipid B as microcapsules for delivering antibiotics, nucleic acids, proteins, dyes, and other substances. These liposomes were also used as biomimetic models to study and characterize biological membranes (Ishigami et al., 1990). Piljac et al. published a patent of rhamnolipids containing formulations that showed significant improvement in skin infections such as acne vulgaris, erythrasma, and pytiriasis versicolor post administration (Piljac & Piljac, 1998). Farmer et al. recently published the use of certain formulations and processes to boost the effectiveness of therapeutic agents. For this purpose, the inventors used an adjuvant composition composed of biosurfactants, i.e., glycolipids

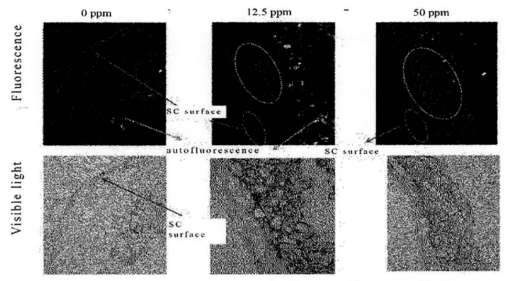

Figure 10.10 Surfactin enhancing skin penetration of gold nanoparticles measured by fluorescence microscope. *From Wang, H.M., Xuan-Rui, X. (2014). Applications of surfactin in cosmetic products. Google Patents US20160030322A1.*

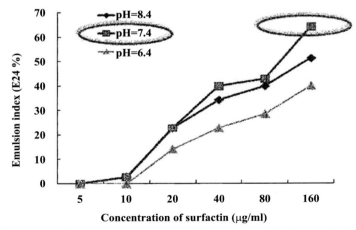

Figure 10.11 The emulsifying power of surfactin at different concentrations with diesel and various pH. *From Wang, H.M., Xuan-Rui, X. (2014). Applications of surfactin in cosmetic products. Google Patents US20160030322A1.*

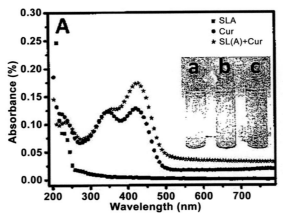

Figure 10.12 Absorption patterns of acidic sophorolipids, curcumin, and acidic sophorolipids plus curcumin complex at different wavelengths. *From Kumar Singh, P., Ashutosh Prabhune, A., Balkrishna Ogale, S. (2015). Curcumin-sophorolipid complex. Google Patents US20170224636A1.*

and lipopeptides. The composition increased the bioavailability, stability, and localization of the drugs and lowered the dosage required (Farmer et al., 2022).

10.8 Unknown aspects of biosurfactants and future directions

Drug adsorption is made possible through the interaction of various receptors present in the plasma membrane. Biosurfactants form micellar structures that interact with the phospholipid bilayers of the plasma membrane making hydrophilic—hydrophilic or hydrophobic—hydrophobic linkages. These unique attributes show that biosurfactants have the potential to advance the drug delivery systems and targeted therapeutic approaches. Geisler et al. reported the interaction of biosurfactants with phospholipids and 1,2-dimyristoyl-sn-glycro-3-phosphocholine (DMPC). Similar interactions are also thought to prevail with other cellular components (Geisler et al., 2020). Due to the emergence of deadly pandemics in recent years, there is a dire need to discover novel biosurfactant receptors for efficient drug adsorption and target drug release.

Various molecular ligands of biosurfactants have been identified recently, and their interactions with the living system are studied through the latest bioinformatics tools. New approaches should be developed to discover cellular-subcellular interactions and unique regulatory pathways to expand the studies related to drug kinetics. The cellular pathways influenced by biosurfactants are still not fully understood; therefore, several intermediate metabolic products have not been reported yet. Unraveling these pathways can improve our understanding and subsequently discovery of new pharmaceutical targets. Vesicle forms of biosurfactants are less studied; hence, research must focus on this unique molecular assembly of biosurfactants concerning drug adsorption and targeted therapies. Their role as drug-delivery agents can be exploited by identifying novel cellular receptors and discovering new metabolic intermediates. Recent scientific advances in bioinformatics can be quite handy in ascertaining the function of ligands formed from biosurfactants and their use in cellular metabolism. Intermediates formed in such processes can be targeted for future therapeutic approaches as possible ailment remedies. This would reduce the use of drugs and minimize the side effects, leading to a better and healthier lifestyle for the people.

10.9 Conclusion and future outlook of biosurfactants in drug development

Despite the increasing demand for biosurfactants in the global market, their use is still limited to a few applications. The remarkable physiochemical properties, low toxicity, bioavailability, and production from cheap and renewable resources can make biosurfactants a good substitute for toxic and hazardous chemical surfactants. Antibacterial, antiviral, antifungal, antiadhesive, and antibiofilm activities are the key attributes of several microbial surfactants, indicating their multipurpose use in the field of pharmaceutical, biomolecular medicine and healthcare sectors. Biosurfactants are quite advantageous for anticancer research owing to their ability to restrict proliferation of cancerous cells. Besides huge potential, the biosurfactants applications in various fields is quite limited because of lacking knowledge and understanding of their self-assembling properties and interactions/behavior at various biological and chemical interfaces. New biosurfactant molecules are being introduced in the market while the potential of previously isolated biosurfactants has not been fully explored. One of the great challenges of the biosurfactants application in the drug delivery system is their purity. The development of novel and advancement of DDS involving different biosurfactants is an area of future research. Comparatively less data are available about the efficacy of biosurfactants as adjuvants in the micro and nanoemulsion formulations. The role of biosurfactants in drug administration processes is not fully understood due to a lack of adequate research on the subject. Until now, only a few biosurfactants have met the criteria set forth by drug regulatory agencies; however, these positive outcomes will pave the way for the use of biosurfactant molecules in various cutting-edge applications.

References

Aguirre-Ramírez, M., Silva-Jiménez, H., Banat, I. M., & Díaz De Rienzo, M. A. (2021). Surfactants: Physicochemical interactions with biological macromolecules. *Biotechnology Letters*, *43*(3), 523–535. Available from https://doi.org/10.1007/s10529-020-03054-1.

Ahuja, K., Singh, S. (2020). *Biosurfactants market size by product*. Global Market Insights (Vol. 564).

Akbari, S., Nour, A. H., Yunus, R. M., & Farhan, A. H. (2018). Biosurfactants as promising multifunctional agent: A mini review. *International Journal of Innovative Research and Scientific Studies, 1*(1), 1–5. Available from https://doi.org/10.53894/ijirss.v1i1.2.

Babu, A., Munshi, A., & Ramesh, R. (2017). Combinatorial therapeutic approaches with RNAi and anticancer drugs using nanodrug delivery systems. *Drug Development and Industrial Pharmacy, 43*(9), 1391–1401. Available from https://doi.org/10.1080/03639045.2017.1313861.

Baccile, N., Babonneau, F., Jestin, J., Pehau-Arnaudet, G., & Van Bogaert, I. (2012). Unusual, pH-induced, self-assembly of sophorolipid biosurfactants. *ACS Nano, 6*(6), 4763–4776. Available from https://doi.org/10.1021/nn204911k.

Baccile, N., Seyrig, C., Poirier, A., Alonso-De Castro, S., Roelants, S. L. K. W., & Abel, S. (2021). Self-assembly, interfacial properties, interactions with macromolecules and molecular modelling and simulation of microbial bio-based amphiphiles (biosurfactants). A tutorial review. *Green Chemistry, 23*(11), 3842–3944. Available from https://doi.org/10.1039/d1gc00097g.

Bernardi, D. S., Pereira, T. A., Maciel, N. R., Bortoloto, J., Viera, G. S., Oliveira, G. C., & Rocha-Filho, P. A. (2011). Formation and stability of oil-in-water nanoemulsions containing rice bran oil: In vitro and in vivo assessments. *Journal of Nanobiotechnology, 9*, 44. Available from https://doi.org/10.1186/1477-3155-9-44.

Bjerk, T. R., Severino, P., Jain, S., Marques, C., Silva, A. M., Pashirova, T., & Souto, E. B. (2021). Biosurfactants: Properties and applications in drug delivery, biotechnology and ecotoxicology. *Bioengineering, 8*(8). Available from https://doi.org/10.3390/bioengineering8080115.

Boonme, P. (2007). Applications of microemulsions in cosmetics. *Journal of Cosmetic Dermatology, 6*(4), 223–228. Available from https://doi.org/10.1111/j.1473-2165.2007.00337.x.

Cao, X. h, Wang, A. h, Wang, C. l, Mao, D. z, Lu, M. f, Cui, Y. q, & Jiao, R. z (2010). Surfactin induces apoptosis in human breast cancer MCF-7 cells through a ROS/JNK-mediated mitochondrial/caspase pathway. *Chemico-Biological Interactions, 183*(3), 357–362. Available from https://doi.org/10.1016/j.cbi.2009.11.027.

Cao, X. H., Zhao, S. S., Liu, D. Y., Wang, Z., Niu, L. L., Hou, L. H., & Wang, C. L. (2011). ROS-Ca2 + is associated with mitochondria permeability transition pore involved in surfactin-induced MCF-7 cells apoptosis. *Chemico-Biological Interactions, 190*(1), 16–27. Available from https://doi.org/10.1016/j.cbi.2011.01.010.

Cao, X., Wang, A. H., Jiao, R. Z., Wang, C. L., Mao, D. Z., Yan, L., & Zeng, B. (2009). Surfactin induces apoptosis and G2/M arrest in human breast cancer MCF-7 cells through cell cycle factor regulation. *Cell Biochemistry and Biophysics, 55*(3), 163–171. Available from https://doi.org/10.1007/s12013-009-9065-4.

Chauhan, S., Singh, K., & Sundaresan, C. N. (2018). Physico-chemical characterization of drug–bio-surfactant micellar system: A road for developing better pharmaceutical formulations. *Journal of Molecular Liquids, 266*, 692–702. Available from https://doi.org/10.1016/j.molliq.2018.07.008.

Chen, J., Song, X., Zhang, H., Qu, Y. B., & Miao, J. Y. (2006). Sophorolipid produced from the new yeast strain Wickerhamiella domercqiae induces apoptosis in H7402 human liver cancer cells. *Applied Microbiology and Biotechnology, 72*(1), 52–59. Available from https://doi.org/10.1007/s00253-005-0243-z.

Chen, W. H., Lecaros, R. L. G., Tseng, Y. C., Huang, L., & Hsu, Y. C. (2015). Nanoparticle delivery of HIF1α siRNA combined with photodynamic therapy as a potential treatment strategy for head-and-neck cancer. *Cancer Letters, 359*(1), 65–74. Available from https://doi.org/10.1016/j.canlet.2014.12.052.

Cheow, W. S., & Hadinoto, K. (2012). Lipid-polymer hybrid nanoparticles with rhamnolipid-triggered release capabilities as anti-biofilm drug delivery vehicles. *Particuology, 10*(3), 327–333. Available from https://doi.org/10.1016/j.partic.2011.08.007.

Chiewpattanakul, P., Phonnok, S., Durand, A., Marie, E., & Thanomsub, B. W. (2010). Bioproduction and anticancer activity of biosurfactant produced by the dematiaceous fungus Exophiala dermatitidis SK80. *Journal of Microbiology and Biotechnology, 20*(12), 1664–1671. Available from https://doi.org/10.4014/jmb.1007.07052.

Chilkoti, A., & Hubbell, J. A. (2005). Biointerface science. *MRS Bulletin, 30*(3), 175–179. Available from https://doi.org/10.1557/mrs2005.48.

Coronel, J. R., Marqués, A., Manresa, Á., Aranda, F. J., Teruel, J. A., & Ortiz, A. (2017). Interaction of the lipopeptide biosurfactant lichenysin with phosphatidylcholine model membranes. *Langmuir*, *33*(38), 9997−10005. Available from https://doi.org/10.1021/acs.langmuir.7b01827.

Daverey, A., Dutta, K., Joshi, S., & Daverey, A. (2021). Sophorolipid: A glycolipid biosurfactant as a potential therapeutic agent against COVID-19. *Bioengineered*, *12*(2), 9550−9560. Available from https://doi.org/10.1080/21655979.2021.1997261.

De Giani, A., Zampolli, J., & Di Gennaro, P. (2021). Recent trends on biosurfactants with antimicrobial activity produced by bacteria associated with human health: Different perspectives on their properties, challenges, and potential applications. *Frontiers in Microbiology*, *12*, 655150. Available from https://doi.org/10.3389/fmicb.2021.655150.

Deore, S. L. (2017). Emulsion micro emulsion and nano emulsion: A review. *Systematic Reviews in Pharmacy*, *8*(1).

Escobar-Díaz, E., López-Martín, E. M., Hernández del Cerro, M., Puig-Kroger, A., Soto-Cerrato, V., Montaner, B., Giralt, E., García-Marco, J. A., Pérez-Tomás, R., & Garcia-Pardo, A. (2005). AT514, a cyclic depsipeptide from Serratia marcescens, induces apoptosis of B-chronic lymphocytic leukemia cells: Interference with the Akt /NF-κB survival pathway. *Leukemia. Official Journal of the Leukemia Society of America, Leukemia Research Fund, U.K*, *19*(4), 572−579. Available from https://doi.org/10.1038/sj.leu.2403679.

Farhood, H., Gao, X., Son, K., Lazo, J. S., Huang, L., Barsoum, J., Bottega, R., & Epand, R. M. (1994). Cationic liposomes for direct gene transfer in therapy of cancer and other diseases. *Annals of the New York Academy of Sciences*, *716*(1), 23−35. Available from https://doi.org/10.1111/j.1749-6632.1994.tb21701.x.

Farias, C. B. B., Almeida, F. C. G., Silva, I. A., Souza, T. C., Meira, H. M., Soares da Silva, R. d C. F., Luna, J. M., Santos, V. A., Converti, A., Banat, I. M., & Sarubbo, L. A. (2021). Production of green surfactants: Market prospects. *Electronic Journal of Biotechnology*, *51*, 28−39. Available from https://doi.org/10.1016/j.ejbt.2021.02.002.

Farmer, S., Alibek, K., & Chen, Y., Locus IP Co LLC. (2022). *Co-culture of myxobacteria and Bacillus for enhanced metabolite production*. U.S. Patent Application 17/439,600.

Fink, J. (2021). *Petroleum engineer's guide to oil field chemicals and fluids* (pp. 1−1046). Elsevier. Available from https://doi.org/10.1016/C2020-0-02705-2.

Fu, S. L., Wallner, S. R., Bowne, W. B., Hagler, M. D., Zenilman, M. E., Gross, R., & Bluth, M. H. (2008). Sophorolipids and their derivatives are lethal against human pancreatic cancer cells. *Journal of Surgical Research*, *148*(1), 77−82. Available from https://doi.org/10.1016/j.jss.2008.03.005.

Furrer, P., Plazonnet, B., Mayer, J. M., & Gurny, R. (2000). Application of in vivo confocal microscopy to the objective evaluation of ocular irritation induced by surfactants. *International Journal of Pharmaceutics*, *207*(1−2), 89−98. Available from https://doi.org/10.1016/S0378-5173(00)00540-8.

Gangwar, M., Singh, R., Goel, R. K., & Nath, G. (2012). Recent advances in various emerging vesicular systems: An overview. *Asian Pacific Journal of Tropical Biomedicine*, *2*(2), S1176−S1188. Available from https://doi.org/10.1016/S2221-1691(12)60381-5.

Gharbavi, M., Manjili, H. K., Amani, J., Sharafi, A., & Danafar, H. (2019). In vivo and in vitro biocompatibility study of novel microemulsion hybridized with bovine serum albumin as nanocarrier for drug delivery. *Heliyon*, *5*(6). Available from https://doi.org/10.1016/j.heliyon.2019.e01858.

Geisler, R., Dargel, C., & Hellweg, T. (2020). The biosurfactant β-Aescin: A review on the physicochemical properties and its interaction with lipid model membranes and Langmuir monolayers. *Molecules (Basel, Switzerland)*, *25*(1), 117. Available from https://doi.org/10.3390/molecules25010117.

Gelperina, S., Kisich, K., Iseman, M. D., & Heifets, L. (2005). The potential advantages of nanoparticle drug delivery systems in chemotherapy of tuberculosis. *American Journal of Respiratory and Critical Care Medicine*, *172*(12), 1487−1490. Available from https://doi.org/10.1164/rccm.200504-613PP.

Google Patents WO2017029175A1. (2016). *Improved lactam solubility*.

Grapentin, C., Barnert, S., Schubert, R., & Bansal, V. (2015). Monitoring the stability of perfluorocarbon nanoemulsions by Cryo-TEM image analysis and dynamic light scattering. *PLoS One*, *10*(6), e0130674. Available from https://doi.org/10.1371/journal.pone.0130674.

Gurpreet, K., & Singh, S. K. (2018). Review of nanoemulsion formulation and characterization techniques. *Indian Journal of Pharmaceutical Sciences*, *80*(5), 781−789. Available from http://www.ijpsonline.com/articles/review-of-nanoemulsion-formulation-and-characterization-techniques.pdf.

He, Z., Zeng, W., Zhu, X., Zhao, H., Lu, Y., & Lu, Z. (2017). Influence of surfactin on physical and oxidative stability of microemulsions with docosahexaenoic acid. *Colloids and Surfaces B: Biointerfaces*, *151*, 232−239. Available from https://doi.org/10.1016/j.colsurfb.2016.12.026.

Heerklotz, H., Wieprecht, T., & Seelig, J. (2004). Membrane perturbation by the lipopeptide surfactin and detergents as studied by deuterium NMR. *Journal of Physical Chemistry B*, *108*(15), 4909−4915. Available from https://doi.org/10.1021/jp0371938.

Huang, W., Lang, Y., Hakeem, A., Lei, Y., Gan, L., & Yang, X. (2018). Surfactin-based nanoparticles loaded with doxorubicin to overcome multidrug resistance in cancers. *International Journal of Nanomedicine*, *13*, 1723−1736. Available from https://doi.org/10.2147/IJN.S157368.

Igarashi, S., Hattori, Y., & Maitani, Y. (2006). Biosurfactant MEL-A enhances cellular association and gene transfection by cationic liposome. *Journal of Controlled Release*, *112*(3), 362−368. Available from https://doi.org/10.1016/j.jconrel.2006.03.003.

Inoh, Y., Kitamoto, D., Hirashima, N., & Nakanishi, M. (2004). Biosurfactant MEL-A dramatically increases gene transfection via membrane fusion. *Journal of Controlled Release*, *94*(2−3), 423−431. Available from https://doi.org/10.1016/j.jconrel.2003.10.020.

Ishigami, Y. G. Y., Nagahora, H., Hongu, T., & Yamaguchi, M. (1990). Rhamnolipid liposomes. *Patent US4902512. Agency Ind Science Techn (JP)*.

Isoda, H., Shinmoto, H., Matsumura, M., & Nakahara, T. (1996). Succinoyl trehalose lipid induced differentiation of human monocytoid leukemic cell line U937 into monocyte-macrophages. *Cytotechnology*, *19*(1), 79−88. Available from https://doi.org/10.1007/bf00749758.

Isoda, H., & Nakahara, T. (1997). Mannosylerythritol lipid induces granulocytic differentiation and inhibits the tyrosine phosphorylation of human myelogenous leukemia cell line K562. *Cytotechnology*, *25* (1−3), 191−195. Available from https://doi.org/10.1023/a:1007982909932.

Isoda, H., Kitamoto, D., Shinmoto, H., Matsumura, M., & Nakahara, T. (1997). Microbial extracellular glycolipid induction of differentiation and inhibition of the protein kinase C activity of human promyelocytic leukemia cell line HL60. *Bioscience, Biotechnology, and Biochemistry*, *61*(4), 609−614. Available from https://doi.org/10.1271/bbb.61.609.

Jaiswal, M., Dudhe, R., & Sharma, P. K. (2015). Nanoemulsion: An advanced mode of drug delivery system. *3 Biotech*, *5*(2), 123−127. Available from https://doi.org/10.1007/s13205-014-0214-0.

Jiang, C., Li, Z., Shi, Y., Guo, D., Pang, B., Chen, X., Shao, D., Liu, Y., & Shi, J. (2020). Bacillus subtilis inhibits Aspergillus carbonarius by producing iturin A, which disturbs the transport, energy metabolism, and osmotic pressure of fungal cells as revealed by transcriptomics analysis. *International Journal of Food Microbiology*, *330*, 108783. Available from https://doi.org/10.1016/j.ijfoodmicro.2020.108783.

Katiyar, S. S., Ghadi, R., Kushwah, V., Dora, C. P., & Jain, S. (2020). Lipid and biosurfactant based core-shell-type nanocapsules having high drug loading of paclitaxel for improved breast cancer therapy. *ACS Biomaterials Science and Engineering*, *6*(12), 6760−6769. Available from https://doi.org/10.1021/acsbiomaterials.0c01290.

Khan, T. A., Mahler, H. C., & Kishore, R. S. K. (2015). Key interactions of surfactants in therapeutic protein formulations: A review. *European Journal of Pharmaceutics and Biopharmaceutics*, *97*, 60−67. Available from https://doi.org/10.1016/j.ejpb.2015.09.016.

Kim, S. y., Kim, J. Y., Kim, S. H., Bae, H. J., Yi, H., Yoon, S. H., Koo, B. S., Kwon, M., Cho, J. Y., Lee, C. E., & Hong, S. (2007). Surfactin from Bacillus subtilis displays anti-proliferative effect via apoptosis induction, cell cycle arrest and survival signaling suppression. *FEBS Letters*, *581*(5), 865−871. Available from https://doi.org/10.1016/j.febslet.2007.01.059.

Kitamoto, D., Morita, T., Fukuoka, T., Konishi, M. a, & Imura, T. (2009). Self-assembling properties of glycolipid biosurfactants and their potential applications. *Current Opinion in Colloid and Interface Science*, *14*(5), 315−328. Available from https://doi.org/10.1016/j.cocis.2009.05.009.

Komiyama, M., Shigi, N., & Ariga, K. (2022). DNA-based nanoarchitectures as eminent vehicles for smart drug delivery systems. *Advanced Functional Materials*, *32*, 2200924. Available from https://doi.org/10.1002/adfm.202200924.

Kulakovskaya, E., & Kulakovskaya, T. (2014). *Extracellular glycolipids of yeasts: Biodiversity, biochemistry, and prospects* (pp. 1−112). Elsevier Inc. Available from https://doi.org/10.1016/C2013-0-12913-3.

Kumar Singh, P., Ashutosh Prabhune, A., Balkrishna Ogale, S. (2015). *Curcumin-sophorolipid complex*. Google Patents US20170224636A1.

Kumar, M., Bishnoi, R. S., Shukla, A. K., & Jain, C. P. (2019). Techniques for formulation of nanoemulsion drug delivery system: A review. *Preventive Nutrition and Food Science*, *24*(3), 225–234. Available from https://doi.org/10.3746/pnf.2019.24.3.225.

Kural, F. H., & Gürsoy, R. N. (2010). Formulation and characterization of surfactin-containing self-microemulsifying drug delivery systems (SF-SMEDDS). *Hacettepe University Journal of the Faculty of Pharmacy*, *30*(2), 171–186. Available from http://www.eczfakder.hacettepe.edu.tr/Arsiv/EskiDergiler/02_2010/04%20makale.pdf.

Lee, J. H., Nam, S. H., Seo, W. T., Yun, H. D., Hong, S. Y., Kim, M. K., & Cho, K. M. (2012). The production of surfactin during the fermentation of cheonggukjang by potential probiotic *Bacillus subtilis* CSY191 and the resultant growth suppression of MCF-7 human breast cancer cells. *Food Chemistry*, *131*(4), 1347–1354. Available from https://doi.org/10.1016/j.foodchem.2011.09.133.

Lewińska, A., Domżał-Kędzia, M., Jaromin, A., & èukaszewicz, M. (2020). Nanoemulsion stabilized by safe surfactin from *Bacillus subtilis* as a multifunctional, custom-designed smart delivery system. *Pharmaceutics*, *12*(10), 953. Available from https://doi.org/10.3390/pharmaceutics12100953.

Li, Y., & Lee, J. S. (2019). Staring at protein-surfactant interactions: Fundamental approaches and comparative evaluation of their combinations - A review. *Analytica Chimica Acta*, *1063*, 18–39. Available from https://doi.org/10.1016/j.aca.2019.02.024.

Liao, Y., Li, Z., Zhou, Q., Sheng, M., Qu, Q., Shi, Y., Yang, J., Lv, L., Dai, X., & Shi, X. (2021). Saponin surfactants used in drug delivery systems: A new application for natural medicine components. *International Journal of Pharmaceutics*, *603*, 120709. Available from https://doi.org/10.1016/j.ijpharm.2021.120709.

Liu, D., Zhang, L., Wang, Y., Li, Z., Wang, Z., & Han, J. (2020). Effect of high hydrostatic pressure on solubility and conformation changes of soybean protein isolate glycated with flaxseed gum. *Food Chemistry*, *333*, 127530. Available from https://doi.org/10.1016/j.foodchem.2020.127530.

Ma, W. D., Zou, Y. P., Wang, P., Yao, X. H., Sun, Y., Duan, M. H., Fu, Y. J., & Yu, B. (2014). Chimaphilin induces apoptosis in human breast cancer MCF-7 cells through a ROS-mediated mitochondrial pathway. *Food and Chemical Toxicology*, *70*, 1–8. Available from https://doi.org/10.1016/j.fct.2014.04.014.

Magalhães, F. F., Nunes, J. C. F., Araújo, M. T., Ferreira, A. M., Almeida, M. R., Freire, M. G., & Tavares, A. P. M. (2021). *Anti-cancer biosurfactants* (pp. 159–196). Springer Science and Business Media LLC. Available from https://doi.org/10.1007/978-981-15-6607-3_8.

Maitani, Y., Yano, S., Hattori, Y., Furuhata, M., & Hayashi, K. (2006). Liposome vector containing biosurfactant-complexed DNA as herpes simplex virus thymidine kinase gene delivery system. *Journal of Liposome Research*, *16*(4), 359–372. Available from https://doi.org/10.1080/08982100600992443.

McClements, D. J. (2012). Nanoemulsions versus microemulsions: Terminology, differences, and similarities. *Soft Matter*, *8*(6), 1719–1729. Available from https://doi.org/10.1039/c2sm06903b.

Md, F. (2012). Biosurfactant: Production and application. *Journal of Petroleum & Environmental Biotechnology*, *03*(04). Available from https://doi.org/10.4172/2157-7463.1000124.

Mnif, I., & Ghribi, D. (2015). Review lipopeptides biosurfactants: Mean classes and new insights for industrial, biomedical, and environmental applications. *Peptide Science*, *104*(3), 129–147. Available from https://doi.org/10.1002/bip.22630.

Moghimipour, E., Salimi, A., & Eftekhari, S. (2013). Design and characterization of microemulsion systems for naproxen. *Advanced Pharmaceutical Bulletin*, *3*(1), 63–71. Available from https://doi.org/10.5681/apb.2013.011.

Molyneux, P., Rhodes, C. T., & Swarbrick, J. (1965). Thermodynamics of micellization of N-alkyl betaines. *Transactions of the Faraday Society*, *61*, 1043–1052. Available from https://doi.org/10.1039/tf9656101043.

Nakanishi, M., Inoh, Y., Kitamoto, D., & Furuno, T. (2009). Nano vectors with a biosurfactant for gene transfection and drug delivery. *Journal of Drug Delivery Science and Technology*, *19*(3), 165–169. Available from https://doi.org/10.1016/S1773-2247(09)50031-7.

Naughton, P. J., Marchant, R., Naughton, V., & Banat, I. M. (2019). Microbial biosurfactants: Current trends and applications in agricultural and biomedical industries. *Journal of Applied Microbiology*, *127*(1), 12–28. Available from https://doi.org/10.1111/jam.14243.

Nazari, M., Kurdi, M., & Heerklotz, H. (2012). Classifying surfactants with respect to their effect on lipid membrane order. *Biophysical Journal, 102*(3), 498−506. Available from https://doi.org/10.1016/j.bpj.2011.12.029.

Nguyen, T. T., & Sabatini, D. A. (2009). Formulating alcohol-free microemulsions using rhamnolipid biosurfactant and rhamnolipid mixtures. *Journal of Surfactants and Detergents, 12*(2), 109−115. Available from https://doi.org/10.1007/s11743-008-1098-y.

Nguyen, T. T., & Sabatini, D. A. (2011). Characterization and emulsification properties of rhamnolipid and sophorolipid biosurfactants and their applications. *International Journal of Molecular Sciences, 12*(2), 1232−1244. Available from https://doi.org/10.3390/ijms12021232.

Oh, H. R., Jo, H. Y., Park, J. S., Kim, D. E., Cho, J. Y., Kim, P. H., & Kim, K. S. (2016). Galactosylated liposomes for targeted co-delivery of doxorubicin/vimentin siRNA to hepatocellular carcinoma. *Nanomaterials, 6*(8). Available from https://doi.org/10.3390/nano6080141.

Ortiz, A., Teruel, J. A., Espuny, M. J., Marqués, A., Manresa, A., & Aranda, F. J. (2009). Interactions of a bacterial biosurfactant trehalose lipid with phosphatidylserine membranes. *Chemistry and Physics of Lipids, 158*(1), 46−53. Available from https://doi.org/10.1016/j.chemphyslip.2008.11.001.

Patel, V., Kukadiya, H., Mashru, R., Surti, N., & Mandal, S. (2010). Development of microemulsion for solubility enhancement of clopidogrel. *Iranian Journal of Pharmaceutical Research, 9*(4), 327−334. Available from http://www.ijpr.ir/?_action = showPDF&article = 898&_ob = 2347562454abf0205 8eb55021cba056c&fileName = full_text.pdf.

Paunovska, K., Loughrey, D., & Dahlman, J. E. (2022). Drug delivery systems for RNA therapeutics. *Nature Reviews. Genetics, 23*(5), 265−280. Available from https://doi.org/10.1038/s41576-021-00439-4.

Peng, Z., Wang, C., Fang, E., Lu, X., Wang, G., Tong, Q., & Rozhkova, E. A. (2014). Co-delivery of doxorubicin and SATB1 shRNA by thermosensitive magnetic cationic liposomes for gastric cancer therapy. *PLoS One, 9*(3), e92924. Available from https://doi.org/10.1371/journal.pone.0092924.

Phale, P. S., Malhotra, H., & Shah, B. A. (2020). Degradation strategies and associated regulatory mechanisms/features for aromatic compound metabolism in bacteria. *Advances in applied microbiology* (Vol. 112, pp. 1−65). Academic Press Inc. Available from https://doi.org/10.1016/bs.aambs.2020.02.002.

Piljac, G., & Piljac, V. (1998). Immunological activity of rhamnolipids. *Comparative Immunology, Microbiology and Infectious Diseases, 1*(21), VIII.

Qu, M. H., Zeng, R. F., Fang, S., Dai, Q. S., Li, H. P., & Long, J. T. (2014). Liposome-based co-delivery of siRNA and docetaxel for the synergistic treatment of lung cancer. *International Journal of Pharmaceutics, 474*(1−2), 112−122. Available from https://doi.org/10.1016/j.ijpharm.2014.08.019.

Rajput, S., Puvvada, N., Kumar, B. N. P., Sarkar, S., Konar, S., Bharti, R., Dey, G., Mazumdar, A., Pathak, A., Fisher, P. B., & Mandal, M. (2015). Overcoming Akt induced therapeutic resistance in breast cancer through siRNA and thymoquinone encapsulated multilamellar gold niosomes. *Molecular Pharmaceutics, 12*(12), 4214−4225. Available from https://doi.org/10.1021/acs.molpharmaceut.5b00692.

Rao, J., & McClements, D. J. (2011). Formation of flavor oil microemulsions, nanoemulsions and emulsions: Influence of composition and preparation method. *Journal of Agricultural and Food Chemistry, 59*(9), 5026−5035. Available from https://doi.org/10.1021/jf200094m.

Ratcliffe, J., Soave, F., Bryan-Kinns, N., Tokarchuk, L., & Farkhatdinov, I. (2021). Extended reality (XR) remote research: A survey of drawbacks and opportunities. *arXiv*. Available from https://arxiv.org.

Ray, S. K., Bano, N., Shukla, T., Upmanyu, N., Pandey, S. P., & Parkhe, G. (2018). Noisomes: As novel vesicular drug delivery system. *Journal of Drug Delivery and Therapeutics, 8*(6), 335−341. Available from https://doi.org/10.22270/jddt.v8i6.2029.

Reddy, A. S., Chen, C. Y., Baker, S. C., Chen, C. C., Jean, J. S., Fan, C. W., Chen, H. R., & Wang, J. C. (2009). Synthesis of silver nanoparticles using surfactin: A biosurfactant as stabilizing agent. *Materials Letters, 63*(15), 1227−1230. Available from https://doi.org/10.1016/j.matlet.2009.02.028.

Renard, P., Canet, I., Sancelme, M., Matulova, M., Uhliarikova, I., Eyheraguibel, B., Nauton, L., Devemy, J., Traïkia, M., Malfreyt, P., & Delort, A.-M. (2019). *Cloud microorganisms, an interesting source of biosurfactants.* IntechOpen. Available from https://doi.org/10.5772/intechopen.85621.

Rodrigues, L. R. (2015). Microbial surfactants: Fundamentals and applicability in the formulation of nano-sized drug delivery vectors. *Journal of Colloid and Interface Science, 449*, 304−316. Available from https://doi.org/10.1016/j.jcis.2015.01.022.

Rodrigues, L., Banat, I. M., Teixeira, J., & Oliveira, R. (2006). Biosurfactants: Potential applications in medicine. *Journal of Antimicrobial Chemotherapy*, 57(4), 609–618. Available from https://doi.org/10.1093/jac/dkl024.

Rofeal, M., El-Malek, F. A., & Qi, X. (2021). In vitro assessment of green polyhydroxybutyrate/chitosan blend loaded with kaempferol nanocrystals as a potential dressing for infected wounds. *Nanotechnology*, 32(37), 375102. Available from https://doi.org/10.1088/1361-6528/abf7ee.

Ruckenstein, E. (1996). Microemulsions, macroemulsions, and the Bancroft rule. *Langmuir*, 12(26), 6351–6353. Available from https://doi.org/10.1021/la960849m.

Sabale, V., & Vora, S. (2012). Formulation and evaluation of microemulsion-based hydrogel for topical delivery. *International Journal of Pharmaceutical Investigation*, 2(3), 140. Available from https://doi.org/10.4103/2230-973x.104397.

Sahnoun, R., Mnif, I., Fetoui, H., Gdoura, R., Chaabouni, K., Makni-Ayadi, F., Kallel, C., Ellouze-Chaabouni, S., & Ghribi, D. (2014). Evaluation of *Bacillus subtilis* SPB1 lipopeptide biosurfactant toxicity towards mice. *International Journal of Peptide Research and Therapeutics*, 20(3), 333–340. Available from https://doi.org/10.1007/s10989-014-9400-5.

Saikia, R. R., Deka, S., & Sarma, H. (2021). Biosurfactants from bacteria and fungi: Perspectives on advanced biomedical applications. *Biosurfactants for a Sustainable Future: Production and Applications in the Environment and Biomedicine* (pp. 293–315). Wiley.

Saini, H. S., Barragán-Huerta, B. E., Lebrón-Paler, A., Pemberton, J. E., Vázquez, R. R., Burns, A. M., Marron, M. T., Seliga, C. J., Gunatilaka, A. L., & Maier, R. M. (2008). Efficient purification of the biosurfactant viscosin from Pseudomonas libanensis strain M9-3 and its physicochemical and biological properties. *Journal of Natural Products*, 71(6), 1011–1015.

Sajid, M., Ahmad Khan, M. S., Singh Cameotra, S., & Safar Al-Thubiani, A. (2020). Biosurfactants: Potential applications as immunomodulator drugs. *Immunology Letters*, 223, 71–77. Available from https://doi.org/10.1016/j.imlet.2020.04.003.

Sen, R., & Swaminathan, T. (2005). Characterization of concentration and purification parameters and operating conditions for the small-scale recovery of surfactin. *Process Biochemistry*, 40(9), 2953–2958. Available from https://doi.org/10.1016/j.procbio.2005.01.014.

Senten, S. M. A., & Engle, C. R. (2020). Impacts of COVID-19 on US aquaculture, aquaponics, and allied businesses. *Journal of the World Aquaculture Society*, 51(3).

Seydlová, G., & Svobodová, J. (2008). Review of surfactin chemical properties and the potential biomedical applications. *Central European Journal of Medicine*, 3(2), 123–133. Available from https://doi.org/10.2478/s11536-008-0002-5.

Shah, P., Bhalodia, D., & Shelat, P. (2010). Nanoemulsion: A pharmaceutical review. *Systematic Reviews in Pharmacy*, 1(1), 24–32. Available from https://doi.org/10.4103/0975-8453.59509.

Sharma, H. B., Vanapalli, K. R., Cheela, V. R. S., Ranjan, V. P., Jaglan, A. K., Dubey, B., Goel, S., & Bhattacharya, J. (2020). Challenges, opportunities, and innovations for effective solid waste management during and post COVID-19 pandemic. *Resources, Conservation and Recycling*, 162, 105052.

Shreve, G. S., & Makula, R. (2019). Characterization of a new rhamnolipid biosurfactant complex from Pseudomonas isolate DYNA270. *Biomolecules*, 9(12), 885. Available from https://doi.org/10.3390/biom9120885.

Smith, M. L., Gandolfi, S., Coshall, P. M., & Rahman, P. K. S. M. (2020). Biosurfactants: A Covid-19 Perspective. *Frontiers in Microbiology*, 11, 1341. Available from https://doi.org/10.3389/fmicb.2020.01341.

Sudo, T., Zhao, X., Wakamatsu, Y., Shibahara, M., Nomura, N., Nakahara, T., Suzuki, A., Kobayashi, Y., Jin, C., Murata, T., & Yokoyama, K. K. (2000). Induction of the differentiation of human HL-60 promyelocytic leukemia cell line by succinoyl trehalose lipids. *Cytotechnology* (Vol. 33, pp. 259–264). Netherlands: Springer Issues 1–3. Available from https://doi.org/10.1023/a:1008137817944.

Sun, H., Liu, K., Liu, W., Wang, W., Guo, C., Tang, B., Gu, J., Zhang, J., Li, H., Mao, X., Zou, Q., & Zeng, H. (2012). Development and characterization of a novel nanoemulsion drug-delivery system for potential application in oral delivery of protein drugs. *International Journal of Nanomedicine*, 7, 5529–5543. Available from https://doi.org/10.2147/IJN.S36071.

Suthar, H., & Nerurkar, A. (2016). Characterization of biosurfactant produced by *Bacillus licheniformis* TT42 having potential for enhanced oil recovery. *Applied Biochemistry and Biotechnology*, 180(2), 248–260. Available from https://doi.org/10.1007/s12010-016-2096-6.

Tayeb, H. H., Felimban, R., Almaghrabi, S., & Hasaballah, N. (2021). Nanoemulsions: Formulation, characterization, biological fate, and potential role against COVID-19 and other viral outbreaks. *Colloid and Interface Science Communications*, *45*, 100533. Available from https://doi.org/10.1016/j.colcom.2021.100533.

Tucker, I. M., Burley, A., Petkova, R. E., Hosking, S. L., Penfold, J., Thomas, R. K., Li, P. X., Webster, J. R. P., Welbourn, R., & Doutch, J. (2021). Adsorption and self-assembly properties of the plant based biosurfactant, Glycyrrhizic acid. *Journal of Colloid and Interface Science*, *598*, 444−454. Available from https://doi.org/10.1016/j.jcis.2021.03.101.

Uchegbu, I. F., Double, J. A., Kelland, L. R., Turton, J. A., & Florence, A. T. (1996). The activity of doxorubicin niosomes against an ovarian cancer cell line and three in vivo mouse tumour models. *Journal of Drug Targeting*, *3*(5), 399−409. Available from https://doi.org/10.3109/10611869608996831.

Ueno, Y., Inoh, Y., Furuno, T., Hirashima, N., Kitamoto, D., & Nakanishi, M. (2007). NBD-conjugated biosurfactant (MEL-A) shows a new pathway for transfection. *Journal of Controlled Release*, *123*(3), 247−253. Available from https://doi.org/10.1016/j.jconrel.2007.08.012.

Varun, T., Sonia, A., Bharat., Patil, V., Kumharhatti, P. O., & Solan, D. (2012). Niosomes and liposomes-vesicular approach towards transdermal drug delivery. *International Journal of Pharmaceutical and Chemical Sciences*, *1*(3), 632−644.

Wang, C. L., Ng, T. B., Yuan, F., Liu, Z. K., & Liu, F. (2007). Induction of apoptosis in human leukemia K562 cells by cyclic lipopeptide from *Bacillus subtilis* natto T-2. *Peptides*, *28*(7), 1344−1350. Available from https://doi.org/10.1016/j.peptides.2007.06.014.

Wang, H. M., Xuan-Rui, X. (2014). *Applications of surfactin in cosmetic products*. Google Patents US20160030322A1.

Warisnoicharoen, W., Lansley, A. B., & Lawrence, M. J. (2000). Nonionic oil-in-water microemulsions: The effect of oil type on phase behaviour. *International Journal of Pharmaceutics*, *198*(1), 7−27. Available from https://doi.org/10.1016/S0378-5173(99)00406-8.

White, D. A., Hird, L. C., & Ali, S. T. (2013). Production and characterization of a trehalolipid biosurfactant produced by the novel marine bacterium Rhodococcus sp., strain PML026. *Journal of Applied Microbiology*, *115*(3), 744−755. Available from https://doi.org/10.1111/jam.12287.

Worakitkanchanakul, W., Imura, T., Morita, T., Fukuoka, T., Sakai, H., Abe, M., Rujiravanit, R., Chavadej, S., & Kitamoto, D. (2008). Formation of W/O microemulsion based on natural glycolipid biosurfactant, mannosylerythritol lipid-A. *Journal of Oleo Science*, *57*(1), 55−59. Available from https://doi.org/10.5650/jos.57.55.

Wu, Y. S., Ngai, S. C., Goh, B. H., Chan, K. G., Lee, L. H., & Chuah, L. H. (2017). Anticancer activities of surfactin potential application of nanotechnology assisted surfactin delivery. *Frontiers in Pharmacology*, *8*, 761. Available from https://doi.org/10.3389/fphar.2017.00761.

Xie, Y., Ye, R., & Liu, H. (2007). Microstructure studies on biosurfactant-rhamnolipid/n-butanol/water/n-heptane microemulsion system. *Colloids and Surfaces A: Physicochemical and Engineering Aspects*, *292*(2−3), 189−195. Available from https://doi.org/10.1016/j.colsurfa.2006.06.021.

Yadav, V., Jadhav, P., Kanase, K., Bodhe, A., & Dombe, S. (2018). Preparation and evaluation of microemulsion containing antihypertensive drug. *International Journal of Applied Pharmaceutics*, *10*(5), 138−146. Available from https://doi.org/10.22159/ijap.2018v10i5.27415.

Yang, H., Ding, Y., Cao, J., & Li, P. (2013). Twenty-one years of microemulsion electrokinetic chromatography (1991-2012): A powerful analytical tool. *Electrophoresis*, *34*(9−10), 1273−1294. Available from https://doi.org/10.1002/elps.201200494.

Yang, T., Li, B., Qi, S., Liu, Y., Gai, Y., Ye, P., Yang, G., Zhang, W., Zhang, P., He, X., Li, W., Zhang, Z., Xiang, G., & Xu, C. (2014). Co-delivery of doxorubicin and Bmi1 siRNA by folate receptor targeted liposomes exhibits enhanced anti-tumor effects in vitro and in vivo. *Theranostics*, *4*(11), 1096−1111. Available from https://doi.org/10.7150/thno.9423.

Yang, Z. Z., Li, J. Q., Wang, Z. Z., Dong, D. W., & Qi, X. R. (2014). Tumor-targeting dual peptides-modified cationic liposomes for delivery of siRNA and docetaxel to gliomas. *Biomaterials*, *35*(19), 5226−5239. Available from https://doi.org/10.1016/j.biomaterials.2014.03.017.

Yao, Y., Su, Z., Liang, Y., & Zhang, N. (2015). pH-sensitive carboxymethyl chitosan-modified cationic liposomes for sorafenib and siRNA co-delivery. *International Journal of Nanomedicine*, *10*, 6185–6198. Available from https://doi.org/10.2147/IJN.S90524.

Zaragoza, A., Teruel, J. A., Aranda, F. J., Marqués, A., Espuny, M. J., Manresa, A., & Ortiz, A. (2012). Interaction of a Rhodococcus sp. trehalose lipid biosurfactant with model proteins: Thermodynamic and structural changes. *Langmuir*, *28*(2), 1381–1390. Available from https://doi.org/10.1021/la203879t.

Zhang, Z., & McClements, D. J. (2018). Overview of nanoemulsion properties: Stability, rheology, and appearance. *Nanoemulsions: Formulation, applications, and characterization* (pp. 21–49). Elsevier Inc. Available from https://doi.org/10.1016/B978-0-12-811838-2.00002-3.

Zhang, Z., Ding, Z. T., Zhong, J., Zhou, J. Y., Shu, D., Luo, D., Yang, J., & Tan, H. (2017). Improvement of iturin A production in *Bacillus subtilis* ZK0 by overexpression of the comA and sigA genes. *Letters in Applied Microbiology*, *64*(6), 452–458. Available from https://doi.org/10.1111/lam.12739.

CHAPTER 11

Recent advancements in biosurfactant-aided adsorption technologies for the removal of pharmaceutical drugs

Jagriti Jha Sanjay[1], Swathi Krishnan Venkatesan[2] and Ramani Kandasamy[2]
[1]Industrial and Environmental Sustainability Laboratory, Department of Biotechnology, SRM Institute of Science and Technology, Kattankulathur, Chengalpattu, Tamil Nadu, India
[2]Industrial and Environmental Sustainability Laboratory, Department of Biotechnology, School of Bioengineering, SRM Institute of Science and Technology, Kattankulathur, Chengalpattu, Tamil Nadu, India

11.1 Introduction

The pharmaceutical industry is one of the ever-growing industries across the globe. Industrial pharmaceutical production has led to their disposal as it is or even complexly modified in the environment. In 2021, the FDA approved 55 new drugs, and more than 2000 drugs are in the pipeline even in the near future, the advancements such as personalized drug production for better health treatment would bring more challenges in dealing with the drugs from an environmental perspective. With the current status, 65% of pharmaceutical production is shared by conventional drugs, which poses a huge number. The complexity and recalcitrance of these pharmaceutical drugs, about 90% of which is released from the body unmetabolized, add up more to their nature, apart from drug disposal (Gadipelly et al., 2014). This further leads to the problem of biomagnification, antibacterial resistance, and ill- effects on the health of living beings that are alarmingly increasing every day, such as endocrine disruption, hormonal imbalance, reduced fecundity, and disruption of feeding. Despite these serious threats, the need for drugs cannot be reduced at any cost. However, the harmful effects are also unendurable; therefore, to combat these serious issues, addressing new and feasible technologies is crucial (Richardson & Bowron, 1985). Adsorption is a technique often sought as a very reliable, energy-, and cost-effective. Activated carbon is often used for the adsorption of pharmaceuticals, but the cost and reusability pose limitations, so few unconventional materials are tested for the same such as clays, biochar, chitosan, agro-industrial waste, and metal-organic framework. Although they are good, each has its own limitation, be it bioavailability, durability, and universality, which limits their use (Ahmed & Hameed, 2018). Biosurfactant as an adsorbent for drugs is a budding concept which, due to its amphiphilic, surface-active, biodegradable, encapsulating nature, made it an appropriate subject for its usage as an adjuvant for drug delivery and the same properties enhance the bioavailability of drugs by even dealing with enhanced membrane permeation, which would also help in removal from the environment, thereby helping in reducing ecotoxicological impact. The high biodegradability profile, low risk of toxicity, easy production from cheap and readily available renewable sources, durability in extreme pH and thermal stress, long physicochemical stability, and enhancing the permeation across membranes all these accolades to its potential candidature for being adsorbent of pharmaceutical drugs. Studies on the application of biosurfactants for the targeted removal of pharmaceutical drugs have been emerging in recent years, and the increase in the research publications on this field was represented in, yet satisfactory field-level implications have not been achieved. This chapter particularly sheds light over the issues of pharmaceutical drugs, the ecological and physiological impact, properties of biosurfactant pertaining to absorbance and its exclusive application in addressing the above-described issues and thus a better understanding of possible niches to be discovered for better application.

11.2 Insight on properties pertaining to ecotoxicological impact of pharmaceutical drugs

Pharmaceutical drugs are complex at a molecular level, making the degradation of residual pharma compounds in effluents very difficult. These pharmaceutical

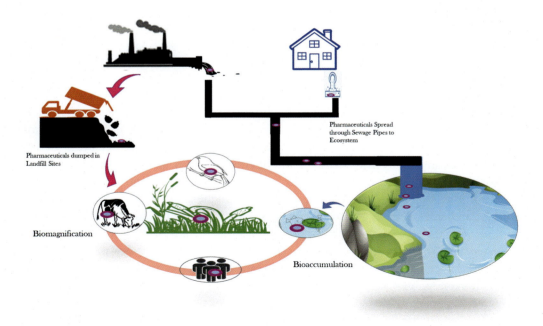

Figure 11.1 *Impacts of pharmaceutical drugs on environment.* Schematic representation of ecotoxicological effects by pharmaceutical drugs.

compounds also have medical implications, which pose a further risk to health as in case the wastewater of the pharmaceutical industries is directly exposed to the outer environment, as depicted in Fig. 11.1. The complexity lies in the molecular structure of these pharmaceutical compounds, which gives the ecotoxicological property to the drugs. This topic discusses in detail the properties causing this effect and the different research works focusing on this study.

11.2.1 Active pharmaceutical ingredients—types

A pharmaceutical drug, in general, consists of many active pharmaceutical ingredients (APIs), organic and inorganic salts, excipients, and additives such as sugars, pigments, and scents. The major categories of pharmaceuticals have been summarized in Fig. 11.2. These belong to a broad range of small-molecular-weight molecules with molecular weight ranging from 200 to 500 Daltons. Some bioengineered classes of pharmaceuticals are synthesized by biotechnological formulation techniques, for example, the genetically engineered human hormone insulin. Pharmaceuticals

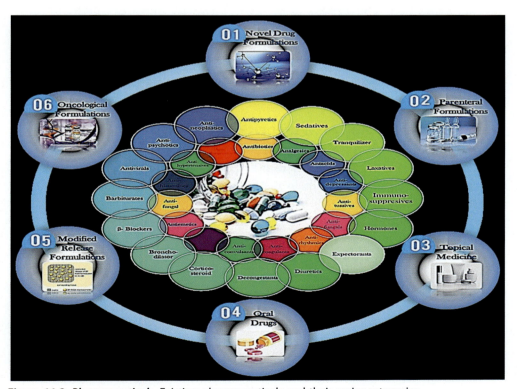

Figure 11.2 *Pharmaceuticals.* Existing pharmaceuticals and their major categories.

are classified generally by their biological functions such as antibiotics, which are used against bacterial infection; analgesics, which are used as pain reliever; and antineoplastic, which is used to treat cancer cells. Bush and Kümmerer classified them into eight primary divisions, namely, anti- inflammatory and analgesics such as diclofenac, antibiotics such as tetracyclines and β-lactams, anti-epileptics such as carbamazepine; antidepressants such as benzodiazepines; lipid-lowering agents such as fibrates; antihistamines such as famotidine; β-blockers such as propranolol; and miscellaneous pharmaceutical compounds such as narcotics, antiseptics, and barbiturates (Bush, 1997; Kümmerer, 2010). Classification with respect to the chemical structure is within the sub-groups like within antibiotics, it is sub-divided into β-lactams, penicillins, cephalosporins, and quinolones. The mode of action is an alternative way of classification of pharmaceuticals, for example, antimetabolites. Sometimes, within the same group, the structures could be chemically very distinct, probably due to their different targets. Certain pharmaceutical compounds are active and fully functional even at very low concentrations, for example, Ethinylsterol, an endocrine-active pharmaceutical hormone, which is the primary API present in birth control pills (Williams et al., 2009). When processed through metabolic pathways in living bodies in certain cases, these pharmaceuticals get biotransformed, while there is no change in other cases. According to a study by Gröning et al. (2007) and Patel et al. (2019) in Germany, 75% of the antibiotics used are excreted unchanged. The transformation can either be completed or could be left incomplete as well. These transformed forms might have new properties, as well as a new chemical entity.

11.2.2 Chemical and physical properties of active pharmaceutical ingredients

Pharmaceutical compounds, contrary to other compounds, are far more complex and resistant to degradation and therefore persist in aqueous ecosystems (Kümmerer, 2009; Rivera-Utrilla et al., 2013). As mentioned by Taylor and Senac (2014), as these pharmaceuticals do not share similar structures and properties, they cannot be classified correctly in groups, which further complicates the designing of universal experiments for drug removal purposes. A majority are polar and specific in nature with peculiar targets, which further diversifies the compounds creating variable chemical structures and thus causing polymorphism. They are heavy as other contaminants and biologically active (Rivera-Utrilla et al., 2013); for instance, clofibric acid could remain intact for several years, while sulfamethoxazole, erythromycin, and naproxen can remain stable for over a year. The pharmaceuticals are also made in enantiomers. In the current scenario, half of them are single enantiomers (Kasprzyk-Hordern & Baker, 2012), while some are made as both single enantiomers and racemates. The enantioselectivity of pharmaceutical compounds further complicates the removal mechanisms as their rectus—sinister shuffling changes periodically as per the reaction concerned; thus, the flexibility in the removal mechanism to switch as per the structure has become the necessity to design in the given aspect, because the enantiomeric catalytic reactions and interactions can further lead to different variable compounds resulting in different therapeutic effects and a different fate in environmental persistence. Based on their biological activity, enantiomers are classified as eutomers possessing a high affinity for receptors and diastomers having less affinity for their receptors (Ariëns et al., 1988). Thus, on this basis, chiral drugs are classified into three distinct classes: one class with drugs with one major bioactive form, a second class with drugs with equal bioactive enantiomers, and a third class of drugs with one enantiomeric form within the body that could be converted to active form through chiral inversion (Kasprzyk-Hordern & Baker, 2012). The physicochemical properties of certain pharmaceuticals may also inhibit their persistence in the ecological environment and may lead it to bioaccumulation. The molecular complexity of the APIs is affected by their stability, ionization capacity, polarity, and solubility, which is consecutively affected by the environment and other surrounding molecules causing the variation. The information about partition coefficient, that is, K_{ow} values and dissociation constant, also helps to determine the occurrence and fate of pharmaceuticals in environment. The pH also determines their distribution because, in solutions, they behave as weak acids or bases depending on their structure. Therefore, k_a (acidity constant), K_b (basicity constant), and ionization determining factors are the parameters that needs to be evaluated accordingly. It has been observed that acidic APIs, as having a positive affinity toward lipids, do not dissociate effectively in an acidic medium, thus making them easily enter the biological system, further spreading their occurrence (Kümmerer, 2010). This can be the possible reason for biomagnification, as can be seen in the case of the bile juice of a few fishes that were analyzed to have PPCPs. In addition, it should be noted that the acidic forms are more soluble in water because they cannot easily dissociate in slightly basic media. Thus, from the above description, it is evident that the fate and occurrence of pharmaceuticals in the environment is also because of the physicochemical property the APIs have and is the reason for the biomagnification in the food chain and web.

11.2.3 Recalcitrance of active pharmaceutical ingredients

Pharmaceutical compounds are generally heterogeneous in nature having varied structural, chemical, and biological properties. Pharmaceutical compounds are complex and refractory in nature (López-Peñalver et al., 2010). The pharmaceutical compounds have molecular weights of less than 500 daltons, although some might be heavier too. Depending on their functionality and sources, they drastically vary in structures even within the same class, as in the case of antibiotics. Due to their polarity and lipophilic characteristics, they are barely soluble in water. As a result, it stays biologically active and stable in the environment for a long time.

The tendency to get adsorbed adds a nuisance to the removal procedures from living and nonliving systems of the environment (Kümmerer, 2009). The wide occurrence of pharmaceuticals in the environment is because of widespread usage for agriculture and animal husbandry, which widely includes steroids, NSAIDS, antibiotics, hospital usage, pharmaceutical companies, and households. The active pharmaceutical ingredients identified in the aquatic system in the environment have been classified as anti-inflammatories, antibiotics, antihistamines, antiepileptic, beta-blockers, antitussives, etc. There are several antimicrobial agents, such as Streptogramins, Glycopeptides, Oxazolidinones, Glycylcyclines, Ketolides, β- lactams, β-lactamase inhibitors, cationic peptides, Inhibitors of Lipid A biosynthesis, quinolone-related agents, and t-RNase synthetase inhibitors (Bush, 1997). The pharmaceutical micropollutants have been mainly exposed to the environment through wastewater treatment plants, hospital wastes, and sewage effluents, while antibiotics are widely spread analgesics (Heberer, 2002; Hughes et al., 2013; Rasheed et al., 2021). With the current pandemic situation, the mass use of specific compounds leads to the wide occurrence of similar compounds in the environment (Slater et al., 2011). This complicates not only drug removal from the environment but also may cause the resistance of microorganisms toward it if immediate action is not taken, leading to further health problems as these drugs would no longer be sufficient. The adsorption capability of pharmaceuticals onto the microbes, solid materials, and the biotransformation characteristics either by host enzymes or by gut microbes could also be the reason for its environmental persistence and introducing new structures of same drug inhibiting the biodegradation procedure (Rasheed et al., 2021). As discussed, the metabolic conversions of the drugs could occur in two distinct phases (Daughton & Ternes, 1999; Khetan & Collins, 2007). Phase one encompassing reactions such as hydrolysis, oxidation, and reduction, while the products from phase I could be conjugated to some different compound in phase II, making them more water soluble and highly polar. The rigid or stable APIs remain in the environment longer while the stability is reduced by metabolic biotransformation and therefore do not persist for long (Löffler et al., 2005). The halogenated pharmaceuticals are highly stable and easily bioavailable, thus posing health risks (Küster & Adler, 2014). The pharmaceutical compounds may even undergo chemical modifications in the water system, and the complexity may increase by its binding with other conjugatory compounds. The complete mineralization of the parent pharmaceutical compounds into nitrates, sulfates, and hydrocarbons makes the RO Reject even more complex and the treatment procedure tedious (Kümmerer, 2009). Thus, from the above reasons, it can be concluded that due to its complex structure, it is difficult to bring the pharmaceutical compounds into complete mineralized form and thus may demand combined and advanced removal methods (Rasheed et al., 2021).

11.2.4 Plausible biosurfactant-mediated treatment approach

The nature of active pharmaceutical ingredients will thus be helpful in planning the biosurfactant-mediated treatment procedure given the knowledge of the interactions that cause adsorption and the factors that stabilize or destabilize these interactions, which has been briefly discussed in this topic. The basic strategy practiced traditionally is represented in Fig. 11.3. The exact mechanisms have been covered in a separate topic further in the chapter. The organic pollutants have an affinity for hydrophobic bonds; therefore, biosurfactants can be used for their removal as they have hydrophobic core pockets that help in binding to those pollutant molecules and in their extrusion. The sole use of biosurfactant for treatment makes the recovery of contaminant- conjugated pollutant removal exceedingly difficult and tedious. But the conjugation of biosurfactant with some efficient adsorbent substratum will help improve the pollutant material surface adsorption. This potential was efficiently used by Asiyeh Kheradmand et al., who laminated the Co/Al- layered double hydroxide composite with rhamnolipid produced by *Pseudomonas aeruginosa*. Ibuprofen's adsorption efficiency was fairly constant at 180 mg/g even after four successive cycles of usage. Similarly, biosurfactants prevent the aggregation of crystal molecules, thus preventing the blockage of the active site of the carrier nanocomposite material. It also leads to better adsorption by increasing the surface area and hydrophobicity (Kheradmand et al., 2021). A similar adsorbent enhancement strategy using biosurfactants was followed by Paskalis Sahaya Murphin Kumar et al. They used glycolipid biosurfactants isolated from rare tropical fruit *Crescentiacujete* biomass to coat Fe_2O_3/C composites. Highest adsorption of diclofenac of about 75.51 mg/g could be achieved. The biosurfactant was found to increase

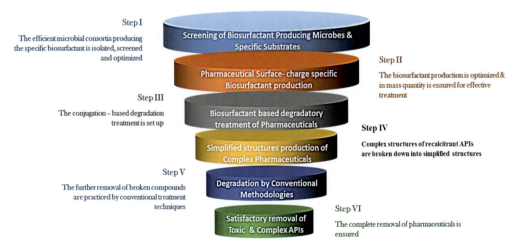

Figure 11.3 *Steps involved in the treatment of pharmaceutical compounds via biosurfactants.* Strategies for biosurfactant-mediated pharmaceutical removal.

elasticity through the formation of inter- lipid hydrogen bonding via polar carbohydrate head (Murphin Kumar et al., 2021). Biosurfactantare particularly characterized by thermostability, tolerance to pH fluctuation, high biodegradability, and durability, which help in their implementation even on an industrial scale (García-Morales et al., 2018). Parameters such as pH and temperature affects the activity of a biosurfactant. Thus, by optimizing the culture conditions the activity of biosurfactant can be enhanced by enhancing the solubility of pollutants, making them more readily available for degradation. In a few studies, it has been found that acidic pH to be favoring for TCS better solubilization in contrary to alkaline pH in rhamnolipid-sediment water system (Jayalatha & Devatha, 2019; Wu et al., 2015). Conventional treatment procedures such as ozonization, advanced oxidation procedures, and chemical treatments can introduce either new contaminants or convert the existing contaminants to other harmful molecular forms. Therefore, in such cases, the use of biosurfactants is helpful as they are of biological origin and easily biodegradable. Biosurfactants recover the chemical pollutant in a single step, even at very low concentration. This method is used in the study by Muthusaravanan et al. (2019), in which they used saponin as a green liquid emulsion to remove pharmaceutical micropollutant norfloxacin from the aqueous solution, whereby norfloxacin concentration was 67.65 mg/L, the removal efficiency was 91.27% (Muthusaravanan et al., 2019). It is difficult to produce these biosurfactants on a large scale because their production cost is very high, and the unconventional sources of these biosurfactants also limit the yield. To overcome this limit, the use of alternative, cheaper, and conventional sources could help to increase the yield of biosurfactants, as in the case of X. Vecino et al. They used corn steep liquor as a biosurfactant source for fluorene removal (Vecino et al., 2015). The microbial biosurfactants have amphiphilic biomolecules that can reduce surface and interfacial tension. They can effectively generate microemulsion-forming micelles, which increase the solubility of hydrophobic pharmaceutical pollutants in water and vice versa (Banat et al., 2000). With such plausible mechanisms, better binding of the drug to the biosurfactant is achieved, making it bioavailable to drug-degrading bacteria and thus simplifying the tedious procedure of drug removal.

11.3 A comprehensive account of the biosurfactants in terms of their types, characteristics, sources, and applications for removing toxic pharmaceutical compounds

11.3.1 Types of biosurfactants

Biosurfactants are amphipathic molecules that form hydrophobic and hydrophilic moieties and are regarded as surface-active agents that help in reducing the interfacial surface tension between solution/surface or interfaces of oil/water or air/water through the formation of micelles. Biosurfactants are generally classified by molecular weight (high or low), source, critical micelle concentration, and mode of action. The comprehensive classification is described as follows (Kumar et al., 2021).

11.3.2 Classification based on molecular weight

The comprehensive description of the classification of biosurfactants and their characteristics are described in detail as given below.

11.3.2.1 High molecular weight

This class of biosurfactants comprises biosurfactants or bioemulsifiers with high molecular mass. They are inherently complex, consisting of proteins, lipoproteins, lipopolysaccharides, and heteropolysaccharides (Bjerk et al., 2021; Uzoigwe et al., 2015). These are amphipathic and very effective in stabilizing oil-in-water emulsion. Although there are many high-molecular-weight bioemulsifiers, the most commercially applied bioemulsifiers are alasan with a CMC of 200 µg/mL and Emulsan with an emulsification activity even at 1% (Rosenberg & Ron, 1999). Bioemulsifiers isolated from *Methanobacterium thermoautotrophium* (Trebbau de Acevedo & McInerney, 1996) and glycolipoprotein-based biosurfactant molecules from *Bacillus stearothermophilus* (Gurjar et al., 1995) have been shown to be stably active at high temperature (Bharali et al., 2011).

11.3.2.2 Low molecular weight

This class of biosurfactants consists of simpler fatty acids, lipids conjugated with a phosphate groups, sugar moiety, and protein molecules, that is, phospholipids, glycolipids, and lipoproteins, respectively. These low-molecular-weight biosurfactants help in lowering the interfacial and surface tension (Kubicki et al., 2019). The most important and most-studied low-molecular-weight biosurfactants are rhamnolipids, sophorolipids, and trehalolipids. It is generally comprise lipopeptides and glycolipids. The detailed description of each sub-type is described as follows.

11.3.3 Classification based on molecular composition

11.3.3.1 Glycolipids

Glycolipids consist of biosurfactants that have lipids, hydroxyl aliphatic saturated or unsaturated as conjugated with ester groups attached to carbohydrate moieties (Drakontis & Amin, 2020). The hydrophilic group includes sugar molecules such as mannose, glucose, glucuronic acid, galactose, and rhamnose while the fatty acids form the hydrophobic tail. These widely used biosurfactants include several biosurfactant sub-types, such as sophorolipids, trehalolipids, and rhamnolipids (Vijayakumar & Saravanan, 2015).

11.3.3.2 Rhamnolipids
This class of glycolipid-based biosurfactants contains one or two rhamnose moieties bonded with one or two hydroxydecanoic acid moieties. Owing to their desirable physicochemical characteristics, such as high surface-active properties, reduction of toxicity, and stability, they are the most desired ones among all biosurfactants (Drakontis & Amin, 2020).

11.3.3.3 Trehalolipids
This class of glycolipid biosurfactants consists of trehalose as the carbohydrate hydrophilic moiety. Trehalose or α-D-glucopyranosyl-(1 1)-α-D−glucopyranoside has a 1−1 α-linkage, which is a stable bond and thus provides stability to the surfactive compound at high temperature and resistance even for acid hydrolysis, have strong surface-active properties, can reduce the surface tension to as low as 30 mN/m, and the interfacial tension to 5 mN/m against hexadecane (Rodrigues et al., 2006; Shao et al., 2021; Singh & Cameotra, 2004).

11.3.3.4 Sophorolipids
This class of glycolipid biosurfactant consists of sophorose as the carbohydrate moiety linked to a long-chain hydroxyl acid by glycosidic linkage. The high cost of production of sophorolipids limits mass production on an industrial scale (Daverey & Pakshirajan, 2009).

11.3.3.5 Lipopeptides and lipoproteins
This crucial class of biosurfactants consists of lipids attached to polypeptide chains and are generally considered to be as non-ionic. The amphiphilic property helps them in creating effective microemulsions (Bharali et al., 2011). They mainly include subclasses such as fatty acids, phospholipids, and neutral lipids. The lipopeptidic biosurfactants have strong surface-active characteristics. Subtilisin or surfactin, a cyclic lipopeptide, produced by *Bacillus subtilis* (Arima et al., 1968), has been regarded as one of the most active biosurfactants with a potential to reduce surface and interfacial tension of up to 27 and 1 mN/m, respectively, and a CMC value of up to 25 mg/L has been synthesized (Cooper et al., 1981).

11.3.3.6 Surfactin
This class of biosurfactants is lipopeptidic and anionic in nature, whereby 3-hydroxyl-1, 3-methyl-tetradecanoic acid is amidated with heptapeptide with an LLDLLDL chiral sequence in the order Glu-Leu-Leu-Val-Asp-Leu-Leu (Arima et al., 1968; Vijayakumar & Saravanan, 2015). This leucine-rich order has both D- and L-isomers. This cyclic lipopeptide is among the most effective biosurfactants (Anuradha, 2010).

11.3.3.7 Lichenysin
This class of biosurfactants includes the highly efficient cyclic lipoheptapeptide biosurfactants that are anionic in nature produced by *Bacillus licheniformis*. They refer to high surface and interfacial tension that reduce the water capacity up to 27 and 0.36 mN/m, respectively. Due to the extreme habitats of this particular species such as *B. licheniformis* BAS50 isolated from deep petroleum reservoirs, the biosurfactants produced thus are highly stable under extreme physiological conditions, such as pH, temperature, salinity, and pressure. This property could be used to treat the Pharmaceutical RO Reject, which has a high saline condition (Javaheri et al., 1985).

11.3.3.8 Polymeric biosurfactants
This class of biosurfactants comprises the eminently variable high-molecular-weight heterogeneous polymers. Such biosurfactants can be classified as either glycolipids or lipoproteins. These biosurfactants have surface-active properties, such as high viscosity, high tensile strength, and trim resistance. Lipomanan, liposan, alasan, and emulsan are the examples of well-studied polymeric biosurfactants (Chakraborty & Das, 2014).

11.3.3.9 Particulate biosurfactants
This class of biosurfactants forms the microemulsion by creating an extracellular partition through membrane vesicle formation. The enormous vesicle produced by self-assembling linoleic acid lipid sophorolipid is one such example of it (Dhasaiyan et al., 2014). It plays a crucial role in the uptake of alkanes by microbes. Sometimes the microbial cells act as biosurfactants (Satpute et al., 2010; Vijayakumar & Saravanan, 2015).

11.3.4 Characteristics of biosurfactants
The biosurfactants produced have some special characteristics that are crucial for determining their physicochemical behavior and their application, which is elucidated in this section (Bjerk et al., 2021).

11.3.5 Self-assembly
Like any ideal synthetic surfactant, biosurfactants can self-assemble spontaneously with the help of weak van der Waal and hydrophobic interactions (Akbari et al., 2018; Lee et al., 2008). According to Arima et al., micelle formation is directly proportional to increase in surfactant concentration above the critical micelle concentration level (Arima et al., 1968). Surfactin and rhamnolipids generally form micelles with a low aggregation number. This property determines the effectiveness of surface activity. The higher the self-assembly strength, the lower the concentration of required biosurfactant. The hydrocarbon chain structure and the sequence of the peptide also determine the supramolecular morphologies of biosurfactants (Liu et al., 2016).

11.3.6 Solubilization

According to Nagarajan et al., surfactants or amphiphiles can solubilize hydrophobic molecules due to the presence of hydrophobic pockets, and this solubility is affected by parameters such as concentration, temperature, salts and additives, and pH (Nagarajan & Ruckenstein, 1991). By increasing the hydrophobicity, the solubilization can be enhanced in rhamnolipids. However, the micelle formation at high pH can negatively impact solubilization. The report shows that even the substrate affects the solubilizing property (Shao et al., 2021).

11.3.7 Surface and interface activity

A good surfactant reduces surface and interfacial tension to the lowest possible level. Cooper et al. reported that surfactin produced by *Bacillus subtilis* could lower the surface tension of water to 25 mN/m and interfacial tension hexadecane/water to less than 1 mN/m (Cooper et al., 1981). Similar results were obtained for rhamnolipids synthesized by *Pseudomonas aeruginosa* (Syldatk et al., 1985).

11.3.8 Temperature and pH tolerance

Depending on the habitability of the microbes, the synthesized biosurfactants show tolerance to different ranges of temperature, pH, and ionic concentration. The biosurfactant lichenysin isolated from *Bacillus licheniformis* proved to be resistant in the pH range from 4.5 to 9, up to a temperature 50°C and a Ca and NaCl concentration of up to 25 and 50 g/L, respectively (McInerney et al., 1990). Singh et al. found that *Arthrobaterprotophormiae* produces biosurfactants that are thermostable over a wide range (30°C–100°C) and very pH stable in a wide range (2–12). These properties help these biosurfactants thrive under harsh industrial conditions such as high pH, temperature, and pressure (Singh & Cameotra, 2004).

11.3.9 Biodegradability

Mulligan et al. (2005) have well described the bioremediation application of biosurfactants and a natural substitute for synthetic chemically derived surfactants because they are easily degraded and cannot further contribute to pollution, unlike their chemical counterparts. Biosurfactants are less toxic than their chemical counterparts. The study by Poremba et al. (1991) shows that Corexit has a 10-fold lower LC50 than rhamnolipids against *Photobacterium phosphoreum*. As described by Campos et al., biosurfactant's property of low toxicity has led to its application in pharma-based industry (Campos et al., 2013).

11.3.10 Anti-adhesive agents

Biosurfactants help alter the hydrophobic bonds and surface charges, thus disrupting the biofilm formation. This shows the special application of biosurfactants, which could be used to prevent membrane fouling and thus help improve wastewater treatment in the pharmaceutical industry (Khalid et al., 2019).

11.3.11 Sources of biosurfactants

Biosurfactants have been successfully isolated from fungi and bacteria and also used in industrial applications. The isolation of biosurfactants from algae has been studied as well. There has been ongoing research to find more conventional and efficient resources. Some of the related findings have been discussed here.

11.3.12 Bacterial biosurfactants

Biosurfactants are produced by bacteria using hydrophobic substrates, including rhamnolipids (Burger et al., 1963; Chong & Li, 2017; Guerra-Santos et al., 1986) or sophorolipids (Elshafie et al., 2015). Bacterial species such as *Rhodococcuserythropolis*, *Arthrobacter* spp, produce trehalolipids (Ristau & Wagner, 1983). The widely used lipopolysaccharide-based biosurfactant Emulsan is produced by Acinetobacter spp. (Kretschmer et al., 1982). Lipoproteins such as Subtilisin and Surfactin are produced by *Bacillus subtilis* (Cooper et al., 1981).

11.3.13 Fungal biosurfactants

Biosurfactants from fungi account for only 195 of the total production compared to bacterial biosurfactants, of which the major part is produced from Ascomycetes (about 12%) and Basidiomycetes (about 7%). Biosurfactant production from *Candida* strains (Casas et al., 1997) and *Aspergillus* spp. (Cortes-Sanchez et al., 2011) is well explored. The main types of biosurfactant produced by fungi includes sophorolipids, cellobiose lipids, xyolipids, mannosylerythreitol lipids, hydrophobins, and lipid polyols (Abdel-Mawgoud & Stephanopoulos, 2018; Garay et al., 2018).

11.3.14 Algal biosurfactants

Algal biosurfactants are still underexplored. Many algal species produce exopolysaccharides, which find application in environmental remediation and, thus, such emulsifiers can be and already are being studied for biosurfactant production, for example, *Dunaliellasalina*, red algae *Porphyridiumcruentum* and *Chlorella spp*. are currently being explored for biosurfactant production (Paniagua-Michel et al., 2014).

11.3.15 Applications: removal of toxic refractory organic pharmaceutical compounds

Biosurfactants are promising potential biomolecules for the removal of toxic refractory and recalcitrant compounds from the ecosystem as their accumulation may pose several health complications, biomagnification, and nuisance for treatment procedures. Earlier, this aspect was underreseached, but as the production and use of pharmaceuticals increased with the growing population, it has become the need of the hour to study this issue and find some satisfactory results. Certain studies have focused on drug removal from wastewater, soil-contaminated sites and water bodies. Biosurfactants have been used as an adjoining factor for adsorbents, thus enhancing drug sorption (Kheradmand et al., 2021; Murphin Kumar et al., 2021). This depicts the various ways in which biosurfactants can be exploited for the removal of drugs. Even the surfactants are used as reliable, nontoxic boosters to build up the interactions, strengthening the interactions and thus making it bioavailable for degradation (Sun et al., 2017). From all aforementioned factors, it is clear that the biosurfactants are suitable for pharmaceutical removal from wastewater and contaminated environment sites and, therefore, further work is needed to remove any other limitations.

11.4 Elucidating biosurfactant drug adsorption properties and mechanisms

11.4.1 Promoting the hydrophobic partitioning/interaction

Hydrophobic interactions help the attachment of PPCPs with low K_{ow} values and other active pharmaceutical ingredients to the hydrophobic tail of micelles. This interaction is very effective in separating the compounds from the aqueous solution. The sorption of diclofenac on the surfactant-modified zeolite adsorbent, especifically on the hydrophobic tails, indicates the role of hydrophobicity in this partitioning effect (Sun et al., 2017). The role of hydrophobic interaction was also described by Cabrera et al. in their study on the adsorption of carbamazepine on surfactant cation conjugated to zeolite, depending on the log K_{ow} values (Cabrera-Lafaurie et al., 2014).

11.4.2 Surface charge exchange

Pharmaceuticals in their anionic forms are also repelled by soil particles as they too are negatively charged. The polar head group binds with the anionic pharmaceutical, as in the case of diclofenac attachment to the polar head group of surfactants. In the mechanism of surface anion exchange, the sorption of Diclofenac is accelerated.

11.4.3 Electrostatic and electron—exchange interaction

The functional groups present in the pharmaceutical molecular structures such as $-C=O$, CH_2CH_3, aromatic rings, and N- and O- donor groups such as $-OH$, $-OOH$, and $-NH_2$ facilitate the interaction with biosurfactant micelles.

11.4.4 Increasing the bioavailability of hydrophobic PPCPs

The hydrophobicity of certain PPCPs is the major reason for its less solubility and prolonged persistence in environment since it is very less available to the microbes for degradation because of its surface sorption capability. The irreversible binding to the surface inhibits biodegradation. As anti-adhesives, biosurfactants can cause desorption of biomass from surfaces and thus promote growth, even apparently even increasing water solubility. Surfactants, which dramatically reduce the interfacial tension, are very efficient in mobilizing the adsorbed hydrophobic molecules and making them bioavailable. As described by Miller and Zhang et al., low-molecular-weight biosurfactants having a low critical micelle concentration incorporate molecules into the micelle hydrophobic cavity and thereby increasing their solubility. This mechanism can be seen in the illustration as follows: the biosurfactant adsorbed over the surface of the adsorbent enhances the adsorption of the drug and helps to make it bioavailable in a modified, simpler form to microbial consortia, thus improving and facilitating the biodegradation procedure (Miller & Zhang, 1997). The biosurfactant-facilitated drug adsorption and microbial biodegradation mechanism is shown in Fig. 11.4.

11.5 Recent advancements in biosurfactant-aided adsorption technologies for removal of drugs from the environment

Several studies have focused on bioremediation using adsorption techniques. In particular, few examples have been provided for drug removal in Table 11.1, which indicates drug removal using conventional and waste-derived adsorbents. Although there are already technologies and studies on drug removal, the limitations involved lead to more diverse research to find better solutions. The biosurfactant overcomes some of these limitations in terms of reproducibility, less toxicity, and predictability, hence studies are slowly progressing on this aspect. This topic particularly deals with more recent studies and the modified approaches specific to the technique are been discussed briefly. The study on removal of ibuprofen by biosurfactant-conjugated nanocomposite showed the potency of biosurfactant as a good adsorbent of pharmaceutical effluents as well (Kheradmand et al., 2021). They constructed a nanocomposite with a layer of a combination of Co-Al, the intercalatory layer is filled by rhamnolipid, and the core shell composed of Fe_3O_4. The adsorption capacity was greater than 180 mg/g even after four successive cycles of use. The underlying functional mechanisms are

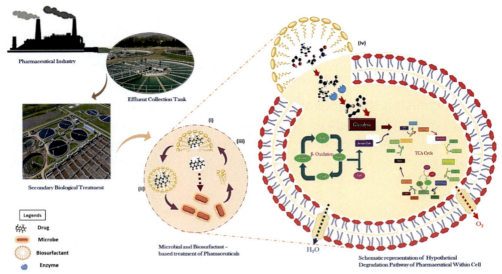

Figure 11.4 *Schematic representation on the mechanism of biosurfactant-aided pharmaceutical compounds removal from effluents.* Biosurfactant-facilitated drug adsorption and microbial biodegradation mechanism [(i) Surface Charge Exchange Adsorption (ii) Hydrophobic Core Adsorption (iii) Electrostatic and Electron−charge exchange interaction (iv) Increasing the bioavailability of hydrophobic PPCPs].

electrostatic attraction, physical adsorption, hydrogen bonding, and anionic exchange. This study therefore shows the proficiency of biosurfactants in pharmaceutical drug removal. The study by Murphin Kumar et al. showed the adsorption-enhancing capacity and highlighted the other potential applications of biosurfactants to address pharmaceutical pollutants. They fabricated the Fe_2O_3/C composites obtained from novel tropical fruit (*Crescentia cujete*)-derived biomass with a glycolipid-based biosurfactant having a high surface area of 466.9 m^2/g through a thermochemical and biofunctional monostep. This biosurfactant-conjugated mesoporous composite could efficiently adsorb pharmaceuticals, especially diclofenac, with nearly 77.51 mg/g as the highest adsorbed concentration. These composites have continued to be used as supercapacitors. The galvanostatic charge-discharge test of the modified electrode showed that it has a specific capacitance of 374 F/g that has current density of 0.2 A/g and could maintain 84% capacitance even after 3000 cycles. This also shows its application in the valuable, resourceful, and renewable energy source (Murphin Kumar et al., 2021). Jayalatha et al. showed that biosurfactant can be very effective in satisfactorily eliminating micropollutants from the environment. Triclosan, the widely used antimicrobial agent, could be degraded to hundred percent over the period of 16 hours by the biosurfactants produced by *Bacillus licheniformis* using crude sunflower oil as a carbon source. This biosurfactant was comparatively thermostable, less toxic, highly

Table 11.1 Techniques used for drug adsorption from aqueous medium.

Sr. No.	Drug adsorbed	Adsorption techniques & matrix	Contact time (min)	Sorption Capacity (mg/g)	Limitations	References
1.	Naproxen	Cu Nanoparticles	60	33.9	Revival of nanoparticles	Husein et al. (2019)
2.	Diclofenac	Cyclamen persicum tubers activated carbon (CTAC)	120	606.78	Recovery of adsorbent, Reduced efficiency on reuse	Jodeh et al. (2016)
3.	Ibuprofen	CO_2 activated Carbonized Precursor (Activated Carbon)	60	78.8	Specific optimization	Mansouri et al. (2015)
4.	Ketoprofen	Olive Waste Cake	11.20	24.69	Low Removal, Inconsistent substrate quality	Baccar et al. (2012)
5.	Tetracycline	Activated Carbon from Apricot shell	24 h	308.33	Regeneration of Used activated Carbon	Marzbali et al. (2016)
6.	Doxycycline Hydrochloride	Pyrolyzed Pumpkin Seed Shell Activated Carbon	10	23.5	Low adsorption Capacity	Kaur et al. (2021)
7.	Amoxicillin	Avocado seed	0.5–1 h	70–325	Recyclability	Leite et al. (2018)
8.	Chlorpheniramine	KOH activated Date palm leaflet activated Carbon	2 h	65–454	Weak—reproducibility	Ali et al. (2019)
9.	Penicillin G	Activated Carbon	24 h	315	Poor recovery	Aksu and Tunç (2005)

(*Continued*)

Table 11.1 (Continued)

Sr. No.	Drug adsorbed	Adsorption techniques & matrix	Contact time (min)	Sorption Capacity (mg/g)	Limitations	References
10.	Enrofloxacin	Cellulose Oxide	24 h	19.3	Less removal	Ötker and Akmehmet-Balcıoğlu (2005)
11.	Ibuprofen	Iron Oxide, Green Synthesis using Black Tea	30 min	0.066.67	Low adsorption	Ali et al. (2016)
12.	Ketoprofen	β-Cyclodextrin	60 min	162.60	Reusability	Skwierawska et al. (2022)
13.	Ibuprofen	Novel Lignosulfonate	30 min	2.318	Low Adsorption	Ciesielczyk et al. (2019)
14.	Dorzolamide	Novel Graphite oxide/polyacrylic acid grafted Chitosan	3 h	334	Reusability	Kyzas et al. (2014)

stable for a long duration and more biodegradable than conventional synthetic counterparts (Jayalatha & Devatha, 2019). Although there are reports of Norfloxacin degradation by biosurfactant producing *Bacillus subtilis* spp.as in a study by Jałowiecki et al., a similar degradation process for tetracycline is not known (Jałowiecki et al., 2017). Therefore, Chun-Xiao Liu et al. focused on the removal of tetracycline from the environment as it is a growing major concern. The biosurfactant-mediated degradation of Axytetracycline, Tetracycline, and Chlortetracycline was widely studied using the indigenously biosurfactant producing *Bacillus amyloliquefaciens* HM 618 and *Bacillus clausii* strains. The maximum removal efficiency was seen with the co-culture system rather than the pure culture system that is 88.9% and 76.6% for Chlortetracycline and Oxytetracycline, respectively. The order of the removal efficiency was CTC> OTC> TC. The surfactant also led to the biotransformation of compounds into minimal toxic forms by demethylation, dehydration, and hydroxylation. This shows the bioremediation potential of surfactants (Liu et al., 2020). Insufficient removal of pharmaceutical compounds by traditional treatment is pushing researchers to seek better alternatives. One such alternative in the form of green surfactant-assisted emulsion liquid membrane (ELM) was proposed by Muthusaravanan et al., in which they used an ELM system to remove Norfloxacin belonging to a class of antibiotics Fluoroquinolone, substantially used for urinary tract infections, from the liquid medium (Muthusaravanan et al., 2019). A major advantage of using ELM system is that they can recover the solutes with very low concentration, as found in the study by Sivarajasekar et al., in which the extraction efficiency of cationic dyes was around 87%, with a solute concentration of less than 20 mg/L (Sivarajasekar et al., 2018). The use of classical synthetic ones such as Span 80 is not only expensive but also hazardous in nature (Sastre et al., 1998); therefore, substituting it with biological sources will be eco-friendly. Muthusaravanan et al. developed an ELM system of natural novel green emulsion-based surfactant from saponin isolated from soapnuts from *Sapindus mukorossi*. The green surfactant thus produced allows satisfactory removal of pharmaceutical-based pollutant Norfloxacin in a single step. The notable extraction efficiency achieved using optimizing strategies opted from Artificial Neural Network (ANN) and Response Surface Methodology-Box Behnken Design (RSM-BBD) of about 91.27%. Hence, these green surfactants could be considered efficient substitutes for chemical-based surfactants (Muthusaravanan et al., 2019). Biosurfactants are good permeation enhancers owing to their hydrophobic nature and, in some cases, also ionic in nature, which alters with the membrane surface and increases permeability through the membrane. Henceforth, this property of biosurfactants could be used to achieve less membrane fouling in wastewater treatment plant, because membrane fouling is the major limit for degrading the efficiency of membrane filtration process by blocking the pores. Lorena Rodriguez Lopez et al. showed the impact of biosurfactant on the permeation of pharmaceutical drugs across a silicone membrane. The biosurfactants studied were produced from corn steep liquor and the effect were analyzed on 10 selected pharmaceutical compounds, namely, benzotriazole, benzocaine, benzoic acid, caffeine, indomethacin, lidocaine, indomethacin, procaine, tetracaine, and salicylic acid. The permeation capacity differed among various selected pharmaceuticals probably because of their molecular size because it was noted that the heavier ones are retarded while the smaller ones are permeated. It was also noted that the biosurfactant and compound interaction favorability also led to the permeation of the compounds (Rodríguez-López et al., 2019). Thus, biosurfactants can help in the transfer of compounds, avoid fouling effects, and enhance the treatment process. The polycyclic aromatic structures of pharmaceutical compounds are refractory and persistent in nature and difficult to degrade by biological action owing to the complexity and toxicity of the compounds such as being mutagenic and carcinogenic. The comparatively high cost than its chemical-derived counterparts drives the research toward cheaper and conventional alternative biosurfactant-synthesizing substrates such as the usage of agricultural feedstocks and byproducts as carbon substrates biosurfactant production from different agricultural wastes by *Lactobacillus pentosus* and the production of lipopeptide-based biosurfactant produced using corn steep liquor as a carbon source (Moldes et al., 2007). Hence, these studies clearly bring out new insights in this field and taking directional cues and further advancements can help in achieving field application of biosurfactants for drug removal.

11.6 Limitations preventing for extensive application of biosurfactant for drug removal from environment and peculiar advantages associated

The production of biosurfactants is very advantageous in some aspects but is not without certain limits, which cannot be ignored. Biosurfactants production is cost-intensive with low productivity. The source are biotic and there is greater chance of batch-to-batch alterations; thus, optimizing the conditions for effective treatment becomes tedious. The production of biosurfactant is scientific; the production plant will require skilled labor. The unconventional sources also pose a problem; however, conventional sources are being researched. The advantages and disadvantages of microbial biosurfactants used in drug adsorption is provided in Table 11.2.

11.7 Future prospects—planning possible strategies to overcome the limited application of biosurfactant over a wide spectrum

Biosurfactant is undoubtedly a suitable alternative for chemical surfactants and finds wide applications in various industrial sectors, pharmacology, ecotoxicology, environmental bioremediation, and immunological implications. However, there are many challenges that

Table 11.2 Advantages and disadvantages of microbial biosurfactants used in drug adsorption.

Sr. No.	Biosurfactant used	Drug adsorbed	Source	Advantages	Disadvantages	References
1.	Surfactin, Iturin, Fengycin	Norfloxacin	*Bacillus subtilis*	Great Interfacial and surface tension reduction, 75% of Drug Removed	Cost-intensive production	Jałowiecki et al. (2017)
2.	Rhamnolipid	Ibuprofen	*Pseudomonas aeruginosa*	Efficient drug removal with 200.9 mg/g, Low toxicity, Durability	Recovery from Spent Effluent	Kheradmand et al. (2022)
3.	Glycolipid	Diclofenac	*Pseudomonas mosselii*	Enhance surface porosity of hydrophobic substrates	Large-scale production	Murphin Kumar et al. (2021)
4.	Lipopeptide	Triclosan	*Bacillus Licheniformis*	Remarkable removal of 100% with 16 h contact time	Unclear Mechanism	Jayalatha and Devatha (2019)
5.	Surfactin, Rhamnolipids	Tetracycline	*Bacillus clausii* T and *Bacillus amyloliquefaciens* HM618	Higher removal efficiency, Lower toxicity of biotransformation products	Growth inhibition of biosurfactant-producing bacteria	Liu et al. (2020)
6.	Saponin	Norfloxacin	*Sapindus mukorossi*	Cost-effective, Higher removal efficiency	Less clear dynamics of removal mechanism	Muthusaravanan et al. (2019)
7.	Rhamnolipid	Rifampin	*Pseudomonas aeruginosa*	Promoter removal efficiency of adsorbent	Large-scale production	Kheradmand et al. (2022)

hinder its large-scale usage. Cost is a very important factor. It is not economically feasible, unless cheaper sources are used, such as corn steep liquor or an industrial waste product. Fermentation and purification are the other costly steps. Studies have recently reported that the cost of production of biosurfactants is 10—12 times more than chemical surfactants, which is not profitable to industries and thus not encouraged in the market. Therefore cost must be reduced by switching to cheaper and renewable resources for biosurfactant production and improving the extraction procedures by cutting short on the steps. The increased yield could also impact the cost. Therefore, the yield can be optimized through strategic parameters. The second aspect is of variability, and this effects on applicability. Because biosurfactant synthesized varies even from strain to strain, it becomes challenging to specifically optimize the parameters, which can be ignored if some universal biosurfactants with wide applications are brought into the market with advanced yield and stability features. The standardized conditions for biosurfactant production need to be stably maintained with a constancy that, at times, becomes tedious to follow and could affect the quality. Therefore, to maintain stringency in quality, the optimized parameters should be scaled in a simplistic and labor-friendly manner. The variability in the source can also lead to variability in yield and thus compromise consistency. Therefore, the choice of sources should be such that there is minimal variation from batch to batch. Thus, following the reasons summarized above, the failure of wide large-scale application can help overcome it.

11.8 Conclusion

This chapter briefly discusses the refractory nature of the pharmaceuticals, which hinders the complete treatment of wastewater and thus accumulates in the environment and causes a risk to the health of living beings. This chapter further discusses the applicability of biosurfactants for drug removal by describing in detail the characteristics, types, and applications, and recent studies being conducted in this field. Using all information gathered, this chapter further describes the underlying mechanism and the prospects for its wide application. With all aspects considered, it can be concluded that that biosurfactants are an excellent substitute for chemical surfactants; with some strategic approach, the limitations can be largely resolved. Drug removal is the need of the hour, and this chapter can help provide the basic necessary concepts to design a study for the same.

References

Abdel-Mawgoud, A. M., & Stephanopoulos, G. (2018). Simple glycolipids of microbes: Chemistry, biological activity and metabolic engineering. *Synthetic and Systems Biotechnology (Reading, Mass.)*, *3*(1), 3—19. Available from https://doi.org/10.1016/j.synbio.2017.12.001.

Ahmed, M. J., & Hameed, B. H. (2018). Removal of emerging pharmaceutical contaminants by adsorption in a fixed-bed column: A review. *Ecotoxicology and Environmental Safety*, *149*, 257—266. Available from https://doi.org/10.1016/j.ecoenv.2017.12.012.

Akbari, S., Abdurahman, N. H., Yunus, R. M., Fayaz, F., & Alara, O. R. (2018). Biosurfactants—A new frontier for social and environmental safety: A mini review. *Biotechnology Research and Innovation*, *2*(1), 81−90. Available from https://doi.org/10.1016/j.biori.2018.09.001.

Aksu, Z., & Tunç, O. (2005). Application of biosorption for penicillin G removal: Comparison with activated carbon. *Process Biochemistry*, *40*(2), 831−847. Available from https://doi.org/10.1016/j.procbio.2004.02.014.

Ali, I., Al-Othman, Z. A., & Alwarthan, A. (2016). Synthesis of composite iron nano adsorbent and removal of ibuprofen drug residue from water. *Journal of Molecular Liquids*, *219*, 858−864. Available from https://doi.org/10.1016/j.molliq.2016.04.031.

Ali, S. N. F., El-Shafey, E. I., Al-Busafi, S., & Al-Lawati, H. A. J. (2019). Adsorption of chlorpheniramine and ibuprofen on surface functionalized activated carbons from deionized water and spiked hospital wastewater. *Journal of Environmental Chemical Engineering*, *7*(1). Available from https://doi.org/10.1016/j.jece.2018.102860.

Anuradha, S. N. (2010). Structural and molecular characteristics of lichenysin and its relationship with surface activity (672, pp. 304−315). Springer Science and Business Media LLC. Available from https://doi.org/10.1007/978-1-4419-5979-9_23.

Ariëns, E. J., Wuis, E. W., & Veringa, E. J. (1988). Stereoselectivity of bioactive xenobiotics. A prepasteur attitude in medicinal chemistry, pharmacokinetics and clinical pharmacology. *Biochemical Pharmacology*, *37*(1), 9−18. Available from https://doi.org/10.1016/0006-2952(88)90749-6.

Arima, K., Kakinuma, A., & Tamura, G. (1968). Surfactin, a crystalline peptidelipid surfactant produced by Bacillus subtilis: Isolation, characterization and its inhibition of fibrin clot formation. *Biochemical and Biophysical Research Communications*, *31*(3), 488−494. Available from https://doi.org/10.1016/0006-291X(68)90503-2.

Baccar, R., Sarrà, M., Bouzid, J., Feki, M., & Blánquez, P. (2012). Removal of pharmaceutical compounds by activated carbon prepared from agricultural by-product. *Chemical Engineering Journal*, *211−212*, 310−317. Available from https://doi.org/10.1016/j.cej.2012.09.099.

Banat, I. M., Makkar, R. S., & Cameotra, S. S. (2000). Potential commercial applications of microbial surfactants. *Applied Microbiology and Biotechnology*, *53*(5), 495−508. Available from https://doi.org/10.1007/s002530051648.

Bharali, P., Das, S., Konwar, B. K., & Thakur, A. J. (2011). Crude biosurfactant from thermophilic Alcaligenes faecalis: Feasibility in petro-spill bioremediation. *International Biodeterioration and Biodegradation*, *65*(5), 682−690. Available from https://doi.org/10.1016/j.ibiod.2011.04.001.

Bjerk, T. R., Severino, P., Jain, S., Marques, C., Silva, A. M., Pashirova, T., & Souto, E. B. (2021). Biosurfactants: Properties and applications in drug delivery, biotechnology and ecotoxicology. *Bioengineering*, *8*(8). Available from https://doi.org/10.3390/bioengineering8080115.

Burger, M. M., Glaser, L., & Burton, R. M. (1963). The enzymatic synthesis of a rhamnose-containing glycolipid by extracts of *Pseudomonas aeruginosa*. *Journal of Biological Chemistry*, *238*(8), 2595−2602. Available from https://doi.org/10.1016/s0021-9258(18)67872-x.

Bush, K. (1997). Antimicrobial agents. *Current Opinion in Chemical Biology*, *1*(2), 169−175. Available from https://doi.org/10.1016/S1367-5931(97)80006-3.

Cabrera-Lafaurie, W. A., Román, F. R., & Hernández-Maldonado, A. J. (2014). Removal of salicylic acid and carbamazepine from aqueous solution with Y-zeolites modified with extra framework transition metal and surfactant cations: Equilibrium and fixed-bed adsorption. *Journal of Environmental Chemical Engineering*, *2*(2), 899−906. Available from https://doi.org/10.1016/j.jece.2014.02.008.

Campos, J. M., Montenegro Stamford, T. L., Sarubbo, L. A., de Luna, J. M., Rufino, R. D., & Banat, I. M. (2013). Microbial biosurfactants as additives for food industries. *Biotechnology Progress*, *29*(5), 1097−1108. Available from https://doi.org/10.1002/btpr.1796.

Casas, J. A., García De Lara, S., & García-Ochoa, F. (1997). Optimization of a synthetic medium for *Candida bombicola* growth using factorial design of experiments. *Enzyme and Microbial Technology*, *21*(3), 221−229. Available from https://doi.org/10.1016/S0141-0229(97)00038-0.

Chakraborty, J., & Das, S. (2014). Biosurfactant-based bioremediation of toxic metals. *Microbial biodegradation and bioremediation* (pp. 168−201). Elsevier Inc. Available from https://doi.org/10.1016/B978-0-12-800021-2.00007-8.

Chong, H., & Li, Q. (2017). Microbial production of rhamnolipids: Opportunities, challenges and strategies. *Microbial Cell Factories*, *16*(1). Available from https://doi.org/10.1186/s12934-017-0753-2.

Ciesielczyk, F., Żółtowska-Aksamitowska, S., Jankowska, K., Zembrzuska, J., Zdarta, J., & Jesionowski, T. (2019). The role of novel lignosulfonate-based sorbent in a sorption mechanism of active pharmaceutical ingredient: Batch adsorption tests and interaction study. *Adsorption*, *25*(4), 865–880. Available from https://doi.org/10.1007/s10450-019-00099-1.

Cooper, D. G., Macdonald, C. R., Duff, S. J. B., & Kosaric, N. (1981). Enhanced production of surfactin from *Bacillus subtilis* by continuous product removal and metal cation additions. *Applied and Environmental Microbiology*, *42*(3), 408–412. Available from https://doi.org/10.1128/aem.42.3.408-412.1981.

Cortes-Sanchez, A., Hernandez-Sanchez, H., & Jaramillo-Flores, M. (2011). Production of glycolipids with antimicrobial activity by Ustilago maydis FBD12 in submerged culture. *African Journal of Microbiology Research*, *5*(17), 2512–2523.

Daughton, C. G., & Ternes, T. A. (1999). Pharmaceuticals and personal care products in the environment: Agents of subtle change? *Environmental Health Perspectives*, *107*(6), 907–938. Available from https://doi.org/10.1289/ehp.99107s6907.

Daverey, A., & Pakshirajan, K. (2009). Production, characterization, and properties of sophorolipids from the yeast Candida bombicola using a low-cost fermentative medium. *Applied Biochemistry and Biotechnology*, *158*(3), 663–674. Available from https://doi.org/10.1007/s12010-008-8449-z.

Dhasaiyan, P., Pandey, P. R., Visaveliya, N., Roy, S., & Prasad, B. L. V. (2014). Vesicle structures from bolaamphiphilic biosurfactants: Experimental and molecular dynamics simulation studies on the effect of unsaturation on sophorolipid self-assemblies. *Chemistry – A European Journal*, *20*(21), 6246–6250. Available from https://doi.org/10.1002/chem.201304719.

Drakontis, C. E., & Amin, S. (2020). Biosurfactants: Formulations, properties, and applications. *Current Opinion in Colloid and Interface Science*, *48*, 77–90. Available from https://doi.org/10.1016/j.cocis.2020.03.013.

Elshafie, A. E., Joshi, S. J., Al-Wahaibi, Y. M., Al-Bemani, A. S., Al-Bahry, S. N., Al-Maqbali, D., & Banat, I. M. (2015). Sophorolipids production by Candida bombicola ATCC 22214 and its potential application in microbial enhanced oil recovery. *Frontiers in Microbiology*, *6*. Available from https://doi.org/10.3389/fmicb.2015.01324.

Gadipelly, C., Pérez-González, A., Yadav, G. D., Ortiz, I., Ibáñez, R., Rathod, V. K., & Marathe, K. V. (2014). Pharmaceutical industry wastewater: Review of the technologies for water treatment and reuse. *Industrial and Engineering Chemistry Research*, *53*(29), 11571–11592. Available from https://doi.org/10.1021/ie501210j.

Garay, L. A., Sitepu, I. R., Cajka, T., Xu, J., Teh, H. E., German, J. B., Pan, Z., Dungan, S. R., Block, D. E., & Boundy-Mills, K. L. (2018). Extracellular fungal polyol lipids: A new class of potential high value lipids. *Biotechnology Advances*, *36*(2), 397–414. Available from https://doi.org/10.1016/j.biotechadv.2018.01.003.

García-Morales, R., García-García, A., Orona-Navar, C., Osma, J. F., Nigam, K. D. P., & Ornelas-Soto, N. (2018). Biotransformation of emerging pollutants in groundwater by laccase from *P. sanguineus* CS43 immobilized onto titania nanoparticles. *Journal of Environmental Chemical Engineering*, *6*(1), 710–717. Available from https://doi.org/10.1016/j.jece.2017.12.006.

Gröning, J., Held, C., Garten, C., Claußnitzer, U., Kaschabek, S. R., & Schlömann, M. (2007). Transformation of diclofenac by the indigenous microflora of river sediments and identification of a major intermediate. *Chemosphere*, *69*(4), 509–516. Available from https://doi.org/10.1016/j.chemosphere.2007.03.037.

Guerra-Santos, L. H., Käppeli, O., & Fiechter, A. (1986). Dependence of *Pseudomonas aeruginosa* continous culture biosurfactant production on nutritional and environmental factors. *Applied Microbiology and Biotechnology*, *24*(6), 443–448. Available from https://doi.org/10.1007/BF00250320.

Gurjar, M., Khire, J. M., & Khan, M. I. (1995). Bioemulsifier production by *Bacillus stearothermophilus* VR-8 isolate. *Letters in Applied Microbiology*, *21*(2), 83–86. Available from https://doi.org/10.1111/j.1472-765X.1995.tb01012.x.

Heberer, T. (2002). Tracking persistent pharmaceutical residues from municipal sewage to drinking water. *Journal of Hydrology*, *266*(3–4), 175–189. Available from https://doi.org/10.1016/S0022-1694(02)00165-8.

Hughes, S. R., Kay, P., & Brown, L. E. (2013). Global synthesis and critical evaluation of pharmaceutical data sets collected from river systems. *Environmental Science and Technology*, 47(2), 661−677. Available from https://doi.org/10.1021/es3030148.

Husein, D. Z., Hassanien, R., & Al-Hakkani, M. F. (2019). Green-synthesized copper nano-adsorbent for the removal of pharmaceutical pollutants from real wastewater samples. *Heliyon*, 5(8). Available from https://doi.org/10.1016/j.heliyon.2019.e02339.

Jałowiecki, è., Żur, J., Płaza, G. A., Kaźmierczak, B., Kutyłowska, M., Piekarska, K., & Trusz-Zdybek, A. (2017). Norfloxacin degradation by *Bacillus subtilis* strains able to produce biosurfactants on a bioreactor scale. *E3S Web of Conferences*, 17, 00033. Available from https://doi.org/10.1051/e3sconf/20171700033.

Javaheri, M., Jenneman, G. E., McInerney, M. J., & Knapp, R. M. (1985). Anaerobic production of a biosurfactant by *Bacillus licheniformis* JF-2. *Applied and Environmental Microbiology*, 50(3), 698−700. Available from https://doi.org/10.1128/aem.50.3.698-700.1985.

Jayalatha, N. A., & Devatha, C. P. (2019). Degradation of triclosan from domestic wastewater by biosurfactant produced from *Bacillus licheniformis*. *Molecular Biotechnology*, 61(9), 674−680. Available from https://doi.org/10.1007/s12033-019-00193-3.

Jodeh, S., Abdelwahab, F., Jaradat, N., Warad, I., & Jodeh, W. (2016). Adsorption of diclofenac from aqueous solution using Cyclamen persicum tubers based activated carbon (CTAC). *Journal of the Association of Arab Universities for Basic and Applied Sciences*, 20, 32−38. Available from https://doi.org/10.1016/j.jaubas.2014.11.002.

Kasprzyk-Hordern, B., & Baker, D. R. (2012). Estimation of community-wide drugs use via stereoselective profiling of sewage. *Science of the Total Environment*, 423, 142−150. Available from https://doi.org/10.1016/j.scitotenv.2012.02.019.

Kaur, G., Singh, N., & Rajor, A. (2021). Adsorption of doxycycline hydrochloride onto powdered activated carbon synthesized from pumpkin seed shell by microwave-assisted pyrolysis. *Environmental Technology & Innovation*, 23, 101601. Available from https://doi.org/10.1016/j.eti.2021.101601.

Khalid, H. F., Tehseen, B., Sarwar, Y., Hussain, S. Z., Khan, W. S., Raza, Z. A., Bajwa, S. Z., Kanaras, A. G., Hussain, I., & Rehman, A. (2019). Biosurfactant coated silver and iron oxide nanoparticles with enhanced anti-biofilm and anti-adhesive properties. *Journal of Hazardous Materials*, 364, 441−448. Available from https://doi.org/10.1016/j.jhazmat.2018.10.049.

Kheradmand, A., Ghiasinejad, H., Javanshir, S., Khadir, A., & Jamshidi, E. (2021). Efficient removal of Ibuprofen via novel core − shell magnetic bio-surfactant rhamnolipid − layered double hydroxide nanocomposite. *Journal of Environmental Chemical Engineering*, 9(5), 106158. Available from https://doi.org/10.1016/j.jece.2021.106158.

Kheradmand, A., Negarestani, M., Kazemi, S., Shayesteh, H., Javanshir, S., & Ghiasinejad, H. (2022). Adsorption behavior of rhamnolipid modified magnetic Co/Al layered double hydroxide for the removal of cationic and anionic dyes. *Scientific Reports*, 12(1). Available from https://doi.org/10.1038/s41598-022-19056-0.

Khetan, S. K., & Collins, T. J. (2007). Human pharmaceuticals in the aquatic environment: A challenge to green chemistry. *Chemical Reviews*, 107(6), 2319−2364. Available from https://doi.org/10.1021/cr020441w.

Kretschmer, A., Bock, H., & Wagner, F. (1982). Chemical and physical characterization of interfacial-active lipids from *Rhodococcus erythropolis* grown on n-alkanes. *Applied and Environmental Microbiology*, 44(4), 864−870. Available from https://doi.org/10.1128/aem.44.4.864-870.1982.

Kubicki, S., Bollinger, A., Katzke, N., Jaeger, K.-E., Loeschcke, A., & Thies, S. (2019). Marine biosurfactants: Biosynthesis, structural diversity and biotechnological applications. *Marine Drugs*, 17(7), 408. Available from https://doi.org/10.3390/md17070408.

Kumar, A., Singh, S. K., Kant, C., Verma, H., Kumar, D., Singh, P. P., Modi, A., Droby, S., Kesawat, M. S., Alavilli, H., Bhatia, S. K., Saratale, G. D., Saratale, R. G., Chung, S. M., & Kumar, M. (2021). Microbial biosurfactant: A new frontier for sustainable agriculture and pharmaceutical industries. *Antioxidants*, 10(9). Available from https://doi.org/10.3390/antiox10091472.

Kümmerer, K. (2009). The presence of pharmaceuticals in the environment due to human use - present knowledge and future challenges. *Journal of Environmental Management*, 90(8), 2354−2366. Available from https://doi.org/10.1016/j.jenvman.2009.01.023.

Kümmerer, Klaus (2010). Pharmaceuticals in the environment. *Annual Review of Environment and Resources*, 35(1), 57−75. Available from https://doi.org/10.1146/annurev-environ-052809-161223.

Küster, A., & Adler, N. (2014). Pharmaceuticals in the environment: Scientific evidence of risks and its regulation. *Philosophical Transactions of the Royal Society B: Biological Sciences*, 369(1656). Available from https://doi.org/10.1098/rstb.2013.0587.

Kyzas, G. Z., Bikiaris, D. N., Seredych, M., Bandosz, T. J., & Deliyanni, E. A. (2014). Removal of dorzolamide from biomedical wastewaters with adsorption onto graphite oxide/poly(acrylic acid) grafted chitosan nanocomposite. *Bioresource Technology*, 152, 399−406. Available from https://doi.org/10.1016/j.biortech.2013.11.046.

Lee, Y. J., Choi, J. K., Kim, E. K., Youn, S. H., & Yang, E. J. (2008). Field experiments on mitigation of harmful algal blooms using a Sophorolipid-Yellow clay mixture and effects on marine plankton. *Harmful Algae*, 7(2), 154−162. Available from https://doi.org/10.1016/j.hal.2007.06.004.

Leite, A. B., Saucier, C., Lima, E. C., dos Reis, G. S., Umpierres, C. S., Mello, B. L., Shirmardi, M., Dias, S. L. P., & Sampaio, C. H. (2018). Activated carbons from avocado seed: Optimisation and application for removal of several emerging organic compounds. *Environmental Science and Pollution Research*, 25(8), 7647−7661. Available from https://doi.org/10.1007/s11356-017-1105-9.

Liu, C. X., Xu, Q. M., Yu, S. C., Cheng, J. S., & Yuan, Y. J. (2020). Bio-removal of tetracycline antibiotics under the consortium with probiotics Bacillus clausii T and Bacillus amyloliquefaciens producing biosurfactants. *Science of the Total Environment*, 710. Available from https://doi.org/10.1016/j.scitotenv.2019.136329.

Liu, G., Liu, J., Tao, X., Li, D. S., & Zhang, Q. (2016). Surfactants as additives make the structures of organic-inorganic hybrid bromoplumbates diverse. *Inorganic Chemistry Frontiers*, 3(11), 1388−1392. Available from https://doi.org/10.1039/c6qi00292g.

Löffler, D., Römbke, J., Meller, M., & Ternes, T. A. (2005). Environmental fate of pharmaceuticals in water/sediment systems. *Environmental Science and Technology*, 39(14), 5209−5218. Available from https://doi.org/10.1021/es0484146.

López-Peñalver, J. J., Sánchez-Polo, M., Gómez-Pacheco, C. V., & Rivera-Utrilla, J. (2010). Photodegradation of tetracyclines in aqueous solution by using UV and UV/H2O2 oxidation processes. *Journal of Chemical Technology and Biotechnology*, 85(10), 1325−1333. Available from https://doi.org/10.1002/jctb.2435.

Mansouri, H., Carmona, R. J., Gomis-Berenguer, A., Souissi-Najar, S., Ouederni, A., & Ania, C. O. (2015). Competitive adsorption of ibuprofen and amoxicillin mixtures from aqueous solution on activated carbons. *Journal of Colloid and Interface Science*, 449, 252−260. Available from https://doi.org/10.1016/j.jcis.2014.12.020.

Marzbali, M. H., Esmaieli, M., Abolghasemi, H., & Marzbali, M. H. (2016). Tetracycline adsorption by H3PO4-activated carbon produced from apricot nut shells: A batch study. *Process Safety and Environmental Protection*, 102, 700−709. Available from https://doi.org/10.1016/j.psep.2016.05.025.

McInerney, M. J., Javaheri, M., & Nagle, D. P. (1990). Properties of the biosurfactant produced by Bacillus licheniformis strain JF-2. *Journal of Industrial Microbiology*, 5(2−3), 95−101. Available from https://doi.org/10.1007/BF01573858.

Miller, R. M., & Zhang, Y. (1997). *Measurement of biosurfactant-enhanced solubilization and biodegradation of hydrocarbons* (Vol. 2, pp. 59−66). Springer Nature. Available from https://doi.org/10.1385/0-89603-437-2:59.

Moldes, A. B., Torrado, A. M., Barral, M. T., & Domínguez, J. M. (2007). Evaluation of biosurfactant production from various agricultural residues by Lactobacillus pentosus. *Journal of Agricultural and Food Chemistry*, 55(11), 4481−4486. Available from https://doi.org/10.1021/jf063075g.

Mulligan, C. N. (2005). Environmental applications for biosurfactants. *Environmental Pollution*, 133(2), 183−198. Available from https://doi.org/10.1016/j.envpol.2004.06.009.

Murphin Kumar, P. S., Ganesan, S., Al-Muhtaseb, A. H., Al-Haj, L., Elancheziyan, M., Shobana, S., & Kumar, G. (2021). Tropical fruit waste-derived mesoporous rock-like Fe2O3/C composite fabricated with amphiphilic surfactant-templating approach showing massive potential for high-tech applications. *International Journal of Energy Research*, 45(12), 17417−17430. Available from https://doi.org/10.1002/er.6798.

Muthusaravanan, S., Vasudha Priyadharshini, S., Sivarajasekar, N., Subashini, R., Sivamani, S., Dharaskar, S., & Dhakal, N. (2019). Optimization and extraction of pharmaceutical micro-pollutant—Norfloxacin using green emulsion liquid membranes. *Desalination and Water Treatment*, *156*, 238−244. Available from https://doi.org/10.5004/dwt.2019.23833.

Nagarajan, R., & Ruckenstein, E. (1991). Theory of surfactant self-assembly: A predictive molecular thermodynamic approach. *Langmuir: The ACS Journal of Surfaces and Colloids*, *7*(12), 2934−2969. Available from https://doi.org/10.1021/la00060a012.

Ötker, H. M., & Akmehmet-Balcıoğlu, I. (2005). Adsorption and degradation of enrofloxacin, a veterinary antibiotic on natural zeolite. *Journal of Hazardous Materials*, *122*(3), 251−258. Available from https://doi.org/10.1016/j.jhazmat.2005.03.005.

Paniagua-Michel, J. d J., Olmos-Soto, J., & Morales-Guerrero, E. R. (2014). Algal and microbial exopolysaccharides: New insights as biosurfactants and bioemulsifiers. In. *Advances in food and nutrition research* (73, pp. 221−257). Academic Press Inc. Available from https://doi.org/10.1016/B978-0-12-800268-1.00011-1.

Patel, M., Kumar, R., Kishor, K., Mlsna, T., Pittman, C. U., & Mohan, D. (2019). Pharmaceuticals of emerging concern in aquatic systems: Chemistry, occurrence, effects, and removal methods. *Chemical Reviews*, *119*(6), 3510−3673. Available from https://doi.org/10.1021/acs.chemrev.8b00299.

Poremba, K., Gunkel, W., Lang, S., & Wagner, F. (1991). Toxicity testing of synthetic and biogenic surfactants on marine microorganisms. *Environmental Toxicology and Water Quality*, *6*(2), 157−163. Available from https://doi.org/10.1002/tox.2530060205.

Rasheed, T., Ahmad, N., Ali, J., Hassan, A. A., Sher, F., Rizwan, K., Iqbal, H. M. N., & Bilal, M. (2021). Nano and micro architectured cues as smart materials to mitigate recalcitrant pharmaceutical pollutants from wastewater. *Chemosphere*, *274*. Available from https://doi.org/10.1016/j.chemosphere.2021.129785.

Richardson, M. L., & Bowron, J. M. (1985). The fate of pharmaceutical chemicals in the aquatic environment. *Journal of Pharmacy and Pharmacology*, *37*(1), 1−12. Available from https://doi.org/10.1111/j.2042-7158.1985.tb04922.x.

Ristau, E., & Wagner, F. (1983). Formation of novel anionic trehalosetetraesters from Rhodococcus erythropolis under growth limiting conditions. *Biotechnology Letters*, *5*(2), 95−100. Available from https://doi.org/10.1007/BF00132166.

Rivera-Utrilla, J., Sánchez-Polo, M., Ferro-García, M. A., Prados-Joya, G., & Ocampo-Pérez, R. (2013). Pharmaceuticals as emerging contaminants and their removal from water. A review. *Chemosphere*, *93*(7), 1268−1287. Available from https://doi.org/10.1016/j.chemosphere.2013.07.059.

Rodrigues, L., Banat, I. M., Teixeira, J., & Oliveira, R. (2006). Biosurfactants: Potential applications in medicine. *Journal of Antimicrobial Chemotherapy*, *57*(4), 609−618. Available from https://doi.org/10.1093/jac/dkl024.

Rodríguez-López, L., Shokry, D. S., Cruz, J. M., Moldes, A. B., & Waters, L. J. (2019). The effect of the presence of biosurfactant on the permeation of pharmaceutical compounds through silicone membrane. *Colloids and Surfaces B: Biointerfaces*, *176*, 456−461. Available from https://doi.org/10.1016/j.colsurfb.2018.12.072.

Rosenberg, E., & Ron, E. Z. (1999). High- and low-molecular-mass microbial surfactants. *Applied Microbiology and Biotechnology*, *52*(2), 154−162. Available from https://doi.org/10.1007/s002530051502.

Sastre, A. M., Kumar, A., Shukla, J. P., & Singh, R. K. (1998). Improved techniques in liquid membrane separations: An overview. *Separation and Purification Methods*, *27*(2), 213−298. Available from https://doi.org/10.1080/03602549809351641.

Satpute, S. K., Banat, I. M., Dhakephalkar, P. K., Banpurkar, A. G., & Chopade, B. A. (2010). Biosurfactants, bioemulsifiers and exopolysaccharides from marine microorganisms. *Biotechnology Advances*, *28*(4), 436−450. Available from https://doi.org/10.1016/j.biotechadv.2010.02.006.

Shao, D., Liu, G., Chen, H., Xu, C., & Du, J. (2021). Combination of surfactant action with peroxide activation for room-temperature cleaning of textiles. *Journal of Surfactants and Detergents*, *24*(2), 357−364. Available from https://doi.org/10.1002/jsde.12471.

Singh, P., & Cameotra, S. S. (2004). Potential applications of microbial surfactants in biomedical sciences. *Trends in Biotechnology*, *22*(3), 142−146. Available from https://doi.org/10.1016/j.tibtech.2004.01.010.

Sivarajasekar, N., Mohanraj, N., Sivamani, S., Prakash Maran, J., Ganesh Moorthy, I., & Balasubramani, K. (2018). Statistical optimization studies on adsorption of ibuprofen onto Albizialebbeck seed pods activated carbon prepared using microwave irradiation. *Materials today: Proceedings, 5*(2), 7264–7274, Elsevier Ltd. Available from https://doi.org/10.1016/j.matpr.2017.11.394.

Skwierawska, A., Nowacka, D., & Kozłowska-Tylingo, K. (2022). Removal of nonsteroidal anti-inflammatory drugs and analgesics from wastewater by adsorption on cross-linked β-cyclodextrin. *Water Resources and Industry, 28*, 100186. Available from https://doi.org/10.1016/j.wri.2022.100186.

Slater, F. R., Singer, A. C., Turner, S., Barr, J. J., & Bond, P. L. (2011). Pandemic pharmaceutical dosing effects on wastewater treatment: No adaptation of activated sludge bacteria to degrade the antiviral drug Oseltamivir (Tamiflu®) and loss of nutrient removal performance. *FEMS Microbiology Letters, 315*(1), 17–22. Available from https://doi.org/10.1111/j.1574-6968.2010.02163.x.

Sun, K., Shi, Y., Wang, X., & Li, Z. (2017). Sorption and retention of diclofenac on zeolite in the presence of cationic surfactant. *Journal of Hazardous Materials, 323*, 584–592. Available from https://doi.org/10.1016/j.jhazmat.2016.08.026.

Syldatk, C., Lang, S., Matulovic, U., & Wagner, F. (1985). Production of four interfacial active rhamnolipids from η-alkanes or glycerol by resting cells of pseudomonas species DSM 2874. *Zeitschrift fur Naturforschung. Section C, Biosciences, 40*(1–2), 61–67. Available from https://doi.org/10.1515/znc-1985-1-213.

Taylor, D., & Senac, T. (2014). Human pharmaceutical products in the environment - The 'problem' in perspective. *Chemosphere, 115*(1), 95–99. Available from https://doi.org/10.1016/j.chemosphere.2014.01.011.

Trebbau de Acevedo, G., & McInerney, M. J. (1996). Emulsifying activity in thermophilic and extremely thermophilic microorganisms. *Journal of Industrial Microbiology, 16*(1), 1–7. Available from https://doi.org/10.1007/bf01569914.

Uzoigwe, C., Burgess, J. G., Ennis, C. J., & Rahman, P. K. S. M. (2015). Bioemulsifiers are not biosurfactants and require different screening approaches. *Frontiers in Microbiology, 6*. Available from https://doi.org/10.3389/fmicb.2015.00245.

Vecino, X., Cruz, J.M., Moldes, A., Aromatic, A.P., & Pahs, H. (2015). Bioremediation of sewage sludge contaminated with fluorene using a lipopeptide biosurfactant. In 17th International Conference on Chemistry and Chemical Engineering, Amsterdam, The Netherlands (Vol. 9, pp. 739–742). https://doi.org/10.5281/zenodo.1108448.

Vijayakumar, S., & Saravanan, V. (2015). Biosurfactants-types, sources and applications. *Research Journal of Microbiology, 10*(5), 181–192. Available from https://doi.org/10.3923/jm.2015.181.192.

Williams, R. J., Keller, V. D. J., Johnson, A. C., Young, A. R., Holmes, M. G. R., Wells, C., Gross-Sorokin, M., & Benstead, R. (2009). A national risk assessment for intersex in fish arising from steroid estrogens. *Environmental Toxicology and Chemistry, 28*(1), 220–230. Available from https://doi.org/10.1897/08-047.1.

Wu, W., Hu, Y., Guo, Q., Yan, J., Chen, Y., & Cheng, J. (2015). Sorption/desorption behavior of triclosan in sediment-water-rhamnolipid systems: Effects of pH, ionic strength, and DOM. *Journal of Hazardous Materials, 297*, 59–65. Available from https://doi.org/10.1016/j.jhazmat.2015.04.078.

CHAPTER 12

Potential of biosurfactants in corrosion inhibition

Qihui Wang and Zhitao Yan
School of Civil Engineering and Architecture, Chongqing University of Science and Technology, Chongqing, P.R. China

12.1 Introduction

Surfactants are used as important industrial chemicals in agriculture (Sachdev & Cameotra, 2013), petroleum (Geetha et al., 2018; Zhang et al., 2021), food (Mnif & Ghribi, 2016), cosmetics (Lourith & Kanlayavattanakul, 2009), biotechnology (Wakamatsu et al., 2001), the pharmaceutical industry, and environmental remediation (Bezza & Chirwa, 2017; Calvo et al., 2009; Nayak et al., 2009; Plaza, 2020). Surfactants decrease the surface and interfacial tension between liquids, solids, and gases. In the mid-nineteenth century, with the industrial production of soap, chemically synthesized surfactants became widely applied in various industries. With increasing awareness of environmental protection, biosurfactants with low toxicity and easy degradation are being considered to replace the existing chemically synthesized surfactants.

Biosurfactants are metabolites secreted by microbes under certain conditions and have a certain surface activity (Benincasa et al., 2002). Biosurfactants are environmentally friendly and can be used as biodegradation aids to promote the degradation of pollutants (Sharma & Sharma, 2020). Biosurfactants are cheap and renewable, conveniently available, produced by microbial metabolism, and have a high regenerative capacity. The structure is complex, and different microbes have different synthetic abilities to produce different biosurfactants and new compounds. Biosurfactants are produced in three major ways: microbial cell metabolism, enzyme catalysis, and natural biological extraction. Microbial cellular metabolism is a multi-enzyme combination of biotransformation processes that perform metabolic activities within the cell. In addition, the use of microbes allows the synthesis of surfactants with new chemical groups that are difficult to synthesize chemically, simplifying the production of biosurfactants and making it suitable for mass production. Enzymatic catalysis is used to synthesize relatively simple surfactants with inexpensive raw materials, specific reactions, and easy separation and purification of the products. Natural biological extraction is a method of obtaining effective biosurfactants from natural biological raw materials, which are easy to produce and abundant but influenced by the raw material.

Biosurfactants have a complex molecular structure (Dagbert et al., 2008) consisting of two main components: polar groups, which are lipophobic and hydrophilic, and non-polar groups, consisting of anions, cations, amino acids, or polysaccharides, such as monosaccharides, polysaccharides, and phosphate groups. Non-polar groups are composed of hydrophobic and lipophilic hydrocarbon chains, such as saturated or unsaturated fatty alcohols and fatty acids. Based on the chemical structure of biosurfactants, they have been classified into five main groups: glycolipids, lipopeptides, phospholipids, fatty acids, and polymeric surfactants. The different chemical structures of the various biosurfactants determine their unique physical and chemical properties. Thus, they exhibit different emulsification, interfacial, and surface tension properties.

Biosurfactants are widely studied in different fields. In the medical field, researchers use biosurfactants to make antibiotics and medical devices. Biosurfactants act as antibiotics. Bluth et al. (2006) found that sophorolipids significantly reduced mortality in septic rats, and sophorolipids acted as antibiotics in the experiment. However, in vitro, sophorolipids do not have antibacterial activity, because biosurfactants have a penetrating effect on cell layers. When combined with proteins, biosurfactants can cause certain proteins to lose their original function and achieve the goal of killing microbial cells. Research on biosurfactants in medical devices has focused on inhibiting biofilm synthesis. Kuiper et al. (2004) produced the lipopeptide surfactants putisolvin I valine and putisolvin II leucine and isoleucine from *Pseudomonas putis strain PCL 1445*, isolated from the roots of plants, which decomposed, formed biofilms, and inhibited the formation of other biofilms. In the field of toiletries, biosurfactants can wet, penetrate, foam, clean, and inhibit bacteria. Therefore, they can be used to produce toiletries instead of chemical surfactants. For example, alkyl sugars, which can be used as raw materials in the production of toiletries such as shower gels and face washes, are soluble, degreasing, and biodegradable and do not cause harm to the environment in the long term. In addition, biosurfactants have a penetrating and wetting effect and can be used as raw materials for producing cosmetics such as lotions and creams to achieve a wetting effect on the skin. Biosurfactants have shown excellent performance in the areas of environmental protection and bioremediation. Sharma et al. (2022) studied the removal of pollutants from textile wastewater by biosurfactants produced by *Staphylococcus sciuri subsp. rodentium strain SE I*. Experiments have shown that strain SE I is a potential biomediator with enhanced decolorization potential for non-ferrous textile wastewater, and a potential for reducing and removing sulfur from heavy metal wastewater. This microbe can be a suitable organism for on-site bioremediation in different industries engaged in decolorization and heavy metals. Dai et al. (2005) found that rhamnolipids have good bioremediation properties, increased the degradation rate of organic matter in the compost to 13.4%, and increased the average content of water-soluble organic carbon by 2.2 g/kg. The results show that adding rhamnolipids can improve the microenvironment of composting treatment and promote the degradation of organic matter for environmental protection (Table 12.1).

In this section, we present a report on the recent research on the use of biosurfactants in corrosion inhibition. The classification of biosurfactants into peptide lipids, glycolipids, and extracellular polymers is based on the chemical structure. According to the biosurfactants' mechanism of action on corrosion inhibition, chemisorption-type, electrostatic attraction adsorption-type, and film-forming mixed type mechanisms of action are proposed. In addition, research has identified a wide range of applications for biosurfactants in corrosion inhibition, such as seawater environments, oil and gas pipelines, highly alkaline concrete environments, industrial pickling, and aviation fuel storage tank channels. The methods of characterization commonly used to detect and analyze the corrosion inhibition of biosurfactants are mainly surface analysis techniques, component characterization techniques, analysis of inhibition effects, and theoretical calculation techniques. This chapter can provide a theoretical basis and useful reference for research on corrosion inhibition by biosurfactants.

12.2 Sources and classification of biosurfactants

There are many types of biosurfactants, and two of the more common methods classify them based on their relative molecular weight and chemical structure. Biosurfactants can be classified into two types: low-molecular-weight and high-molecular-weight biosurfactants, according to their relative molecular weights. Biosurfactants can be classified into five major groups according to their chemical structure, including glycolipids, lipopeptides, phospholipids, fatty acids, and polymers. Table 12.2 demonstrates the classification, sources, and composition of biosurfactants.

Among them, according to the available literature, glycolipids, peptide lipids, and polymers with the ability to inhibit metal corrosion are reported and highlighted in this chapter.

12.2.1 Glycolipids

Glycolipid biosurfactants are one of the more common biosurfactants (Qing Shan et al., 2006). As various glycolipid biosurfactants have different chemical structures, they can be classified as glycolipids, sugar alcohols, and glycosides. Glycolipid biosurfactants are amphiphilic compounds consisting of long-chain fatty acids and hydroxy fatty acids (lipids) at the hydrophobic end and carbohydrates (sugars) at the hydrophilic end. Glycolipid biosurfactants are readily degradable, environmentally friendly, less toxic, and stable surfactants (Aparna et al., 2011). Glycolipid biosurfactants can be produced by bacteria (*Pseudomonas* sp., *Erysipelas* sp., *Bacillus* sp., and *Arthrobacter* sp.) or by fungi (*Pseudostelium* sp., *Clostridium* sp., and *Geobacillus* sp.) (Parthipan et al., 2018).

Table 12.1 Corrosion inhibition properties of biosurfactants.

Sources	Biosurfactant types	Corrosion inhibition properties	References
Pseudomonas aeruginosa	Rhamnolipids	Pre-conditioning of surfaces in Rhamnolipids solutions significantly delays the onset of localized corrosion	Shubina Helbert et al. (2020)
Bacillus sp	Glycolipid	The presentation of biosurfactant F7 during biofilm development or its use as a biofilm eliminator reduces the corrosion rate of mild steel ST37	Purwasena et al. (2019)
Bacillus subtilis, *Pseudomonas stutzeri*, and *Acinetobacter baumannii*	Glycolipid	The biosurfactant production capacity of the bacterial strain plays an important role in solubilizing hydrocarbons in aqueous media, with the highest biodegradation efficiency of 85% for crude oil	Elumalai et al. (2021)
Four indigenous corrosive bacterial strains: *Bacillus subtilis* A1 (KP895564), *Streptomyces parvus* B7 (KP895570), *Pseudomonas stutzeri* NA3 (KU708859) and *Acinetobacter baumannii* MN3 (KU708860)	Glycolipid	The inhibition efficiency was found about 87% for mixed consortia included with biosurfactant system	Parthipan et al. (2018)
The bacterial strain *Pseudomonas cepacia*	Glycolipid	The biosurfactant produced by *Pseudomonas cepacia* CCT 6659 demonstrated stability after being submitted to conservation methods by adding the preservative sodium benzoate and free-flowing steam plus sodium benzoate	da Silva Faccioli et al. (2022)
extracellular rhamnolipids and an extracellular biopolymer	Glycolipid	Biosurfactant anticorrosion efficiency increases with the growth of its concentration, and its addition over critical micelle concentration in corrosion solution does not lead to a significant increase in its protective effect	Zin et al. (2018)
Ethane-1,2-diylbis(N,N-dimethyl-N-hexadecylammoniu-macetoxy)dichloride	Lipopeptide	Corrosion inhibition increases with increasing concentration and solution temperature, with a maximum corrosion inhibition efficiency of 98%	Mobin et al. (2017)
N-dodecyl asparagine and N-dodecyl arginine	Lipopeptide	The addition of a fixed concentration of transition metal cations to the inhibitor significantly improves the inhibition efficiency of B-Surf, with a 90% corrosion inhibition efficiency	Fawzy et al. (2021)

Table 12.2 Types, sources, and composition of biosurfactants.

Class	Biosurfactant	Bacteria	Carbon sources	References
Glycolipids	Rhamnolipid	*Pseudomonas aeruginosa* OG1	Chicken feather	Ozdal et al. (2017)
		Pseudomonas flfluorescens 1895	Olive oil/*n*-hexadecane	Abouseoud et al. (2007)
		Pseudomonas aeruginosa ATCC 10145	Waste frying oil	Wadekar et al. (2012)
		Burkholderia thailandensis	Glycerol	Dubeau et al. (2009)
		Pseudoxanthomonas sp.	Hexadecane	Nayak et al. (2009)
		Ralstonia pickettii SRS, *Alcaligenes piechaudii* SRS	Crude oil	Plaza et al. (2008)
	Sophorolipid	*Candida bombicola* ATCC 22214	Turkish corn oil and honey	Pekin et al. (2005)
		Pichia anomala PY1	Soybean oil	Thaniyavarn et al. (2008)
	Trehalose lipid	*Starnerella bombicola* ATCC 22214	Sweetwater	Wadekar et al. (2012)
		Bordetella hinzii–DAFI	Sucrose/molasses, crude oil	Bayoumi et al. (2010)
Lipopeptides	Iturin	*Bacillus subtilis*	Glucose/rapeseed oil, crude oil	Bayoumi et al. (2010)
	Surfactin	*Bacillus subtilis*	Crude oil	Pereira et al. (2013)
	Fengycin	*B. thuringiensis* CMB26	Glutamic, acid, mannitol, glucose, urea	Kim et al. (2004)
	Lichenysin	*Bacillus licheniformis*	Glucose, sucrose, starch, cane molasses	Joshi et al. (2016)
Phospholipids	Phospholipid	*Bacillus sphaericus* EN3, *Bacillus azotoformans* EN16	Glucose/diesel/crude oil	Adamu et al. (2015)
	Phospholipid	*Klebsiella pneumoniae* IVN51	Dextrose	Astuti et al. (2019)
Fatty acids	Fatty acid	*Fusarium oxysporum*	Crude oil	Santhappan & Pandian (2017)
Polymeric biosurfactants	Polymeric	*Pseudomonas stutzeri*	Diesel	Joshi & Shekhawat (2014)
		Serratia marcescens UCP 1549	Corn waste oil	Araújo et al. (2019)
		Stenotrophomonas maltophilia UCP 1601	Soybean/corn/diesel	Nogueira et al. (2020)
	Complex Carbohydrate/protein/lipid	*Cunninghamella echinulate* UCP	Soybean waste oil, corn steep liquor	Silva et al. (2014)

Rhamnolipids and sophorolipids are the most common glycolipid biosurfactants that can inhibit metal corrosion.

According to available reports, rhamnolipids are the most commonly applied. The hydrophilic group of rhamnolipids consists of one to two rhamnose rings, and the hydrophobic group consists of one to two molecules of saturated or unsaturated fatty acids with different carbon chain lengths, which can be divided into rhamnolipids and bis-rhamnolipids. Our expression "rhamnolipids" is a mixture of many homologous structures, and up to 28 different rhamnolipid structures are known to have been identified (Costa et al., 2011). During biosynthesis, these groups may link to produce multiple congeners with similar chemical structures. Studies have shown that fermentation products generally contain four major rhamnolipids, which are generally denoted R1 (Rha—C10—C10), R2 (Rha—Rha—C10), R3 (Rha—Rha—C10—C10), and R4 (Rha—C10) in academia. The general formula of the structure is shown in Fig. 12.1.

Sophorolipids are also among the biosurfactants that can inhibit metal corrosion, mainly produced by *Pseudomonas aeruginosa* using sugar and vegetable oil as carbon sources and are a secondary metabolite (Zhang et al., 2018). Sophorolipids have many advantages, such as high surface activity, better temperature, acid and salt resistance, environmental friendliness, and special solubilizing ability (Pacwa-Płociniczak et al., 2011). Structurally, the locust lipid molecule consists of a hydrophilic head and a lucid tail, which are usually present as disaccharide locust sugars attached to the terminal or

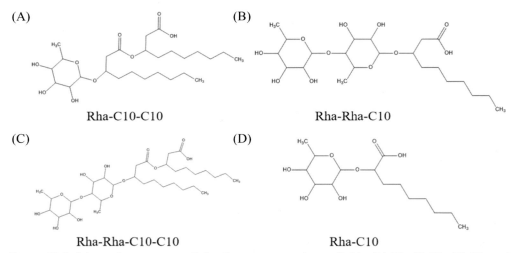

Figure 12.1 Schematic structures of the four common rhamnolipids. (A) R1, (B) R2, (C) R3, and (D) R4. *From Haba, E., Pinazo, A., Jauregui, O., Espuny, M. J., Infante, M. R., & Manresa, A. (2003). Physicochemical characterization and antimicrobial properties of rhamnolipids produced by Pseudomonas aeruginosa 47T2 NCBIM 40044. Biotechnology and Bioengineering, 81(3), 316—322. https://doi.org/10.1002/bit.10474.*

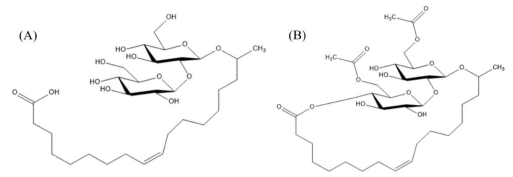

Figure 12.2 Main structures of sophorolipids. (A) Acid sophorolipids and (B) Lactone sophorolipids. From Pratap, A. P., Mestri, R. S., & Mali, S. N. (2021). Waste derived-green and sustainable production of Sophorolipid. Current Research in Green and Sustainable Chemistry, 4. https://doi.org/10.1016/j.crgsc.2021.100209.

sub-terminal of hydroxy fatty acids via β-glycosidic bonds to form amphiphilic biosurfactants (Bisht et al., 1999). Based on whether the sophorolipid molecule undergoes intramolecular esterification, sophorolipids can be divided into two main categories, lactone sophorolipids and acid sophorolipids, with the structures shown in Fig. 12.2.

12.2.2 Lipopeptides

Lipopeptide biosurfactants are surface-active lipopeptide compounds secreted during the metabolic process of microorganisms when they are cultured under certain conditions. Lipopeptides are usually composed of amino or hydroxy fatty acids with peptide chains or rings and are classified into cyclic lipopeptides and linear lipopeptides according to their structural characteristics. Cyclic lipopeptides have a cyclic molecular structure, divided into fatty acid ring formation, attachment, and separation according to the different ring formation methods. Linear lipopeptides are amino acids sequentially linked in a line, and fatty acids are linked to an amino or other hydroxyl group at the N-terminus of the peptide chain without a ring structure. Lipopeptide surfactants have a unique amphiphilic molecular structure with polar hydrophilic peptide bonds and non-polar hydrophobic groups consisting of aliphatic hydrocarbon chains. The available literature indicates that more than 10 lipopeptide biosurfactants have been identified, mainly surfactin, iturin, lichenysin, and fengysin, all of which have the peptide-like structure of proteins (Fig. 12.3).

12.2.3 Polymer class

Extracellular polymeric substances (EPSs) are polymers secreted outside the body by microorganisms, mainly bacteria, under certain environmental conditions.

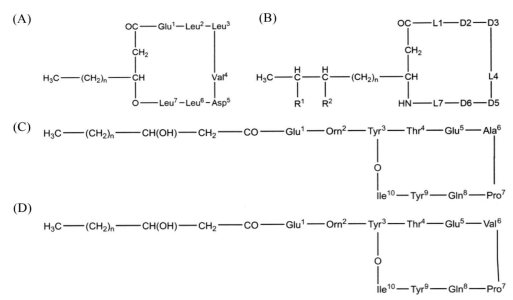

Figure 12.3 Typical structure of lipopeptides. (A) Surfactin, (B) iturin, (C) fengycin A, and (D) fengycin B. *From Denoirjean, T., Doury, G., Poli, P., Coutte, F., & Ameline, A. (2021). Effects of* Bacillus *lipopeptides on the survival and behavior of the rosy apple aphid* Dysaphis plantaginea. *Ecotoxicology and Environmental Safety, 226, 112840. https://doi.org/10.1016/j.ecoenv.2021.112840.*

Its main components are similar to the intracellular components of microorganisms, which are polymers of high-molecular-weight substances such as polysaccharides, proteins, and nucleic acids. EPSs have important physiological functions to adsorb and enrich nutrients from the environment, absorb them into the cells after degradation into small molecules by extracellular enzymes, and reduce the harm of foreign substances to the cells. EPS exhibits a negative surface charge, strong adsorption properties, and biodegradability. It is related to the type of feed water substrate, reactor operating conditions, solution chemistry, and microbial growth stages. EPS is a class of polymers with a complex composition and can exhibit different properties depending on the constituents. Generally, EPSs are divided into soluble EPSs (SEPSs) and bound EPSs (BEPSs). From inside to outside, BEPS can be divided into tightly bound EPS (TB-EPS) and loosely bound EPS (LB-EPS) (Zhu et al., 2009).

EPS contains numerous functional groups that allow EPS to accept or give electrons, so it can adsorb tightly to metals where empty orbitals exist. Because the EPS production process is simple, does not pollute the environment, and is a green corrosion inhibitor. EPS has been reported to adsorb on metal surfaces and form a denser protective film due to its functional groups (Jin et al., 2014).

12.2.4 Phospholipids

Phospholipids are monolipid derivatives containing phosphate roots, which can be divided into two categories according to their molecular structural composition: glycerophospholipids and sphingomgelins (Jun & Yu, 2012). The alcohol that constitute glycerophospholipids is glycerol, which has two hydroxyl groups replaced by fatty acids and the other by phosphoric acid and nitrogen-containing base compounds or inositol. Sphingomgelins are composed of sphingosine, fatty acids, phosphate, and choline. Phospholipid biosurfactants have properties such as surface and interfacial adsorption, as well as the ability to form emulsions and produce liquid crystals and micelles. Nwaguma et al. (2016) isolated *Klebsiella pneumonia* IVN51 from hydrocarbon-polluted soil and confirmed that it was phosphatidylethanolamine that produced the surfactant. Phosphatidylethanolamine is a phospholipid surfactant with strong emulsifying properties. The structures of common phospholipid biosurfactants are shown in Fig. 12.4.

12.2.5 Fatty acids

Fatty acid biosurfactants use carboxylic acid groups as hydrophilic groups and include glycerides, fatty acids, and fatty alcohols. The more common of such biosurfactants is fatty acid monoglycerides. Hua et al. (2000) found that *Candida antarctica* WSH112 can produce glycerol ester biosurfactant when alkanes are used as a carbon source, which has high thermal stability and can reduce the surface tension of water to 38 mN/m.

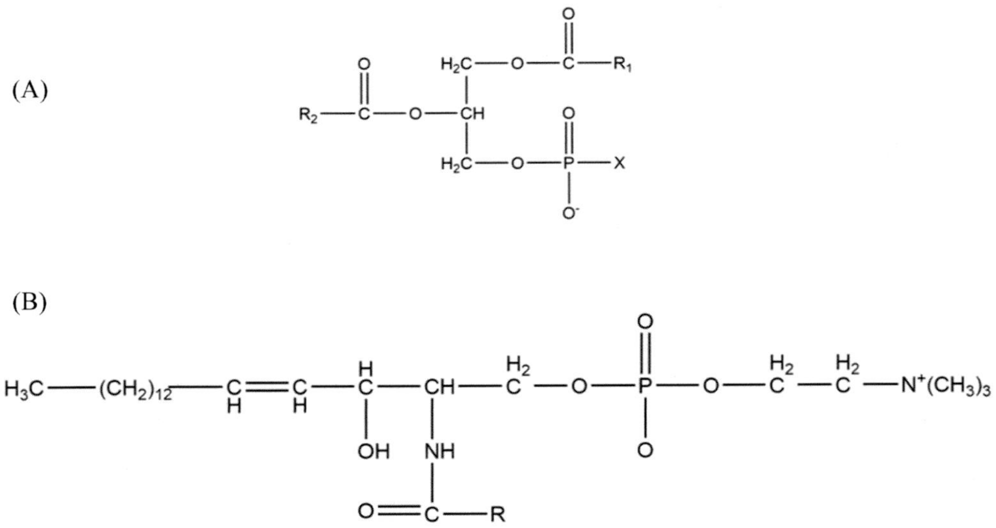

Figure 12.4 Common phospholipid submolecular structures. (A) Glycerophospholipid and (B) sphingomgelins. *From Shi Qian, R., Yun Fen, Z., Shu Ling, X., Xiao Fei, C., Ya, X., Lyu, X., Xiang, J. X., Wei, F., & Chen, H. (2022). Progress on structure, dietary source and nutrition of phospholipid. China Oils and Fats, 47, 68–88.*

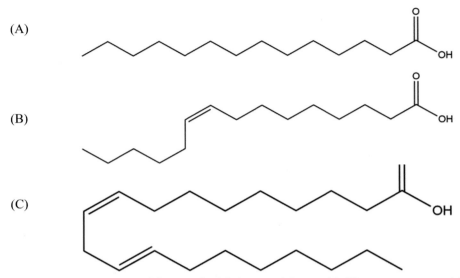

Figure 12.5 Typical structures of fatty acids. (A) Saturated fatty acid, (B) monounsaturated fatty acid, and (C) polyunsaturated fatty acid. *From Esfahanian, M., Nazarenus, T. J., Freund, M. M., McIntosh, G., Phippen, W. B., Phippen, M. E., Durrett, T. P., Cahoon, E. B., & Sedbrook, J. C. (2021). Generating pennycress (Thlaspi arvense) seed triacylglycerols and acetyl-triacylglycerols containing medium-chain fatty acids. Frontiers in Energy Research, 9. https://doi.org/10.3389/fenrg.2021.620118.*

Fatty acids can be classified into saturated, monounsaturated, and polyunsaturated according to whether their molecular structures contain unsaturated carbon bonds and their typical structures are shown in Fig. 12.5.

12.3 Corrosion inhibition mechanism of biosurfactants

In this section, different mechanisms related to the inhibition of metal corrosion by biosurfactants are presented. In addition, since the inhibitory behavior of various active substances secreted by microorganisms varies considerably depending on environmental factors, the specific corrosive media under study are also highlighted in the section. The metabolites with certain surface activity secreted during microbial metabolism can be broadly classified into three types: chemisorbed, electrostatic attraction, and film-forming mixed types.

12.3.1 Chemisorbed type

Biosurfactants produced by bacteria and fungi, such as glycolipids, polysaccharide lipids, and lipopeptides, are all amphiphilic molecules with hydrophilic or hydrophilic portions. The hydrophilic head of the biosurfactant molecule is attached to the metal.

The hydrophobic portion expands outward through the interface of the solution to form a series of hydrophobic tails that provide subsequent protection. Thus, the electrochemical properties of the metal surface are modified.

Parthipan et al. (2018) used *Pseudomonas mosselii* F01 and studied its corrosion inhibition mechanism. The results show that *Pseudomonas mosselii* F01 can form glycolipid-type surfactants. The adsorption of surfactant molecules on the carbon steel surface is due to the higher energy of the interaction between the biosurfactant molecules and the carbon steel surface than between the water and the carbon steel. The mechanism of corrosion inhibition of the glycolipid-type surfactant formed by *Pseudomonas mosselii* F01 can thus be attributed to the adsorption of its bio-surface-active functional groups on the surface. Zhang et al. (2021) fermented *Pseudomonas aeruginosa* and *Burkholderia cepacia* to obtain the secondary *metabolite rhamnolipids*. The study showed that *rhamnolipid* is an anodic-based corrosion inhibitor with an efficiency of more than 90% for X65 steel at 40 mg/L. *Rhamnolipid* is chemisorbed on the steel surface with an optimal adsorption time of 2 h. The results of the surface analysis show that the adsorption mechanism of *rhamnolipid* is mainly a *rhamnopyranose* ring and C=O groups, which provide electrons to bind iron atoms through the hollow orbitals. Xiao et al. (2016) investigated the corrosion inhibition mechanism of exopolysaccharide extracted from *Vibrio Neocaledonicus* sp. on Q235 carbon steel under H_2SO_4. Corrosion begins with the production of Fe^{2+} by anodic oxidation, and the reaction is as follows:

$$Fe \rightarrow Fe^{2+} + 2e^- \tag{12.1}$$

Fe^{2+} continues to oxidize in solution to produce Fe^{3+}, which accelerates the corrosion of carbon steel as follows:

$$Fe^{2+} \rightarrow Fe^{3+} + e^- \tag{12.2}$$

$$Fe + 2Fe^{3+} \rightarrow 3Fe^{2+} \tag{12.3}$$

After the addition of the corrosion inhibitor, the electrochemical reaction changes:

$$Fe + Cl^- \rightarrow (FeCl^-)_{ads} \tag{12.4}$$

$$(FeCl^-)_{ads} + InhH^+ \rightarrow (FeCl^- InhH^+)_{ads} \tag{12.5}$$

From the above reaction, it is evident that the corrosion of carbon steel is contained after the solidification of Fe^{2+}. EPSs are mainly composed of polysaccharides, nucleic acids, and corrosive acids. Fourier transform infrared spectroscopy (FTIR) showed that EPS contains electron-donating groups such as O—H, —COOH, and N—H. EPS is mainly composed of polysaccharides, nucleic acids, proteins, and corrosive acids. Fig. 12.6 shows the mechanistic diagram.

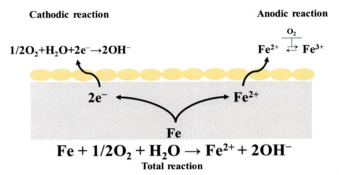

Figure 12.6 Schematic diagram of the electrochemical reaction mechanism on carbon steel.

Chongdar et al. (2005) showed that EPS can create an Fe—EPS layer on carbon steel by giving electron groups that can combine with Fe^{2+} or Fe^{3+}. The bonding of Fe to EPS affects the adsorption strength and densities of Fe—EPS layers on the carbon steel surface. As the concentration of EPS in the solution increases, the amounts of EPS that can be combined with Fe increases, and the Fe—EPS layer formed increases in the denseness of the carbon steel surface, thus increasing the protective force on the carbon steel. Dong et al. (2011) studied the binding of EPS to Fe^{2+} and found that the binding of Fe^{2+} increased with increasing EPS concentration. When EPS reached a specific concentration, the Fe^{2+}—EPS layer was dense on the surface of carbon steel. Above the specific concentration, the adsorption of the Fe^{2+}—EPS layer on the surface of carbon steel reaches dynamic equilibrium, and the adsorption layer is disturbed by the absence of active adsorption sites on the surface. However, EPS will bind Fe^{3+} in the solution, accelerating the dissolution of anode Fe and destroying the density of the protective layer, thus reducing corrosion inhibition. Örnek et al. (2002) concluded from experimental data that *B. subtilis* WB600 and *B. licheniformis* produce *polyaspartate* and *γ—polyglutamate*, respectively, which further reduces the corrosion rate due to their carboxyl acid groups. This functional group can chelate or couple aluminum ions or aluminum oxides present at the metal—solution interface through hydrogen bonds, dipole—dipole interactions, and coupling by Coulombic interactions.

Chemisorbed biosurfactants inhibit the metal corrosion by affecting electrochemical reactions, which has the advantages of high efficiency and fast rate. In particular, corrosion inhibition cases, such as the pickling process and chemisorption-type biosurfactant corrosion inhibitors, can be widely applied.

12.3.2 Electrostatic attraction type

Related studies have confirmed that some microorganisms and their secreted EPS can be adsorbed on metal surfaces by electrostatic attraction, providing a specific gentle microenvironment for the microbial community. The microbial protective film, formed by electrostatic attraction, acts as a barrier, prevents the dispersion of corrosive solvents through the secretion of polymeric biosurfactants and emulsifiers by EPS (Li et al., 2021).

Li et al. (2021) studied the corrosion inhibition mechanism of three bacteria, *Tenacibaculum mesophilum* D−6, *Tenacibaculum litoreum* W−4, and *Bacillus* sp. Y−6. The results of biofilm characterization experiments illustrate that surface hydrophobicity and electrostatic attraction are the attributes that determine the degree of bacterial biofilm adhesion. The conformational change of the EPS permits contact between the hydrophobic groups and the hydrophobic surface, creating a short-range attractive force called hydrophobicity. The uptake of extracellular polysaccharides on metal oxides is generally mediated by hydrogen bonds. The van der Waals interactions and attractive electrostatic forces between EPS and solid surfaces have also been studied. Tamilselvi et al. (2022) used field-emission scanning electron microscopy (FE−SEM), energy dispersive X−ray (EDX), and image processing techniques to investigate the corrosion inhibition mechanism of *Pichia* sp. on mild steel. FE−SEM images show that the metal surface forms smoother planes in the presence of *Pichia* sp. and the elemental composition of the mass spectrometric surface structure of *Pichia* sp. Furthermore, EDX plots show the presence of the elements, as well as mass spectra of mild steel immersed in H_2SO_4. The results show that the direct attack of sulfuric acid is reduced. The corrosion rate is mitigated because *Pichia* sp. form a smooth protective layer on the surface. Meanwhile, FTIR data confirmed the presence of amide and carboxyl groups in *Pichia* sp., indicating that their interaction with the inhibitor during mass spectrometry led to the formation of metal−ligand complexes. The protective layer on the surface of the mass spectrometer is formed by the electrostatic interaction of *Pichia* sp. or its secretion with iron ions.

Biosurfactants formed from the electrostatic attraction are characterized by long inhibition times. However, as EPS consists mainly of polysaccharide lipids and proteins, it reduces the electrostatic force between the cell or subcellular and the metal, leading to inhomogeneous adsorption of biosurfactants. When the metal is in a corrosive medium for a long time, pitting occurs on its surface (Stadler et al., 2010). How to make the active product secreted by microorganisms more uniform by electrostatic attraction adsorption to metal should be further studied.

12.3.3 Film-forming mixed type inhibitor

In most situations, the inhibition mechanism of biosurfactants is a mixture of physical and chemical reactions. The metal is immersed in the solution medium inoculated by bacteria or fungi. Microorganisms colonize the metal surface and secrete heterogeneous EPS, which are promoted by electrostatic attraction between the EPS and the metal surface and adhere to the metal surface to form a biofilm. The main components of biofilms are polysaccharides, which also contain proteins, fatty acids, and other small molecular structures that can form physical barriers and affect electrochemical reactions during corrosion (Li et al., 2021).

Shubina et al. (2016) evaluated the corrosion inhibition efficiency of reinforcement in a simulated concrete pore solution of lipopeptide secreted by *Gram−negative bacteria*. The electrochemical measurements showed that the lipopeptide changed the chemical conditions of the corrosion products by forming an ohmic resistance region between the surface of the carbon steel and the electrolyte, blocking the anode and cathode reaction sites. The regions moderate the galvanic reactions and the loss of electrons from the metal substrate, leading to a slowdown in the corrosion process. The corrosion of reinforcement in this study environment is a mixture of millions of anode and cathode electrodes, and the pore fluid of concrete acts as an electrolyte. The anode chemical reaction formula is as follows:

$$Fe \rightarrow Fe^{2+} + 2e^- \tag{12.6}$$

The oxygen diffuses into the porous system of the concrete, eventually reaching the surface. Oxygen will be reduced to hydroxide ions as described in the following:

$$O_2 + 2H_2O + 4e^- \rightarrow 4OH^- \tag{12.7}$$

$$2H_2O + 2e^- \rightarrow H_2 + 2OH^- \tag{12.8}$$

This reaction may be related to the additional cathodic reactions occurring on the surface of the concrete steel and to the evolution of electrolyte chemistry near the carbon steel. The overall reaction equation is as follows:

$$Fe + 2H^+ \rightarrow Fe^{2+} + H_2 \tag{12.9}$$

SEM showed that Gram−negative bacteria had a strong tendency to adhere to the surface of carbon steel, showing a good corrosion inhibition effect. Furthermore, XPS surface analysis was employed to study the adsorption of lipopeptides on the steel surface. After adding Gram−negative bacteria, the O, N, and Na atoms increase, but the spectral intensity of the Fe core decreases remarkably, indicating that there is an adsorption layer on the surface. Based on electrochemical tests and morphological characterization experiments, lipopeptide biosurfactants secreted by Gram−negative bacteria were a kind of film-forming mixed inhibitor.

The anticorrosion protection of film-forming mixed corrosion inhibitors mainly comes from two aspects. On the one hand, the consumption of O_2 aerobic respiration in the process of microbial secretion of active substances can reduce the participation of O_2 in electrochemical reactions. On the other hand, the EPS secreted by microorganisms can act as a barrier for corrosive substances such as O_2 and chloride ions, which has been confirmed in corrosion protection experiments with abiotic EPS (Gao et al., 2022). The film-forming mixed corrosion inhibitor combines the advantages of chemisorption and electrostatic attraction adsorption, which is a significant research object for future industrial applications.

12.4 Practical application of biosurfactants

Every year, the global direct economic loss resulting from metal corrosion is approximately 700 billion to 1000 billion dollars. Among them, in the United States, the annual corrosion losses amount to more than 300 billion U.S. dollars, accounting for 4.2% of the GDP. The corrosion losses in China in 2014 were approximately RMB 2 trillion, accounting for 3% of the GDP in that year (Wang et al., 2019). The damage of corrosion is obvious, and in addition to economic losses, the damage to society, and the safety of life are incalculable. Therefore, it is important to inhibit corrosion. Common methods are surface protection techniques, environmental treatment, and electrochemical protection. Among them, the addition of corrosion inhibitor technology has become one of the hot spots of research due to the advantages of small dosage, low cost, and environmental protection. The use of biosurfactants as metal corrosion inhibitors in different environments is an efficient and eco-friendly method.

12.4.1 Marine environment

The problem of corrosion is prevalent in the marine environment. In the ocean environment, various types of ships, sea-related exploration equipment, drilling platforms, oil pipelines, and offshore wind power equipment, most structures use various carbon steel, low-alloy steel, stainless steel, and other metal materials. These materials are susceptible to MIC and can cause enormous hazards. As an example, the steel corrosion problem for offshore engineering equipment and ships resulted in an annual economic loss of approximately 100 billion RMB in China, equivalent to 3.34% of the GDP in 2014. Meanwhile, it can pose serious safety threats and pollution problems, thus requiring urgent attention and solutions. The marine environment includes atmospheric environments, seawater environments, sea mud areas, and marine life. Among them, the addition of corrosion inhibitor technology has become one of the hot spots of research due to the advantages of small dosage, low cost, and environmental protection. The use of biosurfactants as metal corrosion inhibitors in different environments is an efficient and eco-friendly method.

For instance, Guo et al. (2017) investigated the adhesion and corrosion ability of *Bacillus subtilis* and *Pseudoalteromonas lipolytica* in a marine environment. It was found that since *Bacillus subtilis* was present, a dense, homogeneous, noncracked, hydrophobic film was formed on the metal surface. However, *Pseudoalteromonas lipolytica* lipids are loose, coarse, inhomogeneous, and hydrophilic. This leads to *Bacillus subtilis* inhibiting steel corrosion, while biofilms of *Bacillus lipolytic* water tend to induce pitting. Ghafari et al. (2013) isolated numerous bacterial strains from various habitats, including soil and water, and subjected them to conditions simulating saltwater that included various bacterial suspensions. The effectiveness of biofilm growth and EPS attachment in 21 bacteria was tested. It was found that *Chryseobacterium indologenes MUT.2* and *Klebsiella pneumonia MUT.1* had good inhibiting effects. In an experiment with simulated seawater, Qu et al. (2015) examined the impact of *Bacillus subtilis* C2 (BS) on the corrosion behavior of cold-rolled steel (CRS). The development of biofilms is closely related to the corrosion of CRS. Once the compact biofilm is fully formed, although initial corrosion is accelerated, later corrosion is significantly inhibited by the BS due to organic acids produced by BS during its physiological activity.

Guo et al. (2022) found that bacterial cellulose (BC) was important in corrosion protection. Endogenous cellulose facilitates the formation of homogeneous biofilms, while exogenous BC provides stable antiseptic activity through biomineralization. Moradi et al. (2015) isolated a new marine inhibitory bacterium (*Vibrio neocaledonicus* sp.). In the presence of these bacteria, the corrosion resistance of carbon steel is increased by more than 60 times. Gao et al. (2021) selected three representative Vibrio species: *V. parahaemolyticus*, *V. alginolyticus*, and *V.* EF187016. Their corrosion inhibition on AA5083 was revealed by electrochemical, surface analysis, and surface characterization techniques. The corrosion of AA5083 was inhibited by each of the three Vibrio species. The biofilm of Vibrio species prevented the pitting of AA5083 and decreased the current density of corrosion. The surface study reveals that the Vibrio species on the surface of AA5083 create a thick, uniform biofilm. The electrochemical results show that the biofilm serves as a barrier against oxygen consumption and the diffusion of corrosive ions. Vibrio biofilms may one day serve as a corrosion-resistant substance for marine environments.

12.4.2 Oil and gas pipelines

The deterioration of metals generated or caused by microbial activity is known as microbiologically influenced corrosion or biocorrosion. Biocorrosion is one of the main problems in the oil and gas industry and has been studied in many industrial systems. As petroleum and its products are complex mixtures containing acids, alkalis, salts, and other corrosive substances, in addition to being exposed to sun and rain, they are prone to corrosion reactions. Moreover, the damage caused by corrosion can easily lead to safety hazards and accidents. It is assumed that more than 75% of corrosion in producing wells and more than 50% of buried pipeline and cable failures are caused by biocorrosion (Purwasena et al., 2019). The annual worldwide corrosion losses from all types of corrosion account for 3%–4% of the total GDP. Inhibiting and resisting corrosion have become important issues for industrial production and the pipeline transmission industry. Owing to its strong corrosion resistance, carbon steel is an ideal material for the gas and oil industries. However, the corrosion resistance has changed with the presence of corrosive microorganisms such as sulfate-reducing bacteria (SRB), acid-producing bacteria (APB), manganese-oxidizing bacteria (MOB), and iron bacteria (IB). Microbial corrosion accounts for approximately 30%—40% of the total corrosion problem in the oil and gas industry (Plaza, 2020).

Silva Faccioli et al. (2022) studied the corrosion inhibition of metals by *Pseudomonas cepacia* CCT 6659 biosurfactant. According to the experimental data, it is a great surface conditioner and can mitigate the growth of biofilms during biocorrosion on metal surfaces. Additionally, the biosurfactant's low toxicity, when mixed with heavy oils dissolved in saltwater, has been proven by ecotoxicological testing. This shows that it is appropriate for both the environment and business. The biosurfactant showed an outstanding ability to promote the biodegradability of petroleum products and demonstrated its usefulness in the bioremediation of contaminated soil and saltwater. Sari et al. (2018) evaluated the effects of switching the carbon source for manufacturing biosurfactants from olive oil to palm oil, a less expensive and fuller vegetable oil. The potential of biosurfactants as corrosion inhibitors for metals was assessed using electrochemical impedance spectroscopy measurements at 303K. The measurements revealed that an inhibition level of 53.23% at a biosurfactant concentration of approximately 200 ppm was the greatest level of inhibition ever recorded.

Rhamnolipid biosurfactant produced by the bacterial strain *Pseudomonas mosselii* F01 was utilized by Parthipan et al. (2018) as an environment-friendly microbial inhibitor. The corrosion inhibition of carbon steel was studied. The results show that biosurfactants have both corrosion inhibitor and biocide properties and can be used as microbial inhibitors to minimize corrosion problems in high-salt environments. Tamilselvi et al. (2022) isolated *Pichia* sp. from *Phyllanthus emblica* and monitored its corrosion inhibition properties through biofilm formation capacity. The results showed that the inhibition efficiency (IE%) tended to increase, and the corrosion rate tended to decrease as the concentration of *Pichia* sp. isolates increased. Finkenstadt et al. (2011) found that a purified exopolysaccharide coating of *Leucococcus enterica* poured from an aqueous solution inhibited the corrosion of mild steel. Dextran-producing bacteria, such as mesophilic bacteria, are specific to a strain in their ability to suppress corrosion. The film's structure, when applied on the substrate, could affect susceptibility of exopolysaccharide to corrosion. Similarly, the coating's bonding to the substrate or the diffusion of corrosive species through the coating is likely to have an impact. Fig. 12.7 illustrates the classification of the various biosurfactant application scenarios into seawater, acidic, chloride, and salt, alkaline, and gas pipeline environments based on the available literature.

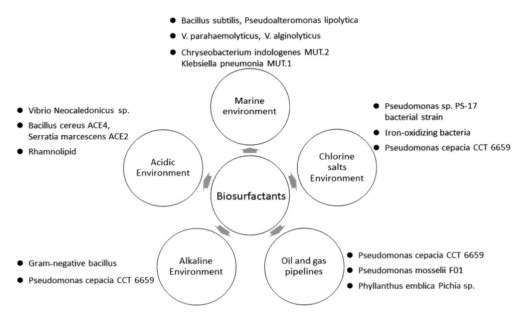

Figure 12.7 Classification chart for biosurfactant application scenarios.

12.4.3 Other environments

Biosurfactants are widely used in addition to inhibiting metal corrosion in marine environments and corrosion in oil and gas pipelines. They are also highly valued in highly alkaline concrete environments, industrial pickling, and aviation fuel storage pipelines.

Pokhmurs'kyi et al. (2014) obtained biosurfactants from the biosynthesis products of *Pseudomonas* sp. PS−17 bacterial strain. The corrosion inhibition performance of D16T aluminum alloy in sodium chloride solution was investigated. The experimental results show that the corrosion inhibition efficiency increases with increasing concentration. Any further increase in the biosurfactant concentration in the corrosive medium will not affect the corrosion protection once it has reached the critical micelle concentration. The basis of the corrosion inhibition mechanism is the adsorption of biosurfactant molecules onto the interface of metal alloys, which causes the development of barrier films. The corrosion behavior and mechanism of carbon steel caused by EPS isolated from iron-oxidizing bacteria cultures in 3.5 wt.% NaCl solution were studied by Liu et al. (2017). At 240 mg/L, 7 days of EPS prevented corrosion while accelerating it at lower and higher concentrations. In addition, EPS facilitated corrosion at 240 mg/L for both 14 and 28 days.

Biosurfactants are also used as corrosion inhibitors in the pickling process. Xiao et al. (2016) investigated the corrosion inhibition of Q235 carbon steel in sulfuric acid by exopolysaccharide of *Vibrio Neocaledonicus* sp. The results show that the Vibrio exopolysaccharides (EPS) can form a dense Fe—EPS layer on the surface of carbon steel. EPS contains electron-bearing groups such as carboxyl and amino groups, which can combine with Fe^{2+} to form an Fe^{2+}—EPS layer adsorbed on the surface of carbon steel. This can effectively inhibit the corrosion of Q235 steel by H_2SO_4 solutions. Zin et al. (2018) investigated the effect of rhamnolipid biosurfactant complexes on the corrosion and repassivation of freshly cut aluminum—copper—magnesium aluminum alloy surfaces. The rhamnolipid biosurfactant was effective in inhibiting alloy corrosion in synthetic acidic rainwater. The corrosion inhibition efficiency increases with increasing biosurfactant concentration.

Biosurfactants are also employed in concrete environments and aeronautical applications. In a new class of chemicals produced by bacterial cells, Shubina et al. (2016) isolated biomolecules of a lipopeptide originating from Gram—negative Bacillus cells. The inhibitory effect in simulated concrete pore solutions was demonstrated using conventional electrochemical measurements, microscopic views, and XPS. This type of biomolecule appears to be a hybrid inhibitor of film formation, according to XPS analysis and SEM observations.

12.4.4 Commercial applications of biosurfactants

Microbial corrosion is one of the main causes of damage and destruction to concrete sewer pipes. The sewage and the sludge layer that accumulates at the bottom of the pipes contain numerous bacteria, including sulfate-reducing and sulfate-oxidizing bacteria, which are very harmful to the concrete. The sulfuric acid produced by their metabolism reacts chemically with the hydration of the cement. On the one hand, the expansion of calcium alumina and gypsum causes micro-cracks in the concrete. On the other hand, the reaction consumes large amounts of $Ca(OH)_2$. Zhengyu et al. (2021) invented a method for upgrading concrete sewage pipes against microbial corrosion using eggshells. Concrete specimens mixed with eggshells are effective in inhibiting the growth and reproduction of harmful bacteria in sewage systems, such as sulfate-reducing and sulfur-oxidizing bacteria, thus reducing the production of biological sulfuric acid and extending the durability of the sewage pipes.

Zhong et al. (2022) invented the rhamnolipid environmental-friendly microbial corrosion inhibitor to inhibit microbial corrosion problems faced by metallic materials in service. Rhamnolipids are used as environmentally friendly microbial corrosion inhibitors. Not only does it inhibit the corrosion of metals itself, but it also effectively inhibits corrosion failure of metallic materials caused by environmental microorganisms. It is also environmentally friendly and non-polluting. Li & Ke, 2012 used electrochemical methods to test the resistance of antimicrobial stainless steel to microbial corrosion. The anti-microbial corrosion properties of the antimicrobial stainless steel were evaluated after different times of interaction with bacteria. This can be used to evaluate the microbial corrosion resistance of antimicrobial stainless steel. The invention solves the problem of being unable to test the microbial corrosion resistance of antimicrobial stainless steel in dry, deep-sea environments. The aim is to provide a more reliable and convenient test of the microbial corrosion resistance of antimicrobial stainless steel.

In addition, biosurfactants are widely used in the composting paper and food industries. For instance, in the medical field, the surfactin biosurfactant produced by *Bacillus subtilis* is an antimicrobial lipopeptide that can be used as an antimicrobial agent. In the cosmetic industry, alkyl sugars have good solubility and detergent ability and are used as raw materials for products such as facial cleansers and body washes.

In conclusion, biosurfactants are very effective in inhibiting metal corrosion and have excellent environmental adaptability in highly alkaline, chloride salt environments, and acidic conditions. Moreover, biosurfactants are aligned with the global strategy of low carbon and green development due to their environmentally friendly synthesis route, low toxicity, rich structure, and superior performance. Biosurfactants, as new green corrosion inhibitors, have become one of the hot spots in the field of corrosion inhibitor research.

12.5 Characterization methods for biosurfactants

With the development and advancement of science and technology, a rich variety of methods are available for the characterization of biosurfactants. The main techniques commonly used for surface analysis of micro-morphological changes induced by biosurfactants are SEM, X—ray photoelectron spectroscopy (XPS), atomic force microscopy (AFM), and contact angle measurement methods (CAM). For component characterization, Fourier infrared spectroscopy (FTIR) and gas chromatography—mass spectrometry (GC—MS) coupling are often used to represent the surfactant components that inhibit corrosion on the material surface. For effectiveness of biosurfactants in corrosion inhibition, electrochemical methods are often used to clarify corrosion behavior and determine the corrosion inhibition mechanism based on EIS and Tafel assay analysis. The stabilized conformations and electron distribution information of the molecules are calculated by theoretical computational techniques.

12.5.1 Surface analysis techniques

For high-resolution micro-area morphology study for various applications, the SEM is a powerful precision tool. On observing the surface of biosurfactant-treated metallic materials by SEM, Guo et al. (2017) found that steel specimens in sterile seawater formed a dense and smooth film with many cracks on the steel surface. Qu et al. (2015) observed that *Bacillus subtilis* caused a tighter surface for bacterial deposition, as seen by SEM, and no significant pitting was found once the film was removed.

CAM refers to a drop of liquid on a solid horizontal surface, the solution phase at the solid–liquid–gas junction point on the solid surface, the gas–liquid interface, and the tangent line of the solid–liquid interface clamped between two angles. Van Oss et al. (1988) determined the energetic properties of stainless steel interfaces with and without adsorbed biosurfactant coatings using the fixed drop technique. Meylheuc et al. (2006) used CAMs on biosurfactant-treated metals. The free energy of the surface computed by contact angle shows that more electrons are supplied to the biosurfactant-treated stainless steel than to the bare stainless steel.

AFM obtains information on the force distribution by scanning the sample and using sensors to detect these changes, obtaining information on the structure of the surface morphology and surface roughness with nanoscale resolution. In their study of the surface morphology of the samples, Meylheuc et al. (2006) observed by AFM that the surface was covered with a layer of biosurfactant after conditioning with *P. fluorescens* 495 biosurfactant.

12.5.2 Component characterization techniques

FTIR is used to obtain the infrared emission and absorption spectra of solids, liquids, or gases, which allows the simultaneous collection of spectral data over a wide range, thus enabling the accurate characterization of molecular functional group information. In the field of biosurfactants, they are widely used to assist in the analysis of their inhibition mechanisms. The effective functional groups in *Halomonas meridiana* BK—AB4 were identified by FTIR by Sari et al. (2018). The biosurfactant's structure is comparable to that of a biosurfactant based on fatty acids.

GC—Ms is a combination of the properties of gas spectrometry and gas chromatography to determine the various substances in a sample. Its main applications are in industrial testing, food safety, environmental protection, and many other areas. The methodology can also be used in applications in the area of biosurfactants for the identification of their main components. Parthipan et al. (2017) detected using GC—Ms that the biosurfactants studied consisted mainly of fatty acids.

XPS is typically used to identify the relative content of elements in chemical bonds, and Shubina et al. (2016) examined carbon steel materials without and with surfactants using XPS. With the blank carbon steel spectra, strong signals were observed for iron (Fe 2p), oxygen (O 1 s), and carbon (C 1 s), with no nitrogen (N 1 s) signal. The Fe (Fe 2p) signal was less significant for the surface of the surfactant-containing steel, and the presence of N 1 s could be clearly observed. This indicates that the surfactant contains elemental N.

12.5.3 Analysis of inhibition effects

The most traditional approach for assessing the effectiveness of corrosion inhibitors is weight reduction. The weight loss technique calculates the average corrosion rate of a metal sample with and without adding corrosion inhibitor. Conversely, the corrosion inhibition performance of the corrosion inhibitor can be judged. The method has the advantages of simple operation, reliable results, and good reproducibility. However, there are also certain disadvantages: the processing of the specimen is more complex, and for the corrosion rate of the smaller systems, the experimental period is longer.

The electrochemical method can accurately assess the inhibition efficiency of the studied biosurfactants on metals. Electrochemical methods include both EIS and Tafel, which are important detection methods in metal corrosion. By assessing the corrosion resistance of metals, EIS offers data on the adsorption of biosurfactants on metal surfaces. Tafel can explain the corrosion behavior on metal electrode surfaces and determine the corrosion inhibition mechanism of biosurfactants.

Two biosurfactants (AsS and ArS) were synthesized by Fawzy et al. (2021). Electrochemical tests showed that at 900 ppm, the maximum corrosion inhibition efficiencies of AsS and ArS were 88% and 90%, respectively. In addition, both AsS and ArS are mixed-type adsorption corrosion inhibitors. Shubina Helbert et al. (2020) investigated the use of rhamnolipids (RLs) derived from *Pseudomonas aeruginosa* as an environmentally friendly corrosion inhibitor for the corrosion protection of steel reinforcement in concrete. Tamilselvi et al. (2022) investigated the corrosion inhibition potential of Pichia sp. against carbon steel corrosion in 1 mol/L H_2SO_4 as assessed by EIS and PP techniques. It was shown that the highest corrosion inhibition efficiency of 90% was achieved at a temperature of 283K and a Pichia sp. concentration of 9×10^6 CFU/mL. As the concentration of Pichia sp. isolate increased, the charge transfer resistance value also increased, and Pichia sp. acted on the carbon steel surface in a mixed type of adsorption.

12.5.4 Theoretical calculation techniques

Quantum chemical computing is the computational processing of information about molecules through simulation techniques using a computer to obtain information about their stable conformation and electron distribution. With the development of science and technology, density functional theory methods have been widely used. This method provides an efficient analysis of the reaction mechanism of molecules, an accurate description of the chemical properties of molecules, and the ability to predict the excited and excess state configurations of molecules accurately. In recent years, it has become an effective method in biology, materials, and pharmaceuticals. Molecular dynamics simulation (MDS) is one of the robust calculation methods used to validate theories and improve models. MDS has been well established in recent years in metal corrosion inhibitor applications. The interaction of corrosion inhibitor molecules with metal substrates can be analyzed using MDS to obtain their stable adsorption conformation, the binding energy of the corrosion inhibitor molecules to the metal substrate, and the diffusion behavior of the molecules in the corrosion medium.

The inhibitory effect of rhamnolipid biosurfactants (RBCs) on the corrosion of aluminum alloy surfaces was investigated by Zin et al. (2018). Quantum chemical results demonstrated that RBCs effectively prevent corrosion on the surface of aluminum alloys. It alters the reduced oxide coating on the alloy's surface. The anion of the biosurfactant is often absorbed in the outer layer of the film. Rhamnolipids combine with aluminum ions to form poorly soluble complexes that may be applied to alloy local anodes and repassivated to create thicker and more durable resistive layers than those seen in regular films.

Biosurfactants as corrosion inhibitors have been characterized in various ways. These characterization methods can be used to identify the components of biosurfactants in order to determine their inhibition efficiency and analyze their mechanism of action. These works provide useful references for the further use of biosurfactants in industry.

12.6 Conclusion and outlook

In recent years, biosurfactants have received increasing attention in corrosion inhibition. Traditional methods of corrosion protection are relatively well established, but they are often expensive and environmentally harmful. Chemically synthesized protective coatings, for example, are notable for their complicated structure, which includes many hazardous and volatile substances. However, these compounds are inevitably discharged into the air, land, and water during manufacturing and usage. As a result, there is a pressing need to develop a low-cost, less hazardous, and eco-friendly corrosion inhibitor to replace the current corrosion protection method, and biosurfactants make ecological and economic sense. This chapter provides a detailed description of the classification of biosurfactants, which are divided into five main categories, peptide phospholipids, mesolipids, glycolipids, sexual lipids, and polymers, with different mechanisms of action for different types of biosurfactants. The mechanism of action of biosurfactants was elucidated from three aspects: chemisorption, electrostatic attraction adsorption, and film-forming mixed type. The application of biosurfactants is elaborated in detail from application scenarios such as seawater environments and oil pipelines, and microbial surface corrosion inhibitors can effectively mitigate material corrosion in each scenario.

The bulk of biosurfactant study results has been achieved under ideal laboratory conditions and mostly focused on understanding any specific biosurfactant's action mechanism. However, because of the complexity and biological variety of the environment, several corrosion mitigation processes are likely to coexist in actual situations and suppress material corrosion through synergistic effects. The adsorption of biosurfactants is influenced by surfactant concentration and amphiphilicity, as well as surface features such as charge, defects, and composition. Surfactant hydrophilicity or hydrophobicity can cause interfacial adsorption and aggregation.

To understand the main processes by which biosurfactants act as corrosion inhibitors, it is essential to understand the microenvironmental changes and the corresponding changes in surface products. This information is generally obtained by microelectrochemical and surface composition analysis. Furthermore, research on biosurfactants in conjunction with projects such as materials science and chemistry was advised to overcome current challenges with new solutions. Harris et al. (2016) described the biological and chemical processes and spatio-temporal development of material biorepairing using SECM ultramicroscopic sensors. The availability of such methods and data is essential for biosurfactant research.

Biosurfactants are imexpensive and environmentally friendly and agree with the development direction of the next generation of corrosion protection technology. To obtain a more stable and long-lasting inhibition effect, several different biosurfactants can be considered for simultaneous use or the use of biosurfactants in combination with other nonbiological corrosion inhibitors. In summary, the research and application of biosurfactants has high potential in material corrosion protection. Compared with traditional corrosion inhibitors, biosurfactants, as corrosion inhibitors, can effectively relieve the pressure on the natural environment and have significant corrosion inhibition efficiency. It can therefore serve as a promising corrosion inhibitor that minimizes the environmental stress caused by existing corrosion inhibitors.

References

Abouseoud, M., Maachi, R., & Amrane, A. (2007). Biosurfactant production from olive oil by *Pseudomonas fluorescens*. *Communicating Current Research and Educational Topics and Trends in Applied Microbiology*, *1*, 340–347.

Adamu, A., Ijah, U. J. J., Riskuwa, M. L., Ismail, H. Y., & Ibrahim, U. B. (2015). Study on biosurfactant production by two Bacillus species. *International Journal of Scientific Research in Knowledge*, *3*(1), 13–20. Available from https://doi.org/10.12983/ijsrk-2015-p0013-0020.

Aparna, A., Srinikethan, G., & Hedge, S. (2011). Effect of addition of biosurfactant produced by Pseudomonas ssp. on biodegradation of crude oil. *International Proceedings of Chemical, Biological & Environmental Engineering*, *6*, 71–75.

Araújo, H. W. C., Andrade, R. F. S., Montero-Rodríguez, D., Rubio-Ribeaux, D., Alves Da Silva, C. A., & Campos-Takaki, G. M. (2019). Sustainable biosurfactant produced by *Serratia marcescens* UCP 1549 and its suitability for agricultural and marine bioremediation applications. *Microbial Cell Factories*, *18*(1). Available from https://doi.org/10.1186/s12934-018-1046-0.

Astuti, D. I., Purwasena, I. A., Putri, R. E., Amaniyah, M., & Sugai, Y. (2019). Screening and characterization of biosurfactant produced by Pseudoxanthomonas sp. G3 and its applicability for enhanced oil recovery. *Journal of Petroleum Exploration and Production Technology*, *9*(3), 2279–2289. Available from https://doi.org/10.1007/s13202-019-0619-8.

Bayoumi, R. A., Haroun, B. M., Ghazal, E. A., & Maher, Y. A. (2010). Structural analysis and characteristics of biosurfactants produced by some crude oil utilizing bacterial strains. *Australian Journal of Basic and Applied Sciences*, *4*(8), 3484–3498. Available from http://www.insipub.com/ajbas/2010/3484-3498.pdf.

Benincasa, M., Contiero, J., Manresa, M. A., & Moraes, I. O. (2002). Rhamnolipid production by *Pseudomonas aeruginosa* LBI growing on soapstock as the sole carbon source. *Journal of Food Engineering*, *54*(4), 283–288. Available from https://doi.org/10.1016/S0260-8774(01)00214-X.

Bezza, F. A., & Chirwa, E. M. N. (2017). The role of lipopeptide biosurfactant on microbial remediation of aged polycyclic aromatic hydrocarbons (PAHs)-contaminated soil. *Chemical Engineering Journal*, *309*, 563–576. Available from https://doi.org/10.1016/j.cej.2016.10.055.

Bisht, K. S., Gross, R. A., & Kaplan, D. L. (1999). Enzyme-mediated regioselective acylations of sophorolipids. *Journal of Organic Chemistry*, *64*(3), 780–789. Available from https://doi.org/10.1021/jo981497m.

Bluth, M. H., Kandil, E., Mueller, C. M., Shah, V., Lin, Y. Y., Zhang, H., Dresner, L., Lempert, L., Nowakowski, M., Gross, R., Schulze, R., & Zenilman, M. E. (2006). Sophorolipids block lethal effects of septic shock in rats in a cecal ligation and puncture model of experimental sepsis. *Critical Care Medicine*, *34*(1), E188. Available from https://doi.org/10.1097/01.CCM.0000196212.56885.50.

Calvo, C., Manzanera, M., Silva-Castro, G. A., Uad, I., & González-López, J. (2009). Application of bioemulsifiers in soil oil bioremediation processes. *Future prospects. Science of the Total Environment*, *407*(12), 3634–3640. Available from https://doi.org/10.1016/j.scitotenv.2008.07.008.

Chongdar, S., Gunasekaran, G., & Kumar, P. (2005). Corrosion inhibition of mild steel by aerobic biofilm. *Electrochimica Acta*, *50*(24), 4655–4665. Available from https://doi.org/10.1016/j.electacta.2005.02.017.

Costa, S. G. V. A. O., Déziel, E., & Lépine, F. (2011). Characterization of rhamnolipid production by Burkholderia glumae. *Letters in Applied Microbiology*, *53*(6), 620–627. Available from https://doi.org/10.1111/j.1472-765X.2011.03154.x.

da Silva Faccioli, Y. E., da Silva, G. O., da Silva, R. d C. F. S., & Sarubbo, L. A. (2022). Application of a biosurfactant from *Pseudomonas cepacia* CCT 6659 in bioremediation and metallic corrosion inhibition processes. *Journal of Biotechnology*, *351*, 109–121. Available from https://doi.org/10.1016/j.jbiotec.2022.04.009.

Dagbert, C., Meylheuc, T., & Bellon-Fontaine, M. N. (2008). Pit formation on stainless steel surfaces pre-treated with biosurfactants produced by *Pseudomonas fluorescens*. *Electrochimica Acta*, *54*(1), 35–40. Available from https://doi.org/10.1016/j.electacta.2008.02.118.

Dai, F., Zeng, G. M., Yuan, X. Z., Wu, X. H., & Shi, J. G. (2005). Application of biosurfactant in composting of agricultural waste. *Huanjing Kexue/Environmental Science*, *26*(4), 181–185.

Dong, Z. H., Liu, T., & Liu, H. F. (2011). Influence of EPS isolated from thermophilic sulphate-reducing bacteria on carbon steel corrosion. *Biofouling*, *27*(5), 487–495. Available from https://doi.org/10.1080/08927014.2011.584369.

Dubeau, D., Déziel, E., Woods, D. E., & Lépine, F. (2009). Burkholderia thailandensis harbors two identical rhl gene clusters responsible for the biosynthesis of rhamnolipids. *BMC Microbiology*, *9*, 263. Available from https://doi.org/10.1186/1471-2180-9-263.

Elumalai, P., Parthipan, P., Huang, M., Muthukumar, B., Cheng, L., Govarthanan, M., & Rajasekar, A. (2021). Enhanced biodegradation of hydrophobic organic pollutants by the bacterial consortium: Impact of enzymes and biosurfactants. *Environmental Pollution*, *289*, 117956. Available from https://doi.org/10.1016/j.envpol.2021.117956.

Fawzy, A., Abdallah, M., Alfakeer, M., Altass, H. M., Althagafi, I. I., & El-Ossaily, Y. A. (2021). Performance of unprecedented synthesized biosurfactants as green inhibitors for the corrosion of mild steel-37-2 in neutral solutions: A mechanistic approach. *Green Chemistry Letters and Reviews*, *14*(3), 488–499. Available from https://doi.org/10.1080/17518253.2021.1943543.

Finkenstadt, V. L., Côté, G. L., & Willett, J. L. (2011). Corrosion protection of low-carbon steel using exopolysaccharide coatings from Leuconostoc mesenteroides. *Biotechnology Letters*, *33*(6), 1093–1100. Available from https://doi.org/10.1007/s10529-011-0539-2.

Gao, Y., Zhang, M., Fan, Y., Li, Z., Cristiani, P., Chen, X., Xu, D., Wang, F., & Gu, T. (2022). Marine Vibrio spp. protect carbon steel against corrosion through secreting extracellular polymeric substances. *Npj Materials Degradation*, *6*(1). Available from https://doi.org/10.1038/s41529-021-00212-2.

Gao, Yu, Feng, D., Moradi, M., Yang, C., Jin, Y., Liu, D., Xu, D., Chen, X., & Wang, F. (2021). Inhibiting corrosion of aluminum alloy 5083 through Vibrio species biofilm. *Corrosion Science*, *180*, 109188. Available from https://doi.org/10.1016/j.corsci.2020.109188.

Geetha, S. J., Banat, I. M., & Joshi, S. J. (2018). Biosurfactants: Production and potential applications in microbial enhanced oil recovery (MEOR). *Biocatalysis and Agricultural Biotechnology*, *14*, 23–32. Available from https://doi.org/10.1016/j.bcab.2018.01.010.

Ghafari, M. D., Bahrami, A., Rasooli, I., Arabian, D., & Ghafari, F. (2013). Bacterial exopolymeric inhibition of carbon steel corrosion. *International Biodeterioration and Biodegradation, 80*, 29−33. Available from https://doi.org/10.1016/j.ibiod.2013.02.007.

Guo, N., Zhao, Q., Hui, X., Guo, Z., Dong, Y., Yin, Y., Zeng, Z., & Liu, T. (2022). Enhanced corrosion protection action of biofilms based on endogenous and exogenous bacterial cellulose. *Corrosion Science, 194*. Available from https://doi.org/10.1016/j.corsci.2021.109931.

Guo, Z., Liu, T., Cheng, Y. F., Guo, N., & Yin, Y. (2017). Adhesion of *Bacillus subtilis* and *Pseudoalteromonas lipolytica* to steel in a seawater environment and their effects on corrosion. *Colloids and Surfaces B: Biointerfaces, 157*, 157−165. Available from https://doi.org/10.1016/j.colsurfb.2017.05.045.

Harris, D., Ummadi, J. G., Thurber, A. R., Allau, Y., Verba, C., Colwell, F., Torres, M. E., & Koley, D. (2016). Real-time monitoring of calcification process by: Sporosarcina pasteurii biofilm. *Analyst, 141*(10), 2887−2895. Available from https://doi.org/10.1039/c6an00007j.

Hua, Z. (2000). Basic characteristics of biosurfactants produced with Candida Antarctica. *Detergent and Cosmetics, 23*(1), 78−80.

Jin, J., Wu, G., Zhang, Z., & Guan, Y. (2014). Effect of extracellular polymeric substances on corrosion of cast iron in the reclaimed wastewater. *Bioresource Technology, 165*(C), 162−165. Available from https://doi.org/10.1016/j.biortech.2014.01.117.

Joshi, S. J., Al-Wahaibi, Y. M., Al-Bahry, S. N., & Elshafie, A. E. (2016). Production, characterization, and application of *Bacillus licheniformis* W16 biosurfactant in enhancing oil recovery. *Frontiers in Microbiology, 7*, 1853.

Joshi, P. A., & Shekhawat, D. B. (2014). Screening and isolation of biosurfactant producing bacteria from petroleum contaminated soil. *Pelagia Research Library, 4*(2248-9215), 164−169.

Jun, X., & Yu, D. (2012). Biological surfactant phospholipids and their application. *Leather and Chemicals, 29*, 31−35.

Kim, P. I., Bai, H., Bai, D., Chae, H., Chung, S., Kim, Y., Park, R., & Chi, Y. T. (2004). Purification and characterization of a lipopeptide produced by *Bacillus thuringiensis* CMB26. *Journal of Applied Microbiology, 97*(5), 942−949. Available from https://doi.org/10.1111/j.1365-2672.2004.02356.x.

Kuiper, I., Lagendijk, E. L., Pickford, R., Derrick, J. P., Lamers, G. E. M., Thomas-Oates, J. E., Lugtenberg, B. J. J., & Bloemberg, G. V. (2004). Characterization of two *Pseudomonas putida* lipopeptide biosurfactants, putisolvin I and II, which inhibit biofilm formation and break down existing biofilms. *Molecular Microbiology, 51*(1), 97−113. Available from https://doi.org/10.1046/j.1365-2958.2003.03751.x.

Li, N., & Ke, Y. (2012). Method for testing the microbial corrosion resistance of antimicrobial stainless steel using electrochemical means. China: CN102590298B,2014-05-14.

Li, Z., Zhou, J., Yuan, X., Xu, Y., Xu, D., Zhang, D., Feng, D., & Wang, F. (2021). Marine biofilms with significant corrosion inhibition performance by secreting extracellular polymeric substances. *ACS Applied Materials and Interfaces, 13*(39), 47272−47282. Available from https://doi.org/10.1021/acsami.1c14746.

Liu, H., Gu, T., Asif, M., Zhang, G., & Liu, H. (2017). The corrosion behavior and mechanism of carbon steel induced by extracellular polymeric substances of iron-oxidizing bacteria. *Corrosion Science, 114*, 102−111. Available from https://doi.org/10.1016/j.corsci.2016.10.025.

Lourith, N., & Kanlayavattanakul, M. (2009). Natural surfactants used in cosmetics: Glycolipids. *International Journal of Cosmetic Science, 31*(4), 255−261. Available from https://doi.org/10.1111/j.1468-2494.2009.00493.x.

Meylheuc, T., Methivier, C., Renault, M., Herry, J. M., Pradier, C. M., & Bellon-Fontaine, M. N. (2006). Adsorption on stainless steel surfaces of biosurfactants produced by gram-negative and gram-positive bacteria: Consequence on the bioadhesive behavior of Listeria monocytogenes. *Colloids and Surfaces B: Biointerfaces, 52*(2), 128−137. Available from https://doi.org/10.1016/j.colsurfb.2006.04.016.

Mnif, I., & Ghribi, D. (2016). Glycolipid biosurfactants: Main properties and potential applications in agriculture and food industry. *Journal of the Science of Food and Agriculture, 96*(13), 4310−4320. Available from https://doi.org/10.1002/jsfa.7759.

Mobin, M., Aslam, R., Zehra, S., & Ahmad, M. (2017). Bio-/environment-friendly cationic gemini surfactant as novel corrosion inhibitor for mild steel in 1 M HCl solution. *Journal of Surfactants and Detergents, 20*(1), 57−74. Available from https://doi.org/10.1007/s11743-016-1904-x.

Moradi, M., Song, Z., & Tao, X. (2015). Introducing a novel bacterium, Vibrio neocaledonicus sp., with the highest corrosion inhibition efficiency. *Electrochemistry Communications*, *51*, 64–68. Available from https://doi.org/10.1016/j.elecom.2014.12.007.

Nayak, A. S., Vijaykumar, M. H., & Karegoudar, T. B. (2009). Characterization of biosurfactant produced by Pseudoxanthomonas sp. PNK-04 and its application in bioremediation. *International Biodeterioration and Biodegradation*, *63*(1), 73–79. Available from https://doi.org/10.1016/j.ibiod.2008.07.003.

Nogueira, I. B., Rodríguez, D. M., Da Silva Andradade, R. F., Lins, A. B., Bione, A. P., Da Silva, I. G. S., Franco, L. D. O., & De Campos-Takaki, G. M. (2020). Bioconversion of agroindustrial waste in the production of bioemulsifier by *Stenotrophomonas maltophilia* UCP 1601 and application in bioremediation process. *International Journal of Chemical Engineering*, *2020*. Available from https://doi.org/10.1155/2020/9434059.

Nwaguma, I. V., Chikere, C. B., & Okpokwasili, G. C. (2016). Isolation, characterization, and application of biosurfactant by *Klebsiella pneumoniae* strain IVN51 isolated from hydrocarbon-polluted soil in Ogoniland, Nigeria. *Bioresources and Bioprocessing*, *3*(1). Available from https://doi.org/10.1186/s40643-016-0118-4.

Örnek, D., Jayaraman, A., Syrett, B., Hsu, C. H., Mansfeld, F., & Wood, T. (2002). Pitting corrosion inhibition of aluminum 2024 by Bacillus biofilms secreting polyaspartate or γ-polyglutamate. *Applied Microbiology and Biotechnology*, *58*(5), 651–657. Available from https://doi.org/10.1007/s00253-002-0942-7.

Ozdal, M., Gurkok, S., & Ozdal, O. G. (2017). Optimization of rhamnolipid production by *Pseudomonas aeruginosa* OG1 using waste frying oil and chicken feather peptone. *3 Biotech*, *7*(2). Available from https://doi.org/10.1007/s13205-017-0774-x.

Pacwa-Płociniczak, M., Płaza, G. A., Piotrowska-Seget, Z., & Cameotra, S. S. (2011). Environmental applications of biosurfactants: Recent advances. *International Journal of Molecular Sciences*, *12*(1), 633–654. Available from https://doi.org/10.3390/ijms12010633.

Parthipan, P., Elumalai, P., Sathishkumar, K., Sabarinathan, D., Murugan, K., Benelli, G., & Rajasekar, A. (2017). Biosurfactant and enzyme mediated crude oil degradation by Pseudomonas stutzeri NA3 and Acinetobacter baumannii MN3. *3 Biotech*, *7*(5). Available from https://doi.org/10.1007/s13205-017-0902-7.

Parthipan, P., Sabarinathan, D., Angaiah, S., & Rajasekar, A. (2018). Glycolipid biosurfactant as an eco-friendly microbial inhibitor for the corrosion of carbon steel in vulnerable corrosive bacterial strains. *Journal of Molecular Liquids*, *261*, 473–479. Available from https://doi.org/10.1016/j.molliq.2018.04.045.

Pekin, G., Vardar-Sukan, F., & Kosaric, N. (2005). Production of sophorolipids from Candida bombicola ATCC 22214 using Turkish corn oil and honey. *Engineering in Life Sciences*, *5*(4), 357–362. Available from https://doi.org/10.1002/elsc.200520086.

Pereira, J. F. B., Gudiña, E. J., Costa, R., Vitorino, R., Teixeira, J. A., Coutinho, J. A. P., & Rodrigues, L. R. (2013). Optimization and characterization of biosurfactant production by *Bacillus subtilis* isolates towards microbial enhanced oil recovery applications. *Fuel*, *111*, 259–268. Available from https://doi.org/10.1016/j.fuel.2013.04.040.

Plaza, G. (2020). Biosurfactants: Eco-friendly and innovative biocides against biocorrosion. *International Journal of Molecular Sciences*, *21*(6), 2152.

Płaza, G. A., èukasik, K., Wypych, J., Nałecz-Jawecki, G., Berry, C., & Brigmon, R. L. (2008). Biodegradation of crude oil and distillation products by biosurfactant-producing bacteria. *Polish Journal of Environmental Studies*, *17*(1), 87–94.

Pokhmurs'kyi, V. I., Karpenko, O. V., Zin', I. M., Tymus', M. B., & Veselivs'ka, H. H. (2014). Inhibiting action of biogenic surfactants in Corrosive Media. *Materials Science*, *50*(3), 448–453. Available from https://doi.org/10.1007/s11003-014-9741-4.

Purwasena, I. A., Astuti, D. I., Ardini Fauziyyah, N., Putri, D. A. S., & Sugai, Y. (2019). Inhibition of microbial influenced corrosion on carbon steel ST37 using biosurfactant produced by Bacillus sp. *Materials Research Express*, *6*(11). Available from https://doi.org/10.1088/2053-1591/ab4948.

Qing Shan, S., Yi Ben, C., & You Sheng, O. (2006). Glycolipids biosurfactants via microorganisms synthesis. *Fine and Specialty Chemicals*, *14*, 1–5.

Qu, Q., He, Y., Wang, L., Xu, H., Li, L., Chen, Y., & Ding, Z. (2015). Corrosion behavior of cold rolled steel in artificial seawater in the presence of Bacillus subtilis C2. *Corrosion Science*, *91*, 321–329. Available from https://doi.org/10.1016/j.corsci.2014.11.032.

Sachdev, D. P., & Cameotra, S. S. (2013). Biosurfactants in agriculture. *Applied Microbiology and Biotechnology*, *97*(3), 1005–1016. Available from https://doi.org/10.1007/s00253-012-4641-8.

Santhappan, R., & Pandian, M. R. (2017). Characterization of novel biosurfactants produced by the strain fusarium oxysporum. *Journal of Bioremediation & Biodegradation*, *08*(06). Available from https://doi.org/10.4172/2155-6199.1000416.

Sari, I. P., Basyiruddin, M. I., & Hertadi, R. (2018). Bioconversion of palm oil into biosurfactant by halomonas meridiana BK-AB4 for the application of corrosion inhibitor. *Indonesian Journal of Chemistry*, *18*(4), 718–723. Available from https://doi.org/10.22146/ijc.27040.

Sharma, P., Rekhi, P., & Debnath, M. (2022). Removal of heavy metal by biosurfactant producing novel halophilic Staphylococcus sciuri subsp. rodentium strain SE I isolated from Sambhar Salt Lake. *ChemistrySelect*, *7*(37). Available from https://doi.org/10.1002/slct.202202970.

Sharma, P., & Sharma, N. (2020). Microbial biosurfactants-an ecofriendly boon to industries for green revolution. *Recent Patents on Biotechnology*, *14*(3), 169–183. Available from https://doi.org/10.2174/1872208313666191212094628.

Shubina Helbert, V., Gaillet, L., Chaussadent, T., Gaudefroy, V., & Creus, J. (2020). Rhamnolipids as an eco-friendly corrosion inhibitor of rebars in simulated concrete pore solution: Evaluation of conditioning and addition methods. *Corrosion Engineering Science and Technology*, *55*(2), 91–102. Available from https://doi.org/10.1080/1478422X.2019.1672008.

Shubina, V., Gaillet, L., Chaussadent, T., Meylheuc, T., & Creus, J. (2016). Biomolecules as a sustainable protection against corrosion of reinforced carbon steel in concrete. *Journal of Cleaner Production*, *112*, 666–671. Available from https://doi.org/10.1016/j.jclepro.2015.07.124.

Silva, N. R. A., Luna, M. A. C., Santiago, A. L. C. M. A., Franco, L. O., Silva, G. K. B., de Souza, P. M., Okada, K., Albuquerque, C. D. C., da Silva, C. A. A., & Campos-Takaki, G. M. (2014). Biosurfactant-and-bioemulsifier produced by a promising Cunninghamella echinulata isolated from caatinga soil in the Northeast of Brazil. *International Journal of Molecular Sciences*, *15*(9), 15377–15395. Available from https://doi.org/10.3390/ijms150915377.

Stadler, R., Wei, L., Fürbeth, W., Grooters, M., & Kuklinski, A. (2010). Influence of bacterial exopolymers on cell adhesion of Desulfovibrio vulgaris on high alloyed steel: Corrosion inhibition by extracellular polymeric substances (EPS). *Materials and Corrosion*, *61*(12), 1008–1016. Available from https://doi.org/10.1002/maco.201005819.

Tamilselvi, B., Bhuvaneshwari, D. S., Padmavathy, S., & Raja, P. B. (2022). Corrosion inhibition of Pichia sp. biofilm against mild steel corrosion in 1 M H2SO4. *Journal of Molecular Liquids*, *359*. Available from https://doi.org/10.1016/j.molliq.2022.119359.

Thaniyavarn, J., Chianguthai, T., Sangvanich, P., Roongsawang, N., Washio, K., Morikawa, M., & Thaniyavarn, S. (2008). Production of sophorolipid biosurfactant by *Pichia anomala*. *Bioscience, Biotechnology, and Biochemistry*, *72*(8), 2061–2068. Available from https://doi.org/10.1271/bbb.80166.

van Oss, C. J., Chaudhury, M. K., & Good, R. J. (1988). Interfacial Lifshitz—van der Waals and Polar Interactions in Macroscopic Systems. *Chemical Reviews*, *88*(6), 927–941. Available from https://doi.org/10.1021/cr00088a006.

Wadekar, S. D., Kale, S. B., Lali, A. M., Bhowmick, D. N., & Pratap, A. P. (2012). Microbial synthesis of rhamnolipids by *Pseudomonas aeruginosa* (ATCC 10145) on waste frying oil as low cost carbon sourCE. *Preparative Biochemistry and Biotechnology*, *42*(3), 249–266. Available from https://doi.org/10.1080/10826068.2011.603000.

Wakamatsu, Y., Zhao, X., Jin, C., Day, N., Shibahara, M., Nomura, N., Nakahara, T., Murata, T., & Yokoyama, K. K. (2001). Mannosylerythritol lipid induces characteristics of neuronal differentiation in PC12 cells through an ERK-related signal cascade. *European Journal of Biochemistry*, *268*(2), 374–383. Available from https://doi.org/10.1046/j.1432-1033.2001.01887.x.

Wang, Z., Li, Y., Xu, W., Yang, L., & Sun, C. (2019). Analysis of global research status and development trends in the field of corrosion and protection: Based on bibliometrics and information visualization analysis. *Journal of the Chinese Society of Corrosion and Protection*, *39*(3), 201−214. Available from https://doi.org/10.11902/1005.4537.2018.123.

Xiao, T., Masoumeh, M., Song, Z., Yang, L., Yan, T., & Hou, L. (2016). Inhibition effect of exopolysaccharide of Vibrio neocaledonicus sp. on Q235 carbon steel in sulphuric acid solution. *Journal of the Chinese Society of Corrosion and Protection*, *36*(2), 150−156. Available from https://doi.org/10.11902/1005.4537.2015.037.

Zhang, J., Wang, M., Liu, C., & Fang, Z. (2021). The corrosion inhibition performance and mechanism of rhamnolipid for X65 steel in CO2-saturated oilfield-produced water. *Journal of Surfactants and Detergents*, *24*(5), 809−819. Available from https://doi.org/10.1002/jsde.12478.

Zhang, Y., Jia, D., Sun, W., Yang, X., Zhang, C., Zhao, F., & Lu, W. (2018). Semicontinuous sophorolipid fermentation using a novel bioreactor with dual ventilation pipes and dual sieve-plates coupled with a novel separation system. *Microbial Biotechnology*, *11*(3), 455−464. Available from https://doi.org/10.1111/1751-7915.13028.

Zhengyu, Z., Hongqiang, C., Yi, X., Jianfeng, W., Yue, G., Mingzhi, G., Xuanlin, L., & Linhua, J. (2021). A method for upgrading concrete sewer pipes to resist microbial corrosion. China: CN114394785A,2022-04-26.

Zhong, L., Dake, X., Xinyi, Y., Zhengtao, L., & Fuhui, W. (2022). A rhamnolipid as an environmentally friendly microbial corrosion inhibitor application as an environmentally friendly microbial corrosion inhibitor. *Corrosion Science*, *204*, 110390.

Zhu, P., Long, G., Ni, J., & Tong, M. (2009). Deposition kinetics of extracellular polymeric substances (EPS) on silica in monovalent and divalent salts. *Environmental Science and Technology*, *43*(15), 5699−5704. Available from https://doi.org/10.1021/es9003312.

Zin, I. M., Pokhmurskii, V. I., Korniy, S. A., Karpenko, O. V., Lyon, S. B., Khlopyk, O. P., & Tymus, M. B. (2018). Corrosion inhibition of aluminium alloy by rhamnolipid biosurfactant derived from pseudomonas sp. PS-17. *Anti-Corrosion Methods and Materials*, *65*(6), 517−527. Available from https://doi.org/10.1108/ACMM-03-2017-1775.

CHAPTER 13

Antimicrobial and anti-biofilm potentials of biosurfactants

John Adewole Alara[1] and Oluwaseun Ruth Alara[2]
[1]St. John of God Accord, Greensborough, VIC, Australia
[2]School of Property, Construction and Project Management, RMIT University, Melbourne, VIC, Australia

13.1 Introduction

Biosurfactants are an extensive range of surface-active structurally different organic molecules formed by several eukaryotic and prokaryotic organisms. These molecules are usually excreted on microbial cell surfaces and contain amphiphilic compounds of both hydrophobic and hydrophilic moieties such as acid, mono-, di- or polysaccharides, or unsaturated or saturated hydrocarbon chains (Gudiña et al., 2013). They are usually classified based on their chemical structure, molecular weight, and mode of action. The most interesting researched biosurfactants are the Lipopeptides (LPs), including fengycin and surfactin, and glycolipids (GLs) such as sophorolipids (SLs) rhamnolipids (RLs), mannosyl erythritol lipids (MELs), and trehalolipid (Abdalsadiq & Zaiton, 2018). Biosurfactants can be effective under various ecological situations and show low toxicity and biodegradable characteristics, which make them be produced from cheap renewable raw materials and owing to their biocompatibility, which makes these molecules gain high interest towards several industrial uses (Satpute et al., 2016).

Biosurfactants have attracted a huge volume of studies in the last two decades as a possible alternative to chemical surfactants in various environmental and industrial uses such as textile, detergent, cosmetic, paint, enhanced oil recovery, food, bioremediation, and agrochemical fields (Ambaye et al., 2021). Nowadays, many studies have resulted in various impacting biochemical actions of biosurfactants, and different medical and pharmaceutical uses have been predicted. Especially their capacity to destabilize membrane integrity and permeability, resulting in leakage and cell lysis, along with their ability to break-up at the interfaces, altering surface features and therefore attacking the microbial adhesion of microbes, which represents the significant roles for antimicrobial and anti-biofilm uses (Fracchia et al., 2015).

The scientific community's attention has been rapidly increased on biosurfactant antimicrobial compounds owing to their intrinsic properties of biosurfactants and antimicrobial agents and their interest regarding the producer strains and less environmental and organismal influence. Glycolipid is one of the powerful antimicrobial biosurfactants derived from the bacteria strain of *Lactobacillus acidophilus* NCIM 2903 associated with human health because they can reduce surface tension to 27 mN/m (Yan et al., 2019). Regarding other bioactive properties of these molecules produced from bacteria-related human health, several studies have proposed that the antimicrobial biosurfactants derived from Lactobacillus species could be utilized to prevent or treat hospital-acquired infections. These novel antimicrobial biosurfactants compounds can produce an antagonistic action against vaginal, urinary, skin, and gastro-intestinal tracts infections; an extremely little cytotoxic activity on lung epithelial cells and an antimicrobial, ant-biofilm, and anti-adhesive effect towards clinical multi-drug resistance strains (Sambanthamoorthy et al., 2014). Besides, antimicrobial biosurfactant agents can be applied on different surfaces of medical devices as anti-biofilm, anti-adhesive and antimicrobial agents or to control microbial overgrowth in the food and nutraceutical company (Mohd Isa et al., 2020) (Satpute et al., 2016).

Current studies have shown that biosurfactants derived from *L. casei* could inhibit the biofilm formation of Staphylococcus aureus (Merghni et al., 2017). Infectious microbial biofilm can cause damaging effects on human health. In today's world, bacterial resistance to antibiotic treatment has become a prominent challenge for physicians. For instance, methicillin-resistant Staphylococcus aureus (MRSA) has been a serious concern due to the reduced effectiveness of antibiotic choices, hospital hygiene, and biofilm-related clinical bacterial resistance strains (Banat et al., 2014). Biofilm can be described as microbial communities that attach to various surfaces and with ability to produce high resistance to several antimicrobial drugs. Some biofilms can also undergo phenotypic alteration due to chemotherapy leading to increased resistance. The intrinsic resistance of biofilms and their extensive participation in implant-associated diseases has given rise to studies for developing anti-biofilm and anti-adhesive agents. Microbial-based biosurfactants can destroy biofilms on clinical implants because of their anti-biofilm, antimicrobial and anti-adhesive potential.

Currently, different methods, which were majorly regarded disturbed with involvement against bacterial attachment, quorum sensing interference, and interference of biofilm structure, have been utilized in preventing the formation of dangerous biofilms. However, the formation of advantageous biofilms is vitalized via manipulating adhesive surfaces, environmental conditions, and quorum-sensing signals (Muhammad et al., 2020). Furthermore, to our best knowledge, biofilms and the potential role of biosurfactants within have become a highly significant topic of study but, yet, have not been the subject of a review article. Thus, this chapter discusses the potential roles of biosurfactants as antimicrobial and anti-biofilm agents, their contribution to the inhibition of biofilms, biofilm characteristic, and monitoring and quantification of this dispersal.

13.2 Biosurfactants as an antimicrobial agent and their mechanisms of action

In recent years, there has been an increase in the number of works that have revealed that biosurfactants accommodate several biological activities which can be exploited by both biomedical and pharmaceutical fields. Biosurfactant's mechanism of action on microbial cell surfaces requires binding to membranes, leading to alteration in surface energy and wettability and surface, resulting in a decrease in hydrophobicity and an increase in permeability via the release of lipopolysaccharide and development of transmembrane pores. Hence, they destroy membrane integrity, resulting in cell lysis and metabolite leakage; lack of membrane functions; and inhibition of protein structures (Cortés-Sánchez et al., 2013; Mandal et al., 2013). The development of basic mechanisms of action for biosurfactants can be important to help discover impacting applications. Biosurfactant glycolipids and lipopeptides are the two biosurfactants with major potent antimicrobial activities, and they show significant sources to identify new antibiotics.

The emergency demand for novel antimicrobial agents in recent times has remained a big issue because of the newly emerging organisms and other local ones, most of which have become almost insensitive to existing antibiotics (Peterson & Kaur, 2018). Microbial metabolites can be regarded as the main source of molecules classified with strong biological effects, among these are biosurfactants which have been demonstrated as potential replacements or adjuvants to synthetic drugs and antimicrobial agents. Besides their capacity to modulate cell interaction (Díaz De Rienzo et al., 2015) with surfaces, biosurfactants still can inhibit microbial adhesion and biofilm formation, a significant and constantly unsafe manifestation on medical equipment and these biofilms consists of bacterial strains that are repeatedly and highly resistant to adverse ecological problems and antibiotics (Ceresa et al., 2021).

Biosurfactants can produce strong antimicrobial and anti-adhesive activities, making them the better choice for applications utilized to fight against diseases. Sambanthamoorthy et al. (2014) have reported the antimicrobial and anti-biofilm potential of BSs isolated from *Lactobacilli* against multi-drug-resistant pathogens. The results of surface effects for both biosurfactants ranged from 6.2 to 25 mg/mL, with clear zones between 7 and 11 cm. Biosurfactants of both *Lactobacillus jensenii* and *L. rhamnosus* indicated antimicrobial effects against *Escherichia coli*, *Staphylococcus aureus*, and *Acinetobacter baumannii* at 25–50 mg/mL. The same pathogens showed anti-biofilm and anti-adhesive activities at 25 and 50 mg/mL concentrations. In addition, an electron microscope demonstrated that both biosurfactants caused damage to the cell wall for *S. aureus* and membrane destruction for *A. baumannii*.

Díaz De Rienzo et al. (2015) have also studied biosurfactants' antibacterial activities and capacity (Sophorolipids, rhamnolipids) and sodium dodecyl sulfate (SDS) to disrupt biofilms of selected Gram- positive bacteria and Gram-negative bacteria. The results indicated that *P. aeruginosa* PAO1was inhibited by sodium dodecyl sulfate and sophorolipids at concentrations less than 5% v/v, and *E. coli* NCTC 10418 was inhibited by sodium dodecyl sulfate and Sophorolipid at 0.1% and less than 5% v/v, respectively. Sophorolipids, rhamnolipids, and sodium dodecyl sulfate inhibited *Bacillus subtilis* NCTC 10400 at a concentration less than 0.5% v/v; similar action was found with *S. aureus* ATCC 9144.

13.2.1 Glycolipid biosurfactants as antimicrobial agents

Regarding the mechanism of action of glycolipid biosurfactants, a study has shown that the subjection of *P. aeruginosa* to RLs can lead to multiple basic responses of the bacterial cells characterized by a decrease in the overall cellular lipopolysaccharides, an improvement in cell hydrophobicity and modifications in membrane proteins and surface morphology. In addition, the antimicrobial action of SLs requires a mode of action that can lead to alteration and destabilization of the permeability of the cellular membrane (Silveira et al., 2021). Another study has revealed the interaction between microbial biosurfactants trehalose with phosphatidylethanolamine and phosphatidylserine membranes (Ortiz et al., 2011). Another popular activity of biosurfactants is their capacity to mitigate bacterial cell adhesion to surfaces, preventing biofilm formation. The rates of initial adhesion and the number of bacteria attaching to the surface can be investigated by interfacial free energies, electrostatic interactions, availability of particular receptors, and the kinds of biosurfactants formed (Krsmanovic et al., 2021). These scholars have also advocated that biosurfactants could decrease hydrophobicity interaction, reducing the hydrophobicity surface that could finally inhibit bacterial adhesions to surfaces and later disrupt biofilm formation.

13.2.1.1 Sophorolipids

It is generally classified as disaccharide sophoroses, which is β-glycosidically attached to fatty acids. Sophorolipids (SLs) can be derived from non-pathogenic fungi such as *Candida bogoriensis*, *C. apicola*, and *C. bombicola*. Sophorlipids can also show antimicrobial properties the same as other biosurfactants. Their mechanisms of action is not limited only to bacteria, but they can also produce antiviral, anti-mycoplasmal, anti-algal, and antifungal properties. According to an investigation by Sleiman et al. (2009), it stated that SLs showed an inhibition effect against *Corynebacterium xerosis*, *Streptococcus faecium*, *S. aureus*, *S. epidermidis*, *Propionibacteriun acnes*, and *B. subtilis* in concentration between 50—29,000 μg/mL. Co-administration of sophorolipids developed by *Candida bombicola* ATCC 22214 can also increase the antimicrobial effect against *S. aureus* (GPB) and *E. coli* (GNB). Joshi-Navare and Prabhune (2013) have revealed that self-assembled sophorolipids can span the bacterial cell membrane and establish the entrance of antimicrobial agents.

13.2.1.2 Oligosaccharide lipids

They can be produced by the strain of Tsukamurella species which produce certain inhibitory effects against some GPB and GNB. Ortiz et al. (2009) have shown that GL-3, GL-2, and GL-1 inhibited *Bacillus maegaterium* growth at 150, 100 and 50 μg/mL concentrations, respectively. Besides, *Tsukamurella* species DM 44370 formed a mixture of glycolipids when cultured in sunflower oil, but culturing them in a marigold oil can increase the production by 60%, and the proportion can be majorly modified on GL-3.

13.2.1.3 Xylolipids

This is derived by probiotic bacteria like *Lactococcus lactis* utilizing paraffin as a carbon source, and it can produce antibacterial properties against the MDR *E. coli* and *S. aureus* such as MRSA. Saravanakumari & Mani (2010) have reported that xylolipids can also be a powerful potential substitute for dermal and oral administration pharmaceuticals.

13.2.1.4 Cellobiose lipids

The cellobiose lipid produced by *Pseudozyma flocculosa* is known as flocculosin. It's made of a C8-hydroxyl acid and two acetyl groups in the cellobiose by-product. Mimee et al. (2009) reported that it has shown in vitro antifungal property against different pathogenic yeasts that leads to human mycoses, such as *Candida albicans*, *C. lusitaniae*, *Trichosporon asahii*, and *Cryptococcus neoformans*. A study has also reported that it can be effective against MDR-resistant *S. aureus* (MRSA), and its antimicrobial property cannot be modified by resistance mechanism against vancomycin and methicillin (Mimee et al., 2009). The Cellobiose lipids (CLs) secreted by *Crptococcus* humicola shows tetra-O-acetyl-β-cellobiosyloxy-2 —hydroxylhexadecanoic acid as the large product and a small of tetra-O-acetyl-β-cellobiosyloxy-2, 15-hydroxylhexadecanoic acid, whereas the CLs secreted by *Pseudozyma fusiformata* produces majorly 16-[6-O-acetyl-20-O-(3-hydroxyhexanoyl)-b-cellobiosyloxy]-2,15-dihydroxyhexadecanoic acid.

13.2.1.5 Rhamnolipis

Haba et al. (2003) reported that Rhamnolipis (RLs) have demonstrated antimicrobial effects against various fungi and microbes in concentrations between 0.4 and 35 μg/mL. Another investigation done by Haba et al. (2003) reported a mixture of rhamnolipids derived by Pseudomonas aeruginosa 47T2 to have shown great antimicrobial activity, and low minimum inhibition concentration (MIC) results were seen in phytopathogenic fungal spp. *Fusarium solani* (75 μg/mL), *Chaetonium globosum* (64 μg/mL), *Gliocadium virens* (32 μg/mL), and *P. funiculosum* (16 μg/mL), and *Staphylococcus epidermidis* and *S. aureus* (32 μg/mL), *Bacillus subtilis* (16 μg/mL), *Enterobacter aerogenes* (8 μg/mL), *S. marcescens* (4 μg/mL), and *Klebsiella pneumoniae* (0.5 μg/mL). Abdel-Megeed et al. (2011) have revealed that BS produced by *Rhodococcus erythropolis* showed great inhibition against

P. aeruginosa, E. coli, A. flavus, and A. niger. Based on the (de Araujo et al., 2011) finding, rhamnolipids derived by P. aeruginosa show between 41% and 71% inhibition effects against bacterial adherence on polystyrene surfaces as tested on Listeria monocytogenes.

13.2.1.6 Mannosylerythritol lipids

Mannosylerythritol lipids (MELs) comprises the hydrophilic group of 4-O-β-D-mannopyr-anosylmesoerythritol and hydrophobic moiety of a fatty acid. The fatty acid can be long, between 10 and 18 carbon atoms, or short, between 2 and 8 carbon atoms. MEL has been revealed to be produced by Ustilago species as a small constituent along with cellobiose (Santos et al., 2017). The mannosylerythritol lipids type differs with the species and the strain of Pseudozyma. Therefore, its development is a significant taxonomic index in identifying Pseudozyma yeasts. The mannosylerythritol can be categorized as MEL-A, MEL-B, MEL-C, and MEL-D were given the level of acetylation at the fourth and sixth carbon positions and their order of outlook on thin-layer chromatography. MEL-A can be regarded as a diacetylated molecule; MEL-B and C are monoacetylated compounds at carbon 6 and 4, respectively, and MEL-D is known to be a perfectly deacetylated compound. In addition, the combination of these four mannosylerythritol lipids, but majorly MEL-A and B, secreted from Candida antarctica T34 demonstrated antimicrobial, surface-active, and antifungal activity against C. albicans (Santos et al., 2017). Santos et al. (2017) have also reported that Schizonellin A and Schizonellin B isolated from S. melanogramma produced effective antimicrobial and antifungal activity against GPB, GNB, and fungi.

13.2.1.7 Trehalose lipids and succinyl trehalose lipids

TL-1 and TL-2 (Trehalose lipids) have been reported to not produce an inhibitory effect against yeast and GNB, although TL-11 shows an antifungal effect by reducing the growth of Glomerella cingulata at a concentration of 300 mg/L (Morita et al., 2009). Santos et al. (2017) investigations have revealed that STL-1 and STL-2 (succinyl trehalose lipids) produced antiviral and antifungal effects against influenza and herpes simplex viruses at 33 mg/L. Another study by Vollbrecht et al. (1999) reported TLs produced by Tsukamurella sp. DSM 44370 combined with tetrasaccharide and trisaccharide lipids demonstrated an incomplete effect against GPB except S. aureus, but there were no inhibitory effects against the GNB.

13.2.2 Lipopeptide biosurfactants as antimicrobial agents

The bactericidal activity of lipopeptides like fengycin and surfactin can be caused by their ability to self-associate and form micellular pore-bearing channels within the lipid membrane. Due to these activities, LPs generally lead to membrane obstruction, increased membrane permeability, metabolite leakages, and cell lysis. In addition, changes in membrane structure and interference of protein conformations can alter important membrane functions such as transport and power generation (Deleu et al., 2008; Liza Kretzschmar & Manefield, 2015). Investigations on daptomycin indicated that the lipopetide oligomer binding that depends on Ca^{2+} can frequently cause the production of pores inside the membranes. These pores can cause the disruption of membranes and cell death due to the influx of transmembrane ions such as Na^+ and K^+. More so, the antimicrobial property of lipopeptides can be increased with the appearance of a lipid tail length of 10–12 carbon atoms, but an increased antifungal property can be produced in lipopeptides with a lipid tail length of 14–16 carbon atoms.

The most active biosurfactant, surfactin, can obstruct the permeability and integrity of membranes by disorganizing them. Surfactin can develop changes in the physical structure of the membrane or protein structures that can alter some main functions of the membrane, such as transport and generation of power (Liza Kretzschmar & Manefield, 2015). One of the important stages for cell membrane destabilization and leakages is the dimerization of surfactin into its bilayer. Regarding antiviral properties, surfactants can directly act on the major viral envelope, leading to complete destruction or leakages, exposing the capsid to loss of infectivity (Liza Kretzschmar & Manefield, 2015).

Other lipopeptide properties and mechanisms of action have been reviewed. Polymyxins majorly produced their powerful antimicrobial actions against Gram-negative bacteria (GNB) by binding to lipopolysaccharide (LPS), obstructing the cellular outer membrane, permeability, and interference of the inner membrane (Moubareck, 2020). Octapeptin A and B can produce a broad-spectrum effect to kill Gram-positive bacteria (GPB) and Gram-negative bacteria (GNB); they can also show antimicrobial effects in some filamentous protozoa, yeast, and fungi because of their capacity to interfere with the cytoplasmic membrane. Moreover, the insulin molecules can produce an antifungal effect via the interaction between the sterol constituents within the fungal membrane, thereby causing an increase in potassium ion (K^+) permeability. The role of surfactant attachment to the GPB or GNB cell wall can possibly be explained due to their selective effect on both kinds of cells, but this fact remains.

13.2.3 Glycoprotein biosurfactants as antimicrobial agent

According to literature retrieval, glycoproteins biosurfactants can be produced only from the strains of Lactobacillus. A study examined the potential of 3 strains of *Lactobacillus* such as *L. plantarum* G88, *L. cellobiosus* TM1, and *L. delbrueckii* N2, to develop biosurfactants during their growth under glycerol or sugar cane molasses (Mouafo et al., 2018). Their extracts showed between 2.40–3.0 g/L on sugar molasses with 49.90%–61.80% of E24 and 2.3–2.81 g/L on glycerol with 41.80%–61.80% of E24. The compounds developed from glycerol growth were made of a larger fragment of lipids than the biosurfactant produced on sugar cane molasses. This result proposed that *Lactobacilli* could manage the glycerol through gluconeogenesis and lipolysis pathways, producing more lipids. However, *Lactobacillus delbrueckii* N2 and *L. cellobiosus* TM1 growths on sugar cane molasses can produce glycoproteins with no single lipid fraction. The bactericidal activity showed that GPB was more sensitive than GNB. For instance, the growth of *Bacillus* sp. BC1 was mostly influenced by the effect of *L. delbrueckii* N2 glycolipid biosurfactant, showing an inhibition zone of 57.5 mm.

13.3 Antimicrobial properties of biosurfactants

Owing to the increase in antibiotic resistance, the demand for discovering new antimicrobials and finding a way to renew recent antibiotics utilized in medicine has become clear. There have been global efforts, internationally and nationally, to solve the problem of antibiotic resistance. Biosurfactants are preferable to solve this challenge based on their applications as bacteriostatic, bactericidal, anti-biofilm, anti-adhesive, synergistic, and adjuvant actions with antibiotics Table 13.1. The antimicrobial properties of sophorolipid depends on its structure and the kind of bacteria investigated. Sophorolipids have been reported to produce virucidal and antibiotic adjuvant properties (Borsanyiova et al., 2016). An investigation reported that SL mixtures with different sugar head groups showed antimicrobial effects against mostly GPB (Naughton et al., 2019). Other scholars have reported that SLs blocked the lethal action of septic shock in rats in a cecal ligation and puncture model of experimental sepsis (Bluth et al., 2006). But, a study has shown that sophorolipids produced from *Starmerella bombicola* can enhance sepsis survival (Hardin et al., 2007). Furthermore, di-rhamnolipid productions have been reported to successfully treat chronic decubitis ulcers and improve the healing of complete-thickness burnt wounds (Naughton et al., 2019).

13.3.1 Antibacterial properties of biosurfactants

An investigation has reported the antibacterial effect of biosurfactants formed by *S. saprophyticus* SBPS 15 against *E. coli*, *Bacillus subtilis*, *Klebsiella pneumonia*, *Vibrio cholera*, and *S. aureus* (Mani et al., 2016). Dusane et al. (2011) reported that rhamnolipid showed biofilm disruptive action against *B. pumilus*. The biosurfactant surfactin has been reported to show an antibacterial effect against *Listeria monocytogenes* growth in food and some GPB, such as *Bacillius pumulis* and *Mariniluteicoccus flavus* (Naughton et al., 2019). Lipopeptides biosurfactants can destroy and gain access to lipids made of negatively charged cell membranes. The rhamnolipids biosurfactants have shown clear proof that they can reduce bacterial growth in the exponential stage and this has suggested that these biosurfactants can affect ordinary cell division. Several studies have also proposed that RLs can be a more effective antibacterial effect against GPB than GNB because of the appearance of an outer membrane in GNB, which assists in excluding

Table 13.1 Antimicrobial properties of different microbial-based biosurfactants against infectious pathogens.

S/N	Types of biosurfactant	Name of pathogen producing biosurfactant	Antimicrobial property		References
			Antibacterial effect	Antifungal effect	
1	Lipopeptide	*Bacillus licheniformis* strain M104	*B. subtilis, B.thuringiensi, B. cereus, S. aureus, E. coli, P. aeruginosa, S. typhimurium, P. vulgaris* and *K. pneumoniae*	*Candida albicans*	Gomaa (2013)
2	Glycolipoprotein	*Lactobacillus acidophilus* NCIM 2903	*E. coli* NCIM 2065, *Proteus vulgaris* NCIM2027, *B. substilis* MTCC2423, *P. putida* MTCC 2467		Satpute et al. (2016)
3	Crude surfactin and rhamnolipid	*Bacillus amyloliquefaciens* ST34 and *Pseudomonas aeruginosa* ST5	Antibiotic resistant *S. aureus* and *E. coli* strains	Pathogenic yeast *C. albicans*	Ndlovu et al. (2017)
4	Lipopepetide	*Bacillus substilis* C 19	It cannot inhibit *Listeria monocytogenes, Salmonella enterica* Typhi, *S. aureus, P. aeruginosa,* and *E. coli*	*Candida albicans*	Yuliani et al. (2018)
5	Sophorolipids, rhamnolipid, and sodium dodecyl sulfate	Obtained from Jeneil Biosurfactant Inc. Saukville, (Wisconsin), MG intobio Co Ltd (Incheon, South Korea)\	Selected GPB and GNB such as *S. aureus* ATCC 9144, *P. aeruginosa* PAO1, *B. subtilis* NCTC 10400 and *E. coli* NCTC 10418		Díaz De Rienzo et al. (2015)
6	Not mentioned	*Bacillus substilis* SPB1	*Pseudomonas stutzeri, P. aeruginosa, E. coli, K. pneumoniae, Enterobacter faecium, Salmonella typhimirium, Brevibacterium flavum, S. aureus,* and *Enterococcus feacalis*	*Saccharomyces cerevisiae, C.* Ghribi et al. (2012), *Rhizopus oryzae, Aspergillus niger, A. oryzae,*	Ghribi et al. (2012)

(*Continued*)

Table 13.1 (Continued)

S/N	Types of biosurfactant	Name of pathogen producing biosurfactant	Antimicrobial property		References
			Antibacterial effect	Antifungal effect	
7	Not mentioned	*Lactobacillus jensenii*, *L. rhamnosus*	Multi-drug resistance pathogens such as MRSA, *E. coli*, and *Acinetobacter baumannii*	*Penicillium italicum*, *P. notatum*, *Puccinia allii*, *Alternaria alternaria*, and *Peronospora destructor*	Sambanthamoorthy et al. (2014)
8	Not mentioned	*Rhodococcus opacus* R7	*S. aureus* ATCC 6538 and *E. coli* ATCC 29522		Zampoli et al. (2022)
9	Not mentioned	*Halomonas* sp. INV PRT125, *Halomonas* sp. INV PRT 124, *Bacillus* sp. INV FIR48	MRSA ATCC 43300		Patiño et al. (2021)
10	Rhamnolipid	Actinomycetes sp NDYS-1, NDYS-3, NDYS-4	Gram-positive and Gram-negative bacteria such as *S. aureus* and *E. coli*		Kourmentza et al. (2021)
11	Not mentioned	*Streptomyces* spp. 8	*S. aureus*, *P. aeruginosa* and *E. coli*		Al-Hulu (2020)
12	Lunasan	*Candida sphaerica* UCP 0995	*Staphylococcus epidermidis*, *Streptococcus oralis*	*Candida albicans*	Luna et al. (2011)
13	Not mentioned	*Lactobacillus acidophilu*, *L. fermentum*, *L. pentosus*, *L. lactic*, *L. plantarum*, and *L. casei*	*P. aeruginosa* 14T28, *P. florescence*, *P. aeruginosa* ATCC2785, *S. typhimurium*, and *E. coli*		Abdalsadiq and Zaiton (2018)

14	Rhamnolipid	P. aeruginosa	Clostridium perfringens, Staphylococcus spp., Listeria. spp., B. subtilis, Enterobacter aerogenes, Salmonella spp., and Escherichia spp.	Fusarium graminearum, Phytophthora spp. Botrytis spp., Mucor spp., and Phytophthoracapsici	Cochis et al. (2012)
15	Glycolipid	Enterobacter cloacae B14	Bacillus subtilis, tetracycline resistance Serratia marcescens		Ekprasert et al. (2020)
16	lipopeptide	Bacillus circulans	MRSA, Alcaligens feacalis, Proteus vulgaris, and other pathogenic MDR		Ghojavand et al. (2019)
17	Rhamnolipid	Pseudomonas aeruginosa Mr01 and MASH1	Only Gram-positive bacteria	Aspergillus niger, Aureobasidium pullulans, Chaetonium globosum, Penicillium chrysogeum, and P. funiculosum	Lotfabad et al. (2012)
18	Not mentioned	Pseudomonas taenensis	E. coli, S. aureus		Borsanyiova et al. (2016)
19	Lipopeptide	Bacillus mojavensis PTCC 1696	Pseudomonas aeruginosa ATCC 27853		Ghojavand et al. (2019)
20	Not mentioned	Pseudomonas aeruginosa RWI, P. putida RWII, P. fluorescens RWIII and Burkholderia cepacia RWIV strains	Klebsiella pneumoniae, E coli	Candida krusei	

biosurfactant compounds (Naughton et al., 2019). Kourmentza et al. (2021) examined lipopeptide biosurfactants' antibacterial activity against foodborne infectious and spoilage pathogens.

13.3.2 Antifungal properties of biosurfactants

Some patients, like those with medical implants, transplants, and immunocompromised individuals, can be highly susceptible to fungal diseases caused by *Candida albicans* and *C. auris* (Schwartz & Patterson, 2018). Investigation done by Haque et al. (2016) reported that sophorolipid formed from *Starmerella bombicola* MTCC1910 can inhibit the hyphal growth and biofilm formation of *C. albican*, in addition to lowering the viability of reserved biofilms. The combined application of sophorolipid with fluconazole or amphotericin B has been reported to act synergistically against both the formation of biofilm and preformed biofilm. Another study has reported that lipopeptide yields showed antifungal action against *Tricoderma atroviride*, *Fusarium oxysporum*, *F. moniliforme*, and *F. solani* (Sarwar et al., 2018). Jumpathong et al. (2022) study showed that biosurfactants derived from *B. velezensi* PW192 can act as an antifungal biological control agent against *Colletotrichum musae* and *C. gloeosporioides*. Matei et al. (2017) recommended that biosurfactant-forming lactic acid bacteria strains can be a substitute antifungal agent in the food industry. An in vitro study had shown biosurfactant rhamnolipids antifungal activity against *Sporisorium scitamineum* disease on sugarcane.

13.3.3 Antiviral properties of biosurfactants

Recently, many emerging and re-emerging viruses have posed a serious global health challenge. The antiviral properties of biosurfactants have been reported majorly against enveloped viruses than non-enveloped viruses. Giugliano et al. (2021) reported the anti-viral property of rhamnolipids mixtures produced from Antarctic strains of *P. gessadii* M15 against coronaviruses and herpes simplex viruses. A group of scholars determined the broad spectrum of antiviral properties of surfactin cyclic LP biosurfactants and antibiotics produced by *B. subtilis* against different viruses such as murine encephalomycocarditis virus, suid herpes virus, simian immunodeficiency virus, vesicular stomatitis virus, feline calicivirus, herpes simplex virus (HSV), and semliki Forest virus (Vollenbroich et al., 1997).

Sophorolipid biosurfactant was reported as a potential antiviral agent against SAR-CoV-2. The result showed that sophorolipid molecules solubilized the lipid envelope of COVID-19 and inactivated it. It also attenuated the cytokine storm produced by COVID-19 during infection to inhibit the progress of SAR-CoV-2 in patients (Daverey et al., 2021). A similar study has evaluated the antiviral effect of sophorolipids and rhamnolipids alginate complex against human immunodeficiency virus and

HSV, respectively (Liza Kretzschmar & Manefield, 2015). In addition, an antiviral effect against Newcastle infection and bursal infection virus has been proposed for lipopeptides-producing by *Bacillus subtilis* (Liza Kretzschmar & Manefield, 2015).

13.3.4 Anticancer property of biosurfactants

Because of their multi-biophysical activities and structural novelty, glycolipids, lipopeptides, and other biosurfactant classes have become potential broad-spectrum agents for cancer chemotherapy and as safe ingredients in drug delivery formulations. Although, several works have indicated that lipopeptides and glycolipids selectively hinder the proliferation of tumor cells and delay cell membranes leading to cell lysis via the apoptosis pathway (Gudiña et al., 2013). Furthermore, lipopeptides and sophorolipids are the most researched biosurfactants with potential anti-tumor properties. Lipopeptide biosurfactants' fatty acid chains and peptides have been reported to have anticancer properties in vitro (Zhao et al., 2013). A report on lipopeptide-producing biosurfactant from *B. substilis* strains such as surfactin, fengycin, and iturin proposed that they have anticancer properties. Dey et al. (2015) have revealed that iturin showed inhibition against the proliferation of MDa-MB-231 cancer cells. Yin et al. (2013) reported that fengycin can disrupt nonsmall cell lung cancer cell 95D and prevent the growth of xeno-grafted 95D cells in nude mice. In addition, the anticancer mode of action of lipopeptides-producing *Bacillus* strains has been widely investigated, but surfactin was revealed to produce an anti-proliferative activity through apoptosis induction, cell cycle arrest, and survival signaling suppression.

13.4 Microbial formation of biofilm

Biofilm is the complex aggregation of microbial cells associated with a different solid surface in almost an irreversible manner to produce extracellular polymers that aid attachment and matrix formation, leading to changes in the organism's phenotype because of gene transcription and growth rate. The discovery of microbial biofilm was first credited to Van Leeuwenhoek, who used his simple microscope to examine the pathogens on the tooth surface (Choudhary et al., 2020). The beginning of biofilm formation needs some conditions as microorganisms must be able to attach themselves to and move on the surface, investigating their cell density and, finally, producing a 3-D mesh of cells enveloped by external polysaccharides. The DNA, extracellular polysaccharides, and cell membrane protein can also play significant roles in biofilm formation, as shown in. The biofilm structure of the extracellular polymeric substance (EPS) matrix comprises one or more proteins, DNA, and extracellular polysaccharides. The channels within the biofilm enable nutrients, air, and water to reach every part of the structure (Rabin et al., 2015).

13.4.1 Stages of biofilm formation

Stage 1. This is known as the attachment stage: At this stage, the conditioning layer is developed, which contains a loose accumulation of proteins and carbohydrates that become attached to minerals in hard water. It captivates the bacterial cells to become attached to the surface.

Stage 2. It is called the irreversible attachment stage: As conditioning layers are adduced, electrical charge aggregates on the surfaces that attract the bacteria with opposite charges, leading to the irreversible attachment of bacterial cells. These charges are weak, and the bacteria can easily be removed using mild sanitizers and cleansers.

Stage 3. This is regarded as the proliferation stage: At this phase, microorganisms become attached to the surface in addition to one another by secreting an extracellular polymeric substance (EPS) that can attract the microbial cells within a glue-like matrix.

Stage 4. The maturation phase: The biofilm surrounding is made of layers rich in nutrients that promote bacterial growth. Complex diffusion channels in matured biofilms help transport oxygen, nutrients, and other constituents needed for microorganisms' growth and remove waste materials and dead cells (Kokare et al., 2009).

Stage 5. This is the dispersion phase: At this phase, the process of dispersal of biofilms in which rapidly growing cells gently get rid of daughter cells. Although fresh nutrients are kept supplied, the biofilm will continue to grow, but if they are deprived of nutrients, they return to their planktonic mode by unbinding themselves from the surface (Choudhary et al., 2020).

13.4.2 Problems associated with resistant bacterial biofilms

Medical settings have faced many remarkable problems associated with microbial pathogenic biofilms. Biofilm-related diseases can be very difficult to treat; they may require strong, usually antibiotic doses (Wu et al., 2015). A higher concentration of antibiotics is needed to kill this bacterial biofilm than that utilized for eradicating its planktonic counterpart. Infectious biofilms can be related to a series of consistent and chronic diseases, including periodontal infection, otitis media infection, non-healing wounds and skin diseases, urogenital diseases, lung diseases in patients with cystic fibrosis, and chronic rhinosinusitis (Madeo & Frieri, 2013; Zhao et al., 2013). When there is a successful biofilm formation in host tissues, it can be associated with poor immune defense to prevent bacterial colonization. In other respects, the prolonged response of the host defense to combat biofilm may destroy the host tissue, where these microbial biofilms are formed; this can progressively influence the patient's quality of life with chronic diseases (Zhao et al., 2013).

Some patients showed a higher risk of getting biofilm diseases because of underlying infections, including diabetes and cystic fibrosis. They are found to be more susceptible to biofilm formation due to the inadequate capacity of their body to mitigate biofilm development. For instance, compromised wound healing in diabetics can promote pathogenic biofilm formation (Hurlow et al., 2015). Individuals with cystic fibrosis have problems coughing up sputum, which makes the lung a normal place for forming biofilm diseases. Other situations that can promote biofilm formation include exposure of internal parts of the body to medical equipment, including catheters and implants, and lack of oral hygiene (Haba et al., 2003). Besides, biofilms may cause more trouble beyond the area where the biofilms are habituated, owing to the dispersion of microbial cells via the formation of molecules that triggers other infections disregarding infections such as autoimmune and cancer diseases (Johnson et al., 2015). A study has reported that oral biofilm in dentures can serve as a reservoir for pneumonia. Yousif et al. (2015) suggested that bacterial cells formed from biofilm in the central nervous system (CNS) can be dispersed, leading to bacteremia. Associations between biofilm compounds, including bacterial DNA and amyloid curly protein during the formation of biofilm may stimulate an immune activation which produces an infectious response in systemic lupus erythematosus, proposing that amyloids developed by pathogens can be responsible for the progress of lupus (Gallo et al., 2015).

13.5 Biosurfactants as antibiofilm agent and their mechanisms of action

Biosurfactant molecules share various activities that can influence their relationship with biofilms. One popular question about biosurfactants' activities against biofilms was why we still have biofilms since biosurfactants have been considered powerful molecules that majorly cause biofilm inhibition? A recent study has suggested that surface-active molecules can serve a prominent role in the maintenance and formation of biofilms partially via the maintenance of water channels via the biofilm that allows movements of gasses and nutrient, which finally results in the detachment of parts of the biofilm into planktonic mobile forms (Banat et al., 2010). Although, different classes of biosurfactants are currently in use to disrupt biofilm in vitro Table 13.2.

The anti-biofilm effect role of BSs can increase abundantly in combination with caprylic acid that prevents biofilm development of *Pseudomonas aeruginosa*, *Bacillus subtilis*, and *E. coli*; fluconazole can harmoniously inhibit the formation of biofilm and preformed biofilm of *Candida albicans* and sodium dodecyl sulfate an anionic surfactant could resulted in killing *P. aeruginosa* PAO1 biofilms (Haque et al., 2016; Nguyen et al., 2020).

13.5.1 Lipopeptide biosurfactants with anti-biofilm property

Lipopeptides can be regarded as one of the biggest classes of biosurfactants that are effective in dispersing microbial biofilms. They are made of more than three varieties of congener compounds, such as surfactins, fusaricidins, fengycins, and polymixins (Banat et al., 2014). Lipopeptides are structurally made of a hydrophilic peptide joined

Table 13.2 Antibiofilm properties of different microbial-based biosurfactants derived from various pathogens to fight against infectious pathogens.

Types of biosurfactant	Name of pathogen-producing BS	Anti-biofilm effectiveness against	References
Lipopeptide	*L. fermentum* B54	Uropathogens	Velraed et al. (2000)
Lipopeptide	*B. subtilis*	*Salmonella enterica*	Kourmentza et al. (2021)
Lipopeptide	*B. tequilensis*	*S. mutans, E. coli*	Pradhan et al. (2013)
Lipopeptide	*Bacillus* strain SW9	Several strains of bacteria	Wu et al. (2015)
Lipopeptide	Coral-associated bacteria	*P. aeruginosa* ATCC10145	Padmavathi and Pandian (2014)
Lipopeptide	*Bacillus* sp.	Various GNP and GNB	Sriram et al. (2011)
Lipopeptide Pseudofactin II	Arctic bacterium *P. fluorescens* BD5	*Proteus mirabilis, E. feacalis, Candida* spp. and *E. coli*	Janek et al. (2012)
Lipopeptide putisolvin 1 and II	*P. putida* PCL1445	Several *Pseudomonas* spp.	Kuiper et al. (2004)
Lipopeptide -Fengycin	*Bacillus licheniformis* and *B. subtilis*	*Salmonella entrica* and *E. coli*	Rivardo et al. (2011)
Glycolipid	Actinobacterium *Brevibacterium casei* MSA19	*C. albicans* FC1, *E. coli* MTCC 2939, *P. mirabilis* PC1, Hemolytic *Streptoococcus* PC2, *K. pneumoniae* PC3, *P. aeruginosa* MTCC 2453, *Vibrio parahaemolyticus* MTCC 451, *V. vulnicus* MTCC 1445, *V. alginolyticus* MTCC, *V. harveyi* MTCC 3438, 4439, *V. alcaligenes* MTCC 4442, *Alteromonas* sp. MMD16, *Pseudoalteromonas* sp. MMD19, *Pseudoalteromonas* sp. MMD18, *Thalassomona* ssp. MMD12, *Ruegeria* sp. MMD27	Rivardo et al. (2011)
Glycolipid	*P. aeruginosa*	*Yarrowia* sp.	Dusane et al. (2011)

Biosurfactant	Producer	Target	Reference
Glycolipid	*Lysinibacillus fusiformis* S9	*Streptococcus mutans*, *E. coli*	Pradhan et al. (2013)
Glycolipid	*Seratia marcescens*	*P. aeruginosa*, *C. albicans* and Marine biofouling *Bacillus pumilus*,	Dusane et al. (2012)
Glycolipid	*Lactobacillus rhamnosus*	*B. subtilis*, *E. coli*, *S. aureus*	Patel et al. (2021)
Glycolipid	*P. aeruginosa*	*B. pumilis*	Banat et al. (2014)
Rhamnolipid	*p. aeruginosa* DS10–129	*Strep. salivarius* GB24/9, *Candida tropicalis*, *S. aureus* GB 9/9, *S. epidermidis* GB 9/6	Rodrigues et al. (2006)
Rhamnolipid	*P. aeruginosa* MA01	*Klebsiella pneumoniae*, *P. aeruginosa*, *Bacillus subtilis*, *S. aureus*	Hajfarajollah et al. (2015)
Rufisan	*C. hypolytica*	*Streptococcus* sp.	Rufino et al. (2011)
Mixed biosurfactant Lunsan	*Strep. thermophiles*, *Pseudomonas lactis*	*Candida* sp., *Strep.* sp., *Staphylococcus* sp.	Rizzo et al. (2021)
Sophorolipid	*C. bombicola* ATCC 22214	*B. subtilis*, *Cupriavdus necator* ATCC 17699, *S. aureus* ATCC 9144	Cho et al. (2022)
Protein-rich biosurfactant, high phosphate, and polysaccharide contents.	*Lactobacillus acidophilus* RC14 and *Lactobacillus fermentum* B54 *L. casei* subsp. *rhamnosus* 36 and ATCC 7469	Urogenital pathogenic *Enterococcus faecalis*	Satpute et al. (2016)
Not specified	*Lactobacillus paracasei* A 20	Filamentous fungi, yeast, and bacteria of different strain	Satpute et al. (2016)

to a fatty acid. These hydrophilic peptides are either aliphatic or cyclic. In addition, the fatty acid chains are different in length and conformation, allowing an extensive diversity of structures. Most recent lipopeptides have been revealed to disperse biofilms derived from *Paenibacillus* (Grady et al., 2016).

13.5.1.1 Polymyxin as potential anti-biofilm agents

Polymyxins can be referred to as a group of non-ribosomal synthesized cyclic lipopeptides. Generally, they can be developed as secondary metabolites of *Bacillus*. They have a typical structurally cyclic polypeptide joined to a fatty acid tail. They are also made of astonishing bacterial amino acids called 2, 4-diaminobutyric acid (DAB). They can be recognized for their low clinical spectrum of inhibition against GNB diseases. Polymyxins have various commercially available formulations, such as polymyxin B, polymyxin E (colistin), and Neosporin (Falagas & Kasiakou, 2005; He et al., 2010). They can also be utilized with trimethoprim to treat eye infections (polytrim), and when combined with neomycinto and bacitracin, they form triple antibiotic ointment neosporin. They can be prescribed as the last drug of choice for treating some diseases due to their toxicity. Banat et al. (2014) have reported that polymyxins can be prescribed as topical creams or powders to treat major cases of multiple drug-resistant *K. pneumoniae, A. baumannii*, and *P. aeruginosa*. Another investigation has revealed can be applied to inhibit biofilms of *Pseudomonas aeruginosa* by 99% within 12 h at a concentration of 20 µg/mL and 100% after 24 h (Banat et al., 2014). Although these results may depend on the bacterial viability and not their dispersal, it was recorded that bacterial cells show changes in their morphology. Colistin has been approved as an early invasive treatment in delaying the onset of chronic *Pseudomonas aeruginosa* disease or the combination of colistin inhalation with oral ciprofloxacin can be used to treat intermittent colonization in cystic fibrosis patients (Doring et al., 2000).

Quinn et al. (2012) stated that polymyxin D1 could be effective against mixed bacterial biofilms. They also reported that this complex of biosurfactants can inhibit the biofilm formation of both GPB, including *B. subtilis, Streptococcus bovis*, and *S. aureus* and GNB, including *Pseudomonas aeruginosa*. Most importantly, these biosurfactants were capable of reducing the development of mixed species biofilms such as self-assembling marine biofilm (SAMB) by 99% in co-incubation assays and initially formed mixed SAMB were disrupted by 72% (Quinn et al., 2012). The mode of activity of polymyxins against microbial biofilms is left mostly unexplained. Although Domingues et al. (2012) proposed that the mechanism of action on planktonic bacteria could be associated with their higher affinity for LPS. This instigated the lipopolysaccharide accumulation, increasing the lipopolysaccharide surface charge, resulting in internalization and attaching to the bacterial phosphatidylglycerol-rich membrane leaflets, which later leak cellular components.

13.5.1.2 Surfactin as a potential anti-biofilm agent

D'Auria et al. (2013) revealed that surfactins are one of the major effective biosurfactants initially derived from *Bacillus subtilis*. They are made of a cyclic peptide heptamer bond to a 13—15-carbon, beta-hydroxy fatty acid chain. However, surfactins could be casually cytotoxic by producing haemolytic effects because of their interactions with cellular membranes. A study reported that surfactin can effectively inhibit S. aureus adhesion and surface biofilm development (Liu et al., 2019). Parreira (2020) stated that after 24 h of incubation, biofilm development was significantly inhibited in the presence of surfactin biosurfactant. The results indicated about an 80% reduction in bacterial counts of *Enterobacter cloacae*, *S. aureus*, and *E. coli* after surfactin application. Mireles et al. (2001) have revealed that surfactin can inhibit biofilm growth formed by Salmonella sp. cultivated on urethral catheters and microtiter wells.

13.5.1.3 Putisolvin as a potential anti-biofilm agent

Rhizobacterium *Pseudomonas putida* strain PCL1445 has been reported for the production of the cyclic lipopeptides putisolvin I and II, and they are biosurfactants that play environmentally significant roles such as swarming motility, biofilm development, and solubility of nutrients and maintenance (Dubern & Bloemberg, 2006). Putisolvin I and II are made of the same structure of a 12 amino acid polar peptide head attached to a fatty acid moiety. Nevertheless, these biosurfactants putisolvin I and II have been revealed as effective inhibitory agents in pre and post-addition to biofilms of other *Pseudomonas* sp. strains (Dubern & Bloemberg, 2006).

13.5.1.4 Fengycin-like lipopeptides as potential anti-biofilm agent

Fengycin is regarded as a bioactive cyclic lipopeptide produced from *Bacillus subtilis* and *B. licheniformis*. They are made of 8—10 amino acids attached to a beta-hydroxy fatty acid. Some studies have reported that fengycin-like lipopeptides could take part in the inhibition of biofilms which can be caused by about 90% dispersion of *Staphylococcus aureus* (GPB) biofilms and around 97% dispersion of *E. coli* biofilm (Banat et al., 2014).

13.5.1.5 Pseudofactin as a potential antibiofilm agent

Pseudofactin is currently recognized as a cyclic lipopeptide biosurfactant formed by *Pseudomonas fluorescens* BD5, the strain isolated from freshwater of the Arctic Archipelago of Svalbard. It is a novel molecule with a palmitic acid connected to the terminal amino group of an 8-amino acid in peptide moiety. The C-terminal carboxylic group of the last amino acid can produce a lactone with the hydroxyl of Thr3 called the third amino acid, threonine. Janek et al. (2012) have stated that pseudofactin II can show about 36 to 90% inhibition against the adhesion of five species, including two *C. albicans* strains, *P. mirabilis*, *S. epidermidis*, *E. coli*, *Enterococcus hirae*, and *E. feacalis* of biofilms on three types of surfaces like glass, silicone and polystyrene. Similarly, they have documented that pseudofactin can show Enterococcus an effective dispersal between 26% and 70% on pre-existing biofilms cultivated on non-treated surfaces. They have been reported for causing significant inhibition of the initial adhesion of *hirae*, *E. feacalis*, *Escherichia coli*, and *Candida albicans* to silicone urethral catheters. The overall growth inhibition of *Staphylococcus epidermidis* was noticed at the highest concentration tested (0.5 mg/mL), which leads to 18%–37% of partial inhibition of other bacteria, 8%–9% inhibition of *Candida albicans* fungi growth and about 99% prevention could be accomplished (Janek et al., 2012).

13.5.2 Glycolipid biosurfactants with anti-biofilm property

Glycolipids are natural amphiphiles that have attracted special interest in their bioactivity, diversity, and biodegradability. They have been studied carefully and extensively for biotechnological applications. Glycolipids are made of a carbohydrate moiety and fatty acid, which are joined glycosidic bond that serves hydrophobic and hydrophilic function, respectively.

13.5.2.1 Sophorolipids as a potential anti-biofilm agent

Sophorolipids (SLs) are typical glycolipid biosurfactants because of their homogeneous product in higher yield that has been deeply studied and commercialized by many industries. Sophorolipids molecular structure is made of a long-chain fatty acid tail of 16 or 18 carbon atoms and a dimer carbohydrate of head sophorose, and they can be produced by *Candida* strains. The synergistic relationship between SLs and antibiotics has been investigated as a possible approach to disrupting biofilm by utilizing the LIVE/DEAD BacLight Bacterial Viability Kits technique. This technique applied two nucleic acid stains, the green-fluorescent SYTO 9 stain and red fluorescent propidium iodide (Di Somma et al., 2020). Studies have shown that SLs could effectively inhibit the growth of caries-causing oral bacteria, such as *L. acidophilus*, *L. fermentum*, *Steptococcus sorbrinus*, *S. salivarius*, and *S. mutans*. It has proposed the important potential of SLs in oral hygiene and health (Solaiman et al., 2017). Another study has demonstrated the actions of SLs to disrupt biofilms from *E. coli* (Joshi-Navare & Prabhune, 2013).

Scholars have investigated the anti-biofilm activities of sophorolipids by showing their ability to cause destructive disruption of *P. aeruginosa* PAO1 biofilm in microfluidic channels. The results showed that SLs impose some damage on the bacteria; they reduced the extracellular polymeric substances (EPS) matrix, causing surface detachment and split-up biofilm (Nguyen et al., 2020). Díaz De Rienzo et al. (2015) have studied the antibacterial and anti-biofilm properties of SLs by utilizing different concentrations of sophorolipid. Their results indicated that at concentrations of 5% v/v of SLs, ATCC 17699 and Gram-positive *B. substilis* BBK006 were inhibited with bactericidal action. Ceresa et al. (2021) investigated the effect of SLs against microbial biofilm formation on medical-grade silicone. The results showed that SLs significantly inhibited the cell attachment of both tested strains of *S. aureus* and *Candida albicans*, and it proposes that sophorolipid molecules could possess a possible function as coating agents on clinical-grade silicone devices to prevent fungi and GPB infections.

13.5.2.2 Rhamnolipids as a potential anti-biofilm agent

Rhamnolipids (RLs) are a group of anionic microbial GLs made of mono or di-rhamnose sugars that bond to β-hydroxyalkanoic acids with different carbon chain lengths. They are initially produced from *Pseudomonas aeruginosa*, and analogs can also be isolated from *Burkholderia* strains, *Tetragenococcus koreensis*, *Cellulomonas cellulans*, *Renibacterium salmoninarum*, and *Nocardioides* (Abdel-Mawgoud et al., 2010; Costa et al., 2011; Dobler et al., 2016). RLs have been reported as an effective anti-biofilm agent against *Bordetella bronchiseptica* (Banat et al., 2014). de Araujo et al. (2011) investigated the anti-biofilm/anti-adhesion and antimicrobial activity of rhamnolipids and surfactin. The result indicated that 0.05% (w/v) crude and 0.50% (w/v) purified tested rhamnolipids significantly reduced the development of biofilm on polystyrene surfaces, where the best inhibition results of about 79% were obtained against *Pseudomonas fluorescens* ATCC13525 and L.monocytogenes ATCC7644 (74% inhibition) respectively. Some scholars have reported that RLs could show an anti-biofilm effect against the dermatophytic fungi *Trichophyton metaggrophytes* and *T. rubrum* (Sen et al., 2020).

Another in vitro investigation has reported that rhamnolipid coating inhibits microbial biofilm development on titanium implants (Tambone et al., 2021). Their results indicated that the titanium discs coated with 4 mg/mL of rhamnolipid biosurfactant (R89BS) reduced biofilm formation of *Staphylococcus aureus* by 7%, 47%, and 99% and of *Staphylococcus epidermidis* by 10%, 29% and 54% at 72, 24 and 48 h, respectively. A similar coating was used on commercial implants' three surfaces, resulting in a biomass inhibition greater than 90% for *Staphylococcus aureus* and around 78% for *Staphylococcus epidermidis* at 24 hours (Tambone et al., 2021). Madeo and Frieri (2013) have compared the anti-biofilm activity of mono-RLs and di-RLs on carbon steel submitted to oil-produced water. Based on their results, it was summarized that RLs are effective molecules that can be utilized to prevent biofilm formation on carbon steel metal when submitted to oil-produced water at a greater percentage of mono-RLs is more demonstrated for this usage. Furthermore, RLs derived from the isolation of *P. aeruginosa* MN1 has been revealed to have higher anti-biofilm and anti-adhesive effect than that surfactin (Abdollahi et al., 2020).

13.5.3 Fungi-based biosurfactants with anti-biofilm properties

Several studies have shown that biosurfactants produced from *Candida bombicola* can inhibit biofilm formation (Banat et al., 2014). It forms SLs that destroy the biofilm formation of *Vibrio cholera*. Luna et al. (2011) also reported *Candida sphaerica*, another strain of fungi that can produce lunasan biosurfactant. It prevents adhesion of *Streptococcus sanguis*, *S. agalactiae*, and *P. aeruginosa* up to 80%–90%. Similarly, *C. lypolytica* can be utilized to produce rufisan that can inhibit biofilm formation at an equal concentration of 0.75 μg/mL or more against *Streptococcus mutans* NS, *S. agalactiae*, and *S. aureus* (Luna et al., 2011).

13.5.4 Combinations of complex biosurfactants with anti-biofilm agent property

Biosurfactants can be rarely seen in isolation, and they are frequently related together with congeners with the same physical and chemical properties, which causes the purification process to be either non-economical or exhaustive. Although these complexes of BSs might possess broader beneficial applications than pure molecules. Combinations of BSs have been isolated from *Nerium oleander* and *Robinia pseudoscscia*. According to (Cochis et al., 2012), these secretions have been reported to show inhibition against the attachment of biofilm formation of *Candida albican* on denture prosthesis and silicon at concentrations of 78 and 156 μg/mL. Moreover, other BSs produced from probiotic *S. thermophiles* and *L. lactis* 53 highly inhibit biofilm formation on preconditioned voice prosthesis in the artificial throat and cause a reduction in the airflow resistance which exists on voice prostheses after biofilm development (Banat et al., 2014).

13.5.5 Surface-active secretions mammals with anti-biofilm property

Looking at the chemotherapeutical point of view, the major fascinating aspects of biosurfactants are those developed by humans. These molecules were known only by just n people, but recently, some researchers have reported palate, lung, nasal epithelium clone (PLUNC) protein has anti—film properties. These compounds are majorly generated as a secretory product of epithelial lining the airway tubes inside mammals, including humans. They can be evolutionarily associated with lipopolysaccharide-binding with protein-lipid transfer or LPB family. PLUNC have been suggested to possess new biologically important surface active activities since they positively mitigated surface tension at the air-to-liquid interface inside the aqueous solution, and on top of that, they inhibited biofilm development within the airways colonizing possible *P. aeruginosa* in vitro at relevant physiological concentrations (Banat et al., 2014; Gakhar et al., 2010).

13.6 Current industrial and medical applications and commercialization of biosurfactant compounds with anti-biofilm and antimicrobial property

Medical devices are often required to be free of microorganisms. Much of the equipment utilized can be disposed, and there may not be any problem of biofilm formation on that equipment. But, in the case of reusable clinical equipment such as endoscopes and surgical equipment, there may be no reason for biofilm formation once the instruments are disinfected. The majority of different types of cleaning agents available now may not be useful as effective anti-biofilm agents (Satpute et al., 2016). There has been a huge attraction toward the possibility of biosurfactant molecules in medically associated usages, such as developing and inhibiting bacteria biofilm (Rodrigues et al., 2006).

Rhamnolipid biosurfactants have been reported to show great potential uses out of all the reported biosurfactants. Cochis et al. (2012) stated the involvement of rhamnolipids in maintaining the structural properties of biofilm and its useful utilization in inhibiting biofilm formation on surfaces of different medical instruments. Sophorolipids type of biosurfactant can also serve as an anti-biofilm agent at various concentrations. Several studies have reported that sophorolipid (5% v/v) can produce a bactericidal effect against *B. subtilis* BBK006 and *Cupriavidus necator* ATCC17699 (Dey et al., 2015). A study conducted by Zezzi do Valle Gomes & Nitschke (2012) revealed that rhamnolipid and surfactin can reduce adhesion and inhibit biofilm formation of pathogenic foodborne pathogens. Similarly, they have also investigated the inhibiting actions of biosurfactant produced by *P. aeruginosa* DSVP20 against *Candida albicans* biofilm. Some commercialized biosurfactant-based formulations have been utilized to inhibit biofilm formation Table 13.3.

An extensive variety of BSs have been found to have the potential for dermatological uses such as wound healing, and a study has investigated the in vivo scavenging properties and wound healing potential of *B. subtilis* SPB1 lipopeptide on excision wounds induced in the experimental mouse. They observed an important increase in the percentage of wound closure compared with the group treated with CICAFLORA and untreated group (Zouari et al., 2016). Another study has evaluated the anti-adhesive and antimicrobial properties of cell-bound BSs secreted by *L. pentosus* (PEB), which can be classified as GL compounds against various pathogens found in human skin flora. The effectiveness of *L. pentosus* was compared against the GLs secreted by *L. paracasei* (PAB). The PEB demonstrated antimicrobial properties against *C. albicans*, *E. coli*, *P. aeruginosa*, *S. pyogenes*, *S. agalactiae*, and *S. aureus*, which was compared to *L. paracasei*. Biosurfactants have been seen in natural environments to help in natural oral care. BSs secreted by *S. mitis* in the oral cavity can help disrupt the adhesion of *S. mutans*. Elshikh et al. (2017) have studied the potential of BSs in the oral cavity. Some groups of researchers have determined the potential of *B. subtilis* SPBI

Table 13.3 Demonstrating some commercial microbial biosurfactants-based formulations recently used as anti-biofilm agents.

Type of biosurfactant	Available commercialized biosurfactant	Producer pathogen	Industrial outline as anti-biofilm against	References
Sophorolipidsus	Rufisan	Candida bypolytica	Streptococcus mutans, S. sanguis, staphylococcus aureus	Rufino et al. (2011)
	Sophorolipid	Candida strains	Escherichia coli	Joshi-Navare and Prabhune (2013)
	Lunasan	C. sphaerica	Streptococcus agalactiae, S. sanguis, P. aeruginosa	Luna et al. (2011)
	Sophorolipid	C. bombicola	V. cholerae	Mukherji and Prabhune (2014)
Surfactin complex	Surfactin plus Fusaricidin and Polymyxin D1	Paenibacillus polymyxa	Biofilm formation of S. aureus and E. coli	Quinn et al. (2012)
	Combination of lipopeptides with other anti-biofilm agents	Bacillus licheniformis	Uropathogen Escherichia coli	Rivardo et al. (2011)
Surfactin	Lipopeptides complexes Surfactin	Bacillus cereus B. subtilis	bio Salmonella strains	Pradhan et al. (2013) Tambone et al. (2021)
Fengycin	Fengycin-like lipopeptide	Bacillus licheniformis and B. subtilis	S. aureus, E. coli	Xu et al. (2013)
	Putisolvin I and putisolvin II (cyclic lipopeptides)	Pseudomonas putida	Other pseudomonas spp.	Dey et al. (2015)
Glycolipids	Rhamnolipid biofungicin	P. aeruginosa	Crops attacked by pathogenic fungi	Davitt and Lavelle (2015)
	Rhamnolipid	P. aeruginosa	Bordetella bronchiseptica	Zouari et al. (2016)

lipopeptide in toothpaste formulation and demonstrated that a lipopeptide-based product produced a significant antimicrobial property against S. *tphimurium* and Enterobacter spp. (Bouassida et al., 2018).

Biosurfactants are currently applied as drug delivery systems (DDS) by providing attractive uses like passive immunization, especially with insufficient treatment methods. For example, treating candidiasis has been proving challenging due to inadequate availability of antifungal drugs, chronic side effects, and toxicity in humans (Naughton et al., 2019). It was also revealed that (Naughton et al., 2019) vesicular drug systems (VDS) such as noisomes and liposomes can be utilized to reduce unwanted side effects and target drug delivery Davitt and Lavelle (2015) have reported that liposomes can be potential candidates with huge applicability based on DDS such as vaccination. Liposomes are known to consist of two hydrophobic tails, and their structures can or cannot be made up of cholesterol, while noisomes can be non–ionic surfactant-based vesicles containing a single hydrophobic chain that creates them adequately appropriate for carrier compounds in DDS (Naughton et al., 2019).

13.7 Future trends and conclusions

Biosurfactants are made of compounds that are of huge attraction for several fields of application. They are a potential substitute for synthetic antimicrobial drugs because no study has been conducted on a possible acquisition of resistance to antimicrobial biosurfactants. Pathogens related to human health can produce antimicrobial and anti-biofilm biosurfactants with significant biomedical potential. For instance, applying anti-biofilm and antimicrobial biosurfactant molecules can treat hospital-acquired diseases. In addition, biosurfactants with ant-biofilm activities can be used to eradicate and preserve urogenital diseases as an alternative to traditional antibiotics. Many studies have revealed that microbial biofilms are found mostly in recalcitrant patients' diseases in hospital surroundings, transmission of airborne organisms, and the fouling of industrial surfaces. These threats are highly made worse by the rise of resistant biofilm communities and the shortage of replacement control strategies. Biosurfactants can be considered emergency therapy with inherent antibacterial, antiviral, and antifungal effects that can effectively disrupt microbial biofilms. Thus, their applications can either be on their own or as adjuvants to other antimicrobial drugs, which may show a possible strategy for eradicating biofilm.

The problems related to further production of biosurfactant uses and exploring their potential remain the huge numbers of strategies and assays to this kind of study. Microbial-based biosurfactants are developed as combinations of congeners, and the amounts of congeners will be different based on the pathogen source, growth medium, and conditions. Because of differences in the congener's properties, the applications of the mixtures in an experiment may cause confusing results. There could also be a problem of endotoxin contamination of BSs secreted by GNB, and fewer researchers have taken a cognizant approach to ensure that their experimental materials are free of these increasingly bioactive compounds. Nevertheless, time-consuming and costly bioactivity must be evaluated with pure single congeners. Several assays recently utilized can give different types of information the mechanism of action and mode of action of BS, either in combination with other therapies or singly against infectious pathogens. However, there may be need for a standardized methodologies and strategies related to biosurfactant work. Another issue that has remained a challenge is the proof of different BSs from many pathogens in various conditions.

13.8 Conclusion

Studies have shown the effectiveness of different biosurfactants in antimicrobial and anti-biofilm properties, and their applications in the biomedical, pharmaceutical, and food industries have increased. The biosurfactant potential uses in healthcare therapeutics can be much more promising due to the value-added antimicrobial, anti-adhesive, and anti-biofilm nature of such products and their suitable fringe benefits to human health. The production cost may seem more favorable in biomedical applications owing to production, which is viable on a small scale. It may be more likely that the natural antimicrobial property of various biosurfactants and their capacity to serve as adjuncts to recent therapeutics in the circumstance of the ever-rising threats of antimicrobial resistance could prove the most advantageous.

Conflict of interest

This article has no conflict of interest.

References

Abdalsadiq, N. K. A., & Zaiton, H. (2018). Biosurfactant and antimicrobial activity of lactic acid bacteria isolated from different sources of fermented foods. *Asian Journal of Pharmaceutical Research and Development, 6*, 60–73.

Abdel-Mawgoud, A. M., Lépine, F., & Déziel, E. (2010). Rhamnolipids: Diversity of structures, microbial origins and roles. *Applied Microbiology and Biotechnology, 86*(5), 1323–1336. Available from https://doi.org/10.1007/s00253-010-2498-2.

Abdel-Megeed, A., Al-Rahma, A. N., Mostafa, A. A., & Husnu Can Baser, K. (2011). Biochemical characterization of anti-microbial activity of glycolipids produced by *Rhodococcus erythropolis*. *Pakistan Journal of Botany, 43*(2), 1323–1334. Available from http://www.pakbs.org/pjbot/PDFs/43%282%29/PJB43%282%291323.pdf.

Abdollahi, S., Tofighi, Z., Babaee, T., Shamsi, M., Rahimzadeh, G., Rezvanifar, H., Saeidi, E., Amiri, M. M., Ashtiani, Y. S., & Samadi, N. (2020). Evaluation of anti-oxidant and anti-biofilm activities of biogenic surfactants derived from bacillus amyloliquefaciens and pseudomonas aeruginosa. *Iranian Journal of Pharmaceutical Research, 19*(2), 115–126. Available from https://doi.org/10.22037/IJPR.2020.1101033.

Al-Hulu, S. M. (2020). Antibacterial activity for biosurfactant produced by streptomyces SPP. Isolated from soil samples. *Asian Journal of Pharmaceutical and Clinical Research*, *13*, 109−110. Available from https://doi.org/10.22159/ajpcr.2020.v13i7.37472.

Ambaye, T. G., Vaccari, M., Prasad, S., & Rtimi, S. (2021). Preparation, characterization and application of biosurfactant in various industries: A critical review on progress, challenges and perspectives. *Environmental Technology & Innovation*, *24*, 102090. Available from https://doi.org/10.1016/j.eti.2021.102090.

Banat, I. M., De Rienzo, M. A. D., & Quinn, G. A. (2014). Microbial biofilms: Biosurfactants as antibiofilm agents. *Applied Microbiology and Biotechnology*, *98*(24), 9915−9929. Available from https://doi.org/10.1007/s00253-014-6169-6.

Banat, I. M., Franzetti, A., Gandolfi, I., Bestetti, G., Martinotti, M. G., Fracchia, L., Smyth, T. J., & Marchant, R. (2010). Microbial biosurfactants production, applications and future potential. *Applied Microbiology and Biotechnology*, *87*(2), 427−444. Available from https://doi.org/10.1007/s00253-010-2589-0.

Bluth, M. H., Kandil, E., Mueller, C. M., Shah, V., Lin, Y. Y., Zhang, H., Dresner, L., Lempert, L., Nowakowski, M., Gross, R., Schulze, R., & Zenilman, M. E. (2006). Sophorolipids block lethal effects of septic shock in rats in a cecal ligation and puncture model of experimental sepsis. *Critical Care Medicine*, *34*(1), E188. Available from https://doi.org/10.1097/01.CCM.0000196212.56885.50.

Borsanyiova, M., Patil, A., Mukherji, R., Prabhune, A., & Bopegamage, S. (2016). Biological activity of sophorolipids and their possible use as antiviral agents. *Folia Microbiologica*, *61*(1), 85−89. Available from https://doi.org/10.1007/s12223-015-0413-z.

Bouassida, M., Fourati, N., Ghazala, I., Ellouze-Chaabouni, S., & Ghribi, D. (2018). Potential application of Bacillus subtilis SPB1 biosurfactants in laundry detergent formulations: Compatibility study with detergent ingredients and washing performance. *Engineering in Life Sciences*, *18*(1), 70−77. Available from https://doi.org/10.1002/elsc.201700152.

Ceresa, C., Fracchia, L., Fedeli, E., Porta, C., & Banat, I. M. (2021). Recent advances in biomedical, therapeutic and pharmaceutical applications of microbial surfactants. *Pharmaceutics*, *13*(4). Available from https://doi.org/10.3390/pharmaceutics13040466.

Cho, W. Y., Ng, J. F., Yap, W. H., & Goh, B. H. (2022). Sophorolipids—Bio-based antimicrobial formulating agents for applications in food and health. *Molecules (Basel, Switzerland)*, *27*, 2−24.

Choudhary, P., Singh, S., & Agarwal, V. (2020). *Microbial biofilms*. IntechOpen.

Cochis, A., Fracchia, L., Martinotti, M. G., & Rimondini, L. (2012). Biosurfactants prevent in vitro Candida albicans biofilm formation on resins and silicon materials for prosthetic devices. *Oral Surgery, Oral Medicine, Oral Pathology and Oral Radiology*, *113*(6), 755−761. Available from https://doi.org/10.1016/j.oooo.2011.11.004.

Cortés-Sánchez, A. d. J., Hernández-Sánchez, H., & Jaramillo-Flores, M. E. (2013). Biological activity of glycolipids produced by microorganisms: New trends and possible therapeutic alternatives. *Microbiological Research*, *168*(1), 22−32. Available from https://doi.org/10.1016/j.micres.2012.07.002.

Costa, S. G. V. A. O., Déziel, E., & Lépine, F. (2011). Characterization of rhamnolipid production by Burkholderia glumae. *Letters in Applied Microbiology*, *53*(6), 620−627. Available from https://doi.org/10.1111/j.1472-765X.2011.03154.x.

D'Auria, L., Deleu, M., Dufour, S., Mingeot-Leclercq, M. P., & Tyteca, D. (2013). Surfactins modulate the lateral organization of fluorescent membrane polar lipids: A new tool to study drug:membrane interaction and assessment of the role of cholesterol and drug acyl chain length. *Biochimica et Biophysica Acta—Biomembranes*, *1828*(9), 2064−2073. Available from https://doi.org/10.1016/j.bbamem.2013.05.006.

Daverey, A., Dutta, K., Joshi, S., & Daverey, A. (2021). Sophorolipid: A glycolipid biosurfactant as a potential therapeutic agent against COVID-19. *Bioengineered*, *12*(2), 9550−9560. Available from https://doi.org/10.1080/21655979.2021.1997261.

Davitt, C. J. H., & Lavelle, E. C. (2015). Delivery strategies to enhance oral vaccination against enteric infections. *Advanced Drug Delivery Reviews*, *91*, 52−69. Available from https://doi.org/10.1016/j.addr.2015.03.007.

de Araujo, L. V., Abreu, F., Lins, U., Anna, L. Md. M. S., Nitschke, M., & Freire, D. M. G. (2011). Rhamnolipid and surfactin inhibit Listeria monocytogenes adhesion. *Food Research International*, *44*(1), 481−488. Available from https://doi.org/10.1016/j.foodres.2010.09.002.

Deleu, M., Paquot, M., & Nylander, T. (2008). Effect of fengycin, a lipopeptide produced by Bacillus subtilis, on model biomembranes. *Biophysical Journal*, *94*(7), 2667−2679. Available from https://doi.org/10.1529/biophysj.107.114090.

Dey, G., Bharti, R., Dhanarajan, G., Das, S., Dey, K. K., Kumar, B. N. P., Sen, R., & Mandal, M. (2015). Marine lipopeptide Iturin A inhibits Akt mediated GSK3β and FoxO3a signaling and triggers apoptosis in breast cancer. *Scientific Reports*, *5*. Available from https://doi.org/10.1038/srep10316.

Di Somma, A., Recupido, F., Cirillo, A., Romano, A., Romanelli, A., Caserta, S., Guido, S., & Duilio, A. (2020). Antibiofilm properties of temporin-L on Pseudomonas fluorescens in static and in-flow conditions. *International Journal of Molecular Sciences*, *21*(22), 8526. Available from https://doi.org/10.3390/ijms21228526.

Díaz De Rienzo, M. A., Banat, I. M., Dolman, B., Winterburn, J., & Martin, P. J. (2015). Sophorolipid biosurfactants: Possible uses as antibacterial and antibiofilm agent. *New Biotechnology*, *32*(6), 720−726. Available from https://doi.org/10.1016/j.nbt.2015.02.009.

Dobler, L., Vilela, L. F., Almeida, R. V., & Neves, B. C. (2016). Rhamnolipids in perspective: Gene regulatory pathways, metabolic engineering, production and technological forecasting. *New Biotechnology*, *33*(1), 123−135. Available from https://doi.org/10.1016/j.nbt.2015.09.005.

Domingues, M. M., Inácio, R. G., Raimundo, J. M., Martins, M., Castanho, M. A. R. B., & Santos, N. C. (2012). Biophysical characterization of polymyxin B interaction with LPS aggregates and membrane model systems. *Biopolymers*, *98*(4), 338−344. Available from https://doi.org/10.1002/bip.22095.

Doring, G., Conway, S. P., Heijerman, H. G. M., Hodson, M. E., Hoiby, N., Smyth, A., & Touw, D. J. (2000). Antibiotic therapy against *Pseudomonas aeruginosa* in cystic fibrosis: A European consensus. *European Respiratory Journal*, *16*(4), 749−767. Available from https://doi.org/10.1034/j.1399-3003.2000.16d30.x.

Dubern, J. F., & Bloemberg, G. V. (2006). Influence of environmental conditions on putisolvins I and II production in *Pseudomonas putida* strain PCL1445. *FEMS Microbiology Letters*, *263*(2), 169−175. Available from https://doi.org/10.1111/j.1574-6968.2006.00406.x.

Dusane, D. H., Pawar, V. S., Nancharaiah, Y. V., Venugopalan, V. P., Kumar, A. R., & Zinjarde, S. S. (2011). Anti-biofilm potential of a glycolipid surfactant produced by a tropical marine strain of Serratia marcescens. *Biofouling*, *27*(6), 645−654. Available from https://doi.org/10.1080/08927014.2011.594883.

Dusane, D. H., Dam, S., Nancharaiah, Y. V., Kumar, A. R., Venugopalan, V. P., & Zinjarde, S. S. (2012). Disruption of Yarrowia lipolytica biofilms by rhamnolipid biosurfactant. *Aquatic Biosystems*, *8*(1). Available from https://doi.org/10.1186/2046-9063-8-17.

Ekprasert, J., Kanakai, S., & Yosprasong, S. (2020). Improved biosurfactant production by enterobacter cloacae B14, stability studies, and its antimicrobial activity. *Polish Journal of Microbiology*, *69*(3), 273−282. Available from https://doi.org/10.33073/pjm-2020-030.

Elshikh, M., Funston, S., Chebbi, A., Ahmed, S., Marchant, R., & Banat, I. M. (2017). Rhamnolipids from non-pathogenic Burkholderia thailandensis E264: Physicochemical characterization, antimicrobial and antibiofilm efficacy against oral hygiene related pathogens. *New Biotechnology*, *36*, 26−36. Available from https://doi.org/10.1016/j.nbt.2016.12.009.

Falagas, M. E., & Kasiakou, S. K. (2005). Colistin: The revival of polymyxins for the management of multidrug-resistant gram-negative bacterial infections. *Clinical Infectious Diseases*, *40*(9), 1333−1341. Available from https://doi.org/10.1086/429323.

Fracchia, L., Banat, J. J., Cavallo, M., Ceresa, C., & Banat, I. M. (2015). Potential therapeutic applications of microbial surface-active compounds. *AIMS Bioengineering*, *2*(3), 144−162. Available from https://doi.org/10.3934/bioeng.2015.3.144.

Gakhar, L., Bartlett, J. A., Penterman, J., Mizrachi, D., Singh, P. K., Mallampalli, R. K., Ramaswamy, S., McCray, P. B., & Kreindler, J. L. (2010). PLUNC is a novel airway surfactant protein with anti-biofilm activity. *PLoS One*, *5*(2), e9098. Available from https://doi.org/10.1371/journal.pone.0009098.

Gallo, P. M., Rapsinski, G. J., Wilson, R. P., Oppong, G. O., Sriram, U., Goulian, M., Buttaro, B., Caricchio, R., Gallucci, S., & Tükel, Ç. (2015). Amyloid-DNA composites of bacterial biofilms stimulate autoimmunity. *Immunity*, *42*(6), 1171−1184. Available from https://doi.org/10.1016/j.immuni.2015.06.002.

Ghojavand, H., Mohammadi Behnazar, M., & Vahabzadeh, F. (2019). Antibacterial activity of the lipopeptide biosurfactant produced by bacillus mojavensis PTCC 1696. *Iranian Journal of Chemistry and Chemical Engineering*, *38*(6), 275−284. Available from http://www.ijcce.ac.ir/article_32436_16f4c802b4fc74d0900c2cb75c8c9ba4.pdf.

Ghribi, D., Abdelkefi-Mesrati, L., Mnif, I., Kammoun, R., Ayadi, I., Saadaoui, I., Maktouf, S., & Chaabouni-Ellouze, S. (2012). Investigation of antimicrobial activity and statistical optimization of Bacillus subtilis SPB1 biosurfactant production in solid-state fermentation. *Journal of Biomedicine and Biotechnology*, *2012*. Available from https://doi.org/10.1155/2012/373682.

Giugliano, R., Buonocore, C., Zannella, C., Chianese, A., Esposito, F. P., Tedesco, P., De Filippis, A., Galdiero, M., Franci, G., & de Pascale, D. (2021). Antiviral activity of the rhamnolipids mixture from the antarctic bacterium pseudomonas gessardii M15 against herpes simplex viruses and coronaviruses. *Pharmaceutics*, *13*(12). Available from https://doi.org/10.3390/pharmaceutics13122121.

Gomaa, E. Z. (2013). Antimicrobial activity of a biosurfactant produced by bacillus licheniformis strain m104 grown on whey. *Brazilian Archives of Biology and Technology*, *56*(2), 259−268. Available from https://doi.org/10.1590/S1516-89132013000200011.

Grady, E. N., MacDonald, J., Liu, L., Richman, A., & Yuan, Z. C. (2016). Current knowledge and perspectives of Paenibacillus: A review. *Microbial Cell Factories*, *15*(1). Available from https://doi.org/10.1186/s12934-016-0603-7.

Gudiña, E. J., Rangarajan, V., Sen, R., & Rodrigues, L. R. (2013). Potential therapeutic applications of biosurfactants. *Trends in Pharmacological Sciences*, *34*(12), 667−675. Available from https://doi.org/10.1016/j.tips.2013.10.002.

Haba, E., Pinazo, A., Jauregui, O., Espuny, M. J., Infante, M. R., & Manresa, A. (2003). Physicochemical characterization and antimicrobial properties of rhamnolipids produced by *Pseudomonas aeruginosa* 47T2 NCBIM 40044. *Biotechnology and Bioengineering*, *81*(3), 316−322. Available from https://doi.org/10.1002/bit.10474.

Hajfarajollah, H., Mehvari, S., Habibian, M., Mokhtarani, B., & Noghabi, K. A. (2015). Rhamnolipid biosurfactant adsorption on a plasma-treated polypropylene surface to induce antimicrobial and antiadhesive properties. *RSC Advances*, *5*(42), 33089−33097. Available from https://doi.org/10.1039/c5ra01233c.

Haque, F., Alfatah, M., Ganesan, K., & Bhattacharyya, M. (2016). Inhibitory efect of sophorolipid on Candida albicans bioflm formation and hyphal growth. *Scientific Report*, *6*.

Hardin, R., Pierre, J., Schulze, R., Mueller, C. M., Fu, S. L., Wallner, S. R., Stanek, A., Shah, V., Gross, R. A., Weedon, J., Nowakowski, M., Zenilman, M. E., & Bluth, M. H. (2007). Sophorolipids improve sepsis survival: Effects of dosing and derivatives. *Journal of Surgical Research*, *142*(2), 314−319. Available from https://doi.org/10.1016/j.jss.2007.04.025.

He, J., Ledesma, K. R., Lam, W. Y., Figueroa, D. A., Lim, T. P., Chow, D. S. L., & Tam, V. H. (2010). Variability of polymyxin B major components in commercial formulations. *International Journal of Antimicrobial Agents*, *35*(3), 308−310. Available from https://doi.org/10.1016/j.ijantimicag.2009.11.005.

Hurlow, J., Couch, K., Laforet, K., Bolton, L., Metcalf, D., & Bowler, P. (2015). Clinical biofilms: A challenging frontier in wound care. *Advances in Wound Care*, *4*(5), 295−301. Available from https://doi.org/10.1089/wound.2014.0567.

Janek, T., Łukaszewicz, M., & Krasowska, A. (2012). Antiadhesive activity of the biosurfactant pseudofactin II secreted by the Arctic bacterium *Pseudomonas fluorescens* BD5. *BMC Microbiology*, *12*. Available from https://doi.org/10.1186/1471-2180-12-24.

Johnson, C. H., Dejea, C. M., Edler, D., Hoang, L. T., Santidrian, A. F., Felding, B. H., Ivanisevic, J., Cho, K., Wick, E. C., Hechenbleikner, E. M., Uritboonthai, W., Goetz, L., Casero, R. A., Pardoll, D. M., White, J. R., Patti, G. J., Sears, C. L., & Siuzdak, G. (2015). Metabolism links bacterial biofilms and colon carcinogenesis. *Cell Metabolism*, *21*(6), 891−897. Available from https://doi.org/10.1016/j.cmet.2015.04.011.

Joshi-Navare, K., & Prabhune, A. (2013). A biosurfactant-sophorolipid acts in synergy with antibiotics to enhance their efficiency. *BioMed Research International*, *2013*, 1−8. Available from https://doi.org/10.1155/2013/512495.

Jumpathong, W., Intra, B., Euanorasetr, J., & Wanapaisan, P. (2022). Biosurfactant-producing Bacillus velezensis PW192 as an anti-fungal biocontrol agent against colletotrichum gloeosporioides and colletotrichum musae. *Microorganisms*, *10*(5). Available from https://doi.org/10.3390/microorganisms10051017.

Kokare, C. R., Chakraborty, S., Khopade, A. N., & Mahadik, K. R. (2009). Biofilm: Importance and applications. *Indian Journal of Biotechnology*, *8*(2), 159–168.

Kourmentza, K., Gromada, X., Michael, N., Degraeve, C., Vanier, G., Ravallec, R., Coutte, F., Karatzas, K. A., & Jauregi, P. (2021). Antimicrobial activity of lipopeptide biosurfactants against foodborne pathogen and food spoilage microorganisms and their cytotoxicity. *Frontiers in Microbiology*, *11*. Available from https://doi.org/10.3389/fmicb.2020.561060.

Krsmanovic, M., Biswas, D., Ali, H., Kumar, A., Ghosh, R., & Dickerson, A. K. (2021). Hydrodynamics and surface properties influence biofilm proliferation. *Advances in Colloid and Interface Science*, *288*. Available from https://doi.org/10.1016/j.cis.2020.102336.

Kuiper, I., Lagendijk, E. L., Pickford, R., Derrick, J. P., Lamers, G. E. M., Thomas-Oates, J. E., Lugtenberg, B. J. J., & Bloemberg, G. V. (2004). Characterization of two *Pseudomonas putida* lipopeptide biosurfactants, putisolvin I and II, which inhibit biofilm formation and break down existing biofilms. *Molecular Microbiology*, *51*(1), 97–113. Available from https://doi.org/10.1046/j.1365-2958.2003.03751.x.

Liu, J., Li, W., Zhu, X., Zhao, H., Lu, Y., Zhang, C., & Lu, Z. (2019). Surfactin effectively inhibits *Staphylococcus aureus* adhesion and biofilm formation on surfaces. *Applied Microbiology and Biotechnology*, *103*(11), 4565–4574. Available from https://doi.org/10.1007/s00253-019-09808-w.

Liza Kretzschmar, A., & Manefield, M. (2015). The role of lipids in activated sludge floc formation. *AIMS Environmental Science*, *2*(2), 122–133. Available from https://doi.org/10.3934/environsci.2015.1.122.

Lotfabad, T., Shahcheraghi, F., & Shooraj, F. (2012). Assessment of antibacterial capability of rhamnolipids produced by two indigenous *Pseudomonas aeruginosa* strains. *Jundishapur Journal of Microbiology*, *6*(1), 29–35. Available from https://doi.org/10.5812/jjm.2662.

Luna, J. M., Rufino, R. D., Sarubbo, L. A., Rodrigues, L. R. M., Teixeira, J. A. C., & DeCampos-Takaki, G. M. (2011). Evaluation antimicrobial and antiadhesive properties of the biosurfactant Lunasan produced by *Candida sphaerica* UCP 0995. *Current Microbiology*, *62*(5), 1527–1534. Available from https://doi.org/10.1007/s00284-011-9889-1.

Madeo, J., & Frieri, M. (2013). Bacterial biofilms and chronic rhinosinusitis. *Allergy and Asthma Proceedings*, *34*(4), 335–341. Available from https://doi.org/10.2500/aap.2013.34.3665.

Mandal, S. M., Barbosa, A. E. A. D., & Franco, O. L. (2013). Lipopeptides in microbial infection control: Scope and reality for industry. *Biotechnology Advances*, *31*(2), 338–345. Available from https://doi.org/10.1016/j.biotechadv.2013.01.004.

Mani, P., Dineshkumar, G., Jayaseelan, T., Deepalakshmi, K., Ganesh Kumar, C., & Senthil Balan, S. (2016). Antimicrobial activities of a promising glycolipid biosurfactant from a novel marine Staphylococcus saprophyticus SBPS 15. *3 Biotech*, *6*(2). Available from https://doi.org/10.1007/s13205-016-0478-7.

Matei, G. M., Matei, S., Matei, A., & Draghici, E. (2017). Antifungal activity of a biosurfactant-producing lactic acid bacteria strain. *The EuroBiotech the Journal*, *1*(3), 212–216. Available from https://doi.org/10.24190/issn2564-615x/2017/03.02.

Merghni, A., Dallel, I., Noumi, E., Kadmi, Y., Hentati, H., Tobji, S., Ben Amor, A., & Mastouri, M. (2017). Antioxidant and antiproliferative potential of biosurfactants isolated from *Lactobacillus casei* and their anti-biofilm effect in oral *Staphylococcus aureus* strains. *Microbial Pathogenesis*, *104*, 84–89. Available from https://doi.org/10.1016/j.micpath.2017.01.017.

Mimee, B., Pelletier, R., & Bélanger, R. R. (2009). In vitro antibacterial activity and antifungal mode of action of flocculosin, a membrane-active cellobiose lipid. *Journal of Applied Microbiology*, *107*(3), 989–996. Available from https://doi.org/10.1111/j.1365-2672.2009.04280.x.

Mireles, J. R., Toguchi, A., & Harshey, R. M. (2001). Salmonella enterica serovar typhimurium swarming mutants with altered biofilm-forming abilities: Surfactin inhibits biofilm formation. *Journal of Bacteriology*, *183*(20), 5848–5854. Available from https://doi.org/10.1128/JB.183.20.5848-5854.2001.

Mohd Isa, M. H., Shamsudin, N. H., Al-Shorgani, N. K. N., Alsharjabi, F. A., & Kalil, M. S. (2020). Evaluation of antibacterial potential of biosurfactant produced by surfactin-producing Bacillus isolated from selected Malaysian fermented foods. *Food Biotechnology*, *34*(1), 1−24. Available from https://doi.org/10.1080/08905436.2019.1710843.

Morita, T., Kitagawa, M., Suzuki, M., Yamamoto, S., Sogabe, A., Yanagidani, S., Imura, T., Fukuoka, T., & Kitamoto, D. (2009). A yeast glycolipid biosurfactant, mannosylerythritol lipid, shows potential moisturizing activity toward cultured human skin cells: The recovery effect of MEL-a on the SDS-damaged human skin cells. *Journal of Oleo Science*, *58*(12), 639−642. Available from https://doi.org/10.5650/jos.58.639.

Mouafo, T. H., Mbawala, A., & Ndjouenkeu, R. (2018). Effect of different carbon sources on biosurfactants' production by three strains of Lactobacillus spp. *BioMed Research International*, *2018*. Available from https://doi.org/10.1155/2018/5034783.

Moubareck, C. A. (2020). Polymyxins and bacterial membranes: A review of antibacterial activity and mechanisms of resistance. *Membranes*, *10*(8), 1−30. Available from https://doi.org/10.3390/membranes10080181.

Muhammad, M. H., Idris, A. L., Fan, X., Guo, Y., Yu, Y., Jin, X., Qiu, J., Guan, X., & Huang, T. (2020). Beyond risk: Bacterial biofilms and their regulating approaches. *Frontiers in Microbiology*, *11*. Available from https://doi.org/10.3389/fmicb.2020.00928.

Mukherji, R., & Prabhune, A. (2014). Novel glycolipids synthesized using plant essential oils and their application in quorum sensing inhibition and as Antibiofilm agents. *The Scientific World Journal*, *2014*. Available from https://doi.org/10.1155/2014/890709.

Naughton, P. J., Marchant, R., Naughton, V., & Banat, I. M. (2019). Microbial biosurfactants: Current trends and applications in agricultural and biomedical industries. *Journal of Applied Microbiology*, *127*(1), 12−28. Available from https://doi.org/10.1111/jam.14243.

Ndlovu, T., Rautenbach, M., Vosloo, J. A., Khan, S., & Khan, W. (2017). Characterisation and antimicrobial activity of biosurfactant extracts produced by Bacillus amyloliquefaciens and *Pseudomonas aeruginosa* isolated from a wastewater treatment plant. *AMB Express*, *7*(1). Available from https://doi.org/10.1186/s13568-017-0363-8.

Nguyen, B. V. G., Nagakubo, T., Toyofuku, M., Nomura, N., & Utada, A. S. (2020). Synergy between Sophorolipid biosurfactant and SDS increases the efficiency of *P. aeruginosa* biofilm disruption. *Langmuir: The ACS Journal of Surfaces and Colloids*, *36*(23), 6411−6420. Available from https://doi.org/10.1021/acs.langmuir.0c00643.

Ortiz, A., Teruel, J. A., Espuny, M. J., Marqués, A., Manresa, A., & Aranda, F. J. (2009). Interactions of a bacterial biosurfactant trehalose lipid with phosphatidylserine membranes. *Chemistry and Physics of Lipids*, *158*(1), 46−53. Available from https://doi.org/10.1016/j.chemphyslip.2008.11.001.

Ortiz, A., Teruel, J. A., Manresa, A., Espuny, M. J., Marqués, A., & Aranda, F. J. (2011). Effects of a bacterial trehalose lipid on phosphatidylglycerol membranes. *Biochimica et Biophysica Acta—Biomembranes*, *1808*(8), 2067−2072. Available from https://doi.org/10.1016/j.bbamem.2011.05.003.

Padmavathi, A. R., & Pandian, S. K. (2014). Antibiofilm activity of biosurfactant producing coral associated bacteria isolated from Gulf of Mannar. *Indian Journal of Microbiology*, *54*(4), 376−382. Available from https://doi.org/10.1007/s12088-014-0474-8.

Parreira, A. G. (2020). Evaluation of the anti-biofilm effects of biosurfactants and silver nanoparticles on biomaterials surfaces. *Archives in Biomedical Engineering & Biotechnology*, *3*(5). Available from https://doi.org/10.33552/abeb.2020.03.000574.

Patel, M., Siddiqui, A. J., Hamadou, W. S., Surti, M., Awadelkareem, A. M., Ashraf, S. A., Alreshidi, M., Snoussi, M., Rizvi, S. M. D., Bardakci, F., Jamal, A., Sachidanandan, M., & Adnan, M. (2021). Inhibition of bacterial adhesion and antibiofilm activities of a glycolipid biosurfactant from lactobacillus rhamnosus with its physicochemical and functional properties. *Antibiotics*, *10*(12). Available from https://doi.org/10.3390/antibiotics10121546.

Patiño, A. D., Montoya-Giraldo, M., Quintero, M., López-Parra, L. L., Blandón, L. M., & Gómez-León, J. (2021). Dereplication of antimicrobial biosurfactants from marine bacteria using molecular networking. *Scientific Reports*, *11*(1). Available from https://doi.org/10.1038/s41598-021-95788-9.

Peterson, E., & Kaur, P. (2018). Antibiotic resistance mechanisms in bacteria: Relationships between resistance determinants of antibiotic producers, environmental bacteria, and clinical pathogens. *Frontiers in Microbiology*, *9*. Available from https://doi.org/10.3389/fmicb.2018.02928.

Pradhan, A. K., Pradhan, N., Mall, G., Panda, H. T., Sukla, L. B., Panda, P. K., & Mishra, B. K. (2013). Application of lipopeptide biosurfactant isolated from a halophile: Bacillus tequilensis ch for inhibition of biofilm. *Applied Biochemistry and Biotechnology*, *171*(6), 1362−1375. Available from https://doi.org/10.1007/s12010-013-0428-3.

Quinn, G. A., Maloy, A. P., McClean, S., Carney, B., & Slater, J. W. (2012). Lipopeptide biosurfactants from *Paenibacillus polymyxa* inhibit single and mixed species biofilms. *Biofouling*, *28*(10), 1151−1166. Available from https://doi.org/10.1080/08927014.2012.738292.

Rabin, N., Zheng, Y., Opoku-Temeng, C., Du, Y., Bonsu, E., & Sintim, H. O. (2015). Biofilm formation mechanisms and targets for developing antibiofilm agents. *Future Medicinal Chemistry*, *7*(4), 493−512. Available from https://doi.org/10.4155/fmc.15.6.

Rivardo, F., Martinotti, M. G., Turner, R. J., & Ceri, H. (2011). Synergistic effect of lipopeptide biosurfactant with antibiotics against *Escherichia coli* CFT073 biofilm. *International Journal of Antimicrobial Agents*, *37*(4), 324−331. Available from https://doi.org/10.1016/j.ijantimicag.2010.12.011.

Rizzo, C., Zammuto, V., Lo Giudice, A., Rizzo, M. G., Spanò, A., Laganà, P., Martinez, M., Guglielmino, S., & Gugliandolo, C. (2021). Antibiofilm activity of antarctic sponge-associated bacteria against *Pseudomonas aeruginosa* and *Staphylococcus aureus*. *Journal of Marine Science and Engineering*, *9*(3), 243. Available from https://doi.org/10.3390/jmse9030243.

Rodrigues, L., Banat, I. M., Teixeira, J., & Oliveira, R. (2006). Biosurfactants: Potential applications in medicine. *Journal of Antimicrobial Chemotherapy*, *57*(4), 609−618. Available from https://doi.org/10.1093/jac/dkl024.

Rufino, R. D., Luna, J. M., Sarubbo, L. A., Rodrigues, L. R. M., Teixeira, J. A. C., & Campos-Takaki, G. M. (2011). Antimicrobial and anti-adhesive potential of a biosurfactant Rufisan produced by *Candida lipolytica* UCP 0988. *Colloids and Surfaces B: Biointerfaces*, *84*(1), 1−5. Available from https://doi.org/10.1016/j.colsurfb.2010.10.045.

Sambanthamoorthy, K., Feng, X., Patel, R., Patel, S., & Paranavitana, C. (2014). Antimicrobial and antibiofilm potential of biosurfactants isolated from lactobacilli against multi-drug-resistant pathogens. *BMC Microbiology*, *14*(1). Available from https://doi.org/10.1186/1471-2180-14-197.

Santos, V. L., Nardi Drummond, R. M., & Dias-Souza, M. V. (2017). *Biosurfactants as antimicrobial and antibiofilm agents. Current developments in biotechnology and bioengineering: Human and animal health applications* (pp. 371−402). Elsevier Inc. Available from https://doi.org/10.1016/B978-0-444-63660-7.00015-2.

Saravanakumari, P., & Mani, K. (2010). Structural characterization of a novel xylolipid biosurfactant from Lactococcus lactis and analysis of antibacterial activity against multi-drug resistant pathogens. *Bioresource Technology*, *101*(22), 8851−8854. Available from https://doi.org/10.1016/j.biortech.2010.06.104.

Sarwar, A., Brader, G., Corretto, E., Aleti, G., Abaidullah, M., Sessitsch, A., & Hafeez, F. Y. (2018). Qualitative analysis of biosurfactants from Bacillus species exhibiting antifungal activity. *PLoS One*, *13*(6). Available from https://doi.org/10.1371/journal.pone.0198107.

Satpute, S. K., Banpurkar, A. G., Banat, I. M., Sangshetti, J. N., Patil, R. H., & Gade, W. N. (2016). Multiple roles of biosurfactants in biofilms. *Current Pharmaceutical Design*, *22*(11), 1429−1448. Available from https://doi.org/10.2174/1381612822666160120152704.

Schwartz, I. S., & Patterson, T. F. (2018). The emerging threat of antifungal resistance in transplant infectious diseases. *Current Infectious Disease Reports*, *20*(3). Available from https://doi.org/10.1007/s11908-018-0608-y.

Sen, S., Borah, S. N., Bora, A., & Deka, S. (2020). Rhamnolipid exhibits anti-biofilm activity against the dermatophytic fungi Trichophyton rubrum and Trichophyton mentagrophytes. *Biotechnology Reports*, *27*. Available from https://doi.org/10.1016/j.btre.2020.e00516.

Silveira, V. A. I., Kobayashi, R. K. T., de Oliveira Junior, A. G., Mantovani, M. S., Nakazato, G., & Celligoi, M. A. P. C. (2021). Antimicrobial effects of sophorolipid in combination with lactic acid against poultry-relevant isolates. *Brazilian Journal of Microbiology*, *52*(4), 1769−1778. Available from https://doi.org/10.1007/s42770-021-00545-9.

Sleiman, J. N., Kohlhoff, S. A., Roblin, P. M., Wallner, S., Gross, R., Hammerschlag, M. R., Zenilman, M. E., & Bluth, M. H. (2009). Sophorolipids as antibacterial agents. *Annals of Clinical and Laboratory Science*, *39*(1), 60−63. Available from http://www.annclinlabsci.org/cgi/reprint/39/1/60.

Solaiman, D. K. Y., Ashby, R. D., & Uknalis, J. (2017). Characterization of growth inhibition of oral bacteria by sophorolipid using a microplate-format assay. *Journal of Microbiological Methods*, *136*, 21−29. Available from https://doi.org/10.1016/j.mimet.2017.02.012.

Sriram, M. I., Kalishwaralal, K., Deepak, V., Gracerosepat, R., Srisakthi, K., & Gurunathan, S. (2011). Biofilm inhibition and antimicrobial action of lipopeptide biosurfactant produced by heavy metal tolerant strain Bacillus cereus NK1. *Colloids and Surfaces B: Biointerfaces*, 85(2), 174–181. Available from https://doi.org/10.1016/j.colsurfb.2011.02.026.

Tambone, E., Bonomi, E., Ghensi, P., Maniglio, D., Ceresa, C., Agostinacchio, F., Caciagli, P., Nollo, G., Piccoli, F., Caola, I., Fracchia, L., & Tessarolo, F. (2021). Rhamnolipid coating reduces microbial biofilm formation on titanium implants: an in vitro study. *BMC Oral Health*, 21(1). Available from https://doi.org/10.1186/s12903-021-01412-7.

Velraed, M. M., Belt-Gritter., Busscher, H., Reid, G., Van., & Mei, H. (2000). Interference in initial adhesion of uropathogenic bacteria and yeasts to silicone rubber by a *Lactobacillus acidophilus* biosurfactant. *World Journal of Urology*, 18, 422–426.

Vollbrecht, E., Rau, U., & Lang, S. (1999). Microbial conversion of vegetable oils into surface-active di-, tri-, and tetrasaccharide lipids (biosurfactants) by the bacterial strain Tsukamurella spec. *Lipid—Fett*, 101(10), 389–394. Available from https://doi.org/10.1002/(SICI)1521-4133(199910)101:10 < 389::AID-LIPI389 > 3.3.CO;2-0.

Vollenbroich, D., Özel, M., Vater, J., Kamp, R. M., & Pauli, G. (1997). Mechanism of inactivation of enveloped viruses by the biosurfactant surfactin from Bacillus subtilis. *Biologicals: Journal of the International Association of Biological Standardization*, 25(3), 289–297. Available from https://doi.org/10.1006/biol.1997.0099.

Wu, H., Moser, C., Wang, H. Z., Høiby, N., & Song, Z. J. (2015). Strategies for combating bacterial biofilm infections. *International Journal of Oral Science*, 7, 1–7. Available from https://doi.org/10.1038/ijos.2014.65.

Xu, Z., Shao, J., Li, B., Yan, X., Shen, Q., & Zhang, R. (2013). Contribution of bacillomycin D in Bacillus amyloliquefaciens SQR9 to antifungal activity and biofilm formation. *Applied and Environmental Microbiology*, 79(3), 808–815. Available from https://doi.org/10.1128/AEM.02645-12.

Yan, X., Gu, S., Cui, X., Shi, Y., Wen, S., Chen, H., & Ge, J. (2019). Antimicrobial, anti-adhesive and anti-biofilm potential of biosurfactants isolated from *Pediococcus acidilactici* and *Lactobacillus plantarum* against *Staphylococcus aureus* CMCC26003. *Microbial Pathogenesis*, 127, 12–20. Available from https://doi.org/10.1016/j.micpath.2018.11.039.

Yin, H., Guo, C., Wang, Y., Liu, D., Lv, Y., Lv, F., & Lu, Z. (2013). Fengycin inhibits the growth of the human lung cancer cell line 95D through reactive oxygen species production and mitochondria-dependent apoptosis. *Anti-Cancer Drugs*, 24(6), 587–598. Available from https://doi.org/10.1097/CAD.0b013e3283611395.

Yousif, A., Jamal, M. A., & Raad, I. (2015). Biofilm-based central line-associated bloodstream infections. *Advances in Experimental Medicine and Biology*, 830, 157–179. Available from https://doi.org/10.1007/978-3-319-11038-7_10.

Yuliani, H., Perdani, M. S., Savitri, I., Manurung, M., Sahlan, M., Wijanarko, A., & Hermansyah, H. (2018). *Antimicrobial activity of biosurfactant derived from Bacillus subtilis C19*. Energy procedia (Vol. 153, pp. 274–278). Elsevier Ltd. Available from https://doi.org/10.1016/j.egypro.2018.10.043.

Zampoli, J., De Giani, A., Di Canito, A., Sello, G., & Gennaro, P. (2022). Identification of a novel biosurfactant with antimicrobial activity produced by Rhodococcus opacus R7. *Microorganisms*, 10, 1–R16.

Zezzi do Valle Gomes, M., & Nitschke, M. (2012). Evaluation of rhamnolipid and surfactin to reduce the adhesion and remove biofilms of individual and mixed cultures of food pathogenic bacteria. *Food Control*, 25(2), 441–447. Available from https://doi.org/10.1016/j.foodcont.2011.11.025.

Zhao, G., Usui, M. L., Lippman, S. I., James, G. A., Stewart, P. S., Fleckman, P., & Olerud, J. E. (2013). Biofilms and inflammation in chronic wounds. *Advances in Wound Care*, 2(7), 389–399. Available from https://doi.org/10.1089/wound.2012.0381.

Zouari, R., Moalla-Rekik, D., Sahnoun, Z., Rebai, T., Ellouze-Chaabouni, S., & Ghribi-Aydi, D. (2016). Evaluation of dermal wound healing and in vitro antioxidant efficiency of *Bacillus subtilis* SPB1 biosurfactant. *Biomedicine and Pharmacotherapy*, 84, 878–891. Available from https://doi.org/10.1016/j.biopha.2016.09.084.

CHAPTER 14

Insecticidal potential of biosurfactants

Natalia Andrade Teixeira Fernandes[1], Luara Aparecida Simões[2],
Angelica Cristina Souza[3] and Disney Ribeiro Dias[4]

[1]Department of Chemistry, University of California, Davis, CA, United States
[2]Biology Department, University of Porto, Porto, Portugal
[3]Department of Biology, Federal University of Lavras, Lavras, Minas Gerais, Brazil
[4]Department of Food Science, Federal University of Lavras, Lavras, Minas Gerais, Brazil

14.1 Introduction

Microbial biosurfactants have unique properties such as low toxicity, greater biodegradability, exceptional surface activity under extreme environmental conditions, low foaming, high efficiency, and regenerative properties (Sachdev & Cameotra, 2013; Shekhar et al., 2015). Due to their potential and safety, biosurfactants are currently used in various fields of application as emulsifiers and stabilizers in the food industry, formulations in the cosmetics industry, biocontrol agents in agriculture or biodegradation, bioremediation in environmental protection systems, and biocontrol of disease-causing insects in humans (Fernandes et al., 2020; Nitschke & Silva, 2017).

In recent years, biosurfactants have been introduced as a green alternative to synthetic chemicals for insect control, and several studies have shown that biosurfactants apply to insect control without causing toxicological harm (Fig. 14.1) (Yang et al., 2017; Zhao et al., 2014). For example, the activity of a biosurfactant produced by a strain of *Bacillus* has been reported against adult mosquitoes, showing that the biosurfactant can penetrate cuticles or spiracular openings, killing mosquitoes (Geetha et al., 2012). The biosurfactants rhamnolipids and orphamide A, produced by *Pseudomonas* strains, have shown insecticidal activity against aphids (Jang et al., 2013; Kim et al., 2011). These studies showed that biosurfactants, as they are amphiphilic molecules (with hydrophobic and hydrophilic portions), can affect insect cuticles.

Studies demonstrating the biotechnological activities of biosurfactants have attracted much attention from agricultural and disease control research programs because biosurfactants are soluble in water and can be used as organic solvents and emulsifying agents, thereby avoiding the use of synthetic chemicals in agricultural products. Thus, this chapter primarily emphasizes the contribution of biosurfactants in managing agricultural insect pests and disease-causing humans, as well as the production and application of biosurfactants, and the way forward.

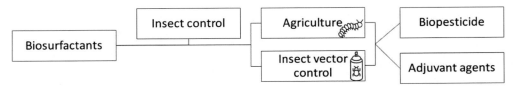

Figure 14.1 Biosurfactants as insects biocontrol.

14.2 Chemical pesticides

Agriculture faces many problems caused by fungi and insects, which can lead to drastically low productivity. Plant pathogens cause substantial economic losses (Savary et al., 2019). Chemical pesticides are used worldwide mainly to contain plant diseases (Popp et al., 2013). Agriculture relies heavily on chemical-based pesticides to destroy plant pathogens and chemical-based fertilizers to increase soil fertility, thereby damaging the environment and humans (Mnif & Ghribi, 2015, 2016).

Pest control is a key component of several strategies aimed at preventing the spread of pests to increase food production. Almost all pest control programs rely on the use of chemical insecticides formulated as direct contact sprays, powders, or baits. Today, commercially available chemical pesticides in modern agriculture are under pressure to be withdrawn from the market due to their hazardous effects on the natural environment (Ghribi, Abdelkefi-Mesrati, et al., 2012).

Although synthetic pesticides can protect crops by suppressing arthropod pests, they pose serious health and environmental risks due to their toxicological properties and harmful side effects (Carvalho, 2017; do Nascimento Silva et al., 2019). The main problems associated with chemical pesticides are their effects on nontarget organisms such as humans, pets, beneficial insects, and wildlife, as well as their often long persistence in nature (Zaller & Brühl, 2019), as shown in Fig. 14.2.

Today, due to human knowledge about the possibility of contamination and the health threats posed by many common synthetic chemical toxins, much attention has been paid to producing biotechnological products to promote sustainable agriculture. Biopesticides can be developed by beneficial microorganisms such as fungi and bacteria. They are also ecological alternatives to sustainable pest control due to their safety for human and nontarget organisms (Sarwar, 2015). Thus, biosurfactants are known for their antimicrobial and biopesticide activity against human and plant pathogens (Mnif & Ghribi, 2015).

14.3 Biosurfactants as agricultural biopesticides

Exploiting microbial metabolites as biocontrol agents is considered an alternative to chemical pesticides in organic agriculture due to their vast potential to increase crop

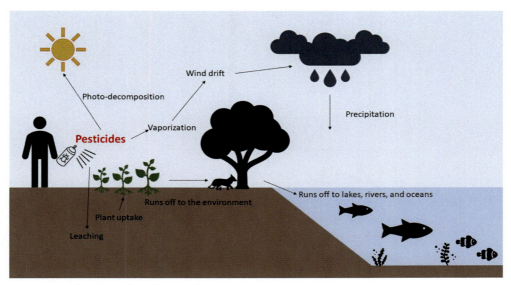

Figure 14.2 The pesticide cycle.

protection and food security. However, the exact mode of action and a field evaluation of the insecticidal activity are necessary to develop a new agricultural biopesticide.

In recent years, the use of biosurfactants as biopesticides has proliferated due to their ecological characteristics and a high degree of degradability (Ben Khedher et al., 2017; Mnif et al., 2013), which increases the interest in studies and applications. Thus, to reduce the adverse effects of synthetic pesticides on the environment and human health, biosurfactants may be a promising alternative choice in the management of agricultural pests.

Synthetic surfactants are popularly used as adjuvants, emulsifiers, dispersants, spreaders, and wetting agents to increase the efficiency of pesticides. Therefore, they are widely used in industries to manufacture pesticides (Mulqueen, 2003). However, these synthetic surfactants are not degradable and accumulate in the soil and groundwater, as well as in agricultural products. As a result, they have adverse effects on the environment and humans. Alternatively, different species of bacteria, yeasts, and fungi have recently been reported to produce biosurfactants (Cuatlayotl-Cottier et al., 2020; Maity et al., 2022; Yang et al., 2017). These biosurfactants have well-known applications in the food industry and the bioremediation of toxic pollutants and have therefore recently received attention in the pesticide and pest control industry (Marcelino et al., 2020; Pradhan et al., 2018), as shown in Table 14.1.

Biosurfactants help solubilize methyl parathion, ethyl parathion, and terifluration (Wattanaphon et al., 2008). They also act as carriers for insecticidal herbicide formulations and, due to the wettability property, keep the leaf surface moist and insecticides remain on the leaf for a longer period of time. Currently, anionic, cationic, and amphoteric

Table 14.1 Biosurfactants used as agricultural biopesticides with insecticidal properties.

Microorganism	Biosurfactant	Activity	Insect	References
Bacillus amyloliquefaciens AG1	Lipopeptide	Larvicidal	*Tuta absoluta*	Ben Khedher et al. (2015)
Bacillus clausii BS02	Surfactin lipopeptide	Adulticidal	*Callosobruchus chinensis* and *Maconellicoccus hirsutus*	Hazra et al. (2015)
Bacillus subtilis Y9	Surfactin	Larvicidal	*Myzus persicae*	Yang et al. (2017)
Bacillus subtilis SPB1	Lipopeptide	Larvicidal	*Ephestia kuehniella*	Ghribi, Elleuch, Abdelkefi-Mesrati, et al. (2012)
Bacillus subtilis SPB1	Lipopeptide	Larvicidal	*Ectomyelois ceratoniae* Zeller	Mnif et al. (2013)
Bacillus subtilis V26	Not given	Larvicidal	*T. absoluta*	Ben Khedher et al. (2020)
Bacillus thuringiensis	Industrial by-product	Adulticidal	*Melanaphis sacchari, Rhophalosiphum maidis, Aphis fabae, Tetranychus urticae*	Cuatlayotl-Cottier et al. (2020)
Pseudomonas aeruginosa PA1	Rhamnolipid	Nymphs and adulticidal	*Bemisia tabaci*	do Nascimento Silva et al. (2019)
Pseudomonas protegens F6	Orfamide A	Larvicidal	*Brassica rapa*, subspecies *pekinensis, and chinensis*	Jang et al. (2013)
Staphylococcus epidermidis	Lipopeptide	Larvicidal	*Tribolium castaneum*	Fazaeli et al. (2021)

biosurfactants are used as insecticides because of defensive properties (Rostás & Blassmann, 2009). Biosurfactant accumulation alters soil texture, color, and plant growth; these also percolate from the soil into groundwater and remain on the surface of fruits and nuts (Blackwell, 2000). Many companies use blends of biosurfactants in various combinations with polymers to create efficient agricultural formulations (Maity et al., 2022).

14.4 Biosurfactant as biocontrol for organic agriculture

Disease and pest management in organic agriculture are primarily based on maintaining soil fertility through crop rotations, especially nitrogen-fixing crops, cover crops, intercropping, compost and manure suplements, and reduced tillage (Hasna et al., 2007; Yogev et al., 2009). In general, crop protection in organic agriculture consists of creating a healthy ecosystem that makes plants resistant to potential pathogens.

Although crop protection in organic agriculture is not aimed at direct pathogen control, biological control is one of the strategies developed for pest and disease control (Letourneau & Bruggen, 2006). Several studies have already highlighted the antagonistic effect of microbial metabolites, such as biosurfactants, against various pathogens and pests (Ben Khedher et al., 2020; Chowdhury et al., 2015; Ghribi, Elleuch, Abdelkefi-Mesrati, et al., 2012). Due to their low toxicity and biodegradability, biosurfactants are considered environmentally safe insecticides (Fig. 14.3). They have received much attention in pest management and are considered green pesticides. For many years, microbial bioinsecticides have been used to control agricultural pests (Maity et al., 2022).

14.4.1 Bacillus biosurfactant as insecticidal

As can be seen in Table 14.1, *Bacillus* species are the best-known and most-studied biosurfactant-producing bacteria. Currently, biosurfactants of this genus have been reported to have applications in pest control. These strains produce broad-spectrum lipopeptides,

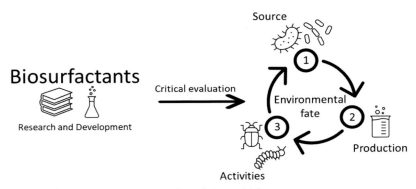

Figure 14.3 Biosurfactant as environmentally safe insecticides.

including surfactin, iturin, bacillomycin, fengycin, and lichenisin (Ghribi, Abdelkefi-Mesrati, et al., 2012; Ghribi, Elleuch, et al., 2012; Mukherjee & Das, 2005). These are bioactive metabolites that cause hemolysis with potent larvicidal activity (Geetha et al., 2011). *Bacillus strains* were considered natural factories of biologically active compounds, such as lipopeptides (Ghribi, Elleuch, et al., 2012), and the importance of their involvement in the control of plant diseases has been demonstrated over the years.

Surfactin produced by *Bacillus amyloliquefaciens* G1 showed insecticidal potential against the green peach aphid, *Myzus persicae*, affecting the aphid cuticles and inducing significant dehydration of the cuticle membrane, and thereby killing the insects (Cheon Yun et al., 2013). Furthermore, Ben Khedher et al. (2015) reported the potential of the biosurfactant produced by *B. amyloliquefaciens* AG1 to control *Tuta absoluta* larvae. The biosurfactant of this bacterium consists of lipopeptides and polyketides. This biosurfactant binds to receptors located on the vesicles of the brush border membrane of larvae. Likewise, the biosurfactants *Bacillus thuringiensis* Vip3Aa16 and *B. amyloliquefaciens* AG1 showed insecticidal potential against *Spodoptera littoralis*. Histopathological examination of the treated larval midgut revealed vacuolization, necrosis, and disintegration of the basement membrane (Ben Khedher et al., 2017).

The insecticidal activity was also shown against storage insect pests. The *B. subtilis* strain SPB1 was shown to produce a lipopeptide biosurfactant and its insecticidal activity evaluated against the locust bean moth, *Ectomielois ceratoniae* (Mnif et al., 2013). The histopathological effects of the biosurfactant *B. subtilis* SPB1 in the midgut of *E. ceratoniae* were also studied to show the formation of vesicles in the apical region of the cells, showing lysis and strong vacuolization of columnar cells. Similar histopathological effects were observed in *Ephestia kuehniella* (Ghribi, Elleuch, Abdelkefi-Mesrati, et al., 2012), *Spodoptera littoralis*, and *Prays oleae* (Ghribi, Abdelkefi-Mesrati, et al., 2012).

Treatment with the partially recovered biosurfactant, identified as a surfactin lipopeptide and produced by *Bacillus clausii* BS02, led to a 50% mortality against *Callosobruchus chinensis* adults and 100% mortality against *Maconellicoccus hirsutus* (Hazra et al., 2015). This study suggested that as the insect crawls on the contact surfaces, the absorption of biosurfactants can occur in two ways: epicuticular waxes present in the host cuticle or through the intersegmental membranes, leading to leakage or dehydration of the cellular components of the host insect.

Bioassays performed with the biosurfactant produced by *Bacillus subtilis* V26 showed that *T. absoluta* larvae were sensitive to the surfactant. The histopathological study of midgut larvae of *T. absoluta* treated by the V26 biosurfactant showed severe damage, indicating that the midgut tissue is a site of biosurfactant action. Compared to untreated larvae, V26 biosurfactant caused necrosis, vacuolization of the cytoplasm, formation of vesicles in the apical region of cells toward the midgut lumen, and detachment of epithelial cells from the basement membrane (Ben Khedher et al., 2020).

Another *Bacillus* species, *B. thuringiensis*, has also been used in agriculture as a biological control agent. This bacterium produces crystalline proteins that are toxic to species of Lepidoptera, Coleoptera, Diptera, and Hymenoptera, among others. Industrial bioprocesses involving different culture media and culture conditions in the development of *B. thuringiensis* strains were studied (Cuatlayotl-Cottier et al., 2020; Kim et al., 2004; Salazar-Magallon et al., 2015); for example, Kim et al. (2004) reported that *B. thuringiensis* produced a lipopeptide with insecticidal activity against the larvae of *Pieris rapae crucivora*.

A recent study evaluated the insecticidal activity of a biosurfactant produced by *B. thuringiensis* against third and fourth instar aphids *Melanaphis sacchari*, *Rhophalosiphum maidis*, *Aphis fabae*, and spider mite, *Tetranychus urticae* (Cuatlayotl-Cottier et al., 2020), in which laboratory mortalities were between 56% and 77%, not reaching mortality rates as high as the control. However, in a greenhouse, it was observed that the biosurfactant had the same effect on the insects evaluated as a chemical insecticide used as a control. This work highlighted that no visual phytotoxic damage was observed in the plants evaluated during the test period, demonstrating potential as alternative insecticides in controlling certain aphids and mites. Cuatlayotl-Cottier et al. (2020) also observed that aphids and mites nymphs showed evidence of cuticle lysis, in which the cuticular membranes became much thinner after treatment, and the membranes were dehydrated and separated from cellular components. These observations suggested that lipopeptides or rhamnolipids can impact the cuticular membrane and kill aphids and mites.

Aphid and mite cell membranes are composed of a complex mixture of alkanes, wax esters, fatty acids and phospholipids and can be lysed by the action of biosurfactants (Dillwith et al., 1993). Due to their amphiphilic nature, biosurfactants can interact with biological membranes composed of amphiphilic lipid bilayers and act as broad-spectrum cytolytic agents, destroying membranes. These differences in biological activity can be attributed to the structural properties of the biosurfactant and the membrane, which determines whether the biosurfactant causes harmful effects on the insect membrane (Ongena & Jacques, 2008).

This shows that the biosurfactants produced by *Bacillus* species target the midgut tissue to cause the death of the treated insects (Fig. 14.4); therefore, oral application method is the effective form of treatment (Edosa et al., 2018). Furthermore, the molecular mechanisms of these biosurfactants should be studied to fully understand their exact modes of action and the large-scale production of new insecticides from this bacterial species needs further investigation.

14.4.2 Biosurfactants as a green pesticide for vectors of diseases

The etiology of vector-borne diseases through mosquitoes is prevalent worldwide. Many fatal diseases are caused by mosquitoes from *Anopheles*, *Culex*, and *Aedes*, which

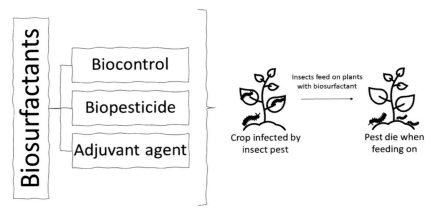

Figure 14.4 Biosurfactants as an effective treatment form for insect pests on crops.

demands economically viable and ecologically correct vector control strategies (Pradhan et al., 2018). Recently, the insecticidal effect of biosurfactants has also been explored against vectors of tropical diseases (Table 14.2), and this application is of extreme interest for public health programs aimed at controlling these urban pests (Silva et al., 2015).

Aedes aegypti has been around the planet for centuries and has been recorded on all continents, mainly in tropical and subtropical countries. This species of mosquito is responsible for transmitting diseases such as dengue, yellow fever, Zika, and chikungunya, causing global concern in recent years (Van De Beek & Brouwer, 2017). Various synthetic insecticides are used to control *A. aegypti*; however, the toxicity of these compounds and the long-term acquired resistance of the mosquito demand the replacement of these insecticides (Braga & Valle, 2007).

With the advancement of microbial biotechnology and green chemistry, studies that evaluate natural substances and methods of eliminating *A. aegypti* have been increasing; for example, these natural substances are applied as larvicides, being one of the widely used means to control the mosquito population (Sun et al., 2014). For example, the biosurfactant produced by the yeast *Wickerhamomyces anomalus* CCMA 0358, using kitchen waste oil as a carbon source, was evaluated against *A. aegypti* larvae and proved to be efficient at the lowest concentration evaluated, in which it eliminated up to 100% of the larvae in 24 h of contact (Fernandes et al., 2020; Silva et al., 2015), evaluated a biosurfactant produced from *Pseudomonas aeruginosa* using sunflower oil as a carbon source against *A. aegypti* larvae, eliminating 100% of the larvae in 48 h.

Another mosquito that poses problems for humans, the Asian tiger *Aedes albopictus*, is one of the most invasive mosquitoes in the world, as it successfully competes with native species of mosquitoes and has a high potential for invasion, in addition to having vector capacity and ecological plasticity (Fischer et al., 2011). The *Ae. albopictus* mosquito is of sociomedical concern due to its diurnal aggressive behavior of biting

Table 14.2 Biosurfactant used for insect vectors of diseases.

Microorganism	Biosurfactant	Activity	Insect	References
Bacillus amyloliquefaciens (VCRC B483)	Not given	Larvicidal and pupicidal	*Culex quinquefasciatus*, *Anopheles stephensi*, *Aedes aegypti*	Geetha et al. (2014)
Bacillus safensis BacI67 and *Bacillus paranthracis* C21	Lipopeptide	Larvicidal	*A. aegypti*	Santos et al. (2021)
Bacillus subtilis (VCRCB471)	Surfactin	Pupicidal	*A. stephensi*	Geetha and Manonmani (2010)
Bacillus subtilis A1 and *Pseudomonas stutzeri* NA3	Lipopeptide	Larvicidal and pupicidal	*A. stephensi*	Parthipan et al. (2018)
Pseudomonas aeruginosa	Rhaminolipid	Larvicidal	*A. aegypti*	Silva et al. (2015)
Pseudomonas fluorescens Migula (VCRC B426)	Di-rhamnolipid	Pupicidal	*C. quinquefasciatus*, *A. stephensi*, and *Aedes aegypti*	Prabakaran et al. (2015)
Salmonella bongori BH11	Rhaminolipid	Larvicidal	*C. quinquefasciatus*	Govindasamy et al. (2018)
Scheffersomyces stipitis NRRL Y-7124	Glycolipid	Larvicidal	*A. aegypti*	Franco Marcelino et al. (2017)
Ustilago maydis	Mannosylerythritol lipids	Larvicidal and pupicidal	*Aedes albopictus*	Ga'al et al. (2021)
Wickerhamomyces anomalus CCMA 0358	Not given	Larvicidal	*A. aegypti*	Fernandes et al. (2020)

humans, which can transmit dengue, chikungunya, and Zika viruses (Sherpa et al., 2018). Recently, the biosurfactant Mannosylerythritol lipids produced from *Ustilago maydis* was evaluated against larvae and pupae of *Ae. albopictus*, obtaining between 50% and 90% of deaths (Ga'al et al., 2021). This study suggests that the biosurfactant can be used as an efficient tool in mosquito vector control strategies.

Anopheles stephensi acts as a vector for *Plasmodium* parasites, responsible for malaria in tropical and subtropical areas worldwide (Benelli & Mehlhorn, 2016). The biosurfactant produced by two bacteria, *Bacillus subtilis* A1 and *Pseudomonas stutzeri* NA3, were evaluated for insecticide applications against malarial mosquitoes and presented between 50% and 90% mortality in all larval and pupal stages (Parthipan et al., 2018). The toxic activity of these biosurfactants on all young instars of *A. stephensi*, as well as their significant impact on adult longevity and fecundity, allows their further consideration for developing insecticides to combat malarial mosquitoes.

The sensibility of larvae of *Anopheles culicifacies* was also evaluated against a cyclic lipopeptide; the biosurfactant showed a reduction of 50% (Pradhan et al., 2018). Likewise, Geetha et al. (2012) tested the surfactin produced by *B. subtilis* against the pupae of *A. stephensi*, the surfactant being active against pupae. The larvicidal activity of another cyclic lipopeptide was also tested in *Culex quinquefasciatus* larvae (Mukherjee & Das, 2005).

According to Christophers (1960), the interaction of biosurfactants with the respiratory siphon, a hydrophobic region, provides greater exposure to the aqueous medium in which water flows through the spiracular cavity of the larvae. In addition, the tracheas, which reach close to the water and facilitate the ascent of the larva to the surface for respiration, are affected by the presence of biosurfactants. Biosurfactants decrease the surface tension of the water, preventing the larvae from staying on the surface and leading to the death of the larvae by drowning.

In addition, mosquito larvae and pupae need atmospheric oxygen to breathe. Due to the reduction of surface tension of water by the action of lipopeptides, larvae and pupae cannot reach oxygen, which could lead to their death (Piper & Maxwell, 1971). Biosurfactants have been reported to potently bind to sulfur groups on DNA, leading to rapid denaturation of organelles and enzymes in the mosquito, followed by decreases in membrane permeability and disturbances in proton motive force that can cause loss of cell function and death (Benelli & Mehlhorn, 2016). However, further investigations on this issue are needed.

14.5 Conclusion and future direction

Biosurfactants have become a promising biopesticide in insect pest management. Although dozens of species of bacteria and fungi have been reported to produce biosurfactants, only a few biosurfactants have been well studied against insect pests and vectors. Most of the biosurfactants explored were produced on a small scale in laboratories; however, the mass production of these biosurfactants needs detailed studies to produce efficient quality products. The targets of most biosurfactants and their modes of action in insects remain elusive. Therefore, advanced biochemical and molecular approaches such as genomic and transcriptomic studies are crucial in the study of their methods of action, leading to the discovery of new biopesticides. Thus, it can be concluded that biosurfactants will be the biopesticides that will replace synthetic pesticides and insecticides used at home and, thus, studies in the areas of molecular biology, biochemistry, microbiology, and environmental sciences are appreciable.

References

Ben Khedher, S., Boukedi, H., Dammak, M., Kilani-Feki, O., Sellami-Boudawara, T., Abdelkefi-Mesrati, L., & Tounsi, S. (2017). Combinatorial effect of *Bacillus amyloliquefaciens* AG1 biosurfactant and *Bacillus thuringiensis* Vip3Aa16 toxin on Spodoptera littoralis larvae. *Journal of Invertebrate Pathology*, *144*, 11−17. Available from https://doi.org/10.1016/j.jip.2017.01.006.

Ben Khedher, S., Boukedi, H., Kilani-Feki, O., Chaib, I., Laarif, A., Abdelkefi-Mesrati, L., & Tounsi, S. (2015). *Bacillus amyloliquefaciens* AG1 biosurfactant: Putative receptor diversity and histopathological effects on *Tuta absoluta* midgut. *Journal of Invertebrate Pathology*, *132*, 42−47. Available from https://doi.org/10.1016/j.jip.2015.08.010.

Ben Khedher, S., Boukedi, H., Laarif, A., & Tounsi, S. (2020). Biosurfactant produced by Bacillus subtilis V26: A potential biological control approach for sustainable agriculture development. *Organic Agriculture*, *10*, 117−124. Available from https://doi.org/10.1007/s13165-020-00316-0.

Benelli, G., & Mehlhorn, H. (2016). Declining malaria, rising of dengue and Zika virus: Insights for mosquito vector control. *Parasitology Research*, *115*(5), 1747−1754. Available from https://doi.org/10.1007/s00436-016-4971-z.

Blackwell, P. S. (2000). Management of water repellency in Australia, and risks associated with preferential flow, pesticide concentration and leaching, *Journal of Hydrology* (Vols. 231−232, pp. 384−395). Elsevier Science B.V. Available from https://doi.org/10.1016/S0022-1694(00)00210-9.

Braga, I. A., & Valle, D. (2007). *Aedes aegypti*: Vigilância, monitoramento da resistência e alternativas de controle no Brasil. *Epidemiologia e Serviços de Saúde*, *16*(4). Available from https://doi.org/10.5123/s1679-49742007000400007.

Carvalho, F. P. (2017). Pesticides, environment, and food safety. *Food and Energy Security*, *6*(2), 48−60. Available from https://doi.org/10.1002/fes3.108.

Cheon Yun, D., Yang, S. Y., Kim, Y. C., Kim, I. S., & Kim, Y. H. (2013). Identification of surfactin as an aphicidal metabolite produced by *Bacillus amyloliquefaciens* G1. *Journal of the Korean Society for Applied Biological Chemistry*, *56*, 751−753. Available from https://doi.org/10.1007/s13765-013-3238-y.

Chowdhury, S. P., Uhl, J., Grosch, R., Alquéres, S., Pittroff, S., Dietel, K., Schmitt-Kopplin, P., Borriss, R., & Hartmann, A. (2015). Cyclic lipopeptides of *Bacillus amyloliquefaciens* subsp. plantarum colonizing the lettuce rhizosphere enhance plant defense responses toward the bottom rot pathogen Rhizoctonia solani. *Molecular Plant-Microbe Interactions*, *28*(9), 984−995. Available from https://doi.org/10.1094/MPMI-03-15-0066-R.

Christophers, S. (1960). Aedes Aegypti *(L.) the Yellow Fever Mosquito: Its Life History, Bionomics and Structure*. Cambridge University Press.

Cuatlayotl-Cottier, R., Huerta-De La, P., Peña-Chora, J., & Magallón, S. (2020). Insecticidal activity of industrial by-products fermented by *Bacillus thuringiensis* strain GP139 against Mites (Prostigmata: Tetranychidae) and Aphids (Hemiptera: Aphidoidea). *Biocontrol Science and Technology*, *32*(1). Available from https://doi.org/10.1080/09583157.2021.1961686.

Dillwith, J. W., Neese, P. A., & Brigham, D. L. (1993). *Insect lipids: Aphids; chemistry, biochemistry, and biology*.

do Nascimento Silva, J., Mascarin, G. M., de Paula Vieira de Castro, R., Castilho, L. R., & Freire, D. M. G. (2019). Novel combination of a biosurfactant with entomopathogenic fungi enhances efficacy against Bemisia whitefly. *Pest Management Science*, *75*(11), 2882−2891. Available from https://doi.org/10.1002/ps.5458.

Edosa, T. T., Hun, Jo, Y., Keshavarz, M., & Soo Han, Y. (2018). Biosurfactants: Production and potential application in insect pest management. *Trends in Entomology*, *14*, 79. Available from https://doi.org/10.31300/tent.14.2018.79-87.

Fazaeli, N., Bahador, N., & Hesami, S. (2021). A study on larvicidal activity and phylogenetic analysis of *Staphylococcus epidermidis* as a biosurfactant-producing bacterium. *Polish Journal of Environmental Studies*, *30*(5), 4511−4519. Available from https://doi.org/10.15244/pjoes/132807.

Fernandes, N. A. T., de Souza, A. C., Simões, L. A., Ferreira dos Reis, G. M., Souza, K. T., Schwan, R. F., & Dias, D. R. (2020). Eco-friendly biosurfactant from *Wickerhamomyces anomalus* CCMA 0358 as larvicidal and antimicrobial. *Microbiological Research*, *241*. Available from https://doi.org/10.1016/j.micres.2020.126571.

Fischer, D., Thomas, S. M., Niemitz, F., Reineking, B., & Beierkuhnlein, C. (2011). Projection of climatic suitability for *Aedes albopictus* Skuse (Culicidae) in Europe under climate change conditions. *Global and Planetary Change*, *78*(1−2), 54−64. Available from https://doi.org/10.1016/j.gloplacha.2011.05.008.

Franco Marcelino, P. R., Da Silva, V. L., Rodrigues Philippini, R., Von Zuben, C. J., Contiero, J., Dos Santos, J. C., & Da Silva, S. S. (2017). Biosurfactants produced by Scheffersomyces stipitis cultured in sugarcane bagasse hydrolysate as new green larvicides for the control of *Aedes aegypti*, a vector of neglected tropical diseases. *PLoS One*, *12*(11). Available from https://doi.org/10.1371/journal.pone.0187125.

Ga'al, H., Yang, G., Fouad, H., Guo, M., & Mo, J. (2021). Mannosylerythritol lipids mediated biosynthesis of silver nanoparticles: An eco-friendly and operative approach against chikungunya vector *Aedes albopictus*. *Journal of Cluster Science*, *32*(1), 17−25. Available from https://doi.org/10.1007/s10876-019-01751-0.

Geetha, I., Arana, R., & Manonmani, A. M. (2014). Mosquitocidal *Bacillus amyloliquefaciens*: Dynamics of growth & production of novel pupicidal biosurfactant. *Indian Journal of Medical Research*, *140* (September), 427−434. Available from http://www.icmr.nic.in/ijmr/2014/september/0915.pdf.

Geetha, I., & Manonmani, A. M. (2010). Surfactin: A novel mosquitocidal biosurfactant produced by Bacillus subtilis ssp. subtilis (VCRC B471) and influence of abiotic factors on its pupicidal efficacy. *Letters in Applied Microbiology*, *51*(4), 406−412. Available from https://doi.org/10.1111/j.1472-765X.2010.02912.x.

Geetha, I., Manonmani, A. M., & Prabakaran, G. (2011). *Bacillus amyloliquefaciens*: A mosquitocidal bacterium from mangrove forests of Andaman & Nicobar Islands, India. *Acta Tropica*, *120*(3), 155−159. Available from https://doi.org/10.1016/j.actatropica.2011.07.006.

Geetha, I., Paily, K. P., & Manonmani, A. M. (2012). Mosquito adulticidal activity of a biosurfactant produced by Bacillus subtilis subsp. subtilis. *Pest Management Science*, *68*(11), 1447−1450. Available from https://doi.org/10.1002/ps.3324.

Ghribi, D., Abdelkefi-Mesrati, L., Boukedi, H., Elleuch, M., Ellouze-Chaabouni, S., & Tounsi, S. (2012). The impact of the Bacillus subtilis SPB1 biosurfactant on the midgut histology of Spodoptera littoralis (Lepidoptera: Noctuidae) and determination of its putative receptor. *Journal of Invertebrate Pathology*, *109*(2), 183−186. Available from https://doi.org/10.1016/j.jip.2011.10.014.

Ghribi, D., Elleuch, M., Abdelkefi, L., & Ellouze-Chaabouni, S. (2012). Evaluation of larvicidal potency of Bacillus subtilis SPB1 biosurfactant against *Ephestia kuehniella* (Lepidoptera: Pyralidae) larvae and influence of abiotic factors on its insecticidal activity. *Journal of Stored Products Research*, *48*, 68−72. Available from https://doi.org/10.1016/j.jspr.2011.10.002.

Ghribi, D., Elleuch, M., Abdelkefi-Mesrati, L., Boukadi, H., & Ellouze-Chaabouni, S. (2012). Histopathological effects of Bacillus subtilis SPB1 biosurfactant in the midgut of Ephestia kuehniella (Lepidoptera: Pyralidae) and improvement of its insecticidal efficiency. *Journal of Plant Diseases and Protection*, *119*(1), 24−29. Available from https://doi.org/10.1007/BF03356415.

Govindasamy, B., Paramasivam, D., Dilipkumar, A., Ramalingam, K. R., Chinnaperumal, K., & Pachiappan, P. (2018). Multipurpose efficacy of the lyophilized cell-free supernatant of Salmonella bongori isolated from the freshwater fish, Devario aequipinnatus: Toxicity against microbial pathogens and mosquito vectors. *Environmental Science and Pollution Research*, *25*(29), 29162−29180. Available from https://doi.org/10.1007/s11356-018-2838-9.

Hasna, M. K., Mårtensson, A., Persson, P., & Rämert, B. (2007). Use of composts to manage corky root disease in organic tomato production. *Annals of Applied Biology*, *151*(3), 381−390. Available from https://doi.org/10.1111/j.1744-7348.2007.00178.x.

Hazra, C., Kundu, D., & Chaudhari, A. (2015). Lipopeptide biosurfactant from *Bacillus clausii* BS02 using sunflower oil soapstock: Evaluation of high throughput screening methods, production, purification, characterization and its insecticidal activity. *RSC Advances*, *5*(4), 2974−2982. Available from https://doi.org/10.1039/c4ra13261k.

Jang, J. Y., Yang, S. Y., Kim, Y. C., Lee, C. W., Park, M. S., Kim, J. C., & Kim, I. S. (2013). Identification of orfamide A as an insecticidal metabolite produced by Pseudomonas protegens F6. *Journal of Agricultural and Food Chemistry*, *61*(28), 6786−6791. Available from https://doi.org/10.1021/jf401218w.

Kim, P. I., Bai, H., Bai, D., Chae, H., Chung, S., Kim, Y., Park, R., & Chi, Y. T. (2004). Purification and characterization of a lipopeptide produced by *Bacillus thuringiensis* CMB26. *Journal of Applied Microbiology*, *97*(5), 942−949. Available from https://doi.org/10.1111/j.1365-2672.2004.02356.x.

Kim, S. K., Kim, Y. C., Lee, S., Kim, J. C., Yun, M. Y., & Kim, I. S. (2011). Insecticidal activity of rhamnolipid isolated from Pseudomonas sp. EP-3 against green peach aphid (*Myzus persicae*). *Journal of Agricultural and Food Chemistry*, *59*(3), 934−938. Available from https://doi.org/10.1021/jf104027x.

Letourneau, D., & Bruggen. (2006). Crop protection. *Organic Agriculture: A Global Perspective* (pp. 93−121). CSIRO.

Maity, S., Acharjee, A., & Saha, B. (2022). Application of biosurfactant as biocontrol agents against soil-borne and root-borne plant pathogens. *Applications of Biosurfactant in Agriculture* (pp. 283−302). Elsevier BV. Available from https://doi.org/10.1016/b978-0-12-822921-7.00015-5.

Marcelino, P. R. F., Gonçalves, F., Jimenez, I. M., Carneiro, B. C., Santos, S. B., & da Silva, S. S (2020). *Sustainable production of biosurfactants and their applications*. *Lignocellulosic Biorefining Technologies*. John Wiley & Sons Ltd. Available from https://doi.org/10.1002/9781119568858.ch8.

Mnif, I., Elleuch, M., Chaabouni, S. E., & Ghribi, D. (2013). Bacillus subtilis SPB1 biosurfactant: Production optimization and insecticidal activity against the carob moth *Ectomyelois ceratoniae*. *Crop Protection*, *50*, 66−72. Available from https://doi.org/10.1016/j.cropro.2013.03.005.

Mnif, I., & Ghribi, D. (2016). Glycolipid biosurfactants: Main properties and potential applications in agriculture and food industry. *Journal of the Science of Food and Agriculture*, *96*(13), 4310−4320. Available from https://doi.org/10.1002/jsfa.7759.

Mnif, I., & Ghribi, D. (2015). Potential of bacterial derived biopesticides in pest management. *Crop Protection*, *77*, 52−64. Available from https://doi.org/10.1016/j.cropro.2015.07.017.

Mukherjee, A. K., & Das, K. (2005). Correlation between diverse cyclic lipopeptides production and regulation of growth and substrate utilization by *Bacillus subtilis* strains in a particular habitat. *FEMS Microbiology Ecology*, *54*(3), 479−489. Available from https://doi.org/10.1016/j.femsec.2005.06.003.

Mulqueen, P. (2003). Recent advances in agrochemical formulation. *Advances in Colloid and Interface Science*, *106*(1−3), 83−107. Available from https://doi.org/10.1016/S0001-8686(03)00106-4.

Nitschke, M., & Silva, S. S. e. (2017). Recent food applications of microbial surfactants. *Critical Reviews in Food Science and Nutrition*, *58*(4), 631−638. Available from https://doi.org/10.1080/10408398.2016.1208635.

Ongena, M., & Jacques, P. (2008). *Bacillus lipopeptides*: Versatile weapons for plant disease biocontrol. *Trends in Microbiology*, *16*(3), 115−125. Available from https://doi.org/10.1016/j.tim.2007.12.009.

Parthipan, P., Sarankumar, R. K., Jaganathan, A., Amuthavalli, P., Babujanarthanam, R., Rahman, P. K. S. M., Murugan, K., Higuchi, A., Benelli, G., & Rajasekar, A. (2018). Biosurfactants produced by Bacillus subtilis A1 and Pseudomonas stutzeri NA3 reduce longevity and fecundity of *Anopheles stephensi* and show high toxicity against young instars. *Environmental Science and Pollution Research*, *25*(11), 10471−10481. Available from https://doi.org/10.1007/s11356-017-0105-0.

Piper, W. D., & Maxwell, K. E. (1971). Mode of action of surfactants on mosquito pupae. *Journal of Economic Entomology*, *64*(3), 601−606. Available from https://doi.org/10.1093/jee/64.3.601.

Popp, J., Pető, K., & Nagy, J. (2013). Pesticide productivity and food security. A review. *Agronomy for Sustainable Development*, *33*(1), 243−255. Available from https://doi.org/10.1007/s13593-012-0105-x.

Prabakaran, G., Hoti, S. L., Prakash Rao, H. S., & Vijjapu, S. (2015). Di-rhamnolipid is a mosquito pupicidal metabolite from Pseudomonas fluorescens (VCRC B426). *Acta Tropica*, *148*, 24−31. Available from https://doi.org/10.1016/j.actatropica.2015.03.003.

Pradhan, A. K., Rath, A., Pradhan, N., Hazra, R. K., Nayak, R. R., & Kanjilal, S. (2018). Cyclic lipopeptide biosurfactant from Bacillus tequilensis exhibits multifarious activity. *3 Biotech*, *8*(6). Available from https://doi.org/10.1007/s13205-018-1288-x.

Rostás, M., & Blassmann, K. (2009). Insects had it first: Surfactants as a defence against predators. *Proceedings of the Royal Society B: Biological Sciences*, *276*(1657), 633−638. Available from https://doi.org/10.1098/rspb.2008.1281.

Sachdev, D. P., & Cameotra, S. S. (2013). Biosurfactants in agriculture. *Applied Microbiology and Biotechnology*, *97*(3), 1005−1016. Available from https://doi.org/10.1007/s00253-012-4641-8.

Salazar-Magallon, J. A., Hernandez-Velazquez, V. M., Alvear-Garcia, A., Arenas-Sosa, I., & Peña-Chora, G. (2015). Evaluation of industrial by-products for the production of *Bacillus thuringiensis* strain GP139 and the pathogenicity when applied to Bemisia tabaci nymphs. *Bulletin of Insectology*, *68*(1), 103−109. Available from http://www.bulletinofinsectology.org/.

Santos, E. M. d. S., Lira, I. R. A. d. S., Meira, H. M., Aguiar, J. D. S., Rufino, R. D., de Almeida, D. G., Casazza, A. A., Converti, A., Sarubbo, L. A., & de Luna, J. M. (2021). Enhanced oil removal by a nontoxic biosurfactant formulation. *Energies*, *14*(2). Available from https://doi.org/10.3390/en14020467.

Sarwar, M. (2015). Biopesticides: An effective and environmental friendly insect-pests inhibitor line of action. *International Journal of Engineering and Advanced Research Technology*, *1*(2), 10−15.

Savary, S., Willocquet, L., Pethybridge, S. J., Esker, P., McRoberts, N., & Nelson, A. (2019). The global burden of pathogens and pests on major food crops. *Nature Ecology & Evolution*, *3*(3), 430−439. Available from https://doi.org/10.1038/s41559-018-0793-y.

Shekhar, S., Sundaramanickam, A., & Balasubramanian, T. (2015). Biosurfactant producing microbes and their potential applications: A review. *Critical Reviews in Environmental Science and Technology*, *45*(14), 1522−1554. Available from https://doi.org/10.1080/10643389.2014.955631.

Sherpa, S., Rioux, D., Pougnet-Lagarde, C., & Després, L. (2018). Genetic diversity and distribution differ between long-established and recently introduced populations in the invasive mosquito *Aedes albopictus*. *Infection, Genetics and Evolution*, *58*, 145−156. Available from https://doi.org/10.1016/j.meegid.2017.12.018.

Silva, V. L., Lovaglio, R. B., Von Zuben, C. J., & Contiero, J. (2015). Rhamnolipids: Solution against *Aedes aegypti*? *Frontiers in Microbiology*, *6*. Available from https://doi.org/10.3389/fmicb.2015.00088.

Sun, D., Williges, E., Unlu, I., Healy, S., Williams, G. M., Obenauer, P., Hughes, T., Schoeler, G., Gaugler, R., Fonseca, D., & Farajollahi, A. (2014). Taming a tiger in the city: Comparison of motorized backpack applications and source reduction against the Asian Tiger Mosquito, *Aedes albopictus*. *Journal of the American Mosquito Control Association*, *30*(2), 99−105. Available from https://doi.org/10.2987/13-6394.1.

Van De Beek, D., & Brouwer, M. C. (2017). CNS Infections in 2016: 2016, the year of Zika virus. *Nature Reviews Neurology*, *13*(2), 69−70. Available from https://doi.org/10.1038/nrneurol.2016.202.

Wattanaphon, H. T., Kerdsin, A., Thammacharoen, C., Sangvanich, P., & Vangnai, A. S. (2008). A biosurfactant from *Burkholderia cenocepacia* BSP3 and its enhancement of pesticide solubilization. *Journal of Applied Microbiology*, *105*(2), 416−423. Available from https://doi.org/10.1111/j.1365-2672.2008.03755.x.

Yang, S. Y., Lim, D. J., Noh, M. Y., Kim, J. C., Kim, Y. C., & Kim, I. S. (2017). Characterization of biosurfactants as insecticidal metabolites produced by Bacillus subtilis Y9. *Entomological Research*, *47*(1), 55−59. Available from https://doi.org/10.1111/1748-5967.12200.

Yogev, A., Raviv, M., Kritzman, G., Hadar, Y., Cohen, R., Kirshner, B., & Katan, J. (2009). Suppression of bacterial canker of tomato by composts. *Crop Protection*, *28*(1), 97−103. Available from https://doi.org/10.1016/j.cropro.2008.09.003.

Zaller, J. G., & Brühl, C. A. (2019). Editorial: Non-target effects of pesticides on organisms inhabiting agroecosystems. *Frontiers in Environmental Science*, *7*. Available from https://doi.org/10.3389/fenvs.2019.00075.

Zhao, P., Quan, C., Wang, Y., Wang, J., & Fan, S. (2014). *Bacillus amyloliquefaciens* Q-426 as a potential biocontrol agent against *Fusarium oxysporum* f. sp. spinaciae. *Journal of Basic Microbiology*, *54*(5), 448−456. Available from https://doi.org/10.1002/jobm.201200414.

CHAPTER 15

Potential of biosurfactants as antiadhesive biological coating

John Adewole Alara[1,2]
[1]Department of Chemical Engineering, College of Engineering, University Malaysia Pahang, Gambang, Pahang, Malaysia
[2]Oyo State Primary Healthcare Board, Ogbomoso, Oyo State, Nigeria

15.1 Introduction

Most pathogens that cause biofilms formation in the natural environment have been structured to be less sensitive to drugs, as revealed by Janek et al. (2012). Several laboratories have been synthesizing new molecules to prevent the development of biofilms (Zhao & Liu, 2010). Adhesion has been regarded as the initial phase of biofilm formation, and this can be the best time for the effect of antibiofilm and anti-adhesive molecules. Microbial biosurfactants (BSs) are found to be potential molecules with anti-adhesive and antimicrobial properties and sometimes penetrate and eliminate mature biofilms (Abdalsadiq et al., 2018). Microbial BSs ranging from lower molecular mass, such as lipopeptides (LPs), glycolipids (GLs), rhamnolipids (RLs), and sophorolipids (SLs), to higher molecular mass, including protein, lipoproteins, and lipopolysaccharides (LPS), can bind to interfaces and prevent the adhesion of pathogens to various surfaces (Das et al., 2009). They can serve as a substitute for synthetic surface-active agents and may be utilized as effective and safe therapeutic agents owing to their lower toxicity, higher biodegradability, and effectiveness at high pH values and temperatures (Haddaji et al., 2022). These BS compounds may be used to coat implant surfaces and preserve biocompatibility because of their lower toxicity. The BSs' amphiphilic structure can affect the adhesion between pathogens and surfaces, inhibiting the bacteria adhesion processes. It is also capable of modifying the permeability of the cell membrane, resulting in the loss of metabolites that later make the cell lyse Table 15.1 (Rodrigues & Teixeira, 2010).

Microbial adhesion and biofilm development on medical instruments are well-known activities that have significant economic and medical consequences. In recent years, the application of permanent or temporary prosthetic instruments fabricated from polymeric biomaterials has significantly increased. Rodrigues (2011) reported that more than five million medical implants are utilized annually within the United States alone. It has also been stated that medical instruments have contributed to about 60%—70% of hospital-acquired infections, especially in severally ill patients. Biofilm

Table 15.1 Anti-adhesive biological coating properties of different biosurfactants.

S/N	Class of biosurfactant	Biosurfactant name	Name of pathogen-producing biosurfactant	Anti-adhesive coating property against	Surface coated	References
1	Cyclic lipopeptide	Pseudofactin II	Pseudomonas fluorescens BD5	E. coli, Enterococcus hirae, E. feacalis, Proteus mirabilis, S. epidermidis, and two C. albicans	Silicone, glass, and polystyrene	Janek et al. (2012)
2	Not specified	Not specified	Lactobacillus brevis and Bacillus sp.	P. aeruginosa, Salmonella typhimurium, E. coli, S. aureus, Streptococcus mutans, Enterococcus faecalis,	Polystyrene	Rodrigues (2011)
3	Glycolipids	Crude biosurfactants	Lactobacillus rhamnosus	E. coli, S. aureus, P. aeruginosa, Bacillus subtilis	Glass	Patel et al. (2021)
4		Rufisan, lunasan	Candida lipolytica, C. sphaerica	Lactobacillus casei, S. epidermidis, E. coli		Rufino et al. (2011)
5	Lipopeptide and glycolipid fractions		Lactobacillus acidophilus and L. pentousus	C. albicans, K. pneumonia, S. pneumoniae, S. aureus, and P. mirabilis	Microtiter plates	Abdalsadiq et al. (2018)
6	Lipopeptides	Fengycin-like family	B. subtilis V9T14, B. licheniformis V19T21	Gram-positive and Gram-negative strains	polystyrene	Rivardo et al. (2009)
7	Trehalolipid		Rhodococcus ruber IEGM 231	B. subtilis, P. fluorescens, C. glutamicum, M. luteus, E. coli	Polystyrene microplates	Kuyukina et al. (2015)
8	Glycoprotein	Not specified	Lactobacillus agilis CCUG31450	P. aeruginosa, S. agalactiae, S. aureus	Metallic surfaces, silicone rubber, polystyrene, glass, and epithelial cell surfaces	Gudiña et al. (2015)
9	Not specified	Rhamnolipid	Not specified	S. aureus, S. epidermidis	Titanium disks	Tambone et al. (2021)
10	Not specified	Biosurfactant	Lactobacillus acidophilus MTCC447	E.coli, Salmonella enteritidis, S. epidermidis, S. aureus	Table eggs	Shawkat et al. (2019)
11	Lipopeptides		Bacillus subtilis VSG4, B. licheniformis VS16	B. cereus ATCC 11778, S. typhimuriun ATCC 19430, S. aureus ATCC 29523	Not specified	Giri et al. (2019)
12	Not specified	Biosurfactant	Lactobacillus acidophilus ATCCC 4356	S. marcescens strains	Not specified	Shokouhfard et al. (2015)
13	Not specified	Rhamnolipids, Sodium dodecyl sulfate (SDS)	Not specified	Yarrowia lipolytica	Microtiter plate wells, polytetrafluoroethylene surfaces	Dusane et al. (2012)

diseases pose a potential clinical threat because of their resistance to antimicrobials and immune defense mechanisms, making it hard to treat many nosocomial diseases such as chronic prostatitis, periodontitis, cystic fibrosis, endocarditis, and otitis media (Zhao & Liu, 2010). Hence, the adhesion of BSs to solid surfaces may constitute new and powerful ways of inhibiting colonization by infectious pathogens and successive biofilm development (Rodrigues et al., 2006).

In the last few years, antimicrobial-resistant (AMR) bacteria have been on the rise owing to overuse and misuse of antibiotics for livestock production and human treatment. AMR bacteria were among the largest health threats because this resistance has become uncontrollable. In addition, it has been revealed that the appearance of the resistant pathogen has constantly risen, increasing multi-drug resistance (MDR) (Ventola, 2015). Tiedje et al. (2019) revealed that public and medical interests have increased in the transmission of MDR human microorganism. Ventola (2016) showed that the prevention of antimicrobial-resistant bacteria has been highly prioritized in hospitals and many other clinical settings. Hence, studies over the last 10 years have focused on biological methods to prevent this resistance and inhibition of microbial colonization within and transmissions among patients.

The anti-adhesive property of BSs in decreasing microbial adhesion and dispersal from various solid surfaces have been conveniently applied in different fields of medicine and many areas of industry (Singh & Cameotra, 2004). BSs produced from *Lactobacillus fermentum* and *L. acidophilus* absorbed on glass were used to reduce the number of attaching uropathogenic cells of *Enterococcus faecalis* to about 77% (Janek et al., 2012). Mireles et al. (2001) stated that using surfactin solution for precoating vinyl urethral catheters before inoculation with media has indicated a reduction in the number of biofilms developed by *Proteus, Escherichia coli, Salmonella enterica*, and *S. tyhimurium*. Rodrigues et al. (2006) developed techniques for the prevention of microbial colonization of silicone rubber voice prostheses. Several studies have shown that RLs display antibiofilm and antimicrobial properties and can inhibit many biological systems, including biofilm layers developed by Gram-positive (GPB), Gram-negative bacteria (GNB), and fungi strains (Peng et al., 2008).

Mandakhalikar et al. (2016) stated that new technologies have been developed for preventing the development of biofilms on medical instruments, including adhesion-resistant and coating surfaces. A studies has revealed the increasing confirmation of the effect of probiotics in prevention and prevention and treatment of device-related biofilms, and an increased focus on encouraging natural methods to improve health has strengthened the study within the area of interest of probiotics and their metabolites in combating infectious biofilms (Azevedo et al., 2017). The de Melo Pereira et al. (2018) study has shown that probiotics have attracted significant interest in terms of their health-promoting activities, receiving the status of Generally Regarded as Safe (GRAS). The commonly used probiotics include species of lactic acid bacteria (LAB), such as *Lactobacillus, Streptococcus, Bifidobacterium, Leuconostoc*, and *Lactococcus*. The major antimicrobial agents secreted by probiotic cells are biosurfactants, hydrogen peroxide and sulfide, bacteriocins, organic acids, carbon dioxide, ethanol, and exopolysaccharides (EPS) (de Melo Pereira et al., 2018).

Lellouche et al. (2012) reported that different kinds of nanomaterials showed antibiofilm and antibacterial properties. These nanomaterials are metal and metal oxide nanoparticles, such as iron, oxide, zinc oxide, copper oxide, and silver oxide, and they have attracted much interest for several years. They have recently been used extensively in the field of biomedical (Kim et al., 2018). Rhamnolipid's antibiofilm and anti-adhesive properties have been integrated with the nanomaterial antimicrobial property of metal and metal oxide nanoparticles (Khalid et al., 2019). The two well-known RLs include mono-rhamnolipid and di-rhamnolipid. Several studies have revealed that rhamnolipid have antimicrobial and antibiofilm activities that inhibit biofilm layers developed by *S. aureus, Listeria monocytogens, B. pumilus*, and *Yarrowia lipolytica* (Kim et al., 2015).

The RLs are considered green antibiofilm molecules because of their anti-adhesive, non-toxicity, biocompatibility, and antifungal properties (Zeraik & Nitschke, 2010). Anti-adhesive and antibiofilm properties of rhamnolipids have been studied using rhamnolipid-coated silver and iron oxide nanoparticles. The results indicated that rhamnolipid-coated silver and iron oxide nanoparticles can be utilized as a potential substitute for reducing disease severity by preventing the formation of biofilm, and they can also show possible biomedical usage for wound dressing and antibacterial coatings.

Hence, this chapter discusses the mechanism of microbial adhesion required for biofilm adhesion, several classes of biosurfactants with anti-adhesive biological coating agents, the development of anti-adhesive and anti-infective biomaterials, and illustrates some outcomes of patents related to biosurfactants potential as anti-adhesive coating effects.

15.2 Microbial adhesion and biofilms

The microbial biofilm can be controlled by several biological, chemical, and physical processes. Adhesion means an attachment of a cell to a substrate, while cohesion means cell-to-cell attachment. The cohesive and adhesive activities a biofilm is determined by the mechanism behind these types of attachment. The first phase in biofilm formation is the adhesion of pathogens to a surface, which is governed by several factors. As the substrate is a key factor in the formation of a biofilm and the knowledge of how substrate activities influence the adhesion of microbial cells may help in designing substrate inhibition to bacterial adhesion (Khalid et al., 2019). Most of these compounds are made of protein constituents such as collagen, fibrogen, and albumin, and some were indicated to influence the imminent microbial adhesion.

The development of a pathogenic biofilm on biomedical material appeared to require many sequential steps. Reid (2000) found that when a device is exposed to body fluids such as urine, saliva, or blood, macromolecular constituents are adsorbed to form a conditioning film. The greatest biosurfactants can be complex compounds consisting of various structures, such as peptides, phospholipids, glycopeptides, glycolipids, and fatty acids. Among several types of BSs, LPs have attracted special attention due to their increased potential antibiotic and surface properties. LPs can also act as antibacterial, anticancer, and antiviral agents and enzyme inhibitors. Based on the study by Rufino et al. (2011), these compounds could decrease or increase the microbial surface hydrophobicity because the surface is more or less hydrophobic. Some studies have characterized and identified a biosurfactant, called arthrofactin, produced by *Arthrobacter* (Golubev et al., 2001). Glycolipids have been regarded as the most common type of BSs, and among these glycolipids are the trehalose lipids produced by *Mycobacterium* sp., which has the most effective surface-active activities, followed by the RLs produced by *Pseudomonas* sp. and the SLs produced by yeasts (Rodrigues et al., 2006).

15.3 The antiadhesive coating property of biosurfactants

Biosurfactants can be seen as microbial amphiphilic compounds involving both hydrophobic and hydrophilic moieties, and they have a unique ability to aggregate at interfaces. Microbial biosurfactants are made of different types of surface-active compounds, and they are seen to exist in different chemical structures (Rodrigues et al., 2006). BSs produced by *Streptococcus thermophiles* A and *L. lactis* 53 have been reported to show anti-adhesive properties against various pathogens isolated from explanted voice prostheses. More than a 90% decrease in the first deposition rate has been reported for most of the bacterial strains investigated. According to Gudiña et al. (2010), BSs formed by *L. paracasei* A20 were evaluated to have anti-adhesive and antimicrobial properties against many pathogens. Another investigation has reported the potential of utilizing BSs derived from *Streptococcus thermophilus* A for preconditioning silicone rubber surfaces in preventing the adhesion of *C. albicans* MFP16−1 and *C. parapsilosis* MFP 16−2 isolated from maxillofacial prostheses. Adhesion methods have shown about 60%−80% reduction in the first deposition rate, and these findings have demonstrated advancement toward developing novel approaches in preventing microbial adhesion to silicone rubber maxillofacial prostheses.

Rivardo et al. (2009) evaluated the anti-adhesive property of two BSs derived by *Bacillus* species against the biofilm development of human bacterial microorganisms. These two strains *B. subtilis* and *B. licheniformis* were grown under higher salinity conditions and produced BSs of about 10% NaCl. They both demonstrated unique anti-adhesive properties by inhibiting the biofilm formation of *S. aureus* ATCC 29213 by 97% and *E. coli* CTF073 by 90%, respectively. Besides, some researchers have studied the anti-adhesive and antimicrobial potential of BS Rufisan produced by C. *lipolytica* UCP0988 (Rufino et al., 2011). The results showed that BS could demonstrate anti-adhesive property against several pathogens tested.

The crude BS can produce anti-adhesive properties against most of the pathogens tested at a minimum concentration of 0.75 mg/L utilized. The highest anti-adhesive activity was found against *Lactobacillus casei* at 91—99% with the lowest concentration used, while the lowest anti-adhesive specificity was seen against *S. agalactiae*, *S. epidermidis*, and *S. aureus*. Janek et al. (2012) also investigated the anti-adhesive property of BS pseudofactin II produced by *Pseudomonas fluorescens* BDS And showed that pseudofactin II reduced the adhesion to three surfaces, such as glass, silicon, and polystyrene, of five bacterial strains: *E. coli*, *S. epidermidis*, *P. mirabilis*, *Enterococcus hirae*, *E. feacalis*, and two *C. albicans* strains. The pretreatment of a polystyrene surface (PS) using 0.5 mg/mL of pseudofactin II resulted in 30%—90% reduction in bacterial adhesion, and *C. albicans* is reduced by 92%—99%. A similar concentration of this biosurfactant also removed between 26% and 70% of preexisting biofilms developed on initially untreated surface. It also significantly reduced the previous *E. coli*, *E. hirae*, *E. feacalis*, and *Candida albicans* strains on silicone urethra catheters.

Moreover, with antiviral, antibacterial, and antifungal properties, BSs are also powerful inhibitors of bacterial adhesion and biofilm development. For instance, a study by Rufino et al. (2012) on BSs produced from *Streptococcus mitis* had an anti-adhesive effect against *S. mutans*. In addition, *L. fermentum* RC-14 produces biosurfactant molecules that show anti-adhesive properties against uropathogenic pathogens such as *E. faecalis*. Another has evaluated the anti-adhesive coating of biosurfactant produced by *Lactobacillus acidophilus* MTCC 447 for table eggs and its use in food protection and safety (Shawkat et al., 2019). The results indicated that biosurfactant was effective against *S. enteritidis*, *E. coli*, *S. aureus*, and *S. epidermidis*. It can inhibit bacterial adhesion to eggs and prevent spoiling.

Kuyukina et al. (2015) investigated the activities of a trehalolipid BS secreted by *Thodococcus ruber* IEGM 231 against bacterial biofilm development and adhesion on the surface of polystyrene microplates. The adhesion of GPB such as *Bacillus subtilis*, *Mircococcus luteus*, Brevibacterium *linens*, *Corynebacterium glutamicum*, and *Arthrobacter* simplex and GNB such as *Pseudomonas fluorescens* and *E. coli* were correlated in different ways with the surface charge and cell hydrophobicity. The results indicated higher anti-adhesive properties of the BS actively against *P. fluorescence* (11%—30%), *M. luteus* (15%—20%), *E. coli* (49%), *C. glutamicum* (36%), and *B. subtilis* (76%) cells, which helped in evaluating the BS concentrations, effectively reducing the bacterial biofilm development on polystyrene at a concentration of 10 mg/L.

15.4 Antiadhesive property of glycolipids biosurfactants

RLs and many other plant-based surface-active oil extracts have been investigated by Quinn et al. (2013), and they showed that these biosurfactants play an important role in preventing complex biofilms and can also serve as adjuvants to enhance selected antibiotics microbial inhibitors. Singh et al. (2013) investigated glycolipid BSs produced by *P. aeruginosa* DSVP20 to determine their capacity to inhibit *C. albicans* biofilm. The polystyrene surfaces adhesion of *C. albicans* can be positively reduced when treated with di-rhamnolipid at concentrations between 0.04 and 5.0 mg/mL in a dose-dependent manner. Results showed nearly a 50% decrease in the number of adherent cells after 2 hours of treatment with 0.16 mg/mL of RL-2, and this continued to rise until a total reduction of adherence was achieved at a concentration of 5 mg/mL. In addition, the biofilm formation of *C. albicans* on the polystyrene surface was reduced to about 70% and 90% with di-rhamnolipid treatment at concentrations of 2.5 and 5.0 mg/mL, respectively. A novel GL produced by *Lysinibacillus fusiformis* S9 is observed to produce a remarkable antibiofilm effect against pathogenic *S. mutans* and *E. coli* without influencing the microbial cell viability. The results indicated that the biosurfactant showed complete inhibition of biofilm development at a concentration of 40 μg/mL.

Some authors have studied the antibiofilm effects of rhamnolipid BSs on the prevention of biofilm formation of GPB and GNB on polystyrene surfaces.

The RL molecule produced by *Pseudomonas aeruginosa* isolated from a patient suffering from cystic fibrosis was used at 500 μg/mL. The results indicated that RL significantly inhibited the biofilm development capacities of *S. epidermidis* (GPB) by 75% and *E. coli* (GNB) by 82%, respectively. About 31% reductions were also observed in three related strains of *P. aeruginosa*. It has been seen that RL plays a significant role in various phases of *P. aeruginosa*, and their properties can depend on the concentration. Another study by Nickzad and Déziel (2014) showed that small amounts of RLs enhance the first adherence of cells to a surface and the development of microbial biofilms. The presence of increased concentrations in the anti-adhesion assays can reduce binding of the cells, further microbial development of biofilms, and also decrease the formation of biofilm. The interesting results observed against strains producing biofilms showed that rhamnolipid biosurfactants are better anti-adhesive agents in preventing the adhesion to plastic surfaces. Bruce et al. (2000) revealed a patent approved for biosurfactants synthesized by *Lactobacillus* sp. with the capacity to prevent bacterial adhesion and biofilm formation on medical devices to inhibit urogenital disease in mammals.

Padmapriya and Suganthi (2013) showed that they had fairly purified two BSs secreted from *C. albicans* and *C. tropicalis* and evaluated their anti-adhesive property against various urinary tract infections. Their results indicated a decrease in the adhesion of cells on the surface of the urinary catheter precoated with BS produced by *Candida albicans*. Moreover, Tahmourespour et al. (2011) studied the activity of biosurfactant synthesized by *L. acidophilius* DSM 20079 on adherence and on the expression stage of the gene *gtf* B and *gtf* C in *Streptococcus mutans*. The results showed that BS produced by *L. acidophilius* could prevent the adhesion and biofilm development of

S. mutans on a glass slide and resulted in shortening the chain development. In addition, different activities of *S. mutans* cells, such as gene expression, surface activities, biofilm development, and adhesive capacity, were changed because of the treatment with BS secreted by *L. acidophilius*.

15.5 Antiadhesive property of lipopeptides biosurfactants

Hajfarajollah et al. (2014) showed new anti-adhesive properties of LP biosurfactants produced from the probiotic strain *Propionibacterium freudenreichii*. Their results indicated a significant anti-adhesive effect against large numbers of pathogenic fungi and bacteria. The greatest adhesive inhibition of about 67.1% was reported for *P. aeruginosa* at a concentration of 40 mg/mL, while low actions were reported at a similar concentration for *E. coli* (47.7%), *Bacillus cereus* (39.1%), and *S. aureus* (32.1%), respectively. Some studies revealed that a lipopeptide BS secreted by *B. subtilis* V9T14 in combination with antibiotics symbiotically improved the effectiveness of antibiotics against adhesion and biofilm development of pathogenic *E. coli* CFT073 (Rivardo et al., 2011). Ceri et al. (2010) received an international patent on this usage.

In another study, Quinn et al. (2012) evaluated the anti-adhesive activity of lipopeptide BS produced by *Paenibacillus polymyxa* against single and mixed strains of biofilm. The BS complex, mainly composed of polymyxin D1 and fusaricidin B, inhibited the biofilm biomass of *S. bovis*, *Micrococcus luteus*, *S. aureus*, *B. subtilis*, and *P. aeruginosa*. A study has also evaluated a lipopeptide BS secreted by a soil strain *B. cereus* resistant to heavy metals such as zinc, iron, and lead (Sriram et al., 2011). The highest inhibition of 57% was seen against *S. epidermidis* at a concentration of 15 mg/mL. Some studies showed the anti-adhesive properties of *L. monocytogenes*, *M. luteus*, and *S. aureus* on PS at different temperatures when treated with surfactin and rhamnolipid (Zeraik & Nitschke, 2010). The results indicated that RLs slowly reduced adhesion of *S. aureus*, while surfactin effectively prevented the attachment of the tested strains under all conditions, with improved effect as temperature decreased, with the highest decrease in adhesion of 63%—66% at 4°C.

Cochis et al. (2012) studied the anti-adhesive effect of lipopeptide BS secreted by *Bacillus* sp. against *Candida albicans* biofilm development on acrylic resins and silicone disks to denature prostheses. The application of biosurfactants for precoating can provide a higher reduction of biofilm and the number of viable cells than chlorhexidine disinfectant. In addition, Pradhan et al. (2013) investigated the anti-adhesive property of lipopeptide BS derived by *B. tequilensis* CH (CHBS) against biofilm development of pathogenic bacteria on hydrophobic and hydrophilic surfaces. *Streptococcus mutans* and *Escherichia coli* biofilm were cultured with BS of various concentrations on PVC surfaces. The result indicated biofilm formation of *S. mutans* and *E. coli* on the surfaces co-incubated with CHBS concentration between 0 and 25 μg/mL, but there was no biofilm formation on the surfaces incubated with CHBS at a concentration of 50–70 μg/mL. The results also showed that CHBS did not reduce the formation of S. mutans and E. coli planktonic cells under all tested concentrations, and this has also indicated that CHBS is not acting as a bactericidal agent but only differentiated microbial adhesion to various surfaces.

Ceresa et al. (2015) showed that BS secreted by L. brevis CV8LAC can significantly inhibit the biofilm development and adhesion of Candida albicans on silicone elastomeric disks. Their results have shown that co-incubated CV8LAC biosurfactant positively inhibited biofilm development by 90%, while precoated silicone disks showed 60% inhibition of fungi adhesion. Fracchia et al. (2015) reported a significant decrease in the biofilm formation of bacteria on polystyrene coated with LP biosurfactant secreted by B. subtilis. The results showed that the biofilms of three P. aeruginosa decreased by 70%–90%, while a 70% reduction of biofilm was found in S. epidermidis and E. coli. Moreover, a similar LP biosurfactant demonstrated the capacity to significantly decrease the growth of C. albicans on silicone elastomeric disks coated with BS.

15.6 Production of antiadhesive and antiineffective biomaterials

The development of anti-infective and anti-adhesive implants arose due to a significant resistance of microbial biofilms to traditional antibiotic treatment methods and are produced as follows: (1) as mechanical design substitutes; (2) alteration of molecule surface properties such as brushes, and biosurfactant; (3) anti-infective molecules that bind to the surface of the material such as biosurfactant, synthetic antibiotics, quaternary ammonium, and silver; and (4) soluble toxic substances such as chlorhexidine antibiotics (Bryers, 2008; Rodrigues, 2011). Bryers (2008) reported that mechanical design substitutes showed only marginal breakthroughs and can be used only for short-term indwelling catheters. Coatings are created to inhibit microbial adhesion by changing the physiochemical characteristics of the materials to prevent the formation of biofilm or bacterial–substrate relationships. These coatings can be passive with modified surfaces with brushes, poly (ethylene oxide), and poly (ethylene glycol) (Kaper et al., 2003; Kingshott et al., 2003).

Hetrick and Schoenfisch (2006) reported the effectiveness of passive coatings in mitigating bacterial adhesion to be limited and highly varied based on the bacteria strains. The physical and chemical characteristics of the coating are covered with an adsorbed conditioning film, reducing its effectiveness. In addition, Bryers (2008) revealed that surface-bound anti-infective agents can be toxic only to the initial wave of incoming bacteria and have little residual effect once the layers of dead cells aggregate and cause inflammation. A good instance is the production of anti-adhesive silicone rubber surfaces utilized for voice prostheses. These voice prostheses can be constantly exposed to drinks, food, and saliva, combined with the oropharyngeal microflora, which are responsible for valve failure, and there is a need for frequent replacement of the implant (Rodrigues, 2011). Another investigation has reported several strategies used to modify the silicone rubber surface in preventing biofilm development and thus extend the lifespan of voice prostheses (Rodrigues, 2011).

A current replacement strategy to mitigate microbial adhesion is based on a coating that can produce potent antibacterial agents. Anderson (2001) revealed that such active coatings were developed to temporarily produce large initial fluxes of antibacterial agents during the post-implementation period to prevent bacterial adhesion. A study also reported progress on antibiotic-eluting medical instruments for various uses (Zilberman & Elsner, 2008). Gollwitze et al. (2003) evaluated biodegradable polymeric coatings produced from polylactic acid and glycolic acid copolymer. Schmidmaier et al. (2006) used an easy dip-coating method to show the additional benefits of these coatings used by the polymer for both plastics and alloys with porous or polished surfaces. Teller et al. (2007) demonstrated in vitro production of antibiotics from hydroxyapatite-coated implants. The conventional plasma spraying method for hydroxyapatite-coated implants can be related to high processing temperatures and thus prevent the integration of antibiotics during the process. Overall, despite the active or passive drug release technique utilized, the production of a toxic substance from a biological material of a soluble anti-infective agent is automatically interrupted once the entrapped agent is consumed.

15.7 Results of some patents related to antiadhesive biological coating property of biosurfactants

Wesal et al. (2022) invented a technique to coat the surface of substrate material with a biosurfactant to increase its reactivity with a silane killer, thereby oxidizing the surface of the substrate materials, functionalizing the surface of the substrate material with a silane linker, and reacting the functionalized substrate material surface with the modified biosurfactant, resulting in covalent bonding of the biosurfactant to the surface of the substrate material. This substrate can be chosen from the group consisting of ferrous metal, stainless, polymers, PVC, and high-density polyethylene. Before coating, the surface of this substrate material was susceptible to biofilm formation, and it has generated activities needed for material chosen from the group made up of water distribution apparatus. This apparatus can be configured with food, medical device apparatus, marine, and shipping apparatus. The invention also states that the biosurfactant can be synthesized by a minimum of a single strain of a species of pathogen chosen from the group consisting of *B. amyioiiquefaciens*, *P. aeruginosa*, and *S. marcescens*.

In another study, an invention aimed at isolating a strain of *Lactobacillus* BSs and the process for producing it was presented. This invention is reported to be directed toward techniques that can be used to prevent urogenital infection in mammals using the isolated *Lactobacillus* strain BS. In addition, this invention showed techniques for reducing microbial biofilm formation utilizing the isolated strains of *Lactobacillus* BS to prevent the development of bacterial biofilms and preventing bacteria with the ability to form biofilms from adhering to surfaces. This patent has solved problem associated with biomaterial devices such as catheters. They also found that the BS produced by the Lactobacillus sp. after the dialysis phase was highly effective. The inventors revealed that diluted Lactobacillus BS showed effective reduction by inhibiting adhesion of *Enterococcus* in vitro.

Ceri et al. (2010) also invented a new *B. licheniformis* strain V9T14 that synthesizes a novel BS composition that is effective in preventing biofilm formation or colonization of bacteria. This novel BS composition may be utilized together with biocides to eradicate bacteria grown planktonically or as a biofilm formation, particularly on abiotic or biotic surfaces. The BS composition secreted by V9T14 strain consists of surfactin and fengycin, which were detected by mass spectrometry and infra spectra analysis. The result obtained through spectra analysis investigated on V9T14 BS confirmed that this was an LP molecule, while the mass spectrometry analysis of V9T14 BS indicated two major groups of compounds belonging to fengycin and surfactin metabolites and differing for isoform and homologue compositions. The depth of adhesion of four bacteria species on polystyrene in the absence or presence of the V9T14 BS was evaluated using the Calgary Biofilm Device under shear forces. The outcomes after precoating the pegs before inoculation showed that the effectiveness was the same as when BS was included in the growth medium. Biosurfactant V9T14 showed a 97% reduction in *Escherichia coli* CF073 biofilm (Wang & Mulligan, 2004).

15.8 Challenges and future perspective

It is essential to emphasize that inadequate information and studies on BSs' toxicity to humans and their cost of production have led to their use been discouraged in many fields of applications. BSs can demonstrate an attractive approach because of their potential to modify the surface activities while acting as anti-adhesive and antimicrobial agents. In the future, problems related to the structural function and genetic associations of biosurfactants, and techniques of adhering them to surfaces will again promote study in this area. Further study on natural microbiota and human cell may be needed to confirm the application of biosurfactants in various health-related fields. Several inventors have suggested that in the future there will be a need for a method directed toward inhibiting microbial biofilm formation consisting of coating a biomaterial for insertion into a mammal containing pathogenically inhibitory amount of *Lactobacillus* BS. Moreover, RL-coated nanoparticles have been suggested to further improve the prevention of microbial films because of the synergistic property of RLs as anti-adhesive agents and silver and iron oxide as antimicrobial substances.

15.9 Conclusion

Biosurfactants can play a vital role in the production of anti-adhesive coatings for many biomaterials such as silicone rubber as they efficiently reduce microbial adhesion and stop biofilm development. Hence, bulk and surface modification methods, laser-induced surface grafting, and the sequential technique used in interpenetrating polymer networks can be employed to bind the BSs more firmly to the biomaterial surfaces, therefore preventing them from leaching from the surfaces and increasing their activity. Besides, as an appropriate substitute for antimicrobial agents, BS may be utilized as an effective and safe therapeutic agent. However, the efficiency of these coatings was seen to be greatly reduced and varied by bacteria species, mainly because of the different environments in which the instruments are placed and the way the pathogens colonize the surfaces. There may be a need to produce substitutes for the conventional surface-modified strategies that can sufficiently focus on the antimicrobial coating of instruments and the use of antibiotics. In addition, BSs can be used as an anti-adhesive biological coating for medical implants, hence preventing hospital-acquired diseases and limiting the application of chemicals and synthetic drugs. Thus, BSs make an attractive substitute for their chemical equivalents, mainly because of their lower toxicity and biodegradability.

Reference

Abdalsadiq, N., Hassan, Z., & Lani, M. (2018). Antimicrobial, anti-adhesion and anti-biofilm activity of biosurfactant isolated from Lactobacillus spp. *Research Journal of Life Science and Bioinformatic, 4*, 280–292.

Anderson, J. (2001). Biological responses to materials. *Annual Review Material Research, 31*, 81–110.

Azevedo, A. S., Almeida, C., Melo, L. F., & Azevedo, N. F. (2017). Impact of polymicrobial biofilms in catheter-associated urinary tract infections. *Critical Reviews in Microbiology, 43*(4), 423–439. Available from https://doi.org/10.1080/1040841X.2016.1240656.

Bruce, A., Busscher, H., & Reid, G. (2000). *Lactobacillus therapies*, U.S. Patent US6051552A.

Bryers, J. (2008). Medical biofilms. *Biotechnology and Bioengineering, 100*, 1–18.

Ceresa, C., Tessarolo, F., & Caola, I. (2015). Inhibition of Candida albicans adhesion on medical-grade silicone by a Lactobacillus-derived biosurfactant. *Journal of Applied Microbiology, 18*, 1116–1125.

Ceri, H., Turner, R., Martinotti, M.G., Rivardo, F., & Allegrone, G. (2010). Biosurfactant composition produced by a new *Bacillus licheniformis* strain, uses and products thereof. Available at: https://patentscope.wipo.int/search/en/detail.jsf?docId = WO2010067245.

Cochis, A., Fracchia, L., Martinotti, M., & Rimondini, L. (2012). Biosurfactants prevent in vitro Candida albicans biofilm formation on resins and silicon materials for prosthetic devices. *Oral Surgery, Oral Medicine, Oral Pathogen, and Oral Radiology, 113*, 755–761.

Das, P., Mukherjee, S., & Sen, R. (2009). Antiadhesive action of a marine microbial surfactant. *Colloids Surfaces B Biointerfaces, 71*, 183–186. Available from https://doi.org/10.1016/j.colsurfb.2009.02.004.

Dusane, D., Dam, S., Nancharaiah, Y., Kumar, A., Venugopalan, V., & Zinjarde, S. (2012). Disruption of Yarrowia lipolytica biofilms by rhamnolipid biosurfactant. *Aquatic Biosystems, 8*(1), 17.

Fracchia, L., Banat, J., Cavallo, M., Chiara, C., & Banat, I. (2015). Potential therapeutic applications of microbial surface-active compounds. *Bioengineering, 2*(3), 144–162.

Giri, S., Ryu, E., Sukumaran, V., & Park, S. (2019). Antioxidant, antibacterial, and anti-adhesive activities of biosurfactants isolated from Bacillus strains. *Microbiology Pathology, 132*, 66–72. Available from https://doi.org/10.1016/j.micpath.2019.04.035.

Gollwitze, H., Ibrahim, K., Meyer, H., Mittelmeier, W., Busch, R., & Stemberger, A. (2003). Antibacterial poly(D,L-lactic acid) coating of medical implants using a biodegradable drug delivery technology. *Journal of Antimicrobial Chemotherapy*, *51*(3), 585–591.

Golubev, W., Kulakovskaya, T., & Golubeva, W. (2001). The yeast *Pseudozyma fusiformata* VKM Y-2821 producing an antifungal glycolipid. *Microbiological Research*, *70*, 553–556.

Gudiña, E. J., Rodrigues, A. I., Alves, E., Domingues, M. R., Teixeira, J. A., & Rodrigues, L. R. (2015). Bioconversion of agro-industrial by-products in rhamnolipids toward applications in enhanced oil recovery and bioremediation. *Bioresource Technology*, *177*, 87–93. Available from https://doi.org/10.1016/j.biortech.2014.11.069.

Gudiña, E. J., Teixeira, J. A., & Rodrigues, L. R. (2010). Isolation and functional characterization of a biosurfactant produced by *Lactobacillus paracasei*. *Colloids and Surfaces B: Biointerfaces*, *76*(1), 298–304. Available from https://doi.org/10.1016/j.colsurfb.2009.11.008.

Haddaji, N., Mahdhi, A., Bouali, N., Ghorbel, M., Bechambi, O., Leban, N., Ameur, M., & Mzoughi, R. (2022). Biosurfactants as inhibitors of the adhesion of *Pathogenic bacteria*. *Emirates Journal of Food and Agriculture*, *34*(1), 36–43. Available from https://doi.org/10.9755/ejfa.2022.v34.i1.2803.

Hajfarajollah, H., Mokhtarani, B., & Noghabi, K. (2014). Newly antibacterial and antiadhesive lipopeptide biosurfactant secreted by a probiotic strain, *Propionibacterium freudenreichii*. *Applied Biochemistry and Biotechnology*, *174*, 2725–2740.

Hetrick, E., & Schoenfisch, M. (2006). Reducing implant-related infections: Active release strategies. *Chemical Society Reviews*, *35*, 780–789.

Janek, T., Lukaszewicz, M., & Krasowska, A. (2012). Antiadhesive activity of the biosurfactant pseudofactin II secreted by the Arctic bacterium *Pseudomonas fluorescens* BD5. *BMC Microbiology*, *12*, 24.

Kaper, H., Busscher, H., & Norde, W. (2003). Characterization of poly(ethylene oxide) brushes on glass surfaces and adhesion of *Staphylococcus epidermidis*. *Journal of Biomaterial Science and Polymer Education*, *14*(4), 313–324.

Khalid, H. F., Tehseen, B., Sarwar, Y., Hussain, S. Z., Khan, W. S., Raza, Z. A., Bajwa, S. Z., Kanaras, A. G., Hussain, I., & Rehman, A. (2019). Biosurfactant coated silver and iron oxide nanoparticles with enhanced anti-biofilm and anti-adhesive properties. *Journal of Hazardous Materials*, *364*, 441–448. Available from https://doi.org/10.1016/j.jhazmat.2018.10.049.

Kim, L., Jung, Y., Yu, H., Chae, K., & Kim, I. (2015). Physicochemical interactions between rhamnolipids and *Pseudomonas aeruginosa* biofilm layers. *Environmental Science and Technology*, *49*(6), 3718–3726. Available from https://doi.org/10.1021/es505803c.

Kim, M. J., Ko, D., Ko, K., Kim, D., Lee, J. Y., Woo, S. M., Kim, W., & Chung, H. (2018). Effects of silver-graphene oxide nanocomposites on soil microbial communities. *Journal of Hazardous Materials*, *346*, 93–102. Available from https://doi.org/10.1016/j.jhazmat.2017.11.032.

Kingshott, P., Wei, J., Bagge-Ravn, D., Gadegaard, N., & Gram, L. (2003). Covalent attachment of poly (ethylene glycol) to surfaces, critical for reducing bacterial adhesion. *Langmuir: The ACS Journal of Surfaces and Colloids*, *19*, 6912–6921.

Kuyukina, M. S., Ivshina, I., Baeva, T., Chereshnev, V., Kochina, O., Gein, V., & Sergey, A. (2015). Trehalolipid biosurfactants from nonpathogenic *Rhodococcus actinobacteria* with diverse immunomodulatory activities. *New Biotechnology*, *32*(6), 559–568. Available from https://doi.org/10.1016/j.nbt.2015.03.006.

Lellouche, J., Friedman, A., Gedanken, A., & Banin, E. (2012). Antibacterial and antibiofilm properties of yttrium fluoride nanoparticles. *International Journal of Nanomedicine*, *7*, 5611–5624. Available from https://doi.org/10.2147/IJN.S37075.

Mandakhalikar, K. D., Chua, R. R., & Tambyah, P. A. (2016). New technologies for prevention of catheter associated urinary tract infection. *Current Treatment Options in Infectious Diseases*, *8*(1), 24–41. Available from https://doi.org/10.1007/s40506-016-0069-5.

de Melo Pereira, G., Coelho, B. D., Júnior, A. I., Thomaz-Soccol, V., & Soccol, C. (2018). How to select a probiotic? A review and update of methods and criteria. *Biotechnology Advances*, *36*, 2060–2076.

Mireles, J., Toguchi, A., & Harshey, R. (2001). Salmonella enteric se-rovar typhimurium swarming mutants with altered biofilm forming abilities: Surfactin inhibits biofilm formation. *Journal of Biotechnology*, *183*(20), 5848–5854. Available from https://doi.org/10.1128/JB.183.20.5848-5854.2001.

Nickzad, A., & Déziel, E. (2014). The involvement of rhamnolipids in microbial cell adhesion and biofilm development — an approach for control? *Letter of Applied Microbiology, 58*, 447–453.

Padmapriya, B., & Suganthi, S. (2013). Antimicrobial and anti adhesive activity of purified biosurfactants produced by Candida species. *Middle East Journal of Scientific Research, 14*, 1359–1369.

Patel, M., Siddiqui, A., Hamadou, W., Surti, M., Awadelkareem, A., Ashraf, S., & Adnan, M. (2021). Inhibition of bacterial adhesion and antibiofilm activities of a glycolipid biosurfactant from *Lactobacillus rhamnosus* with its physicochemical and functional properties. *Antibiotics, 10*(12), 2–25.

Peng, F., Wang, Y., Sun, F., Liu, Z., Lai, Q., & Shao, Z. (2008). A novel lipopeptide produced by a Pacific Ocean deep-sea bacterium, Rhodococcus sp. *Journal of Applied Microbiology, 105*, 698–705.

Pradhan, A., Pradhan, N., Mall, G., Panda, H., Sukla, L., Panda, P., & Mishra, B. (2013). Application of lipopeptide biosurfactant isolated from a halophile: Bacillus tequilensis CH for inhibition of biofilm. *Applied Biochemistry and Biotechnology, 171*, 1362–1375.

Quinn, G., Maloy, A., McClean, S., Carney, B., & Slater, J. (2012). Lipopeptide biosurfactants from Paenibacillus polymyxa inhibit single and mixed species biofilms. *Biofouling, 28*, 1151–1166.

Quinn, G. A., Maloy, A. P., Banat, M. M., & Banat, I. M. (2013). A comparison of effects of broad-spectrum antibiotics and biosurfactants on established bacterial biofilms. *Current Microbiology, 67*(5), 614–623. Available from https://doi.org/10.1007/s00284-013-0412-8.

Reid, G. (2000). In vitro testing of Lactobacillus acidophilus NCFM(TM) as a possible probiotic for the urogenital tract. *International Dairy Journal, 10*(5–6), 415–419. Available from https://doi.org/10.1016/S0958-6946(00)00059-5.

Rivardo, F., Martinotti, M., Turner, R., & Ceri, H. (2011). Synergistic effect of lipopeptide biosurfactant with antibiotics against *Escherichia coli* CFT073 biofilm. *International Journal of Antimicrobial Agents, 37*, 324–331.

Rivardo, F., Turner, R. J., Allegrone, G., Ceri, H., & Martinotti, M. G. (2009). Anti-adhesion activity of two biosurfactants produced by Bacillus spp. prevents biofilm formation of human bacterial pathogens. *Applied Microbiology and Biotechnology, 83*(3), 541–553. Available from https://doi.org/10.1007/s00253-009-1987-7.

Rodrigues, L., Banat, I. M., Teixeira, J., & Oliveira, R. (2006). Biosurfactants: Potential applications in medicine. *Journal of Antimicrobial Chemotherapy, 57*(4), 609–618. Available from https://doi.org/10.1093/jac/dkl024.

Rodrigues, L. R. (2011). Inhibition of bacterial adhesion on medical devices. *Advances in Experimental Medicine and Biology, 715*, 351–367. Available from https://doi.org/10.1007/978-94-007-0940-9_22.

Rodrigues, L. R., & Teixeira, J. A. (2010). Biomedical and therapeutic applications of biosurfactants. *Advances in Experimental Medicine and Biology, 672*, 75–87. Available from https://doi.org/10.1007/978-1-4419-5979-9_6.

Rufino, R., Luna., Sarubbo, L. A., Marona, L., Teixeira, J. A., & Campos-Takaki. (2012). Antimicrobial and anti-adhesive potential of a biosurfactants produced by Candida species. *Practical application in biomedical engineering* (pp. 246–256). IntechOpen.

Rufino, R. D., Luna, J. M., Sarubbo, L. A., Rodrigues, L. R. M., Teixeira, J. A. C., & Campos-Takaki, G. M. (2011). Antimicrobial and anti-adhesive potential of a biosurfactant Rufisan produced by *Candida lipolytica* UCP 0988. *Colloids and Surfaces B: Biointerfaces, 84*(1), 1–5. Available from https://doi.org/10.1016/j.colsurfb.2010.10.045.

Schmidmaier, G., Lucke, M., Wildemann, B., Haas, N., & Raschke, M. (2006). Prophylaxis and treatment of implant-related infections by antibiotic-coated implants: A review. *Injury, 37*(2), S105–S112.

Shawkat, S. M., Al-Jawasim, M., & Hussain, Z. F. (2019). Production of anti-adhesive coating agent for table eggs by lactobacillus acidophilus MTCC 447. *Plant Archives, 19*, 1795–1798. Available from http://plantarchives.org/SPL%20ISSUE%20SUPP%202,2019/311%20(1795-1798).pdf.

Shokouhfard, M., Kasra Kermanshahi, R., Vahedi Shahandashti, R., Feizabadi, M. M., & Teimourian, S. (2015). The inhibitory effect of a Lactobacillus acidophilus derived biosurfactant on Serratia marcescens biofilm formation. *Iranian Journal of Basic Medical Sciences, 18*(10), 1001–1007. Available from http://ijbms.mums.ac.ir/pdf_5464_ac1dd2d5fb936edd661c99fc3812d23a.html.

Singh, N., Pemmaraju, S. C., Pruthi, P. A., Cameotra, S. S., & Pruthi, V. (2013). Candida biofilm disrupting ability of di-rhamnolipid (RL-2) produced from *Pseudomonas aeruginosa* DSVP20. *Applied Biochemistry and Biotechnology, 169*(8), 2374–2391. Available from https://doi.org/10.1007/s12010-013-0149-7.

Singh, P., & Cameotra, S. S. (2004). Potential applications of microbial surfactants in biomedical sciences. *Trends in Biotechnology, 22*(3), 142–146. Available from https://doi.org/10.1016/j.tibtech.2004.01.010.

Sriram, M., Kalishwaralal, K., Deepak, V., Gracerosepat, R., Srisakthi, K., & Gurunathan, S. (2011). Biofilm inhibition and antimicrobial action of lipopeptide biosurfactant produced by heavy metal tolerant strain *Bacillus cereus* NK1. *Colloid Surface B Biointerfaces, 85*, 174–181.

Tahmourespour, A., Salehi, R., & Kermanshahi, R. (2011). Lactobacillus acidophilus-derived biosurfactant effect on gtfB and gtfC expression level in Streptococcus mutans biofilm cells. *Brazilian Journal of Microbiology, 42*, 330–339.

Tambone, E., Bonomi, E., Ghensi, P., Maniglio, D., Ceresa, C., Agostinacchio, F., Caciagli, P., Nollo, G., Piccoli, F., Caola, I., Fracchia, L., & Tessarolo, F. (2021). Rhamnolipid coating reduces microbial biofilm formation on titanium implants: An in vitro study. *BMC Oral Health, 21*(1). Available from https://doi.org/10.1186/s12903-021-01412-7.

Teller, M., Gopp, U., Neumann, H., & Kuhn, K. (2007). Release of gentamicin from bone regenerative materials: An in vitro study. *Journal of Biomedical Material Research Part B, 81*(1), 23–29.

Tiedje, J. M., Wang, F., Manaia, C. M., Virta, M., Sheng, H., Ma, L., Zhang, T., & Topp, E. (2019). Antibiotic resistance genes in the human-impacted environment: A one health perspective. *Pedosphere, 29*(3), 273–282. Available from https://doi.org/10.1016/S1002-0160(18)60062-1.

Ventola, C. L. (2015). The antibiotic resistance crisis: Part 1: Causes and threats. *P and T, 40*(4), 277–283. Available from http://www.ptcommunity.com/system/files/pdf/ptj4004277.pdf.

Ventola, C. L. (2016). Immunization in the United States: Recommendations, barriers, and measures to improve compliance. *Pharmacy and Therapeutics, 41*(8), 492–506. Available from https://ptcommunity.com/journal/article/full/2016/8/492/immunization-united-states-recommendations-barriers-and-measures.

Wang, S., & Mulligan, C. N. (2004). An evaluation of surfactant foam technology in remediation of contaminated soil. *Chemosphere, 57*, 1079.

Wesal, K., Thando, N., Lee, C., & Begum, M. (2022). *Coating of materials with biosurfactant compounds*. Retrieved from: https://www.sumobrain.com/patents.

Zeraik, A. E., & Nitschke, M. (2010). Biosurfactants as agents to reduce adhesion of pathogenic bacteria to polystyrene surfaces: Effect of temperature and hydrophobicity. *Current Microbiology, 61*(6), 554–559. Available from https://doi.org/10.1007/s00284-010-9652-z.

Zhao, T., & Liu, Y. (2010). N-acetylcysteine inhibit biofilms produced by *Pseudomonas aeruginosa*. *BMC Microbiology, 10*(1), 140. Available from https://doi.org/10.1186/1471-2180-10-140.

Zilberman, M., & Elsner, J. (2008). Antibiotic-eluting medical devices for various applications. *Journal Control Release, 130*, 202–215.

CHAPTER 16

Advantages of biosurfactants over petroleum-based surfactants

Angelica Cristina de Souza[1], Monique Suela Silva[2], Luara Aparecida Simões[3], Natalia Andrade Teixeira Fernandes[4], Rosane Freitas Schwan[1] and Disney Ribeiro Dias[2]

[1]Department of Biology, Federal University of Lavras, Lavras, Minas Gerais, Brazil
[2]Department of Food Science, Federal University of Lavras, Lavras, Minas Gerais, Brazil
[3]Biology Department, University of Porto, Porto, Portugal
[4]Department of Chemistry, University of California, Davis, CA, United States

16.1 Introduction

The term surfactant covers various compounds with numerous properties and functions, both synthetic and biological. These properties allow surfactants to be used as moderators of immiscible phases to form emulsions, foams, suspensions, and microemulsions or to provide wetting, liquid film formation, and surface detergency (Johnson et al., 2021; Kosaric & Sukan, 2014).

Surfactants or surfactant agents are amphipathic compounds. They have both hydrophilic and hydrophobic groups in the same molecule (Fig. 16.1). The apolar portion, also called the tail, consists of one or two carbonic, or fluorocarbon, or siloxane chains, while the polar portion, or head, can have ionic groups (cations or anions), nonionic or amphoteric groups (Drakontis & Amin, 2020; Joshi & Tejas, 2017). This composition is related to the ability of these molecules to reduce the surface tension at the oil–water interface, increasing the solubility of immiscible substances in water (Nitschke & Pastore, 2002). These versatile amphiphilic molecules have applications in various industrial sectors, including food, pharmaceuticals, cosmetics, detergents, paints, petroleum, and water treatment (Adetunji & Olaniran, 2021; Akbari et al., 2018; Drakontis & Amin, 2020).

The two main characteristics of surfactants are surface tension and the ability to form micelles in solutions, where the critical micelle concentration (CMC) is the minimum surfactant concentration required to achieve the lowest surface tension. On reaching CMC, the amphipathic molecules are aggregated and the hydrophilic portions are positioned toward the outside of the molecule and the hydrophobic portions toward the inside, leading to the formation of micelles. Thus, the efficiency of these compounds can be determined by their ability to reduce surface tension. Good surfactants reduce the surface tension of water from 72 to 30 mN/m and the interfacial

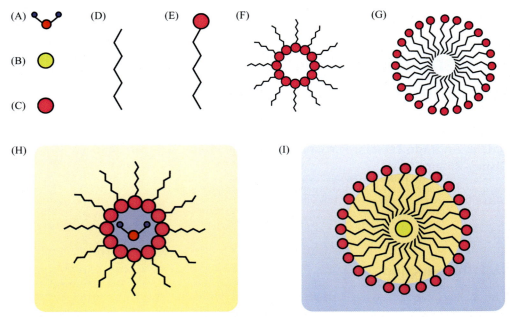

Figure 16.1 Surfactant molecule and schematic illustration of micelle formation. (A) Water molecule (H_2O), (B) Oil molecule, (C) Hydrophilic head, (D) Hydrophobic tail, (E) Biosurfactant monomers, (F and G) Micellar systems, (H) Water-in-oil emulsions, (I) Oil-in-water emulsions.

tension of water against n-hexadecane from 40 to 1 mN/m (De et al., 2015). In this sense, lower surface tensions facilitate interactions between molecules of different polar natures (Gecol, 2007; Moutinho et al., 2021).

Surfactants can be divided into three distinct categories according to their origin and composition: (1) synthetic, (2) bio-based, and (3) microbial (Kosaric & Sukan, 2014; Moldes et al., 2021). The former comprises the majority of commercially used surfactants synthesized from petroleum derivatives (Sarubbo et al., 2022), 70%—75% of surfactants consumed in industrialized countries are of petrochemical origin and, therefore, can be produced at low cost and high yield (Moldes et al., 2021; Nitschke & Pastore, 2002). The second category comprises chemically synthesized surfactants obtained from renewable sources and which have vitamins, sugars or amino acids in their structure, snedo generally referred to as green surfactants (Farias et al., 2021; Moldes et al., 2021). The third group of surfactants, produced by various microorganisms, is called microbial biosurfactants with different molecular structures (Drakontis & Amin, 2020; Farias et al., 2021). They are considered more biocompatible and biodegradable than synthetic surfactants (Moldes et al., 2021).

The use of synthetic surfactants in various industrial and domestic processes causes environmental damage associated with their production and disposal. Another environmental issue is the use of nonrenewable raw materials, mainly petroleum derivatives. In addition to environmental issues, most of these synthetic compounds are potentially associated with toxicological problems owing to their recalcitrant and persistent nature (De et al., 2015). Due to growing environmental concerns, there is a need to produce safe, biodegradable, and less toxic surfactants than those derived from petroleum (Akbari et al., 2018; Jahan et al., 2020; Mondal et al., 2022; Singh et al., 2019).

Natural surfactants or biosurfactants are amphiphilic biological compounds, generally extracellular, obtained from the secondary metabolism of microorganisms such as bacteria, yeasts, and filamentous fungi, and produced by biotechnological routes using various substances, including waste materials (Shekhar et al., 2015). Microbial biosurfactants share the same mechanisms as synthetic surfactants to reduce interfacial and surface tension but have shown great acceptability greater biocompatibility, structural diversity, ability to be used at different temperatures, salinity and pH, as well as low CMC due to their low toxic effect (Mondal et al., 2022).

The demand for new sustainable formulations is increasingly becomingimportant in all sectors, and these are widely used in different industries such as cosmetics, food, pharmaceuticals, cleaning products, petroleum, agriculture, and medicine. As surfactants are used in various processes, they must be effective and environmentally friendly. Thus, biosurfactants, surfactant agents produced by microorganisms, have shown promise in several aspects, mainly related to the environment. This chapter will discuss the main advantages of microbial biosurfactants over petroleum-based surfactants.

16.2 Chemical classification of biosurfactants

Biosurfactants are classified into microbial origin, chemical composition, or molecular weight (Drakontis & Amin, 2020). The molecular mass can range from 500 to 1500 Da. Low-molecular-weight biosurfactants are more efficient in reducing the surface tension at the water–air interface and the water–oil interface (e.g., glycolipids, lipopeptides, and phospholipids), while high-molecular-weight compounds are more effective as stabilizers of oil-in-water emulsions (polymeric and particulate surfactants) (De et al., 2015; Sarubbo et al., 2022). According to their chemical composition, biosurfactants are classified into glycolipids, lipopeptides, lipoproteins, fatty acids, phospholipids, polymeric surfactants, and particulate surfactants (Table 16.1).

Table 16.1 Major classes/subclasses of microbial biosurfactants according to molecular structure.

Class	Subclass	Microbial source	References
Glycolipids	Rhamnolipds	Pseudomonas aeruginosa	Araújo et al. (2018), Eslami et al. (2020)
		Pseudomonas cepacia	Soares da Silva et al. (2017)
		Lysinibacillus sphaericus	Gaur et al. (2019)
		Serratia rubidaea	Nalini and Parthasarathi (2013)
	Trehalolipids	Nocardia farcinica	Christova et al. (2015)
		Rhodococcus erythropolis	Pirog et al. (2013)
	Sophorolipids	Candida bombicola	Luna et al. (2016), Pinto et al. (2022)
		Starmerella bombicola	De Graeve et al. (2018), Van Renterghem et al. (2018), Wang et al. (2019)
		Candida sphaerica	Luna et al. (2015)
		Candida magnoliae	Mnif and Ghribi (2016)
		Torulopsis petrophilum	Cooper and Paddock (1983)
		Torulopsis apicola	Celligoi et al. (2014)
	Mannosylerythritol lipids	Cutaneotrichosporon mucoides	Jiménez et al. (2020), Marcelino et al. (2019)
		Pseudozyma aphidis	Morita et al. (2007)
Lipopetides	Surfactin	Bacillus subtilis	Ciurko et al. (2022), Cruz Mendoza et al. (2022), Hu et al. (2021)
		Bacillus nealsonii	Phulpoto et al. (2020)
		Kocuria marina	Sarafin et al. (2014)
	Licheysin	Bacillus licheniformis	Gudiña and Teixeira (2022), Liu et al. (2021)
		Pseudomonas putida	Biktasheva et al. (2022)
		Thiobacillus thiooxidans	Sarubbo et al. (2022)
Phospholipids, fatty acids, and neutral lipids	Fatty acids	Corynebacterium lepus	Cooper et al. (1979)
	Neutral lipids	Nocardia erythropolis	Gayathiri et al. (2022), Yang et al. (2019)
	Phospholipids	Acidithiobacillus thiooxidans	Vandana and Singh (2018)
Polymeric biosurfactants	Rufisan	Candida lipolytica	Janek et al. (2020), Sarubbo et al. (2022)
	Liposan	Candida lipolytica	Farias et al. (2021), Sarubbo et al. (2022)
	Emulsan	Acinetobacter lwoffii	Liu et al. (2020)
	Biodispersan	Acinetobacter calcoaceticus	Moshtagh et al. (2021), Mujumdar et al. (2019)
	Alasan	Acinetobacter radioresistens	Mujumdar et al. (2019)
Particulate surfactants	Vesicles	Acinetobacter calcoaceticus	Mujumdar et al. (2019)
	Cells	Cyanobacteria	Mehjabin et al. (2020)

16.3 Microbial biosurfactant production

Biosurfactants are produced by microorganisms in response to environmental conditions and can be associated with distinct growth stages (Moldes et al., 2021). From the microorganism perspective, some physiological functions are associated with biosurfactant production, such as improvement in the ability to solubilize and emulsify water-insoluble compounds, facilitating the use of these substrates; formation of polysaccharide—fatty acid complexes on the cell surface of certain microorganisms, which are involved in the transport of hydrocarbons; use of biosurfactants adhered to the cell wall for attachment to surfaces; antimicrobial action, mainly by certain classes of lipopeptides and glycopeptides; and improve competitiveness and search for nutrients (Nitschke & Pastore, 2002).

Most of these compounds are produced extracellularly as lipopeptides and glycolipids, but they can also be associated with the plasma membrane; these compounds are the so-called microbial cell-bound biosurfactants (glycolipopeptides) (Moldes et al., 2021). The main species or genera that produce biosurfactants are *Bacillus subtilis*, *Pseudomonas aeruginosa*, *Arthrobacter calcoaceticus*, *Candida lipolytica*, and *Candida bombicola* (Campos et al., 2015).

The use of different substrate sources by microorganisms can result in different structures with altered surfactant properties (Moldes et al., 2021). Not only the species used but also the cultivation conditions can change the type of biosurfactant and its concentration in the medium. Physicochemical parameters such as the nature of the carbon and nitrogen source (the C:N ratio), as well as temperature, pH, and aeration rate, can result in products with different compositions (Gayathiri et al., 2022; Md, 2012; Sarubbo et al., 2022).

The nature of the carbon source has a strong influence on the microbial growth rate and biosurfactant production by various microorganisms (Santos et al., 2016). Diesel oil, petroleum, glucose, sucrose and glycerol have been shown to be promising substrate sources (Rahman & Gakpe, 2008). The production of rhamnolipid by *Pseudomonas* sp. reaches higher yield when using petroleum-type carbon sources, whichrepresents an alternative to the use of renewable and cheap sources such as lignocellulosic waste and palm oil refinery industry (Santos et al., 2016). It was observed for yeasts *Torulopsis bombicola*, *Starmerella bombicola*, and *Candida lipolytica*, when grown on a medium containing different hydrophobic sources or a combination of hydrophilic and hydrophobic sources, improved the productivity of biosurfactants as concentrations above 5% were used (e.g., glucose and vegetable oil) (Sarubbo et al., 2022).

Nitrogen is essential for microbial growth for the synthesis of proteins and enzymes. Some compounds commonly used for biosurfactant production are urea, peptone, yeast extract, ammonium sulfate, sodium nitrate, meat extract, and malt (De et al., 2015). In addition, the C:N ratio can alter productivity. Maximum rhamnolipid production by *Pseudomonas aeruginosa* and *Rhodococcus* sp. was observed after nitrogen limitation at a

C:N ratio of 16:1−18:1, while no surfactant was produced below a C/N ratio of 11:1. Some metal ions are important enzyme cofactors, and their concentration can be a determinant for the synthesis of certain biosurfactants. For the production of surfactants, for example, the presence of Fe^{2+} ions is important (Thimon et al., 1992).

The incubation time is another limiting factor for biosurfactant production because each microorganism requires a different time interval depending on the type of substrate used. *Candida bombicola* shows maximum biosurfactant production after 68 h of incubation in a medium containing animal fat (Bhardwaj et al., 2013). For *Aspergillus ustus*, the maximum production of biosurfactant was observed after 5 days of incubation (Santos et al., 2016). In addition, environmental factors are also important for production and the characteristics of the biosurfactant. Therefore, production can be optimized through changes in temperature, pH, aeration, and agitation speed (Desai & Banat, 1997).

The optimum pH for biosurfactant production by most microorganisms varies between 5.7−7.8 (Jimoh & Lin, 2019). The rhamnolipid carboxyl groups that confer anionic character to the molecules are affected by the pH of the solution. Most carboxyl groups are dissociated at pH 6.8 to form carboxylate groups, while at pH 5, the protonated form prevails. Thus, higher efficiency in hydrolysis and acidification of activated sludge waste in the presence of rhamnolipids is found at alkaline pH than at acidic pH or near neutrality, since the rhamnosyl group is dissociated at high pH (pH > 11) (Özdemir et al., 2004).

The variation, even if small, in temperature can modulate the composition of the biosurfactants produced (Sarubbo et al., 2022). The optimum temperature for biosurfactant production by yeasts of the genus *Candida* is 27°C−30°C (Santos et al., 2016). While rhamnolipid synthesis by *Pseudomonas aeruginosa* occurs at 37°C (Rocha et al., 1992), other bacterial strains isolated from environmental sources such as the ocean have an optimum temperature between 22°C and 30°C (Tripathi et al., 2019).

Studies also show that the agitation and aeration rate can influence the production of compounds (Jahan et al., 2020; Santos et al., 2016; Sarubbo et al., 2022). Optimization of oxygen transfer has been shown to improve the process of scaling up surfactin production by *Bacillus subtilis* (Desai & Banat, 1997). Increased stirring speed favors the accumulation of biosurfactants synthesized by *P. aeruginosa* UCP 0992 grown on glycerol (Silva et al., 2010). Changes in stirring speed from 50 to 200 rpm for *Pseudomonas alcaligenes* grown on palm oil favored the reduction of the surface tension of the broth to 27.6 mN/m (Oliveira et al., 2009). The opposite, however, was observed for the biosurfactant produced by *Serratia* sp. SVGG16 growing in a medium containing hydrocarbons; the increased stirring speed negatively affected the surface tension (Cunha et al., 2004).

Strategies used to optimize the cultivation process and production of biosurfactants are shake flask, batch, fed-batch, continuous, and integrated microbial/enzymatic process. Bioreactors are commonly used for continuous or fed-batch fermentation, but they induce heavy foaming due to agitation and aeration, making the process more expensive. An alternative to reducing the costs of the large-scale production process of biosurfactants is the choice of substrate as they represent about 50% of the costs of the final product (Farias et al., 2021).

16.4 Renewable natural resources used in biosurfactant production

The generation of agro-industrial by-products is growing rapidly. In 2019, bioethanol, animal slaughter, cassava processing, palm oil, and dairy industries generated more than 4 billion liters of wastewater (Martinez-Burgos et al., 2021). Increasing the sustainability of these agro-industrial processes requires reducing waste or using these by-products or effluents to enhance other products, such as biosurfactants (Santos et al., 2016).

The substitution of synthetic culture media by agro-industrial residues or renewable substrates is interesting from both economic and environmental perspectives (Lorena de Oliveira & Sandra de Cássia, 2017). There are several researches involving the use of by-products for biosurfactant production, such as vegetable oils, oily effluents (Batista et al., 2010; Sarubbo et al., 2007), starch-rich effluents (Fox & Bala, 2000), fatty acids of animal (Maneerat, 2005; Santos et al., 2013) and vegetables (de Gusmão et al., 2010), vegetable oil waste (Alcantara et al., 2010; Batista et al., 2010), soap (Benincasa et al., 2002; Maneerat, 2005), molasses (Makkar & Cameotra, 1997; Maneerat, 2005), dairy waste (whey) (Alkan et al., 2019), corn liquor (Santos et al., 2018), cassava flour wastewater (Gaur, Sharma, Sirohi, et al., 2022), petroleum distillery waste (Rufino et al., 2007), and glycerol (Gaur, Sharma, Sirohi, et al., 2022; Silva et al., 2010). Some commonly used industrial by-products for biosurfactant production are further detailed.

16.4.1 Oil wastes

Waste vegetable oils from refineries and the food industry are used as substrates to produce surfactants by microorganisms (De et al., 2015). Sunflower seed oil and oleic acid have been used successfully by *Thermus thermophilus* HB8 to produce rhamnolipids (Pantazaki et al., 2010). Olive oil mill effluent (OOME) waste presents environmentally challenging phenols, usually nitrogen compounds (12–14 g/L), sugars (20–80 g/L), residual oil (0.3–5 g/L), and organic acids (5–15 g/L) (Santos et al., 2016). The application of this residue for the production of rhaminolipids by *Pseudomonas* sp (Mercadé et al., 1993) has shown satisfactory results. Surfactin production by *Bacillus* sp. growing in a culture medium supplemented with 2% cooking oil waste reached a concentration of 9.5 g/L within 7 days of incubation (Md Badrul Hisham et al., 2019).

16.4.1.1 Animal fat
Animal fat obtained from meat processing has been commonly used for food preparation; however, the search for healthier products by consumers has led to vegetable oils gaining more market space (Banat et al., 2014). This by-product by *Candida bombicola* has been shown to stimulate the production of sophorolipids (Deshpande & Daniels, 1995). Similarly, Santos et al. (2013, 2014) achieved maximum glycolipid production by the yeast *C. lipolytica* UCP 0988 when grown on animal fat and corn liquor.

16.4.1.2 Dairy whey
Whey, a waste product from the dairy industry, represents an environmental problem; this is especially important for countries based on this economy (Banat et al., 2014). For every 1 kg of cheese, about 6 L of waste are produced (De et al., 2015). Whey is an excellent source for microbial growth and production of surfactants and is composed of approximately 75% lactose, vitamins, and organic acids (Santos et al., 2016). Studies have shown a high concentration of sophorolipids by *Cryptococcus curvatus* ATCC 20509 when grown on this substrate using a two-stage cultivation process (two-stage) (Daniel et al., 1999).

16.4.1.3 Molasses
Molasses are by-products of the sugar industry, coming from sugar beets or sugar cane (Patel & Desai, 1997). They are cheap substrates with high sugar content (48%–56%), 2.5% protein, 1.5%–5.0% potassium and 1% magnesium, phosphorus, and calcium. It has been used as a substrate source for cultivating microorganisms to produce different metabolites (Santos et al., 2016), e.g., *Bacillus subtilis* (Makkar & Cameotra, 1997).

16.4.1.4 Starchy substrates
Starchy by-products are the main raw materials from the potato industry and are a cheap source for biosurfactant production. They are rich in carbon, starch and sugars, nitrogen, sulfur, inorganic minerals, trace elements, and vitamins (Fox & Bala, 2000). There are other starchy sources for surfactant production such as wastewater from the processing of cassava, rice, and cereals (Mukherjee et al., 2006). For example, the production of surfactant by *Bacillus subtilis* from cassava wastewater has been demonstrated (Santos et al., 2016).

16.4.1.5 Soap stock
Soap stock is a by-product of the vegetable oil refinery industry, with high content of saponified fatty acids (Lopes et al., 2021). It has been used for biosurfactant synthesis by some microorganisms, such as rhamnolipids-producing *P. aeruginosa* (Benincasa et al., 2002). The production of biosurfactants on an industrial scale is an area under development, and some criteria should be considered to improve the feasibility of the process,

such as the raw material used, microorganisms, design of the bioreactor, purification processes of the product, and fermentation time (Farias et al., 2021). The main problem in using waste in biotechnological processes lies in the selection of a substrate that contains the right balance of nutrients to support both cell growth and the production of the compound of interest (Miranda-Durán et al., 2020). In general, residues with high carbohydrate or lipid content are ideal for use as substrate (Santos et al., 2016).

16.5 Properties and advantages of biosurfactants

Despite having similar functions as amphipathic molecules, synthetic surfactants and biosurfactants differ significantly with respect to their origins. The biosurfactants produced by microorganisms have shown promise in different aspects because they have several advantages, such as surface and interface activity, emulsifying and demulsifying capacity, biodegradability and low toxicity, production using renewable raw materials, temperature, pH, and ionic strength tolerance. Therefore, biosurfactants are considered "green" alternatives to synthetically derived surfactants (De et al., 2015; Ribeiro et al., 2020; Saranraij et al., 2022).

Biosurfactants are complex molecules with specific functional groups and often have specific actions. This is of particular interest because they demonstrate superior performance to petroleum-based surfactants, making them more attractive. The main properties and advantages are further discussed.

16.5.1 Surface and interface activity

Efficiency and effectiveness are essential characteristics of a good surfactant. While the measurement of the CMC value refers to efficiency, surface and interfacial tensions are related to effectiveness (Santos et al., 2016). Most biosurfactants have lower CMC and surface and interfacial tension values than their synthetic counterparts, making them more efficient and effective when used in similar applications (Sarubbo et al., 2022). In general, a good biosurfactant can reduce the surface tension of water from 72 to 35 mN/m and the interfacial tension of water against hexadecane from 40 to 1 mN/m (Mulligan, 2005). The CMC of biosurfactants ranges from 1 to 2000 mg/L, while the interfacial (oil/water) and surface tensions are approximately 1 and 30 mN/m, respectively (Santos et al., 2016). For example, surfactin, produced mainly by the bacterium *Bacillus* sp., is considered one of the most effective biosurfactants ever described in the literature due to its ability to reduce the surface tension of water from 72 to 27 mN/m (Banat, 1993; Cooper & Goldenberg, 1987).

Furthermore, the CMCs of microbial surfactants are about 10–40 times lower than that of synthetic surfactants, implying that a smaller amount of the product is required to lower the surface tension to a set level (De et al., 2015; Desai & Banat, 1997). Biosurfactant from *Bacillus subtilis* DS03, using a medium with molasses, reduced the water surface tension from 72 to 35 mN/m, reaching a CMC of 24.66 ppm (Cruz Mendoza et al., 2022). The calculated CMC for the crude biosurfactant produced by *Wickerhamomyces anomalus* CCMA 0358 in media containing glucose (1 g/L) and olive oil (20 g/L) as carbon sources was 0.9 mg/mL, and the minimum surface tension value was 31.2 ± 0.4 mN/m (Souza et al., 2017).

16.5.1.1 Emulsification

Biosurfactants can stabilize (emulsify) or destabilize (de-emulsify) emulsions (De et al., 2015; Satpute et al., 2010). Bioemulsifiers, referring to high-molecular-weight biosurfactants, are complex mixtures of heteropolysaccharides, lipopolysaccharides, lipoproteins, and proteins. These molecules can efficiently emulsify two immiscible liquids and stabilize emulsions, even at low concentrations, without necessarily reducing the surface tension (STF) (Satpute et al., 2010; Uzoigwe et al., 2015). This property of biosurfactants is especially useful for the cosmetic, food, and petroleum industries as microbial enhanced oil recovery (MEOR) and bioremediation of contaminated soils (Pandey et al., 2022).

Emulsan is a lipopolysaccharide bioemulsifier with a molecular weight of 1000 kDa produced by *Acinetobacter calcoaceticus* RAG-1. It is one of the most widely studied bacterial emulsifiers. In the pure form, emulsan shows emulsifying activity at low concentrations (0.01%−0.001%) (Uzoigwe et al., 2015). Bioemulsifier produced by bacteria *Geobacillus thermodenitrificans* NG80−2, under anaerobic conditions at a temperature of 60°C using lignocellulosic biomass as carbon source, efficiently reduced the viscosity of heavy oil and extra-heavy oil. In addition, it emulsified various aliphatic and aromatic hydrocarbons and exhibited excellent emulsifying stability over a wide temperature range (4°C−100°C), salinity (NaCl, 0−100 g/L), and pH (3−10) (Li et al., 2023). The yeast strain *Yarrowia lipolytica* produced a potent bioemulsifier under saline conditions in the presence of diesel oil and exhibited high emulsification values with hydrocarbons, with no reduction in surface tension (Souza et al., 2012). The emulsifying activity of biosurfactants produced by *Bacillus licheniformis* and *Bacillus subtilis* strains isolated from Egyptian oil fields show significant emulsifying activity against long-chain hydrocarbons, with emulsification rates of 50.2% and 63.7%, respectively (Aboelkhair et al., 2022).

16.5.1.2 Biodegradability and low toxicity

Excessive use of any surfactants and their disposal in the environment, especially in aquatic bodies, could seriously affect the ecosystem. If the concentrations of synthetic surfactants in the environment reach high concentrations, toxicity can be accumulated in animals through the food chain, reaching humans through food consumption. Biosurfactants, because of their biological origin, are significantly less toxic to aquatic flora and fauna and are more easily biodegraded by microorganisms in water, soil, or in treatment plants without producing harmful end products, making these compounds adequate for bioremediation and waste treatment (Ivanković & Hrenović, 2010; Sarubbo et al., 2022). A low degree of toxicity allows the use of biosurfactants in foods, cosmetics, and pharmaceuticals (Abbot et al., 2022) and is also of fundamental importance to environmental applications (Johnson et al., 2021).

The cytotoxicity of biosurfactants produced from *Lactobacillus paracasei* was evaluated and compared with other emulsions formulated with the chemical surfactant sodium dodecyl sulfate (SDS) using a mouse fibroblast cell line. Solutions containing 5 g/L of biosurfactant presented cell proliferation values of 97%, whereas 0.5 g/L of SDS showed a strong inhibitory effect (Fei et al., 2020; Ferreira et al., 2017) reported the performance of surfactin from *Bacillus subtilis* HSO121 as a low toxic and nonirritant ingredient for detergent formations. In addition, surfactin showed other advantages, such as excellent surface and interfacial properties in terms of emulsification, wettability, foamability, high compatibility, and stability in a wide range of temperatures. Several personal care products, such as skin creams and lotions, shampoo, conditioner, liquid soap, toothpaste, sunscreen, shaving foam and gel, hair gel, deodorants, and cosmetics, can be prepared or even replaced by biosurfactants as long as they ensure safety for consumers (Ambaye et al., 2021).

Bacillus cereus UCP 1615 was cultivated in a mineral medium composed of 2% frying oil and 0.12% peptone to produce a biosurfactant characterized as a lipopeptide. Toxicity tests showed survival rates of the fish *Poecilia vivipara* and the bivalve *Anomalocardia brasiliana* higher than 90% and 55%, respectively. Moreover, biosurfactants stimulated the growth of autochthonous microorganisms independently of the presence of motor oil in bioassays performed in seawater. Biosurfactants with a low production cost, high productivity, and biocompatibility have the potential for industrial-scale production and application to bioremediation of oil spills–polluted marine environment (Durval et al., 2020). Microbial surfactants such as lipopeptides are less toxic, eco-friendly, with more stability in harsh environments, and highly biodegradable than their synthetic counterparts (Umar et al., 2021).

16.5.1.3 Temperature, pH, and ionic strength tolerance

The physicochemical properties of surfactants are strongly affected by temperature, pH, and salt concentrations. In general, salinity influences surfactant adsorption due to the interactions that arise between salt ions and surfactant molecules. The increase in temperature generally leads to a considerable decrease in the adsorption of surfactants due to an increase in the effects of kinetic and different pH on surfactant adsorption depending on the surfactant charge, which interacts with the charges available at the surface (Belhaj et al., 2020). On the other hand, many biosurfactants can be used at high temperatures, and pH values ranging from 2 to 12 also tolerate a salt concentration up to 10%, whereas 2% NaCl is enough to inactivate synthetic surfactants (Santos et al., 2016).

The biosurfactant produced by *Streptomyces* sp. DPUA1566, isolated from lichens from the Brazilian Amazon region, characterized as a lipoprotein, and denominated bioelan, was effective in the pH range from 2 to 12, at temperatures of 4°C–120°C, salt concentration 2%–12%, and heating time at 90°C for 10–120 min (Santos et al., 2019).

Umar and co-workers (Umar et al., 2021) used a cost-effective media for lipopeptide production by *Bacillus subtilis* SNW3. The lipopeptides obtained exhibited potential emulsifying and surface tension-reducing capabilities with strong stability at a wide range of pH (1–11), salinity (1%–8%), temperature (20°C–121°C), and even after autoclaving.

Bacillus licheniformis DS1 indigenous bacteria isolated from an oil reservoir produced lipopeptide in SMSSe medium supplemented with 2% (v/v) crude oil as the sole carbon source and incubated at 50°C, shaken at 120 rpm for 72 h. The biosurfactant exhibited good stability in maintaining emulsification activity at pH 4–10, high temperatures up to 120°C, and NaCl concentration up to 10% (w/v) (Purwasena et al., 2019).

The emulsifying capacity of the biosurfactant produced by *Lactobacillus pentosus* CECT-4023 T (ATCC-8041) for gasoline/water emulsions was evaluated under different conditions of salinity, pH, and temperature. The maximum emulsifying capacity of the biosurfactant and the stability of gasoline/water emulsions by the biosurfactant were achieved at temperatures below 30°C, salinity below 3%, and pH higher than 5, and at low values, salinity and temperature had a synergistic effect on the emulsifying properties of the biosurfactant (Vecino Bello et al., 2012).

16.6 Production using renewable raw materials

Large-scale production of biosurfactants can be expensive (De et al., 2015). It is estimated that the availability and type of raw material account for about 10%–30% of the total cost of a biotechnology project (Soares da Silva et al., 2019). However, this problem can be overcome using renewable raw materials and agro-industrial waste, contributing to help manage environmental waste and reduce greenhouse gas emissions and reduce production costs (Mgbechidinma et al., 2022). The use of abundantly available raw materials has been fundamental for the production of commercially and economically viable biosurfactants (Pardhi et al., 2022) (see Section Renewable natural resources used in biosurfactant production).

The yeast *Wickerhamomyces anomalus* CCMA 0358 was used to produce eco-friendly glycolipid from a low-cost substrate such as kitchen waste oil (KWO) and showed larvicidal, antibacterial, antiadhesive, and antifungal activities (de Andrade Teixeira Fernandes et al., 2020). Santos et al. (2019) produced a novel biosurfactant called bioelan from *Streptomyces* sp. DPUA1566 using 10 g/L soybean frying waste oil and 20 g/L corn maceration liquor. The biosurfactant reduced water surface tension from 72 to 28 mN/m, with a CMC of 0.08%, and was effective over wide ranges of temperature, pH, and salt concentration.

16.7 Challenges that limit the production of biosurfactants

The global market for microbial surfactants has increased significantly in recent decades, owing to the outstanding functional properties of these biomolecules (Adetunji & Olaniran, 2021). Despite several advantages, some challenges limit the production and large-scale industrial application of biosurfactants (Fig. 16.2).

16.7.1 Price and low productivity

The main limitation in the wide applicability of biosurfactants is the cost-effectiveness of production (Gaur, Sharma, Gupta, et al., 2022; Gaur, Sharma, Sirohi, et al., 2022; Kosaric & Sukan, 2014). The high price of biosurfactants is due to factors, such as lower yields, higher recovery and purification costs, energy requirements for sterilization, and maintenance of the biological culture (Farias et al., 2021).

While the average price of synthetic surfactants, such as SDS and plant-based surfactants, is $1–4/kg, the average price of sophorolipids, which are the most viable of the microbial biosurfactants, can range from US$ 20 to US$22.7/mg for rhamnolipids manufactured by Sigma-Aldrich, US$ 2.5 to 6.3/kg offered as SophoronTM from Saraya (Japan) and US$ 13.94/mg for surfactin (98% purity) available from the Sigma Chemical

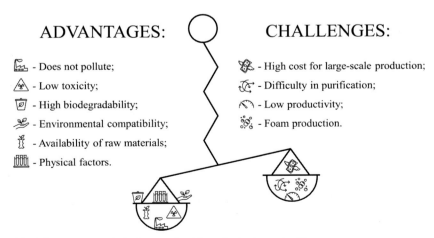

Figure 16.2 Advantages and challenges for biosurfactant production.

Company (Farias et al., 2021; Sarubbo et al., 2022; Soares da Silva et al., 2019). The key to successful biosurfactant production is the development of a cost-effective process that uses low-cost materials and offers high productivity (Rufino et al., 2014).

16.7.1.1 Recovery and purification

The recovery and purification of biosurfactants is challenging and represents 60%–80% of the total production costs (Mgbechidinma et al., 2022). The choice of the most suitable biosurfactant recovery processes depends, among other things, on the ionic charge, the water solubility and the location of the biosurfactant synthesis (intracellular or extracellular). These processes as the concentration of the biosurfactant, are decisive in determining the cost of biosurfactant production and are of prime industrial interest (Sant'Ana et al., 2017). The main methods for recovery of biosurfactants are acid precipitation, crystallization, organic solvent extraction, ammonium sulfate precipitation, centrifugation, foam fractionation, adsorption, membrane ultrafiltration, tangential flow filtration, and ion exchange chromatography. Industries have been investing in the study of different methods of extraction of biosurfactants and in the optimization of processes to achieve efficiency and low cost (Pardhi et al., 2022).

16.7.1.2 Foam formation

Most of the known biosurfactant-producing microorganisms require aerobic conditions for efficient production. However, the use of conventional submerged aeration can lead to foaming, a colloidal mixture formed by the dispersion of a gas in a liquid, causing serious operational problems (De et al., 2015; Nurfarahin et al., 2018). The high foaming is further increased by extracellular proteins, resulting in significant costs for process control, making the production process impractical (de Kronemberger et al., 2007; Sodagari et al., 2018). Foaming drags the culture medium out of the bioreactor where the cultivation occurs, causing losses and favoring contamination during the process. In addition, foaming would also restrict the extent of aeration and, as a result, limit the maximum cell concentration usable in the process (Sodagari et al., 2018). Mechanical foam breakers are not very efficient, and chemical defoaming agents can change the product quality and the bioreactor's the final effluent's potential to pollute (de Kronemberger et al., 2007).

16.8 Concluding remarks

Currently, the market demands for surfactants are met by numerous synthetic, mainly petroleum-based, chemical surfactants, which are usually toxic to the environment and nondegradable. Nevertheless, the growing interest of consumers who value eco-friendly products is a factor that has increasingly influenced the production and commercialization of biosurfactants. Various microorganisms, such as bacteria, yeasts, and filamentous fungi, are capable of producing biosurfactants with different molecular structures from renewable raw materials and less environmental impact during the production and disposal process. The chemical structure of biosurfactants is unique and exhibits excellent structural diversity, including glycolipids, lipopeptides, phospholipids, neutral fatty acids and lipids, particulate surfactants, and polymeric surfactants.

Despite the technical challenges, global biosurfactant consumption is estimated to increase further and global production is expected to reach US$ 28.8 billion by 2023. Global biosurfactants market size was more than USD 1.75 billion in 2020 and is estimated to grow at over 5.5% Compound Annual Growth Rate (CAGR) between 2021 and 2027 owing to growing consumer awareness regarding the hazardous effects of petroleum-based or synthetic surfactants on the environment and human health. Growing environmental concerns related to the use of synthetic surfactants and increasing consumer demand for natural household cleaners and cosmetics will increase the market share of biosurfactants (Global Market Insights Inc., 2022).

In terms of regional demand, the global market of biosurfactants is extended and prevails from Middle East Africa, North, South, and Central America, Asia Pacific, and Europe to Asia. Key players in the biosurfactants industry are BASF Cognis (Germany), Evonik Industries (Germany), Ecover, Lion Corporation (Belgium), Croda International PLC (British), Biotensidon GmbH (Germany), AkzoNobel N.V. (Netherlands), Saraya Co., Ltd. (Japan), Jeneil Biotech, Inc. (USA), and Givaudan SA (Soliance) (Switzerland) (Ambaye et al., 2021; Global Market Insights Inc., 2022).

Despite the current limitations to commercial production and the minimal amount of the market that is represented by biosurfactants, there is great interest in these materials as they offer some advantages over chemical surfactants, such as higher biodegradability, lower toxicity, temperature, pH and ionic strength tolerance, production using renewable raw materials, and better environmental compatibility. Therefore, several studies can be found in the literature aiming at lower-cost production processes and establishing biosurfactants as economical commercial compounds, mainly using renewable sources, including agro-industrial residues, as culture media.

References

Abbot, V., Paliwal, D., Sharma, A., & Sharma, P. (2022). A review on the physicochemical and biological applications of biosurfactants in biotechnology and pharmaceuticals. *Heliyon*, *8*(8), e10149. Available from https://doi.org/10.1016/j.heliyon.2022.e10149.

Aboelkhair, H., Diaz, P., & Attia, A. (2022). Environmental comparative study of biosurfactants production and optimization using bacterial strains isolated from Egyptian oil fields. *Journal of Petroleum Science and Engineering*, *216*, 796. Available from https://doi.org/10.1016/j.petrol.2022.110796.

Adetunji, A. I., & Olaniran, A. O. (2021). Production and potential biotechnological applications of microbial surfactants: An overview. *Saudi Journal of Biological Sciences*, *28*(1), 669–679. Available from https://doi.org/10.1016/j.sjbs.2020.10.058.

Akbari, S., Abdurahman, N. H., Yunus, R. M., Fayaz, F., & Alara, O. R. (2018). Biosurfactants—a new frontier for social and environmental safety: A mini review. *Biotechnology Research and Innovation*, *2*(1), 81−90. Available from https://doi.org/10.1016/j.biori.2018.09.001.

Alcantara, V. A., Pajares, I. G., Simbahan, J. F., Villarante, N. R., & Rubio, M. L. D. (2010). Characterization of biosurfactant from saccharomyces cerevisiae 2031 and evaluation of emulsification activity for potential application in bioremediation. *Philippine Agricultural Scientist*, *93*(1), 22−30. Available from http://www.pas-uplbca.edu.ph/002_Alcantara.pdf.

Alkan, Z., Erginkaya, Z., Konuray, G., & Ünal Turhan, E. (2019). Production of biosurfactant by lactic acid bacteria using whey as growth medium. *Turkish Journal of Veterinary and Animal Sciences*, *43*(5), 676−683. Available from https://doi.org/10.3906/vet-1903-48.

Ambaye, T. G., Vaccari, M., Prasad, S., & Rtimi, S. (2021). Preparation, characterization and application of biosurfactant in various industries: A critical review on progress, challenges and perspectives. *Environmental Technology and Innovation*, *24*, 102090. Available from https://doi.org/10.1016/j.eti.2021.102090.

Araújo, J., Rocha, J., Filho, M. O., Matias, S., Júnior, S. O., Padilha, C., & Santos, E. (2018). Rhamno lipids biosurfactants from *Pseudomonas aeruginosa*—A review. *Biosciences Biotechnology Research Asia*, *15*(4), 767−781. Available from https://doi.org/10.13005/bbra/2685.

Banat, I. M. (1993). The isolation of a thermophilic biosurfactant producing Bacillus SP. *Biotechnology Letters*, *15*(6), 591−594. Available from https://doi.org/10.1007/BF00138546.

Banat, I.M., Satpute, S.K., Cameotra, S.S., Patil, R., & Nyayanit, N.V. (2014). Cost effective technologies and renewable substrates for biosurfactants' production. Frontiers in Microbiology, 5, 697. https://doi.org/10.3389/fmicb.2014.00697

Batista, R. M., Rufino, R. D., Luna, J. M., De Souza, J. E. G., & Sarubbo, L. A. (2010). Effect of medium components on the production of a biosurfactant from Candida tropicalis applied to the removal of hydrophobic contaminants in soil. *Water Environment Research*, *82*(5), 418−425. Available from https://doi.org/10.2175/106143009X12487095237279.

Belhaj, A. F., Elraies, K. A., Mahmood, S. M., Zulkifli, N. N., Akbari, S., & Hussien, O. S. (2020). The effect of surfactant concentration, salinity, temperature, and pH on surfactant adsorption for chemical enhanced oil recovery: A review. *Journal of Petroleum Exploration and Production Technology*, *10*(1), 125−137. Available from https://doi.org/10.1007/s13202-019-0685-y.

Benincasa, M., Contiero, J., Manresa, M. A., & Moraes, I. O. (2002). Rhamnolipid production by *Pseudomonas aeruginosa* LBI growing on soapstock as the sole carbon source. *Journal of Food Engineering*, *54*(4), 283−288. Available from https://doi.org/10.1016/S0260-8774(01)00214-X.

Bhardwaj, G., Cameotra, S. S., & Chopra, H. K. (2013). Utilization of oleo-chemical industry byproducts for biosurfactant production. *AMB Express*, *3*, 68. Available from https://doi.org/10.1186/2191-0855-3-68.

Biktasheva, L., Gordeev, A., Selivanovskaya, S., & Galitskaya, P. (2022). Di- and monorhamnolipids produced by the *Pseudomonas putida* PP021 isolate significantly enhance the degree of recovery of heavy oil from the romashkino oil field (Tatarstan, Russia). *Processes*, *10*(4). Available from https://doi.org/10.3390/pr10040779.

Campos, J. M., Stamford, T. L. M., Rufino, R. D., Luna, J. M., Stamford, T. C. M., & Sarubbo, L. A. (2015). Formulation of mayonnaise with the addition of a bioemulsifier isolated from Candida utilis. *Toxicology Reports*, *2*, 1164−1170. Available from https://doi.org/10.1016/j.toxrep.2015.08.009.

Celligoi, M. A. C., Marcos, R. D. O., Camilios-Neto, D., Baldo, C., Magri, A., Pedrine, M. A., & Celligoi, C. (2014). Biosynthesis And Production of Sophorolipids. *International Journal of Scientific & Technology Research*, *3*(11), 132−143.

Christova, N., Lang, S., Wray, V., Kaloyanov, K., Konstantinov, S., & Stoineva, I. (2015). Production, structural elucidation, and in vitro antitumor activity of trehalose lipid biosurfactant from Nocardia farcinica strain. *Journal of Microbiology and Biotechnology*, *25*(4), 439−447. Available from https://doi.org/10.4014/jmb.1406.06025.

Ciurko, D., Czyżnikowska, Ż., Kancelista, A., èaba, W., & Janek, T. (2022). Sustainable production of biosurfactant from Agro-Industrial Oil wastes by *Bacillus subtilis* and its potential application as antioxidant and ACE inhibitor. *International Journal of Molecular Sciences*, *23*(18). Available from https://doi.org/10.3390/ijms231810824.

Cooper, D. G., & Goldenberg, B. G. (1987). Surface-active agents from two Bacillus species. *Applied and Environmental Microbiology*, *53*(2), 224−229. Available from https://doi.org/10.1128/aem.53.2.224-229.1987.

Cooper, D. G., & Paddock, D. A. (1983). Torulopsis petrophilum and surface activity. *Applied and Environmental Microbiology*, *46*(6), 1426−1429. Available from https://doi.org/10.1128/aem.46.6.1426-1429.1983.

Cooper, D. G., Zajic, J. E., & Gerson, D. F. (1979). Production of surface-active lipids by *Corynebacterium lepus*. *Applied and Environmental Microbiology*, *37*(1), 4−10. Available from https://doi.org/10.1128/aem.37.1.4-10.1979.

Cruz Mendoza, I., Villavicencio-Vasquez, M., Aguayo, P., Coello Montoya, D., Plaza, L., Romero-Peña, M., Marqués, A. M., & Coronel-León, J. (2022). Biosurfactant from *Bacillus subtilis* DS03: Properties and application in cleaning out place system in a pilot sausages processing. *Microorganisms*, *10*(8). Available from https://doi.org/10.3390/microorganisms10081518.

Cunha, C. D., Do Rosário, M., Rosado, A. S., & Leite, S. G. F. (2004). Serratia sp. SVGG16: A promising biosurfactant producer isolated from tropical soil during growth with ethanol-blended gasoline. *Process Biochemistry*, *39*(12), 2277−2282. Available from https://doi.org/10.1016/j.procbio.2003.11.027.

Daniel, H. J., Otto, R. T., Binder, M., Reuss, M., & Syldatk, C. (1999). Production of sophorolipids from whey: Development of a two-stage process with *Cryptococcus curvatus* ATCC 20509 and Candida bombicola ATCC 22214 using deproteinized whey concentrates as substrates. *Applied Microbiology and Biotechnology*, *51*(1), 40−45. Available from https://doi.org/10.1007/s002530051360.

de Andrade Teixeira Fernandes, N., de Souza, A. C., Simões, L. A., Ferreira dos Reis, G. M., Souza, K. T., Schwan, R. F., & Dias, D. R. (2020). Eco-friendly biosurfactant from Wickerhamomyces anomalus CCMA 0358 as larvicidal and antimicrobial. *Microbiological Research*, *241*, 126571. Available from https://doi.org/10.1016/j.micres.2020.126571.

De Graeve, M., De Maesenejre, S. L., Roelants, S. L. K. W., & Soetaert, W. (2018). Starmerella bombicola, an industrially relevant, yet fundamentally underexplored yeast. *FEMS Yeast Research*, *18*(7). Available from https://doi.org/10.1093/femsyr/foy072.

de Gusmão, C. A. B., Rufino, R. D., & Sarubbo, L. A. (2010). Laboratory production and characterization of a new biosurfactant from Candida glabrata UCP1002 cultivated in vegetable fat waste applied to the removal of hydrophobic contaminant. *World Journal of Microbiology and Biotechnology*, *26*(9), 1683−1692. Available from https://doi.org/10.1007/s11274-010-0346-2.

de Kronemberger, F. A., Anna, L. M. M. S., Fernandes, A. C. L. B., de Menezes, R. R., Borges, C. P., & Freire, D. M. G. (2007). *Oxygen-controlled biosurfactant production, A bench scale bioreactor* (147, pp. 401−413). Springer Science and Business Media LLC Issues 1−3. Available from https://doi.org/10.1007/978-1-60327-526-2_39.

De, S., Malik, S., Ghosh, A., Saha, R., & Saha, B. (2015). A review on natural surfactants. *RSC Advances*, *5*(81), 65757−65767. Available from https://doi.org/10.1039/c5ra11101c.

Desai, J. D., & Banat, I. M. (1997). Microbial production of surfactants and their commercial potential. *Microbiology and Molecular Biology Reviews*, *61*(1), 47−64. Available from https://doi.org/10.1128/0.61.1.47-64.1997.

Deshpande, M., & Daniels, L. (1995). Evaluation of sophorolipid biosurfactant production by Candida bombicola using animal fat. *Bioresource Technology*, *54*(2), 143−150. Available from https://doi.org/10.1016/0960-8524(95)00116-6.

Drakontis, C. E., & Amin, S. (2020). Biosurfactants: Formulations, properties, and applications. *Current Opinion in Colloid and Interface Science*, *48*, 77−90. Available from https://doi.org/10.1016/j.cocis.2020.03.013.

Durval, I. J. B., Mendonça, A. H. R., Rocha, I. V., Luna, J. M., Rufino, R. D., Converti, A., & Sarubbo, L. A. (2020). Production, characterization, evaluation and toxicity assessment of a *Bacillus cereus* UCP 1615 biosurfactant for marine oil spills bioremediation. *Marine Pollution Bulletin*, *157*, 111357. Available from https://doi.org/10.1016/j.marpolbul.2020.111357.

Eslami, P., Hajfarajollah, H., & Bazsefidpar, S. (2020). Recent advancements in the production of rhamnolipid biosurfactants by *Pseudomonas aeruginosa*. *RSC Advances*, *10*(56), 34014−34032. Available from https://doi.org/10.1039/d0ra04953k.

Farias, C. B. B., Almeida, F. C. G., Silva, I. A., Souza, T. C., Meira, H. M., Soares da Silva, R. d C. F., Luna, J. M., Santos, V. A., Converti, A., Banat, I. M., & Sarubbo, L. A. (2021). Production of green surfactants: Market prospects. *Electronic Journal of Biotechnology*, *51*, 28–39. Available from https://doi.org/10.1016/j.ejbt.2021.02.002.

Fei, D., Zhou, G. W., Yu, Z. Q., Gang, H. Z., Liu, J. F., Yang, S. Z., Ye, R. Q., & Mu, B. Z. (2020). Low-toxic and nonirritant biosurfactant surfactin and its performances in detergent formulations. *Journal of Surfactants and Detergents*, *23*(1), 109–118. Available from https://doi.org/10.1002/jsde.12356.

Ferreira, A., Vecino, X., Ferreira, D., Cruz, J. M., Moldes, A. B., & Rodrigues, L. R. (2017). Novel cosmetic formulations containing a biosurfactant from *Lactobacillus paracasei*. *Colloids and Surfaces B: Biointerfaces*, *155*, 522–529. Available from https://doi.org/10.1016/j.colsurfb.2017.04.026.

Fox, S. L., & Bala, G. A. (2000). Production of surfactant from *Bacillus subtilis* ATCC 21332 using potato substrates. *Bioresource Technology*, *75*(3), 235–240. Available from https://doi.org/10.1016/S0960-8524(00)00059-6.

Gaur, V. K., Bajaj, A., Regar, R. K., Kamthan, M., Jha, R. R., Srivastava, J. K., & Manickam, N. (2019). Rhamnolipid from a *Lysinibacillus sphaericus* strain IITR51 and its potential application for dissolution of hydrophobic pesticides. *Bioresource Technology*, *272*, 19–25. Available from https://doi.org/10.1016/j.biortech.2018.09.144.

Gaur, V. K., Sharma, P., Gupta, S., Varjani, S., Srivastava, J. K., Wong, J. W. C., & Ngo, H. H. (2022). Opportunities and challenges in omics approaches for biosurfactant production and feasibility of site remediation: Strategies and advancements. *Environmental Technology and Innovation*, *25*, 102132. Available from https://doi.org/10.1016/j.eti.2021.102132.

Gaur, V. K., Sharma, P., Sirohi, R., Varjani, S., Taherzadeh, M. J., Chang, J. S., Yong Ng, H., Wong, J. W. C., & Kim, S. H. (2022). Production of biosurfactants from agro-industrial waste and waste cooking oil in a circular bioeconomy: An overview. *Bioresource Technology*, *343*, 126059. Available from https://doi.org/10.1016/j.biortech.2021.126059.

Gayathiri, E., Prakash, P., Karmegam, N., Varjani, S., Awasthi, M. K., & Ravindran, B. (2022). Biosurfactants: Potential and eco-friendly material for sustainable agriculture and environmental safety—A review. *Agronomy*, *12*(3). Available from https://doi.org/10.3390/agronomy12030662.

Gecol, H. (2007). The basic theory. *Chemistry and technology of surfactants* (pp. 24–45). Blackwell Publishing Ltd. Available from https://doi.org/10.1002/9780470988596.ch2.

Gudiña, E. J., & Teixeira, J. A. (2022). Bacillus licheniformis: The unexplored alternative for the anaerobic production of lipopeptide biosurfactants? *Biotechnology Advances*, *60*, 108013. Available from https://doi.org/10.1016/j.biotechadv.2022.108013.

Hu, J., Luo, J., Zhu, Z., Chen, B., Ye, X., Zhu, P., & Zhang, B. (2021). Multi-scale biosurfactant production by bacillus subtilis using tuna fish waste as substrate. *Catalysts*, *11*(4). Available from https://doi.org/10.3390/catal11040456.

Ivanković, T., & Hrenović, J. (2010). Surfactants in the environment. *Archives of Industrial Hygiene and Toxicology*, *61*(1), 95–110. Available from https://doi.org/10.2478/10004-1254-61-2010-1943.

Jahan, R., Bodratti, A. M., Tsianou, M., & Alexandridis, P. (2020). Biosurfactants, natural alternatives to synthetic surfactants: Physicochemical properties and applications. *Advances in Colloid and Interface Science*, *275*, 102061. Available from https://doi.org/10.1016/j.cis.2019.102061.

Janek, T., Mirończuk, A. M., Rymowicz, W., & Dobrowolski, A. (2020). High-yield expression of extracellular lipase from Yarrowia lipolytica and its interactions with lipopeptide biosurfactants: A biophysical approach. *Archives of Biochemistry and Biophysics*, *689*, 108475. Available from https://doi.org/10.1016/j.abb.2020.108475.

Jiménez, I. M., Chandel, A. K., Marcelino, P. R. F., Anjos, V., Batesttin Costa, C., Jose, V., Bell, M., Pereira, B., & da Silva, S. S. (2020). Comparative data on effects of alkaline pretreatments and enzymatic hydrolysis on bioemulsifier production from sugarcane straw by *Cutaneotrichosporon mucoides*. *Bioresource Technology*, *301*, 122706. Available from https://doi.org/10.1016/j.biortech.2019.122706.

Jimoh, A. A., & Lin, J. (2019). Biosurfactant: A new frontier for greener technology and environmental sustainability. *Ecotoxicology and Environmental Safety*, *184*, 109607. Available from https://doi.org/10.1016/j.ecoenv.2019.109607.

Johnson, P., Trybala, A., Starov, V., & Pinfield, V. J. (2021). Effect of synthetic surfactants on the environment and the potential for substitution by biosurfactants. *Advances in Colloid and Interface Science*, *288*102340. Available from https://doi.org/10.1016/j.cis.2020.102340.

Joshi, P., & Tejas. (2017). A short history and preamble of surfactants. *International Journal of Applied Chemistry*, *13*(2), 283–292.

Kosaric, N., & Sukan, F. V. (2014). *Biosurfactants versus chemically synthesized surface-active agents* (pp. 48–59). Informa UK Limited. Available from https://doi.org/10.1201/b17599-5.

Li, M., Yu, J., Cao, L., Yin, Y., Su, Z., Chen, S., Li, G., & Ma, T. (2023). Facultative anaerobic conversion of lignocellulose biomass to new bioemulsifier by thermophilic *Geobacillus thermodenitrificans* NG80-2. *Journal of Hazardous Materials*, *443*130210. Available from https://doi.org/10.1016/j.jhazmat.2022.130210.

Liu, Q., Niu, J., Yu, Y., Wang, C., Lu, S., Zhang, S., Lv, J., & Peng, B. (2021). Production, characterization and application of biosurfactant produced by *Bacillus licheniformis* L20 for microbial enhanced oil recovery. *Journal of Cleaner Production*, *307*, 127193. Available from https://doi.org/10.1016/j.jclepro.2021.127193.

Liu, Y., Wan, Y. Y., Wang, C., Ma, Z., Liu, X., & Li, S. (2020). Biodegradation of n-alkanes in crude oil by three identified bacterial strains. *Fuel*, *275*, 117897. Available from https://doi.org/10.1016/j.fuel.2020.117897.

Lopes, P. R. M., Montagnolli, R. N., Cruz, J. M., Lovaglio, R. B., Mendes, C. R., Dilarri, G., Contiero, J., & Bidoia, E. D. (2021). Production of rhamnolipids from soybean soapstock: Characterization and comparison with synthetics surfactants. *Waste and Biomass Valorization*, *12*(4), 2013–2023. Available from https://doi.org/10.1007/s12649-020-01159-2.

Lorena de Oliveira, F., & Sandra de Cássia, D. (2017). Surfactantes sintéticos e biossurfactantes: Vantagens e desvantagens. *Química Nova Na Escola*, *39*(3). Available from https://doi.org/10.21577/0104-8899.20160079.

Luna, J. M., Filho, A. S. S., Rufino, R. D., & Sarubbo, L. A. (2016). Production of biosurfactant from Candida bombicola URM 3718 for environmental applications. *Chemical Engineering Transactions*, *49*, 583–588. Available from https://doi.org/10.3303/CET1649098.

Luna, J. M., Rufino, R. D., Jara, A. M. A. T., Brasileiro, P. P. F., & Sarubbo, L. A. (2015). Environmental applications of the biosurfactant produced by Candida sphaerica cultivated in low-cost substrates. *Colloids and Surfaces A: Physicochemical and Engineering Aspects*, *480*, 413–418. Available from https://doi.org/10.1016/j.colsurfa.2014.12.014.

Makkar, R. S., & Cameotra, S. S. (1997). Utilization of molasses for biosurfactant production by two Bacillus strains at thermophilic conditions. *Journal of the American Oil Chemists' Society*, *74*(7), 887–889. Available from https://doi.org/10.1007/s11746-997-0233-7.

Maneerat, S. (2005). Production of biosurfactants using substrates from renewable-resources. *Songklanakarin Journal of Science and Technology*, *27*(3), 675–683.

Marcelino, P. R. F., Peres, G. F. D., Terán-Hilares, R., Pagnocca, F. C., Rosa, C. A., Lacerda, T. M., dos Santos, J. C., & da Silva, S. S. (2019). Biosurfactants production by yeasts using sugarcane bagasse hemicellulosic hydrolysate as new sustainable alternative for lignocellulosic biorefineries. *Industrial Crops and Products*, *129*, 212–223. Available from https://doi.org/10.1016/j.indcrop.2018.12.001.

Martinez-Burgos, W. J., Bittencourt Sydney, E., Bianchi Pedroni Medeiros, A., Magalhães, A. I., de Carvalho, J. C., Karp, S. G., Porto de Souza Vandenberghe, L., Junior Letti, L. A., Thomaz Soccol, V., de Melo Pereira, G. V., Rodrigues, C., Lorenci Woiciechowski, A., & Soccol, C. R. (2021). Agro-industrial wastewater in a circular economy: Characteristics, impacts and applications for bioenergy and biochemicals. *Bioresource Technology*, *341*, 125795. Available from https://doi.org/10.1016/j.biortech.2021.125795.

Md Badrul Hisham, N. H., Ibrahim, M. F., Ramli, N., & Abd-Aziz, S. (2019). Production of biosurfactant produced from used cooking oil by Bacillus sp. HIP3 for heavy metals removal. *Molecules*, *24*(14). Available from https://doi.org/10.3390/molecules24142617.

Md, F. (2012). Biosurfactant: Production and application. *Journal of Petroleum & Environmental Biotechnology*, *03*(04). Available from https://doi.org/10.4172/2157-7463.1000124.

Mehjabin, J. J., Wei, L., Petitbois, J. G., Umezawa, T., Matsuda, F., Vairappan, C. S., Morikawa, M., & Okino, T. (2020). Biosurfactants from marine cyanobacteria collected in Sabah, Malaysia. *Journal of Natural Products*, *83*(6), 1925–1930. Available from https://doi.org/10.1021/acs.jnatprod.0c00164.

Mercadé, M. E., Manresa, M. A., Robert, M., Espuny, M. J., de Andrés, C., & Guinea, J. (1993). Olive oil mill effluent (OOME). New substrate for biosurfactant production. *Bioresource Technology*, *43*(1), 1–6. Available from https://doi.org/10.1016/0960-8524(93)90074-L.

Mgbechidinma, C. L., Akan, O. D., Zhang, C., Huang, M., Linus, N., Zhu, H., & Wakil, S. M. (2022). Integration of green economy concepts for sustainable biosurfactant production – A review. *Bioresource Technology*, *364*, 128021. Available from https://doi.org/10.1016/j.biortech.2022.128021.

Miranda-Durán, S., Porras-Reyes, L., & Schmidt-Durán, A. (2020). Evaluation of agro-industrial residues produced in Costa Rica for a low-cost culture medium using *Bacillus subtilis* 168. *Revista Tecnología En Marcha*, *33*(4), 15–25. Available from https://doi.org/10.18845/tm.v33i4.4807.

Mnif, I., & Ghribi, D. (2016). Glycolipid biosurfactants: Main properties and potential applications in agriculture and food industry. *Journal of the Science of Food and Agriculture*, *96*(13), 4310–4320. Available from https://doi.org/10.1002/jsfa.7759.

Moldes, A. B., Rodríguez-López, L., Rincón-Fontán, M., López-Prieto, A., Vecino, X., & Cruz, J. M. (2021). Synthetic and bio-derived surfactants versus microbial biosurfactants in the cosmetic industry: An overview. *International Journal of Molecular Sciences*, *22*(5), 1–23. Available from https://doi.org/10.3390/ijms22052371.

Mondal, S., Acharjee, A., & Saha, B. (2022). Utilization of natural surfactants: An approach towards sustainable universe. *Vietnam Journal of Chemistry*, *60*(1), 1–14. Available from https://doi.org/10.1002/vjch.202100137.

Morita, T., Konishi, M., Fukuoka, T., Imura, T., & Kitamoto, D. (2007). Physiological differences in the formation of the glycolipid biosurfactants, mannosylerythritol lipids, between Pseudozyma antarctica and Pseudozyma aphidis. *Applied Microbiology and Biotechnology*, *74*(2), 307–315. Available from https://doi.org/10.1007/s00253-006-0672-3.

Moshtagh, B., Hawboldt, K., & Zhang, B. (2021). Biosurfactant production by native marine bacteria (Acinetobacter calcoaceticus P1-1A) using waste carbon sources: Impact of process conditions. *The Canadian Journal of Chemical Engineering*, *99*(11), 2386–2397. Available from https://doi.org/10.1002/cjce.24254.

Moutinho, L. F., Moura, F. R., Silvestre, R. C., & Romão-Dumaresq, A. S. (2021). Microbial biosurfactants: A broad analysis of properties, applications, biosynthesis, and techno-economical assessment of rhamnolipid production. *Biotechnology Progress*, *37*(2). Available from https://doi.org/10.1002/btpr.3093.

Mujumdar, S., Joshi, P., & Karve, N. (2019). Production, characterization, and applications of bioemulsifiers (BE) and biosurfactants (BS) produced by Acinetobacter spp.: A review. *Journal of Basic Microbiology*, *59*(3), 277–287. Available from https://doi.org/10.1002/jobm.201800364.

Mukherjee, S., Das, P., & Sen, R. (2006). Towards commercial production of microbial surfactants. *Trends in Biotechnology*, *24*(11), 509–515. Available from https://doi.org/10.1016/j.tibtech.2006.09.005.

Mulligan, C. N. (2005). Environmental applications for biosurfactants. *Environmental Pollution*, *133*(2), 183–198. Available from https://doi.org/10.1016/j.envpol.2004.06.009.

Nalini, S., & Parthasarathi, R. (2013). Biosurfactant production by Serratia rubidaea SNAU02 isolated from hydrocarbon contaminated soil and its physico-chemical characterization. *Bioresource Technology*, *147*, 619–622. Available from https://doi.org/10.1016/j.biortech.2013.08.041.

Nitschke, M., & Pastore, G. M. (2002). Biossurfactantes: Propriedades e aplicações. *Quimica Nova*, *25*(5), 772–776. Available from https://doi.org/10.1590/S0100-40422002000500013.

Nurfarahin, A., Mohamed, M., & Phang, L. (2018). Culture medium development for microbial-derived surfactants production—An overview. *Molecules*, *23*(5), 1049. Available from https://doi.org/10.3390/molecules23051049.

Oliveira, F. J. S., Vazquez, L., de Campos, N. P., & de França, F. P. (2009). Production of rhamnolipids by a Pseudomonas alcaligenes strain. *Process Biochemistry*, *44*(4), 383–389. Available from https://doi.org/10.1016/j.procbio.2008.11.014.

Özdemir, G., Peker, S., & Helvaci, S. S. (2004). Effect of pH on the surface and interfacial behavior of rhamnolipids R1 and R2. *Colloids and Surfaces A: Physicochemical and Engineering Aspects*, *234*(1−3), 135−143. Available from https://doi.org/10.1016/j.colsurfa.2003.10.024.

Pandey, R., Krishnamurthy, B., Singh, H. P., & Batish, D. R. (2022). Evaluation of a glycolipopepetide biosurfactant from Aeromonas hydrophila RP1 for bioremediation and enhanced oil recovery. *Journal of Cleaner Production*, *345*, 131098. Available from https://doi.org/10.1016/j.jclepro.2022.131098.

Pantazaki, A. A., Dimopoulou, M. I., Simou, O. M., & Pritsa, A. A. (2010). Sunflower seed oil and oleic acid utilization for the production of rhamnolipids by *Thermus thermophilus* HB8. *Applied Microbiology and Biotechnology*, *88*(4), 939−951. Available from https://doi.org/10.1007/s00253-010-2802-1.

Pardhi, D. S., Panchal, R. R., Raval, V. H., Joshi, R. G., Poczai, P., Almalki, W. H., & Rajput, K. N. (2022). Microbial surfactants: A journey from fundamentals to recent advances. *Frontiers in Microbiology*, *13*, 982603. Available from https://doi.org/10.3389/fmicb.2022.982603.

Patel, R. M., & Desai, A. J. (1997). Biosurfactant production by *Pseudomonas aeruginosa* GS3 from molasses. *Letters in Applied Microbiology*, *25*(2), 91−94. Available from https://doi.org/10.1046/j.1472-765X.1997.00172.x.

Phulpoto, I. A., Yu, Z., Hu, B., Wang, Y., Ndayisenga, F., Li, J., Liang, H., & Qazi, M. A. (2020). Production and characterization of surfactin-like biosurfactant produced by novel strain *Bacillus nealsonii* S2MT and it's potential for oil contaminated soil remediation. *Microbial Cell Factories*, *19*(1). Available from https://doi.org/10.1186/s12934-020-01402-4.

Pinto, M. I. S., Guerra, J. M. C., Meira, H. M., Sarubbo, L. A., & de Luna, J. M. (2022). A biosurfactant from Candida bombicola: Its synthesis, characterization, and its application as a food emulsions. *Foods*, *11*(4). Available from https://doi.org/10.3390/foods11040561.

Pirog, T., Sofilkanych, A., Shevchuk, T., & Shulyakova, M. (2013). Biosurfactants of *Rhodococcus erythropolis* IMV Ac-5017: Synthesis intensification and practical application. *Applied Biochemistry and Biotechnology*, *170*(4), 880−894. Available from https://doi.org/10.1007/s12010-013-0246-7.

Purwasena, I. A., Astuti, D. I., Syukron, M., Amaniyah, M., & Sugai, Y. (2019). Stability test of biosurfactant produced by *Bacillus licheniformis* DS1 using experimental design and its application for MEOR. *Journal of Petroleum Science and Engineering*, *183*106383. Available from https://doi.org/10.1016/j.petrol.2019.106383.

Rahman, P. K. S. M., & Gakpe, E. (2008). Production, characterisation and applications of biosurfactants—Review. *Biotechnology*, *7*(2), 360−370. Available from https://doi.org/10.3923/biotech.2008.360.370.

Ribeiro, B. G., Guerra, J. M. C., & Sarubbo, L. A. (2020). Biosurfactants: Production and application prospects in the food industry. *Biotechnology Progress*, *36*(5). Available from https://doi.org/10.1002/btpr.3030.

Rocha, C., San-Blas, F., San-Blas, G., & Vierma, L. (1992). Biosurfactant production by two isolates of *Pseudomonas aeruginosa*. *World Journal of Microbiology & Biotechnology*, *8*(2), 125−128. Available from https://doi.org/10.1007/BF01195830.

Rufino, R. D., de Luna, J. M., de Campos Takaki, G. M., & Sarubbo, L. A. (2014). Characterization and properties of the biosurfactant produced by *Candida lipolytica* UCP 0988. *Electronic Journal of Biotechnology*, *17*(1), 34−38. Available from https://doi.org/10.1016/j.ejbt.2013.12.006.

Rufino, R. D., Sarubbo, L. A., & Campos-Takaki, G. M. (2007). Enhancement of stability of biosurfactant produced by Candida lipolytica using industrial residue as substrate. *World Journal of Microbiology and Biotechnology*, *23*(5), 729−734. Available from https://doi.org/10.1007/s11274-006-9278-2.

Sant'Ana, G. C. F., Silva, K. A., Coelho, M. A. Z., Biossurfactante, C., Ribeiro, B. D., & Pereira, K. S. (2017). *Microbiologia industrial: Bioprocessos. do Nascimento*. Elsevier.

Santos, D. K. F., Brandão, Y. B., Rufino, R. D., Luna, J. M., Salgueiro, A. A., Santos, V. A., & Sarubbo, L. A. (2014). Optimization of cultural conditions for biosurfactant production from Candida lipolytica. *Biocatalysis and Agricultural Biotechnology*, *3*(3), 48−57. Available from https://doi.org/10.1016/j.bcab.2014.02.004.

Santos, D. K. F., Rufino, R. D., Luna, J. M., Santos, V. A., Salgueiro, A. A., & Sarubbo, L. A. (2013). Synthesis and evaluation of biosurfactant produced by Candida lipolytica using animal fat and corn steep liquor. *Journal of Petroleum Science and Engineering*, *105*, 43−50. Available from https://doi.org/10.1016/j.petrol.2013.03.028.

Santos, D. K. F., Rufino, R. D., Luna, J. M., Santos, V. A., & Sarubbo, L. A. (2016). Biosurfactants: Multifunctional biomolecules of the 21st century. *International Journal of Molecular Sciences, 17*(3). Available from https://doi.org/10.3390/ijms17030401.

Santos, E. F., Teixeira, M. F. S., Converti, A., Porto, A. L. F., & Sarubbo, L. A. (2019). Production of a new lipoprotein biosurfactant by Streptomyces sp. DPUA1566 isolated from lichens collected in the Brazilian Amazon using agroindustry wastes. *Biocatalysis and Agricultural Biotechnology, 17*, 142–150. Available from https://doi.org/10.1016/j.bcab.2018.10.014.

Santos, F. F., Freitas, K. M. L., Da Costa Neto, J. J. G., Fontes-Sant'Ana, G., Rocha-Leão, M. H. M., & Amaral, P. F. F. (2018). Tiger nut (Cyperus esculentus) milk byproduct and corn steep liquor for biosurfactant production by *Yarrowia lipolytica*. *Chemical Engineering Transactions, 65*, 331–336. Available from https://doi.org/10.3303/CET1865056.

Sarafin, Y., Donio, M. B. S., Velmurugan, S., Michaelbabu, M., & Citarasu, T. (2014). Kocuria marina BS-15 a biosurfactant producing halophilic bacteria isolated from solar salt works in India. *Saudi Journal of Biological Sciences, 21*(6), 511–519. Available from https://doi.org/10.1016/j.sjbs.2014.01.001.

Saranraij, P., Sivasakthivelan, P., Hamzah, K. J., Hasan, M. S., & Al-Tawaha, A. R. M. (2022). *Microbial fermentation technology for biosurfactants production* (pp. 25–43). Informa UK Limited. Available from https://doi.org/10.1201/9781003247739-2.

Sarubbo, L. A., Farias, C. B. B., & Campos-Takaki, G. M. (2007). Co-utilization of canola oil and glucose on the production of a surfactant by Candida lipolytica. *Current Microbiology, 54*(1), 68–73. Available from https://doi.org/10.1007/s00284-006-0412-z.

Sarubbo, L. A., Silva, M. d G. C., Durval, I. J. B., Bezerra, K. G. O., Ribeiro, B. G., Silva, I. A., Twigg, M. S., & Banat, I. M. (2022). Biosurfactants: Production, properties, applications, trends, and general perspectives. *Biochemical Engineering Journal, 181*, 108377. Available from https://doi.org/10.1016/j.bej.2022.108377.

Satpute, S. K., Banpurkar, A. G., Dhakephalkar, P. K., Banat, I. M., & Chopade, B. A. (2010). Methods for investigating biosurfactants and bioemulsifiers: A review. *Critical Reviews in Biotechnology, 30*(2), 127–144. Available from https://doi.org/10.3109/07388550903427280.

Shekhar, S., Sundaramanickam, A., & Balasubramanian, T. (2015). Biosurfactant producing microbes and their potential applications: A review. *Critical Reviews in Environmental Science and Technology, 45*(14), 1522–1554. Available from https://doi.org/10.1080/10643389.2014.955631.

Silva, S. N. R. L., Farias, C. B. B., Rufino, R. D., Luna, J. M., & Sarubbo, L. A. (2010). Glycerol as substrate for the production of biosurfactant by Pseudomonas aeruginosa UCP0992. *Colloids and Surfaces B: Biointerfaces, 79*(1), 174–183. Available from https://doi.org/10.1016/j.colsurfb.2010.03.050.

Singh, P., Patil, Y., & Rale, V. (2019). Biosurfactant production: Emerging trends and promising strategies. *Journal of Applied Microbiology, 126*(1), 2–13. Available from https://doi.org/10.1111/jam.14057.

Soares da Silva, R. d C. F., Almeida, D. G., Meira, H. M., Silva, E. J., Farias, C. B. B., Rufino, R. D., Luna, J. M., & Sarubbo, L. A. (2017). Production and characterization of a new biosurfactant from *Pseudomonas cepacia* grown in low-cost fermentative medium and its application in the oil industry. *Biocatalysis and Agricultural Biotechnology, 12*, 206–215. Available from https://doi.org/10.1016/j.bcab.2017.09.004.

Soares da Silva, R. d C. F., de Almeida, D. G., Brasileiro, P. P. F., Rufino, R. D., de Luna, J. M., & Sarubbo, L. A. (2019). Production, formulation and cost estimation of a commercial biosurfactant. *Biodegradation, 30*(4), 191–201. Available from https://doi.org/10.1007/s10532-018-9830-4.

Sodagari, M., Invally, K., & Ju, L. K. (2018). Maximize rhamnolipid production with low foaming and high yield. *Enzyme and Microbial Technology, 110*, 79–86. Available from https://doi.org/10.1016/j.enzmictec.2017.10.004.

Souza, F. A. S. D., Salgueiro, A. A., & Albuquerque, C. D. C. (2012). Production of bioemulsifiers by *Yarrowia lipolytica* in sea water using diesel oil as the carbon source. *Brazilian Journal of Chemical Engineering, 29*(1), 61–67. Available from https://doi.org/10.1590/S0104-66322012000100007.

Souza, K. S. T., Gudiña, E. J., Azevedo, Z., de Freitas, V., Schwan, R. F., Rodrigues, L. R., Dias, D. R., & Teixeira, J. A. (2017). New glycolipid biosurfactants produced by the yeast strain Wickerhamomyces anomalus CCMA 0358. *Colloids and Surfaces B: Biointerfaces, 154*, 373–382. Available from https://doi.org/10.1016/j.colsurfb.2017.03.041.

Thimon, L., Peypoux, F., & Michel, G. (1992). Interactions of surfactin, a biosurfactant from Bacillus subtilis, with inorganic cations. *Biotechnology Letters*, *14*(8), 713–718. Available from https://doi.org/10.1007/BF01021648.

Tripathi, L., Twigg, M. S., Zompra, A., Salek, K., Irorere, V. U., Gutierrez, T., Spyroulias, G. A., Marchant, R., & Banat, I. M. (2019). Biosynthesis of rhamnolipid by a Marinobacter species expands the paradigm of biosurfactant synthesis to a new genus of the marine microflora. *Microbial Cell Factories*, *18*(1). Available from https://doi.org/10.1186/s12934-019-1216-8.

Umar, A., Zafar, A., Wali, H., Siddique, M. P., Qazi, M. A., Naeem, A. H., Malik, Z. A., & Ahmed, S. (2021). Low-cost production and application of lipopeptide for bioremediation and plant growth by *Bacillus subtilis* SNW3. *AMB Express*, *11*(1). Available from https://doi.org/10.1186/s13568-021-01327-0.

Uzoigwe, C., Burgess, J. G., Ennis, C. J., & Rahman, P. K. S. M. (2015). Bioemulsifiers are not biosurfactants and require different screening approaches. *Frontiers in Microbiology*, *6*. Available from https://doi.org/10.3389/fmicb.2015.00245.

Van Renterghem, L., Roelants, S. L. K. W., Baccile, N., Uyttersprot, K., Taelman, M. C., Everaert, B., Mincke, S., Ledegen, S., Debrouwer, S., Scholtens, K., Stevens, C., & Soetaert, W. (2018). From lab to market: An integrated bioprocess design approach for new-to-nature biosurfactants produced by *Starmerella bombicola*. *Biotechnology and Bioengineering*, *115*(5), 1195–1206. Available from https://doi.org/10.1002/bit.26539.

Vandana, P., & Singh, D. (2018). Review on biosurfactant production and its application. *International Journal of Current Microbiology and Applied Sciences*, *7*(8), 4228–4241. Available from https://doi.org/10.20546/ijcmas.2018.708.443.

Vecino Bello, X., Devesa-Rey, R., Cruz, J. M., & Moldes, A. B. (2012). Study of the synergistic effects of salinity, pH, and temperature on the surface-active properties of biosurfactants produced by *Lactobacillus pentosus*. *Journal of Agricultural and Food Chemistry*, *60*(5), 1258–1265. Available from https://doi.org/10.1021/jf205095d.

Wang, H., Roelants, S. L. K. W., To, M. H., Patria, R. D., Kaur, G., Lau, N. S., Lau, C. Y., Van Bogaert, I. N. A., Soetaert, W., & Lin, C. S. K. (2019). *Starmerella bombicola*: Recent advances on sophorolipid production and prospects of waste stream utilization. *Journal of Chemical Technology and Biotechnology*, *94*(4), 999–1007. Available from https://doi.org/10.1002/jctb.5847.

Yang, R., Liu, G., Chen, T., Li, S., An, L., Zhang, G., Li, G., Chang, S., Zhang, W., Chen, X., Wu, X., & Zhang, B. (2019). Characterization of the genome of a Nocardia strain isolated from soils in the Qinghai-Tibetan Plateau that specifically degrades crude oil and of this biodegradation. *Genomics*, *111*(3), 356–366. Available from https://doi.org/10.1016/j.ygeno.2018.02.010.

CHAPTER 17

Commercialization of biosurfactants

Ruby Aslam[1], Jeenat Aslam[2] and C.M. Hussain[3]
[1]School of Civil Engineering and Architecture, Chongqing University of Science and Technology, Chongqing, P.R. China
[2]Department of Chemistry, College of Science, Taibah University, Yanbu, Al-Madina, Saudi Arabia
[3]Department of Chemistry and Environmental Sciences, New Jersey Institute of Technology, Newark, NJ, United States

17.1 Introduction

Surfactants are soluble compounds that reduce the surface tension of liquids or interfacial tension between two liquids or a liquid and a solid (Aslam et al., 2021; Mittal & Lindmann, 1984). These compounds have various desirable characteristics, such as their ability to reduce surface tension and improve the wetting and detergency power, and low critical micelle concentration. They are useful in various industries, such as food, petroleum, and textiles. Besides their applications in these sectors, they are also used in various other areas, such as agriculture, antiscalants, and anticorrosive applications (Kosaric et al., 1987; Mobin et al., 2017, Mobin Zehra et al., 2017). As many of the chemicals used in the production of various products have a harmful environmental impact, the availability of nontoxic alternatives such as biosurfactants and bio-emulsifiers is desirable. Compared with chemically produced surfactants, biosurfactants offer several advantages. These include their low toxicity, biodegradability, and broad structural range. They also have physical properties ideal for applications such as water-based and non-water-based cleaning and dispersing agents. In addition, they can be modified through the use of biochemical or biological techniques, and they can be tailored to meet specific needs. The broad functional properties of microbial surface-active compounds have led to their potential applications in various industries. These include the reduction of the viscosity of heavy crude oil, foaming, wetting, de-emulsification, corrosion inhibition, and phase separation (Araujo et al., 2022). They have been identified to have applications in various areas, such as pulp and paper, textiles, ceramics, uranium ore processing, and personal health care (Banat et al., 2000). Owing to their high production cost, the use of biosurfactants in industrial applications is limited. This is mainly due to their poor strain productivity and the need to use expensive substrates. In addition, other factors, such as inefficient downstream processing technologies and the lack of process technology, can also affect the producer's competitiveness and the high production cost of the substrates. This can be improved by utilizing cheaper substrates and improving the process technology (Banat et al., 2014; Hayes et al., 1986).

In the industrial sector, companies are currently replacing synthetic chemicals with biosurfactants. These are nontoxic and natural and can be produced by microorganisms grown on various substrates. Some chemical structures comprising biosurfactants include amino acids, phospholipids, lipids, and hydrocarbons (http://www.biosurfactant.com/rhamnolipidproducts.html). The initial studies on the use of biosurfactants started during the 1980s. The main advantages of these substances are their low toxicity and stability when exposed to harsh environmental conditions (Peele Karlapudi et al., 2018). Currently, emulsan is the only industrial-scale bio-emulsifiers (BE)/biosurfactants (BS) produced by microorganisms that can be used in the production of oil and other liquid products. This is a polyionic lipopolysaccharide that can be used to clean oil-contaminated vessels and to recover oil from various oil spills. It is marketed by the Netherlands-based Petroleum Fermentations. Rhamnols, on the other hand, are derived from Jeneil Bioproduct (Anjum et al., 2016).

The potential of biosurfactants to address various environmental issues has been shown. For instance, they can be used to bioremediate heavy-metal contamination and enhance the oil recovery (Marchant & Banat, 2012; Sarma & Prasad, 2015; Soares da Silva et al., 2017). For these environmental applications, biosurfactants are required in huge quantities; thus, scale-up processes must be optimized. This chapter aims to provide a comprehensive overview of the various studies conducted on the commercialization of biosurfactants. It also explores the outlook for the industry's future growth.

17.2 Global biosurfactant market and their impact on the COVID-19 pandemic

The biosurfactants market was valued at approximately $1.2 billion in 2022 (https://www.marketsandmarkets.com/Market-Reports/biosurfactant-market-163644922.html). It is expected to grow at a CAGR of 11% and reach $1.9 million by 2027. Among the various kinds of biosurfactants, the market for glycolipids is expected to grow at a high rate during the next few years. The personal care industry is expected to be the fastest-growing application area of the biosurfactants market during the next few years. Europe is expected to be the largest consumer of biosurfactants.

Biosurfactants are currently used across an extensive range of industrial and medical processes, and their innate versatility opens up their use for a large variety of coronavirus-related applications (Fig. 17.1) (Aytar Çelik et al., 2021). The emergence of the COVID-19 pandemic has shown a positive influence on biosurfactants. Amphiphilic biosurfactants have a hydrophobic domain, which allows them to interact with the surface of the virus while also interacting with water and other hydrophilic substances. Consequently, they can disrupt the structure of the virus and deactivate it. One of the most interesting characteristics of biosurfactant materials is their ability to

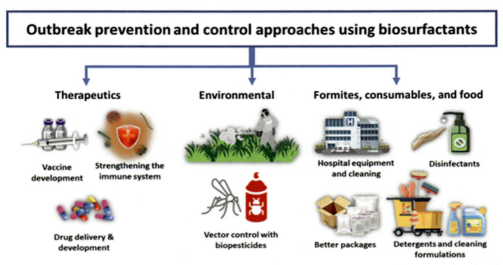

Figure 17.1 Use of biosurfactants in the prevention of the COVID-19 pandemic. *Reproduced with permission Aytar Çelik, P., Blaise Manga, E., Çabuk, A., & Banat, I. M. (2021). Biosurfactants' potential role in combating COVID-19 and similar future microbial threats. Applied Science, 11(1), 334. Use of biosurfactants in prevention of Covid-19.*

create structures around their crucial micelle concentration, which varies depending on the type of biosurfactant. This structural component will be used in drug delivery and the direct attack against the virus. These structures could be used as liposomes that can deliver drugs to an infected area. They can also function as a protective barrier against harsh environmental conditions that severely affect their function. They are now being used in various ways to prevent the spread and treatment of the virus.

In addition to being used in handwashes, biosurfactants are also being utilized to deliver drugs and relieve the symptoms of a person suffering from infections. One of the most important examples of their use is the production of antiviral facemasks made from biosurfactants.

17.3 Factors affecting scale up of biosurfactants

Different biosurfactant-producing organisms and their environments can lead to different patterns of production. Physical and environmental parameters, as well as nutrients, can also affect the process. It is believed that sufficient mechanical agitation and dissolved oxygen can improve the production of lipopeptides (Davis et al., 2001). Mass transfer and oxygen are required to meet the demands of the fermentation process. The vigorous agitation and consequent reduction in the volume of fluids can result in inefficient and unstable operations. Therefore, the optimal strategies for mass transfer and agitation must be optimized (Davis et al., 1999; Kim et al., 1997). Bioreactors should also be designed to handle various foaming issues that can occur during the production of biosurfactants (Yeh et al., 2006a).

High production costs are among the main factors that prevent biosurfactant companies from achieving commercial success. Various methods can be used to reduce these costs. These include optimizing the medium composition, using cheap substrates, and bioreactor operation (Rao et al., 2008; Yeh et al., 2006b). A shake flask has been widely used in plant cultivation to reduce the time and cost of operation. In addition to the efficiency of the process, various factors such as the power input, agitation rate, and tip speed are also considered to determine the scale up of the bioreactor (Yezza et al., 2004). Before the start of a fermentation process, various criteria are thoroughly considered. This can help determine the optimal technology for the production of biosurfactants. Conducting tests in bioreactors and shake flasks can help develop the necessary technology for commercialization. One of the most important criteria that can be considered when it comes to increasing the scale up of aerobic fermentation is maintaining the same volumetric oxygen transfer coefficient (KLa) between the bioreactor and the shake flask (Garcia-Ochoa & Gomez, 2009).

17.4 Scale-up studies of biosurfactant production

There is a lack of knowledge about the scale up of biosurfactant manufacturing. To develop effective technology for commercialization, studies on bioreactors and shake flasks were performed. The study compared the results of the two processes (da Silva et al., 2013). The studies were conducted using a semi-industrial 50-L bioreactor and a flask. They discovered that the latter's ability to control temperature, agitation, and aeration resulted in a significant increase in the biosurfactant's yield. This is because bioreactors, which are closed systems, tend to favor greater cell growth. In addition, the cultivation of biosurfactant using a flask involves orbital shaking. On the other hand, a bioreactor uses mechanical agitation to continuously supply oxygen, which allows for a higher concentration of the substance (Luna et al., 2015). Almeida et al. (2017) noted that the yield of the biosurfactant produced by *Candida tropicalis* was increased by 40% and 75% in experiments conducted in two different bioreactors, i.e., 2-L bench and 50-L pilot bioreactors, respectively. In another study, Zhu et al. (2007) found that the rhamnolipid concentration reached 12.47 g/L after the cultivation of *P. aeruginosa* using waste frying oil.

A reactor with an integrated foam collector was designed for biosurfactant production using *Bacillus subtilis* isolated from agricultural soil (Amani, 2010). The yield of biosurfactant on biomass (Y(p/x)), biosurfactant on sucrose (Y(p/s)), and the volumetric production rate (Y) for shake flask of about 0.45, 0.18 g/g, and 0.03 g/L/h, respectively, were obtained. The best condition for bioreactor was 300 rpm and 1.5 vvm, giving Y(x/s), Y(p/x), Y(p/s), and Y of 0.42, 0.595, 0.25 g/g, and 0.057 g/L/h, respectively. The biosurfactant maximum production, 2.5 g/L, was reached in 44 hours of growth, which was 28% better than the shake flask. Under optimal conditions, the K(L)a-based oxygen transfer coefficient values were obtained from the shake flask and a bioreactor. The comparison of these values revealed that biosurfactants could be efficiently scaled up from their initial state to the bioreactor at a very accurate scale. After water flooded the sand pack, almost 8% of the original oil in the area was recovered using a biosurfactant.

The fermentative production of biosurfactant by *Bacillus licheniformis* R2 strain in a bench-scale bioreactor under different dissolved oxygen tensions was investigated (Joshi et al., 2013). Mini mal media was utilized for the production of biosurfactant products, which mainly utilized glucose as a carbon source. Different concentrations of DO, such as 30%, 50%, 70%, and 100%, were fed during the batch fermentations. The media was continuously maintained during the process through a cascading mode. Adjusting the dissolved oxygen level to 100% saturation during the fermentation process significantly increased the biosurfactant's production. This process also resulted in the reduction of the interfacial and surface tension by about 50% in less than 10 hours. The results of this study suggest that the strain could be utilized to develop sustainable biosurfactants. The process was conducted under different agitation and aeration rates to determine the optimal conditions for the production of biosurfactant products using *B. licheniformis* R2. The objective of this study was to develop a feasible and commercially viable process for the production of biosurfactants.

Silva et al. (2019) reported that the commercial production of a biosurfactant from *Pseudomonas cepacia* CCT6659 grown on industrial waste was investigated in a semi-industrial 50-L bioreactor for use in the removal of hydrocarbons from oily effluents. The researchers noted that the 40.5 g/L concentration achieved during the scale-up process reduced the surface tension to 29 mN/m, which is a significant factor in making biosurfactant economically viable. They noted that biosurfactants could be economically viable for a large-scale application. They estimated that the cost of formulating and isolating a biosurfactant could be approximately US$ 0.01–0.02/g and US$ 0.14 to 0.15/L, respectively.

The use of a cell-free biosurfactant developed by *C. bombicola* for industrial purposes was evaluated by Freitas and colleagues (Freitas et al., 2016). The cost of the product was estimated at about 0.1–0.22 dollars.

17.5 Patents related to the biosurfactants

The following section focuses on the patents granted on the compounds. Table 17.1 shows the results of these granted worldwide. They highlight the various types of microorganisms used to synthesize the biosurfactants.

Table 17.1 Some available patents on biosurfactants.

S. No.	Organism	Patent title	Inventor	Patent No.	Date
1.	Pseudzyma sp. TM-453	Glycolipid containing mannose and mannitol	Matsura F, Ota M, Tamai M, Tamura K	JP2005104837	21-04-2005
2.	Acinetobacter spp. RAG-1. ATCC 31012	Production of a-emulsans	Gutnick D L, Rosenberg E, Belsky I, Zosim Z	US 4395354	17-07-1980
3.	Acinetobacter spp. RAG-1. ATCC 31012	Apo b-emulsans	Gutnick D L, Rosenberg E, Belsky I, Zosim Z	US 4311829	19-1-1982
4.	Acinetobacter spp. RAG-1 ATCC 31012	Proemulsans	Gutnick D L, Rosenberg E, Belsky I, Zosim Z	US 4311832	19-1-1982
5.	Acinetobacter spp. RAG-1 ATCC 31012	Psi emulsans	Gutnick D L, Rosenberg E, Belsky I, Zosim Z	US 4380504	19-4-1983
6.	A. calcoaceticus	Novel A. calcoaceticus and novel biosurfactant	Fukui T, Negi T, Tanaka Y	EP0401700	12-12-1990
7.	calcoaceticus 217 strain (FERM BP-2905)	Novel A. calcoaceticus and novel biosurfactant	Tanaka Y, Fukui T, Negi T	JP 03130073	03-6-1991
8.	A. calcoaceticus CL(KCTC 0081BP)	Microorganism producing biosurfactant	Hwang K A, Kim Y S	KR9706157	24-4-1997
9.	Acinetobacter spp. KRCK4	New lipopolysaccharide biosurfactant	Prosperi G, Camilli M, Crescenzi F, Fascetti E, Porcelli F, Sacceddu P	JP 11269203	05-10-1999
10.	A. calcoaceticus	New lipopolysaccharide biosurfactant	Crescenzi F, Sacceddu P, Prosperi G, Camilli M, Fascetti E, Porcelli F	JP11269203	23-6-1999
11.	B. subtilis ATCC 55033	Mutant of B. subtilis and production of surfactin by use of the mutant	Carrera P, Cosmina P, Grandi G	J04299981	23-10-1992
12.	Bacillus spp.	Method for concentrating biosurfactant	Sakurai S, Imanaka T, Morikawa M	JP 05211876	24-08-1993
13.	B. subtilis	Controlling agent against plant disease and injury	Shoda M, Sato H	JP 06135811	17-5-1994
14.	Bacillus spp. TY-8 (FERM P-13666) & TY-34 (FERM P-13665)	New microorganism assimilating petroleum and waste oil	Okura I, Kubo M, Hasumi F, Yamamoto E	JP 07008271	13-01-1995

15.	*Bacillus* spp.	Straight-chain surfactin	Mohamado O, Ishigami Y, Ishizuka Y	J08092279	09-04-1996
16.	*Bacillus* spp.	Methods for producing polypeptides, in surfactin mutants of Bacillus cells	Adams L F, Brown S, Sloma A, Sternberg D	WO9822598	28-5-1998
17.	*B. subtilis* AQ713	A novel strain of Bacillus for controlling plant diseases and corn rootworm	Manker D C, Mccoy R J, Heins S D, Jimenez D R, Orjala J E	WO9850422	12-11-1998
18.	*B. subtilis* AQ713	Compositions and methods for controlling, plant pests	Jimenez D R, Mccoy R J, Heins S D, Manker D C, Orjala J E, Marrone P G	WO0029426	25-5-2000
19.	*Bacillus* spp.	Production process of surfactin	Tsuzuki T, Furuya K, Miyota Y, Yoneda T	WO0226961	4-4-2002
20.	*Bacillus* spp.	It can be used to improve the efficiency and quality of compost	Shi J, Zeng G, Huang G	CN 1431314	23-7-2003
Pseudomonas					
21.	*Pseudomonas* strains	Pharmaceutical preparation based on rhamnolipid	Piljac G, Piljac V	US 5455232	3-10-1995
22.	*P. aeruginosa* DSM 7107, DSM 7108, and their mutants	*P. aeruginosa* and its use in a process for the biotechnological preparation of L-rhamnose	Giani C, Wullbrandt D, Rothert R, Meiwes J	US 5501966	26-3-1996
23.	*P. fluorescens* R.sub.4	Microbially produced rhamnolipids (biosurfactants) for the control of plant pathogenic zoosporic fungi	Stanghellini M E, Miller R M, Rasmussen S L, Kim D H, Zhang Y	WO 97/25866	17-1-1997

(Continued)

Table 17.1 (Continued)

S. No.	Organism	Patent title	Inventor	Patent No.	Date
24.	*P. rubescens*	Improvement in cellulosic fiber and composition	Masuoka K, Obata T, Fukuda Y, Taharu N	JP 10096174	14-04-1998
25.	*P. fluorescens* R.sub.4	Microbially produced rhamnolipids (biosurfactants) for the control of plant pathogenic zoosporic fungi	Stanghellini M E, Miller R M, Rasmussen S L, Kim D H, Zhang Y	US 5767090	16-6-1998
26.	*P. aeruginosa* (USBCS1)	Production of oily emulsions mediated by a microbial tenso-active agent	Rocha C A, Gonzalez D, Iturralde M L, Lacoa U L, Morales F A	US 5866376	2-2-1999
27.	*P. aeruginosa* (USBCS1)	Production of oily emulsions mediated by a microbial tenso-active agent	Rocha C A, Gonzalez D, Iturralde M L, Lacoa U L, Morales F A	US 6060287	9-5-2000
28.	*Pseudomonas* 11 C, *Bacillus* 60 A	Method for inhibiting eukaryotic protein Kinases	Davies J E, Waters B, Saxena G	US 6319898	20-11-2001
29.	*P. aeruginosa*	Used for preparing compost from lift garbage to improve efficiency and quality of compost	Shi J, Yuan X, Zeng G	CN 1431036	23-7-2003
30.	*P. fluorescens* KPM-018 P	Used for controlling plant insect pests	Mayama S, Tosa Y, Otsu Y, Toyada H, Matsuda K, Nonomura T.	JP 2005102510	21-4-2005
31.	*Corynebacteria SalvinicumSFC*	Microbiological production of novel biosurfactants	Zajic, J E, Panchal C, Gerson R K, Gerson D F	US 4355109	19-10-1982

32.	Sophorolipid producer	Biodegradable low-foaming detergent compositions, which sustain favorable detergency over a wide temperature range	Furuta T, Igarashi K, Hirata Y	WO 03002700	9-1-2003
33.	Sophorolipid	Can be used as slimming agents and/or active agents stimulating leptin synthesis through adipocytes, in the producer, manufacture of a cosmetic composition, for reducing the subcutaneous fat, overload	Pellicier F, Andre P	WO 2004108063	16-12-2004
34.	*C. bombicola*	Acts as a spermicidal and/or antiviral, agent, and/or anti-spermidical agents	Gross R A, Shah V, Doncel G F	WO 2005089522, US 2004242501, US 2005164955, AU 2003299557	29-9-2005
35.		OLIGOMERIC BIOSURFACTANTS	Owen, Donald Baton Rouge, Fan Lili, Baton Rouge	US 7198508 P	29-05-2008
36.	*Bacillus subtilis*	Method of producing biosurfactant	Michael Paul Bralkowski, Sarah Ashley Brooks, Stephen M. Hinton, David Matthew Wright, Shih-Hsin Yang	US20150037302A1	05-02-2015
37.	B. subtilis	Method of producing biosurfactants	Michael Paul Bralkowski, Sarah Ashley Brooks, Stephen M. Hinton, David Matthew Wright, Shih-Hsin Yang	US 2015/0037302 A1, WO2013110132A1	05-02-2015

17.6 Challenges and future outlook

One of the most effective ways to reduce the costs of biosurfactant production is by replacing the substrates commonly used for manufacturing biosurfactants with low-cost raw materials such as industrial wastes. This can be done through the use of residue from various sources, such as oil and grease, starch, and sugar cane. The availability of raw materials and their type can significantly affect the cost of biosurfactant manufacturing. It is estimated that nearly 10%–30% of the overall cost of biotechnological products is due to the availability of raw materials, while purification accounts for 60%. However, in certain cases, the cost of purification can be minimized by avoiding it when the biosurfactant has been applied in its crude form.

17.7 Conclusion

Due to the environmental issues caused by chemical surfactants, we must use biosurfactants in various industries. The different types of biosurfactants include oligopeptides, glycolipids, lipopeptides, neutral lipids, and fatty acid phospholipids. The pH, temperature, and salinity conditions, as well as the culture conditions, are some of the factors that can influence the physicochemical properties and production of biosurfactants. Although numerous patents have been issued for these products, only a few have been commercialized. Biosurfactant applications are mainly focused on the oil sector. However, due to the cost of production, the oil industry cannot adopt them for microbially enhanced recovery. This will lead to the development of new biosurfactant technologies that can be used in the oil recovery industry. Various methods that can be used to reduce the costs of biosurfactants. Some of these include using cheap substrates, bioreactor operation, and optimizing medium composition.

References

Almeida, D. G., Soares da Silva, R. C. F., Luna, J. M., et al. (2017). Response surface methodology for optimizing the production of biosurfactant by Candida tropicalis on industrial waste substrates. *Frontiers in Microbiology, 8*, 157.

Amani, H., Mehrnia, M. R., Sarrafzadeh, M. H., Haghighi, M., & Soudi, M. R. (2010). Scale up and application of biosurfactant from *Bacillus subtilis* in enhanced oil recovery. *Applied Biochemistry and Biotechnology, 162*(2), 510–523.

Anjum, F., Gautam, G., Edgard, G., & Negi, S. (2016). Biosurfactant production through Bacillus sp. MTCC 5877 and its multifarious applications in food industry. *Bioresource Technology, 213*, 262–269.

Araujo, J., Monteiro, J., Silva, D., Alencar, A., Silva, K., Coelho, L., Pacheco, W., Silva, D., Silva, M., Silva, L., et al. (2022). Surface-active compounds produced by microorganisms: Promising molecules for the development of antimicrobial, anti-inflammatory, and healing agents. *Antibiotics, 11*, 1106.

Aslam, R., Mobin, M., Aslam, J., Aslam, A., Zehra, S., & Masroor, S. (2021). Application of surfactants as anticorrosive materials: A comprehensive review. *Advances in Colloid and Interface Science, 295*, 102481.

Aytar Çelik, P., Blaise Manga, E., Çabuk, A., & Banat, I. M. (2021). Biosurfactants' potential role in combating COVID-19 and similar future microbial threats. *Applied Science, 11*(1), 334.

Banat, Á. M., Makkar, R. S., & Cameotra, S. S. (2000). Cameotra, commercial applications of microbial surfactants. *Applied Microbiology and Biotechnology, 53*(495), 508.

Banat, I. M., Satpute, S. K., Cameotra, S. S., Patil, R., & Nyayanit, N. V. (2014). Cost effective technologies and renewable substrates for biosurfactants' production. *Frontiers in Microbiology, 5*, 697.

da Silva, R. C. F. S., Rufino, R. D., Luna, J. M., et al. (2013). Enhancement of biosurfactant production from *Pseudomonas cepacia* CCT6659 through optimisation of nutritional parameters using response surface methodology. *Tenside Surfactants Detergents, 50*, 137—142.

Davis, D. A., Lynch, H. C., & Varley, J. (2001). The application of foaming for the recovery of surfactin from *B. subtilis* ATCC 21332 cultures. *Enzyme Microbial Technol, 28*, 346—354.

Davis, D. A., Lynch, H. C., & Varley, J. (1999). The production of surfactin in batch culture by *Bacillus subtilis* ATCC 21332 is strongly influenced by the conditions of nitrogen metabolism. *Enzyme Microbial Technology, 25*, 322—329.

Freire Soares da Silva, R. d C., de Almeida, D. G., Ferreira Brasileiro, P. P., Rufino, R. D., de Luna, J. M., & Asfora Sarubbo, L. (2019). Production, formulation and cost estimation of a commercial biosurfactant. *Biodegradation, 30*, 191—201.

Freitas, B. G., Brito, J. G. M., Brasileiro, P. P. F., et al. (2016). Formulation of a commercial biosurfactant for application as a dispersant of petroleum and by-products spilled in oceans. *Frontiers in Microbiology, 7*, 1646.

Garcia-Ochoa, F., & Gomez, E. (2009). Bioreactor scale-up and oxygen transfer rate in microbial processes: An overview. *Biotechnology Advances, 27*, 153—176.

Hayes, M. E., Nestaas, E., & Hrebenar, K. R. (1986). Microbial surfactants. *Chemtech, 4*(239), 243.

Joshi, S. J., Geetha, S. J., Yadav, S., & Desai, A. J. (2013). Optimization of bench-scale production of biosurfactant by *Bacillus licheniformis* R2. *APCBEE Procedia, 5*(232), 236.

Kim, H. S., Yoon, B. D., Lee, C. H., Suh, H. H., Oh, H. M., Katsuragi, T., & Tani, Y. (1997). Production and properties of a lipopeptide biosurfactant from *Bacillus subtilis* C9. *Journal of Fermentation and Bioengineering, 84*, 41—46.

Kosaric, N., Cairns, W. L., & Gray, N. C. (1987). *Bioesurfactants and biotechnology* (Vol. 25). New York: Marcel Dekker, Inc.

Luna, J. M., Rufino, R. D., Jara, A. M. A. T., et al. (2015). Environmental applications of the biosurfactant produced by Candida sphaerica cultivated in low-cost substrates. *Colloids Surfaces A, 480*, 413—418.

Marchant, R., & Banat, I. M. (2012). Microbial biosurfactants: Challenges and opportunities for future exploitation. *Trends in Biotechnology, 30*, 558—565. Available from https://doi.org/10.1016/j.tibtech2012.07.003.

Mittal, K., & Lindmann, B. (1984). *Surfactants in solution* (Vol. 1). New York: Plenum Press.

Mobin, M., Aslam, R., & Aslam, J. (2017). Non-toxic biodegradable cationic gemini surfactants as novel corrosion inhibitor for mild steel in hydrochloric acid medium and synergistic effect of sodium salicylate: Experimental and theoretical approach. *Materials Chemistry and Physics, 91*, 151.

Mobin, M., Aslam, R., Zehra, S., & Ahmad, M. (2017). Bio-/environment-friendly cationic gemini surfactant as novel corrosion inhibitor for mild steel in 1 M HCl solution. *Journal of Surfactants and Detergents, 20*, 57.

Peele Karlapudi, A., Venkateswarulu, T. C., Tammineedi, J., Kanumuri, L., KumarRavuru, B., Ramu Dirisala, V., & Prabhakar Kodali, V. (2018). Petroleum role of biosurfactants in bioremediation of oil pollution-a review. *Petroleum, 4*(3), 241—249.

Rao, R. S., Kumar, C. G., Prakasham, R. S., & Hobbs, P. J. (2008). The Taguchi methodology as a statistical tool for biotechnological applications: A critical appraisal. *Biotechnology Journal, 3*(4), 510—523.

Sarma, H., & Prasad, M. N. V. (2015). *Plant-Microbe association-assisted removal of heavy metals and degradation of polycyclic aromatic hydrocarbons. Petroleum Geosciences: Indian Contexts* (pp. 219—236). Springer International Publishing.

Soares da Silva, R. C. F., Almeida, D. G., Meira, H. M., et al. (2017). Production and characterization of a new biosurfactant from *Pseudomonas cepacia* grown in low-cost fermentative medium and its application in the oil industry. *Biocatalysis and Agricultural Biotechnology, 12*, 206—215.

Yeh, M. S., Wei, Y. W., & Chang, J. S. (2006a). Bioreactor design for enhanced carrier-assisted surfactin production with *B. subtilis*. *Process Biochemistry, 41*, 1799-5.

Yeh, M. S., Wei, Y. H., & Chang, J. S. (2006b). Bioreactor design for enhanced carrier-assisted surfactin production with *Bacillus subtilis*. *Process Biochemistry, 41,* 1799−1805.

Yezza, A., Tyagi, R. D., Valero, J. R., Surampalli, R. Y., & Smith, J. (2004). Scale-up of biopesticide production processes using wastewater sludge as a raw material. *Journal Industrial Microbiology and Biotechnology, 31*(12), 545−552.

Zhu, Y., Gan, J., Zhang, G., et al. (2007). Reuse of waste frying oil for production of rhamnolipids using Pseudomonas aeruginosa zju.u1M. *Journal of Zhejiang University-Science A, 8,* 1514−1520.

Further reading

Joshi, S., Bharucha, C., Jha, S., Yadav, S., Nerurkar, A., & Desai, A. J. (2008). Biosurfactant production using molasses and whey under thermophilic conditions. *Bioresource Technology, 99,* 195−199.

CHAPTER 18

Biosurfactants for environmental health and safety

Luara Aparecida Simões[1], Natalia Andrade Teixeira Fernandes[2], Angelica Cristina de Souza[3] and Disney Ribeiro Dias[4]

[1]Biology Department, University of Porto, Porto, Portugal
[2]Department of Chemistry, University of California, Davis, CA, United States
[3]Department of Biology, Federal University of Lavras, Lavras, Minas Gerais, Brazil
[4]Department of Food Science, Federal University of Lavras, Lavras, Minas Gerais, Brazil

18.1 Introduction

Significant population development, associated with the excessive consumption of natural resources, increase in environmental pollution, and climate change, is challenging the current economy for a structural transformation. Therefore, both legislation and industry, in general, are challenged to promote environmental health in a sustainable manner (Ni'matuzahroh, Sari et al., 2020). In this context, the use of biosurfactants has attracted much attention due to the gradual emergence of public policies that promote world economic progress with great concern for reducing environmental impact (Nunes et al., 2022).

Due to the influence of the synthetic surfactants in the environment and also by the consumers' current preference for more natural products, the possible substitution of synthetic surfactants with biosurfactants is being investigated (Jimoh & Lin, 2019; Mulligan et al., 2014). These biosurfactants present better biodegradability and low toxicity that avoid adverse environmental effects caused by synthetic surfactants (Johnson et al., 2021). The main advantages of biosurfactants compared to synthetic surfactants are presented in Fig. 18.1.

Despite the immense advantages of using biosurfactants compared with synthetic surfactants, the industrial production of biosurfactants is currently still underexplored as the various production processes, such as cultivation of microorganisms, are associated with high costs, purification, and recovery of biosurfactants (Schultz & Rosado, 2020; Varjani & Upasani, 2019). To overcome the problem of production costs of biosurfactants, an alternative to reduce production costs, which also helps reduce the environmental impact, is the use of highly available and low-cost substrates such as agro-industrial wastes, making biosurfactants sufficiently competitive in the market, in addition to their production contributing to environmental sustainability (Singh et al., 2019).

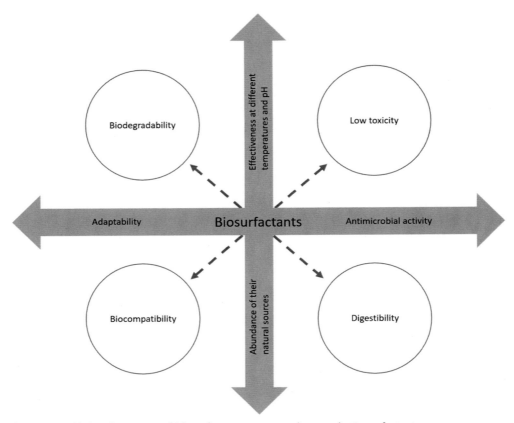

Figure 18.1 Main advantages of biosurfactants compared to synthetic surfactants.

Implementing the waste use strategy results in improved biosurfactant production, which establishes a competitive market, together with waste bioconversion and revenue generation. This adoption of the "green process" aligns with the sustainability desired by the world (Gaur et al., 2022). This chapter discusses the impacts of synthetic surfactants on aquatic and terrestrial environments and the role of biosurfactants in environmental health and safety.

18.2 Environmental effect of synthetic surfactants

18.2.1 Effect on aquatic environment

Over the past years, due to environmental problems caused by toxic pollutants, these hazardous compounds in water have attracted certain attention. Such contamination occurs through the disposal of products, or by industrial processes, and even by direct or indirect unintentional events, as is the case of wastewater treatment. Therefore, due to the lack of great toxicity of the surfactants, the biodegradability of the biosurfactants is in fact a great benefit, because even if the biosurfactants are dumped in the environment, these compounds will break down, not persisting in the environment, which does not cause the accumulation (Johnson et al., 2021). In addition, biosurfactants have a low toxicity to flora and fauna, presenting less threat to the environment, when compared to the toxic effect of synthetic surfactants (Hogan et al., 2019; Santos, Meira, et al., 2017; Santos, Resende, et al., 2017).

Surfactants in the aquatic environment can exhibit toxicity behavior in different ways. The damage caused by the presence of surfactants in the aquatic environment is summarized in Fig. 18.2. The presence of surfactant results in a foaming or settling

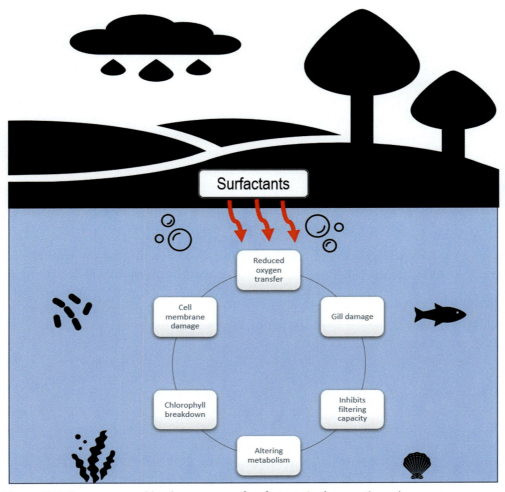

Figure 18.2 Damage caused by the presence of surfactants in the aquatic environment.

process, which directly interferes with water quality, reduces oxygen transfer, and decreases the self-cleaning capacity of the aquatic environment (Gheorghe et al., 2013). In addition, surfactants increase the solubility of persistent organic pollutants in the aqueous phase, significantly impacting the quality of the aquatic environment (Olkowska et al., 2014). It is also worth mentioning that surfactants directly affect the physiological characteristics and biochemical activities of aquatic organisms, altering metabolism, growth, damaging the cell membrane, and breaking down the chlorophyll—protein complex (Koparal et al., 2006).

The presence of synthetic surfactants negatively affects both fish and shellfish. Synthetic surfactants damage fish gills and destroy the mucus-producing outer layers, which protect against parasitic bacteria (Effendi et al., 2017). In addition, surfactants inhibit the filtering capacity of oysters and mussels (Ostroumov, 2003). Several works confirm the toxic effect of surfactants in the aquatic environment, typically when testing the effects on aquatic organisms, the authors use the median effective concentration (EC50) and the median concentration of growth inhibition (IC50) to determine the risk and toxicity of surfactants on microorganisms, animals, and plants from the aquatic environment.

Sütterlin et al. (2008) investigated the toxicity of surfactants often found in wastewater, such as linear anionic surfactant alkylbenzene sulfonate (LAS), naphthalene sulfonic acid (NSA), sodium dodecylsulfonate (SDS), and benzenesulfonic acid (BSA). To determine the toxicity of the compounds, the authors tested the inhibitory activity of *Pseudomonas putida* and *Vibrio fischeri*. All tested surfactants showed a toxic effect on the two microorganisms studied; for the *P. putida*, the EC50 for LAS, NSA, SDS, and BSA were, respectively, 33.4, >160, >160, >1000 mg/L, and for *V. fischeri*, the EC50 for LAS, NSA, SDS, and BSA were 109.7, >167, 8.2, >1000 mg/L, respectively.

Three types of surfactants were studied by Masakorala et al. (2011), where the toxicities of hexadecyltrimethylammonium bromide, sodium dodecyl sulfate, and Triton X-100 were tested on marine macroalgae; as a result, the authors found that all tested surfactants were able to reduce the photochemical energy conversion efficiency of microalgae. Of the surfactants studied, the quaternary ammonium cationic compound (HDTMA) showed a greater effect in reducing the photosynthetic response (EC50 of 2.4 mg/L).

The toxicity of amine oxide-based surfactants was tested in *Photobacterium phosphoreum* and *Daphnia magna* (García et al., 2007). Surfactants had more effect on aquatic bacteria *P. phosphoreum*, showing an EC50 from 0.11 to 11 mg/L; for the crustacea *Daphnia magna*, the IC50 ranged from 6.8 to 45 mg/L. The same aquatic models *D. magna* and *P. phosphoreum* were studied by García et al. (2001), who showed the toxic effect of the alkyl chain of quaternary ammonium-based surfactants (alkyl trimethyl ammonium and alkyl benzyl dimethyl ammonium halides) and found the EC50 values in the range of 0.1 ± 1 mg/L for the two cationic surfactants. They also showed that the substitution of a benzyl group for a methyl group increases the toxicity.

The anionic surfactants, consisting of alkyl sulfates (AS), alkylbenzene sulphonates (LAS), and alkylpolyoxy-ethylene sulfates (AES), and the nonionic surfactants, consisting of polyoxyethylene alkyl ethers (AE) and polyoxylethylene alkyl-phenyl ethers (APE), were tested in three different toxicity assays on aquatic organisms, namely, *Physa acuta Draparnaud*, *Artemia salina*, and *Raphidocelis subcapitata* (Liwarska-Bizukojc et al., 2005). All tested anionic surfactants were harmful to aquatic organisms (LC50 between 10 and 100 mg/L); on the other hand, nonionic surfactants also showed toxicity (EC50 between 1 and 10 mg/L) or even high toxicity (EC50 below 1 mg/L).

The ecotoxicity of commercial surfactants, well-known pollutants in surface waters, and constituents of cleansing hair products (shampoos) was tested by Pavlić et al. (2005) in freshwater green algae *Pseudokirchneriella subcapitata* and *Scenedesmus subspicatus* and marine diatoms microalgae *Skeletonema costatum* and *Phaeodactylum tricornutum*. Considering the classification supported by the environmental effects, all tested surfactants had a toxic effect on freshwater green alga *P. subcapitata*. Surfactants were more toxic to marine diatoms (EC50 0.14–1.7 mg l/L) than freshwater green algae (EC50 0.32–4.4 mg/L).

The toxicity of linear alkylbenzene sulfonate (LAS) and quaternary alkylammonium chloride: alkyl trimethyl ammonium chloride (TMAC), dialkyl dimethyl ammonium chloride (DADMAC), and alkyl benzyl dimethyl ammonium chloride (BDMAC) was evaluated by Utsunomiya et al. (1997) on their study on unicellular green alga *Dunaliella* sp. Based on toxicity, the surfactants were classified in the following order TMAC (EC50 = 0.79 mg/L) > BDMAC (EC50 = 1.3 mg/L) > LAS (EC50 = 3.5 mg/L) > DADMAC (EC50 = 18 mg/L).

The toxicity of alkyl ethoxysulphate surfactant, widely used in commercial and industrial applications, was tested by Sibila et al. (2008) on microalgae *Nannochloropsis gaditana*, *Isochrysis galbana*, *Chaetoceros gracilis*, *Dunaliella salina*, and *Tetraselmis chuii* and the invertebrate *Artemia franciscana*. Alkyl ethoxysulphate surfactants showed toxicity by inhibiting the growth of microalgae, the EC50 values ranged from 4.68 g/L (*D. salina*) to 24.02 mg/L (*I. galbana*), and the LC50 value for *A. franciscana* were 38.30 mg/L.

18.2.1.1 Effect on the terrestrial environment

Cationic surfactants are widely used as disinfectants and antiseptics in various products because these compounds have inhibitory properties that prevent or delay the growth of microorganisms. However, these cationic surfactants, in addition to causing toxic effects when present in the aquatic environment, can cause irritation or burns in humans in different parts of the body, such as the eyes, skin, and respiratory system (Ying, 2006).

The in vitro effect of the cationic surfactants cetyl trimethylammonium bromide (CTAB), anionic surfactant sodium dodecyl sulfate (SDS), and dodecyl trimethylammonium bromide (DTAB) was investigated on mature human serum albumin (HAS) fibrils by Pandey et al. (2013); cationic surfactant solutions added to the formed HSA fibrils demonstrated the ability to disintegrate fibrillar species.

The disturbance induced by various surfactants in human skin *in vivo* was demonstrated by Mizushima et al. (2000), who used three types of surfactants, namely, anionic (SLS), cationic (MSAC), and amphoteric (HEA), and concluded that HEA has the lowest irritant potential. This result was confirmed in the work of (Tabohashi et al., 1998), where HEA was less irritating than MSAC in a skin irritation assay.

Surfactants have a proven toxic effect on mammalian cells (Grant et al., 1992; Thorsteinsson et al., 2003). In a study (Inácio et al., 2013), quaternary ammonium surfactants were studied in Madin-Darby canine kidney (MDCK) cells, and the authors found that toxicity is mediated mainly by mitochondrial dysfunction, characterized by mitochondrial fragmentation and reduced cellular energy load; even at very low concentrations, surfactants can inhibit mitochondrial respiration, causing a reduction in cellular energy load, which can induce apoptosis.

Chen et al. (2018) showed that organosilicon surfactants had a toxic effect on bees and demonstrated that organosilicon surfactants had a severe impact on bee survival rates in oral toxicity tests. The study by Ciarlo et al. (2012) tested organosilicon surfactants demonstrated to harm the honey olfactory learning of bees. The toxic effect of surfactants on bees is problematic as bees are responsible for much of the pollination; therefore, the environmental effect can go far beyond insects, leading to shortages of different plant species.

Therefore, surfactants have evident toxic effects, in both the terrestrial environment, including human toxicity, and the aquatic environment. Thus, there is a great worldwide concern in replacing the use of synthetic surfactants with biosurfactants, substances less aggressive to the environment, which are a great bet for a sustainable and eco-friendly world future.

18.3 Role of biosurfactants in environmental pollution

Pollution from heavy metals, oils, and petroleum hydrocarbons is becoming a serious problem due to the increasing use of crude oil or related products in various fields. This pollution leads to ecological damage in terrestrial, aquatic, and marine environments, affecting the entire ecosystem. Therefore, different strategies have been proposed to solve this problem and clean the environment; several techniques, such as the use of chemical surfactants and other complex technologies, have already been tested; however, recently, much attention has been given to biosurfactant compounds, because they are considered an eco-friendly solution for remediation technology (Jimoh & Lin, 2019).

Heavy metals are natural and often essential compounds present in small amounts in the terrestrial environment, which in high concentrations, cause severe intoxication. Heavy metals contaminate rivers and oceans, normally through the wastewater dumped into them (Hazra et al., 2012). On the other hand, soil contamination by heavy metals is usually due to different industrial activities (mining, incineration, production of batteries for vehicles) and even as a result of industrial waste deposits (Sarubbo et al., 2015).

Heavy metals in the environment causes serious problems because these compounds are non-biodegradable and more persistent than organic contaminants, such as petroleum derivatives and pesticides, which leads to the contamination of biological systems and groundwater by leaching. On reaching different ecosystems, heavy metals become bioavailable to living organisms, thus reacting with biological molecules in living beings, resulting in the formation of extremely stable and dangerous biotoxic compounds (Santona et al., 2006).

Paper industries are a concern for the environment. The sludge waste may contain toxic compounds (such as metallic Pb, Zn, and Cu) from soluble ink waste. In this context, biosurfactants enter the remediation process. In the study by Hidayati and Surtiningsih (2014), the biosurfactant produced by *Pseudomonas putida*, *Acinetobacter* sp. *Bacillus subtilis*, *Actinobacillus* sp., in the removal of heavy metals, lead, zinc, and copper was evaluated in a batch wash test. The highest removal of Zn (6.5%) and Cu (2.01%) was observed for the biosurfactant produced by *P. putida*; on the other hand, the biosurfactant produced by *Acinetobacter* sp. showed the highest removal of Pb (14.04%). Therefore, the authors proved that the biosurfactants studied have the capacity as a washing agent in the removal of heavy metals.

The capacity of the chemical surfactant Triton X-100 to remove heavy metals was compared with a biosurfactant rhamnolipid in contaminated sandy soil (Wang & Mulligan, 2004). The chemical surfactant in the concentration of 0.5% removed 52.8% of the cadmium (Cd) and 45.2% of the nickel (Ni), while the same concentration of rhamnolipid removed 73.2% of the Cd and 68.1% of the Ni, that is, an increase of approximately 38% and 50% removal of heavy metals Cd and Ni, respectively. Therefore, this study demonstrates that the application of biosurfactants can be more effective and safer for the environment than synthetic surfactants.

The lipopeptide biosurfactant produced by a strain of *Bacillus* sp. was evaluated by Ravindran et al. (2020) for the effectiveness of bioremediation of heavy metals, showing a removal of 99.93% cadmium, 75.5% mercury, 89.5% manganese, and 97.73% lead. Interestingly, the authors also tested the ability to remove heavy metals from the biosurfactants on the surface of carrots, fresh cabbage, and lettuce, and the biosurfactant proved to be effective. The heavy metals lead (Pb) and mercury (Hg) were removed by 62.50% Pb and 50.20% Hg from intertidal marine sediments by rhamnolipid biosurfactants produced by *Pseudomonas aeruginosa* strain (Chen et al., 2021).

In the study by Santos, Resende, et al. (2017), crude biosurfactant produced by *Candida lipolytica* UCP 0988 showed some efficiency in the removal of heavy metals, the cell-free crude extract removed about 30%–40% of lead and copper from the sand, which indicates that the application of the crude biosurfactant has the potential for treating soils contaminated with heavy metals. Rufino et al. (2012) studied the proprieties in lipopeptide biosurfactant produced by *Candida lipolytica* (UCP 0988) in the bioremediation of heavy metals. The biosurfactant showed the ability to remove 96% zinc and copper and could to reduce the concentration of lead, iron, and cadmium.

Biosurfactants can emulsify hydrocarbons, enhancing their water solubility, decreasing surface tension, and increasing the displacement of oil substances from soil particles (Geetha et al., 2018; Montero-Rodríguez et al., 2015; Santos et al., 2016). The advancement in microbial-enhanced oil recovery (MEOR) and use of cheap substrate results in better yield, economical production, and better use for oil recovery (Luna et al., 2016; Santos, Meira, et al., 2017).

The enhanced oil recovery by biosurfactants can be achieved by two strategies: in situ and ex situ (Araújo et al., 2019; Geetha et al., 2018). In in situ strategies, the biosurfactants are produced in the reservoir together with the oil by adding either nutrients or native strains. The biosurfactants adsorb quickly at the oil—water interface and on the rock surface, which can decrease the oil—water interfacial tension, changing emulsification patterns. On the other hand, in ex situ strategies, biosurfactants are produced in fermenters and then injected into the reservoir where the oil is present, in that if there is no need to consider the influence of reservoir conditions on the metabolism of the strain; therefore, a high yield of biosurfactant is always used (Santos et al., 2019).

The oil industry is responsible for most of the processes conducted in the marine environment. In some cases, some of the oil may accidentally reach the seawater. To remedy this problem, the use of biosurfactants becomes a hypothesis due to the dispersing power, a process in which the hydrocarbon is dispersed in the aqueous phase as small emulsions (Almeida et al., 2017). A matter of concern in relation to oil contamination is also related to mechanical workshops, which produce a high amount of used engine oils that are released without any treatment, where a liter of used motor oil can pollute a million gallon of fresh water. Therefore, remediation of used motor oil is mandatory for environment security (Adelowo et al., 2006).

The oil displacement capacity of the lipoprotein biosurfactant produced by *Streptomyces* sp. DPUA1566 was evaluated in the work of (Santos et al., 2019), this displacement test measures the surface activity of the biosurfactant in the oil, and the higher the diameter of the clear zone, the stronger the surface activity, in this study the clear zone area was up to 94.9 cm^2, with a dispersion rate of 95.0% of the initial diameter of the oil. Therefore, the studied biosurfactant showed high surface activity and can be applied in the bioremediation process of sites contaminated with hydrocarbons.

The ability of microorganisms isolated from contaminated sites has been studied by some authors and in some cases the microorganisms themselves can be used to treat the same place from which they were isolated (Chaprão et al., 2015; Vollenbroich et al., 1997). In the work of Moldes et al. (2011), the biosurfactant produced by *Lactobacillus pentosus* showed 63% efficiency of octane biodegradation after 15 days of treatment in the soil in which this microorganism was isolated.

18.4 Biosurfactants and sustainability

Generally, the most industrially used biosurfactants are produced on water-insoluble carbon substrates such as oils, fats, and hydrocarbons (Pacwa-Płociniczak et al., 2011). Such substrates directly affect the cost of production, with the raw material accounting for 10%—30% of the total cost of producing biosurfactants. Therefore, several studies have focused on the production of biosurfactants from industrial waste as the treatment and the removal of these residues involve a high cost for different industries and often pollute the environment (Marchant & Banat, 2012). Therefore, the strategy of using these residues is in accordance with environmental protection and safety, and several studies have reported the production of biosurfactants with renewable and low-cost agro-industrial substrates, such as glycerin, residual vegetable oils from fried foods, dairy by-products, sugar cane molasses, and distilleries residues (Banat et al., 2014).

Microorganisms can use different substrates for the production of biosurfactants, for example, biosurfactant produced by *Bacillus pumilus* using vinasse and residual frying oil (Oliveira & Garcia-Cruz, 2013); *C. lipolytica* UCP 0988 (Luna et al., 2015), *P. aeruginosa* Mr01 (Bagheri Lotfabad et al., 2017), and *Pseudomonas aeruginosa* PACL (De Lima et al., 2009) using medium containing inexpensive soybean oil refinery residue; *Saccharomyces cerevisiae* URM 6670 using soybean frying oil and corn steep liquor (Schultz & Rosado, 2020); *Pseudomonas aeruginosa* ATCC27853 using waste cooking oil and coffee wastewater (Yañez-Ocampo et al., 2017); *Achromobacter* sp. using rice straw and hydrolyzed corncob (Ni'matuzahroh, Sari, et al., 2020); and *Lactobacillus paracasei* using vineyard pruning waste (Vecino et al., 2017). Table 18.1 lists studies on the different renewable by-products used for the production of biosurfactants.

18.5 Beneficial effects on plants

Biosurfactants have great applications in agriculture and can be inoculated directly into the soil by immersion of roots and leaves or inoculated into seeds (O'Callaghan, 2016). The presence of biosurfactants can protect plants directly due to the antimicrobial properties against phytopathogenic pathogens proven in different studies (Akladious et al., 2019; Khare & Arora, 2021; Shalini et al., 2017) and indirectly through the stimulation of plant defense systems through the "induced systemic resistance" (ISR) mechanism; when inducing this mechanism, biosurfactants can make plants more resistant and less susceptible to attack by pathogens (Ongena et al., 2007).

Table 18.1 Use of renewable by-products for the production of biosurfactant.

Biosurfactant-producing microorganism	Biosurfactant type	Renewable by-products Agro-waste as substrate	Properties	References
Bacillus sp. HIP3	Lipopeptide	Used Cooking Oil	Removing copper (13.6%), lead (12.7%), zinc (2.9%), chromium (1.7%), and cadmium (0.7%)	Md Badrul Hisham et al. (2019)
Candida sphaerica UCP 0995	Glycolipid	Ground-nut oil refinery and corn steep liquor	Recovered 95% of motor oil adsorbed in a sand sample	Luna et al. (2012)
Pantoea sp.	Glycolipid	Corn steep liquor and pineapple peel residue	Removed between 64 and 92% of oils from the soil	De Almeida et al. (2015)
Candida lipolytica UCP 0988	Glycolipid	Animal fat and	Recovering up to 70% of the residual oil from oil-saturated sand	Santos et al. (2013)
Pseudomonas cepacia CCT6659	Glycolipid	Soybean waste frying oil and corn steep liquor	Biodegradation of motor oil activity (83%)	Silva et al. (2014)
Meyerozyma guilliermondii	Glycolipid	Soybean oil used	Solubilize 15.9% of cadmium from the sewage sludge	Camargo et al. (2018)
Serratia marcescens UCP 1549	—	Cassava flour wastewater	Remove 94% of motor oil contained in sand	Araújo et al. (2019)
Ochrobactrum anthropi HM-1 and Citrobacter freundii HM-2	Glycolipid	Waste frying oil	Recover 70% and 67% of the trapped oil	Ibrahim. (2018)
Bacillus licheniformis K51, B. subtilis 20B, B. subtilis R1 and Bacillus strain HS3	Lipopeptides	Molasses or cheese whey	Recover 25%–33% of residual oil	Joshi et al. (2008)
Bacillus subtilis strains viz. DM-03 and DM-04	Lipopeptide	Potato peels	Releasing kerosene or crude oil from saturated sand pack column	Das and Mukherjee (2007)
Corynebacterium aquaticum Corynebacterium spp. CCT 1968	—	Fish waste and sugarcane bagasse	Paint removal process	Martins and Martins (2018)

Aspergillus niger	—	Wheat bran, sugarcane molasses, and corncob	Contaminant degradation (biodiesel) of 64%	Kreling et al. (2020)
Pseudomonas azotoformans AJ15	Rhamnolipid	Bagasse and potato peels	Recovering up to 36% of trapped oil under saline condition	Das and Kumar (2018)
Pseudomonas indica MTCC 3714	Di-rhamnolipids	Rice-bran oil industry residues	Recover kerosene up to 70%	Bhardwaj et al. (2015)
Candida tropicalis UCP0996	—	Sugarcane molasses, corn steep liquor, waste frying oil	Motor oil spreading efficiency of 75%	Almeida et al. (2017)
Bacillus subtilis TD4	Lipopeptide	Palm oil effluent	Biodegradation of oil industry waste	Louhasakul et al. (2020)
Bacillus subtilis SPB1	Lipopeptide	Cooking tuna fish water and sesame peel flour	Diesel solubility enhancement	Mnif et al. (2013)
Halobacteriaceae archaeon AS65	Lipopeptide	Banana peel	Removal of used motor lubricating oil	Chooklin et al. (2014)

Another factor to be considered in the application of biosurfactants in environmental health is increase in the concentration of micronutrients present in the soil (Kumar et al., 2021) and increased fertility due to the ability to provide a better environment for seed development, which results in the promotion of plant growth, as proven by El-Sheshtawy et al. (2022). They studied the impact of a lipopeptide biosurfactant produced by *B. megaterium* on lettuce growth parameters. The leaf and root size increased when lettuce seeds were subjected to a treatment with a high concentration of biosurfactant, proving the ability of biosurfactants to promote plant growth.

In the study by Araújo et al. (2019), biosurfactant produced by *S. marcescens* UCP 1549 had a positive effect on the germination and growth of cabbage seeds, where the presence of roots and secondary leaves for all tested biosurfactant solutions could be observed. This positive property, combined with the high environmental compatibility of biosurfactants, can lead to the effective application of these compounds in agriculture, reducing the use of toxic agrochemicals that do not contribute to environmental health.

Biosurfactants can also contribute to environmental health and safety due to their effect against phytopathogenic microorganisms that lead to huge worldwide economic losses (Savary et al., 2019). Thus, biosurfactants emerge as a new sustainable technique to combat diseases and pests that affect crops, serving as promising biocontrol agents and an ecologically suitable tool (Singh & Rale, 2022).

The antifungal activity and possible application as a biocontrol of the biosurfactant produced by *Pseudomonas guariconensis* LE3 were presented by Khare and Arora (2021). The use of the biosurfactant the occurrence of the sunflower charcoal rot disease caused by the fungus *Macrophomina phaseolina* decreases by about 54.95%. Therefore, treatment with biosurfactants could protect the plant, showing a reduction in the disease caused by the fungus studied.

18.6 Concluding remarks and future perspectives

Concerns about the ecotoxicity of surfactants arise due to the significant exploitation in the daily life across the world. Therefore, to replace the use of these synthetic surfactants with natural biosurfactants, which contribute to environmental health and safety, regulatory measures should be considered to expand the biosurfactant market. To overcome the problem associated with the cost and production of biosurfactants competing with synthetic surfactants, there is a need to study the use of cheap substrates and efficient biosurfactant-producing microorganisms. The production of biosurfactants by reusing industrial waste has been a viable strategy that reduces waste disposal in the environment, contributes to the circular bioeconomy, and the commercialization of biosurfactants on a large scale.

Biosurfactants play important roles in a wide range of processes that contribute to the environment, such as the biodegradation of heavy metals and hydrocarbons, and have benefits in the relationship between the plant and the soil, in addition to presenting antimicrobial activity against pathogenic microorganisms. Further, to leverage the use of biosurfactants, collaborations with academics, regulatory bodies, and biosurfactant producers are needed to contribute to greater synergy between the environmental and economic aspects of the biosurfactant market.

References

Adelowo, O. O., Alagbe, S. O., & Ayandele, A. A. (2006). Time-dependent stability of used engine oil degradation by cultures of *Pseudomonas fragi* and *Achromobacter aerogenes*. *African Journal of Biotechnology*, *5*(24), 2476—2479. Available from http://www.academicjournals.org/AJB/PDF/pdf2006/18Dec/Adelowo%20et%20al.pdf.

Akladious, S. A., Gomaa, E. Z., & El-Mahdy, O. M. (2019). Efficiency of bacterial biosurfactant for biocontrol of Rhizoctonia solani (AG - 4) causing root rot in faba bean (*Vicia faba*) plants. *European Journal of Plant Pathology*, *153*(1), 15—35. Available from https://doi.org/10.1007/s10658-018-01639-1.

Almeida, D. G., da Silva, R. d C. F. S., Luna, J. M., Rufino, R. D., Santos, V. A., & Sarubbo, L. A. (2017). Response surface methodology for optimizing the production of biosurfactant by Candida tropicalis on industrial waste substrates. *Frontiers in Microbiology*, *8*. Available from https://doi.org/10.3389/fmicb.2017.00157.

Araújo, H. W. C., Andrade, R. F. S., Montero-Rodríguez, D., Rubio-Ribeaux, D., Alves Da Silva, C. A., & Campos-Takaki, G. M. (2019). Sustainable biosurfactant produced by *Serratia marcescens* UCP 1549 and its suitability for agricultural and marine bioremediation applications. *Microbial Cell Factories*, *18*(1). Available from https://doi.org/10.1186/s12934-018-1046-0.

Bagheri Lotfabad, T., Ebadipour, N., Roostaazad, R., Partovi, M., & Bahmaei, M. (2017). Two schemes for production of biosurfactant from *Pseudomonas aeruginosa* MR01: Applying residues from soybean oil industry and silica sol—gel immobilized cells. *Colloids and Surfaces B: Biointerfaces*, *152*, 159—168. Available from https://doi.org/10.1016/j.colsurfb.2017.01.024.

Banat, I. M., Satpute, S. K., Cameotra, S. S., Patil, R., & Nyayanit, N. V. (2014). Cost effective technologies and renewable substrates for biosurfactants' production. *Frontiers in Microbiology*, *5*. Available from https://doi.org/10.3389/fmicb.2014.00697.

Bhardwaj, G., Cameotra, S. S., & Chopra, H. K. (2015). Utilization of oil industry residues for the production of rhamnolipids by *Pseudomonas indica*. *Journal of Surfactants and Detergents*, *18*(5), 887—893. Available from https://doi.org/10.1007/s11743-015-1711-9.

Camargo, F. P., de Menezes, A. J., Tonello, P. S., Dos Santos, A. C. A., & Duarte, I. C. S. (2018). Characterization of biosurfactant from yeast using residual soybean oil under acidic conditions and their use in metal removal processes. *FEMS Microbiology Letters*, *365*(10). Available from https://doi.org/10.1093/femsle/fny098.

Chaprão, M. J., Ferreira, I. N. S., Correa, P. F., Rufino, R. D., Luna, J. M., Silva, E. J., & Sarubbo, L. A. (2015). Application of bacterial and yeast biosurfactants for enhanced removal and biodegradation of motor oil from contaminated sand. *Electronic Journal of Biotechnology*, *18*(6), 471—479. Available from https://doi.org/10.1016/j.ejbt.2015.09.005.

Chen, J., Fine, J. D., & Mullin, C. A. (2018). Are organosilicon surfactants safe for bees or humans. *Science of the Total Environment*, *612*, 415—421. Available from https://doi.org/10.1016/j.scitotenv.2017.08.175.

Chen, Q., Li, Y., Liu, M., Zhu, B., Mu, J., & Chen, Z. (2021). Removal of Pb and Hg from marine intertidal sediment by using rhamnolipid biosurfactant produced by a *Pseudomonas aeruginosa* strain. *Environmental Technology & Innovation*, *22*, 101456. Available from https://doi.org/10.1016/j.eti.2021.101456.

Chooklin, C. S., Maneerat, S., & Saimmai, A. (2014). Utilization of banana peel as a novel substrate for biosurfactant production by *Halobacteriaceae archaeon* AS65. *Applied Biochemistry and Biotechnology*, *173*(2), 624−645. Available from https://doi.org/10.1007/s12010-014-0870-x.

Ciarlo, T. J., Mullin, C. A., Frazier, J. L., & Schmehl, D. R. (2012). Learning impairment in honey bees caused by agricultural spray adjuvants. *PLoS One*, *7*(7). Available from https://doi.org/10.1371/journal.pone.0040848.

Das, A. J., & Kumar, R. (2018). Utilization of agro-industrial waste for biosurfactant production under submerged fermentation and its application in oil recovery from sand matrix. *Bioresource Technology*, *260*, 233−240. Available from https://doi.org/10.1016/j.biortech.2018.03.093.

Das, K., & Mukherjee, A. K. (2007). Comparison of lipopeptide biosurfactants production by *Bacillus subtilis* strains in submerged and solid state fermentation systems using a cheap carbon source: Some industrial applications of biosurfactants. *Process Biochemistry*, *42*(8), 1191−1199. Available from https://doi.org/10.1016/j.procbio.2007.05.011.

De Almeida, F. C. G., Silva, T., Garrard, I., Asfora, L., Sarubbo, G., & Tambourgi, E. B. (2015). Optimization and evaluation of biosurfactant produced by Pantoea sp. using pineapple peel residue, vegetable fat and corn steep liquor. *Journal of Chemistry and Chemical Engineering*, *9*(2015), 269−279.

De Lima, C. J. B., Ribeiro, E. J., Sérvulo, E. F. C., Resende, M. M., & Cardoso, V. L. (2009). Biosurfactant production by *Pseudomonas aeruginosa* grown in residual soybean oil. *Applied Biochemistry and Biotechnology*, *152*(1), 156−168. Available from https://doi.org/10.1007/s12010-008-8188-1.

Effendi, I., Nedi, S., Ellizal, E., Nursyirwani, N., Feliatra, F., Fikar, F., Tanjung, T., Pakpahan, R., & Pratama, P. (2017). Detergent Disposal into Our Environmentand Its Impact on Marine Microbes. In IOP Conference Series: Earth and Environmental Science (Vol. 97, Issue 1). Institute of Physics Publishing. https://doi.org/10.1088/1755-1315/97/1/012030.

El-Sheshtawy, H. S., Mahdy, H. M., Sofy, A. R., & Sofy, M. R. (2022). Production of biosurfactant by *Bacillus megaterium* and its correlation with lipid peroxidation of Lactuca sativa. *Egyptian Journal of Petroleum*, *31*(2), 1−6. Available from https://doi.org/10.1016/j.ejpe.2022.03.001.

García, M. T., Ribosa, I., Guindulain, T., Sánchez-Leal, J., & Vives-Rego, J. (2001). Fate and effect of monoalkyl quaternary ammonium surfactants in the aquatic environment. *Environmental Pollution*, *111*(1), 169−175. Available from https://doi.org/10.1016/s0269-7491(99)00322-x.

García, M., Campos, E., & Ribosa, I. (2007). Biodegradability and ecotoxicity of amine oxide based surfactants. *Chemosphere*, *69*(10), 1574−1578.

Gaur, V. K., Sharma, P., Sirohi, R., Varjani, S., Taherzadeh, M. J., Chang, J. S., Yong Ng, H., Wong, J. W. C., & Kim, S. H. (2022). Production of biosurfactants from agro-industrial waste and waste cooking oil in a circular bioeconomy: An overview. *Bioresource Technology*, *343*. Available from https://doi.org/10.1016/j.biortech.2021.126059.

Geetha, S. J., Banat, I. M., & Joshi, S. J. (2018). Biosurfactants: Production and potential applications in microbial enhanced oil recovery (MEOR. *Biocatalysis and Agricultural Biotechnology*, *14*, 23−32. Available from https://doi.org/10.1016/j.bcab.2018.01.010.

Gheorghe, S., Lucaciu, I., Paun, I., Stoica, C., & Stanescu, E. (2013). Ecotoxicological behavior of some cationic and amphoteric surfactants (biodegradation, toxicity and risk assessment). *Biodegradation-Life of Science*, *83*, 114.

Grant, R. L., Yao, C., Gabaldon, D., & Acosta, D. (1992). Evaluation of surfactant cytotoxicity potential by primary cultures of ocular tissues: I. Characterization of rabbit corneal epithelial cells and initial injury and delayed toxicity studies. *Toxicology*, *76*(2), 153−176. Available from https://doi.org/10.1016/0300-483X(92)90162-8.

Hazra, C., Kundu, D., & Chaudhari, A. (2012). *Biosurfactant-assisted bioaugmentation in bioremediation*. In Microorganisms in environmental management: microbes and environment (Vol. 9789400722293, pp. 631−664). Netherlands: Springer. Available from https://doi.org/10.1007/978-94-007-2229-3_28.

Hidayati, N., & Surtiningsih, T. (2014). Removal of heavy metals Pb, Zn and Cu from sludge waste of paper industries using biosurfactant. *Journal of Bioremediation and Biodegradation*, *5*(7).

Hogan, D. E., Tian, F., Malm, S. W., Olivares, C., Palos Pacheco, R., Simonich, M. T., Hunjan, A. S., Tanguay, R. L., Klimecki, W. T., Polt, R., Pemberton, J. E., Curry, J. E., & Maier, R. M. (2019). Biodegradability and toxicity of monorhamnolipid biosurfactant diastereomers. *Journal of Hazardous Materials*, *364*, 600−607. Available from https://doi.org/10.1016/j.jhazmat.2018.10.050.

Ibrahim, H. M. M. (2018). Characterization of biosurfactants produced by novel strains of Ochrobactrum anthropi HM-1 and Citrobacter freundii HM-2 from used engine oil-contaminated soil. *Egyptian Journal of Petroleum, 27*(1), 21−29. Available from https://doi.org/10.1016/j.ejpe.2016.12.005.

Inácio, A. S., Costa, G. N., Domingues, N. S., Santos, M. S., Moreno, A. J. M., Vaz, W. L. C., & Vieiraa, O. V. (2013). Mitochondrial dysfunction is the focus of quaternary ammonium surfactant toxicity to mammalian epithelial cells. *Antimicrobial Agents and Chemotherapy, 57*(6), 2631−2639. Available from https://doi.org/10.1128/AAC.02437-12.

Jimoh, A. A., & Lin, J. (2019). Biosurfactant: A new frontier for greener technology and environmental sustainability. *Ecotoxicology and Environmental Safety, 184*. Available from https://doi.org/10.1016/j.ecoenv.2019.109607.

Johnson, P., Trybala, A., Starov, V., & Pinfield, V. J. (2021). Effect of synthetic surfactants on the environment and the potential for substitution by biosurfactants. *Advances in Colloid and Interface Science, 288*. Available from https://doi.org/10.1016/j.cis.2020.102340.

Joshi, S., Bharucha, C., Jha, S., Yadav, S., Nerurkar, A., & Desai, A. J. (2008). Biosurfactant production using molasses and whey under thermophilic conditions. *Bioresource Technology, 99*(1), 195−199. Available from https://doi.org/10.1016/j.biortech.2006.12.010.

Khare, E., & Arora, N. K. (2021). Biosurfactant based formulation of *Pseudomonas guariconensis* LE3 with multifarious plant growth promoting traits controls charcoal rot disease in Helianthus annus. *World Journal of Microbiology and Biotechnology, 37*(4). Available from https://doi.org/10.1007/s11274-021-03015-4.

Koparal, A. S., Önder, E., & Öütveren, Ü. B. (2006). Removal of linear alkylbenzene sulfonate from a model solution by continuous electrochemical oxidation. *Desalination, 197*(1−3), 262−272. Available from https://doi.org/10.1016/j.desal.2005.12.024.

Kreling, N. E., Simon, V., Fagundes, V. D., Thomé, A., & Colla, L. M. (2020). Simultaneous production of lipases and biosurfactants in solid-state fermentation and use in bioremediation. *Journal of Environmental Engineering, 146*(9). Available from https://doi.org/10.1061/(asce)ee.1943-7870.0001785.

Kumar, A., Singh, S. K., Kant, C., Verma, H., Kumar, D., Singh, P. P., Modi, A., Droby, S., Kesawat, M. S., Alavilli, H., Bhatia, S. K., Saratale, G. D., Saratale, R. G., Chung, S. M., & Kumar, M. (2021). Microbial biosurfactant: A new frontier for sustainable agriculture and pharmaceutical industries. *Antioxidants, 10*(9). Available from https://doi.org/10.3390/antiox10091472.

Liwarska-Bizukojc, E., Miksch, K., Malachowska-Jutsz, A., & Kalka, J. (2005). Acute toxicity and genotoxicity of five selected anionic and nonionic surfactants. *Chemosphere, 58*(9), 1249−1253. Available from https://doi.org/10.1016/j.chemosphere.2004.10.031.

Louhasakul, Y., Cheirsilp, B., Intasit, R., Maneerat, S., & Saimmai, A. (2020). Enhanced valorization of industrial wastes for biodiesel feedstocks and biocatalyst by lipolytic oleaginous yeast and biosurfactant-producing bacteria. *International Biodeterioration & Biodegradation, 148*, 104911. Available from https://doi.org/10.1016/j.ibiod.2020.104911.

Luna, J. M., Filho, A. S. S., Rufino, R. D., & Sarubbo, L. A. (2016). Production of biosurfactant from *Candida bombicola* URM 3718 for environmental applications. *Chemical Engineering Transactions, 49*, 583−588. Available from https://doi.org/10.3303/CET1649098.

Luna, J. M., Rufino, R. D., Campos-Takaki, G. M., & Sarubbo, L. A. (2012). *Properties of the biosurfactant produced by Candida sphaerica cultivated in low-cost substrates*. In *Chemical engineering transactions* (Vol. 27, pp. 67−72). Italian Association of Chemical Engineering—AIDIC. Available from https://doi.org/10.3303/CET1227012.

Luna, J. M., Rufino, R. D., Jara, A. M. A. T., Brasileiro, P. P. F., & Sarubbo, L. A. (2015). Environmental applications of the biosurfactant produced by *Candida sphaerica* cultivated in low-cost substrates. *Colloids and Surfaces A: Physicochemical and Engineering Aspects, 480*, 413−418. Available from https://doi.org/10.1016/j.colsurfa.2014.12.014.

Marchant, R., & Banat, I. M. (2012). Biosurfactants: A sustainable replacement for chemical surfactants. *Biotechnology Letters, 34*(9), 1597−1605. Available from https://doi.org/10.1007/s10529-012-0956-x.

Martins, P. C., & Martins, V. G. (2018). Biosurfactant production from industrial wastes with potential remove of insoluble paint. *International Biodeterioration and Biodegradation, 127*, 10−16. Available from https://doi.org/10.1016/j.ibiod.2017.11.005.

Masakorala, K., Turner, A., & Brown, M. T. (2011). Toxicity of synthetic surfactants to the marine macroalga, Ulva lactuca. *Water, Air, and Soil Pollution, 218*(1−4), 283−291. Available from https://doi.org/10.1007/s11270-010-0641-4.

Md Badrul Hisham, N. H., Ibrahim, M. F., Ramli, N., & Abd-Aziz, S. (2019). Production of biosurfactant produced from used cooking oil by Bacillus sp. HIP3 for heavy metals removal. *Molecules (Basel, Switzerland), 24*(14). Available from https://doi.org/10.3390/molecules24142617.

Mizushima, J., Kawasaki, Y., Tabohashi, T., Kitano, T., Sakamoto, K., Kawashima, M., Cooke, R., & Maibach, H. I. (2000). Effect of surfactants on human stratum corneum: Electron paramagnetic resonance study. *International Journal of Pharmaceutics, 197*(1−2), 193−202. Available from https://doi.org/10.1016/S0378-5173(00)00323-9.

Mnif, I., Ellouze-Chaabouni, S., & Ghribi, D. (2013). Economic production of *Bacillus subtilis* SPB1 biosurfactant using local agro-industrial wastes and its application in enhancing solubility of diesel. *Journal of Chemical Technology and Biotechnology, 88*(5), 779−787. Available from https://doi.org/10.1002/jctb.3894.

Moldes, A. B., Paradelo, R., Rubinos, D., Devesa-Rey, R., Cruz, J. M., & Barral, M. T. (2011). Ex situ treatment of hydrocarbon-contaminated soil using biosurfactants from lactobacillus pentosus. *Journal of Agricultural and Food Chemistry, 59*(17), 9443−9447. Available from https://doi.org/10.1021/jf201807r.

Montero-Rodríguez, D., Andrade, R. F., Ribeiro, D. L. R., Rubio-Ribeaux, D., Lima, R. A., Araújo, H., & Campos-Takaki, G. M. (2015). Bioremediation of petroleum derivative using biosurfactant produced by *Serratia marcescens* UCP/WFCC 1549 in low-cost medium. *International Journal of Current Microbiology and Applied Sciences, 4*(7), 550−562.

Mulligan, C. N., Sharma, S. K., & Mudhoo, A. (2014). *Biosurfactants research trends and applications* (p. 34) Boca Raton: CRC Press.

Ni'matuzahroh, Sari, S. K., Trikurniadewi, N., Ibrahim, S. N. M. M., Khiftiyah, A. M., Abidin, A. Z., Nurhariyati, T., & Fatimah. (2020). Bioconversion of agricultural waste hydrolysate from lignocellulolytic mold into biosurfactant by Achromobacter sp. BP(1)5. *Biocatalysis and Agricultural Biotechnology, 24*. Available from https://doi.org/10.1016/j.bcab.2020.101534.

Nunes, H. M. A. R., Vieira, I. M. M., Santos, B. L. P., Silva, D. P., & Ruzene, D. S. (2022). Biosurfactants produced from corncob: A bibliometric perspective of a renewable and promising substrate. *Preparative Biochemistry and Biotechnology, 52*(2), 123−134. Available from https://doi.org/10.1080/10826068.2021.1929319.

O'Callaghan, M. (2016). Microbial inoculation of seed for improved crop performance: Issues and opportunities. *Applied Microbiology and Biotechnology, 100*(13), 5729−5746. Available from https://doi.org/10.1007/s00253-016-7590-9.

Oliveira, J. G. d, & Garcia-Cruz, C. H. (2013). Properties of a biosurfactant produced by *Bacillus pumilus* using vinasse and waste frying oil as alternative carbon sources. *Brazilian Archives of Biology and Technology, 56*(1), 155−160. Available from https://doi.org/10.1590/S1516-89132013000100020.

Olkowska, E., Ruman, M., & Polkowska, Z. (2014). Occurrence of surface active agents in the environment. *Journal of Analytical Methods in Chemistry, 2014*. Available from https://doi.org/10.1155/2014/769708.

Ongena, M., Jourdan, E., Adam, A., Paquot, M., Brans, A., Joris, B., Arpigny, J. L., & Thonart, P. (2007). Surfactin and fengycin lipopeptides of Bacillus subtilis as elicitors of induced systemic resistance in plants. *Environmental Microbiology, 9*(4), 1084−1090. Available from https://doi.org/10.1111/j.1462-2920.2006.01202.x.

Ostroumov, S. A. (2003). Studying effects of some surfactants and detergents on filter-feeding bivalves. *Hydrobiologia, 500*, 341−344. Available from https://doi.org/10.1023/A:1024604904065.

Pacwa-Płociniczak, M., Płaza, G. A., Piotrowska-Seget, Z., & Cameotra, S. S. (2011). Environmental applications of biosurfactants: Recent advances. *International Journal of Molecular Sciences, 12*(1), 633−654. Available from https://doi.org/10.3390/ijms12010633.

Pandey, N. K., Ghosh, S., & Dasgupta, S. (2013). Effect of surfactants on preformed fibrils of human serum albumin. *International Journal of Biological Macromolecules, 59*, 39−45. Available from https://doi.org/10.1016/j.ijbiomac.2013.04.014.

Pavlić, Z., Vidaković-Cifrek, Z., & Puntarić, D. (2005). Toxicity of surfactants to green microalgae *Pseudokirchneriella subcapitata* and *Scenedesmus subspicatus* and to marine diatoms Phaeodactylum tricornutum and Skeletonema costatum. *Chemosphere*, *61*(8), 1061−1068. Available from https://doi.org/10.1016/j.chemosphere.2005.03.051.

Ravindran, A., Sajayan, A., Priyadharshini, G. B., Selvin, J., & Kiran, G. S. (2020). Revealing the efficacy of thermostable biosurfactant in heavy metal bioremediation and surface treatment in vegetables. *Frontiers in Microbiology*, *11*. Available from https://doi.org/10.3389/fmicb.2020.00222.

Rufino, R. D., Luna, J. M., Campos-Takaki, G. M., Ferreira, S. R. M., & Sarubbo, L. A. (2012). *Application of the biosurfactant produced by* Candida lipolytica *in the remediation of heavy metals*. In *Chemical engineering transactions* (Vol. 27, pp. 61−66). Italian Association of Chemical Engineering—AIDIC. Available from https://doi.org/10.3303/CET1227011.

Santona, L., Castaldi, P., & Melis, P. (2006). Evaluation of the interaction mechanisms between red muds and heavy metals. *Journal of Hazardous Materials*, *136*(2), 324−329. Available from https://doi.org/10.1016/j.jhazmat.2005.12.022.

Santos, D. K. F., Meira, H. M., Rufino, R. D., Luna, J. M., & Sarubbo, L. A. (2017). Biosurfactant production from *Candida lipolytica* in bioreactor and evaluation of its toxicity for application as a bioremediation agent. *Process Biochemistry*, *54*, 20−27. Available from https://doi.org/10.1016/j.procbio.2016.12.020.

Santos, D. K. F., Resende, A. H. M., de Almeida, D. G., da Silva, R. d C. F. S., Rufino, R. D., Luna, J. M., Banat, I. M., & Sarubbo, L. A. (2017). *Candida lipolytica* UCP0988 biosurfactant: Potential as a bioremediation agent and in formulating a commercial related product. *Frontiers in Microbiology*, *8*. Available from https://doi.org/10.3389/fmicb.2017.00767.

Santos, D. K. F., Rufino, R. D., Luna, J. M., Santos, V. A., & Sarubbo, L. A. (2016). Biosurfactants: Multifunctional biomolecules of the 21st century. *International Journal of Molecular Sciences*, *17*(3). Available from https://doi.org/10.3390/ijms17030401.

Santos, D. K. F., Rufino, R. D., Luna, J. M., Santos, V. A., Salgueiro, A. A., & Sarubbo, L. A. (2013). Synthesis and evaluation of biosurfactant produced by *Candida lipolytica* using animal fat and corn steep liquor. *Journal of Petroleum Science and Engineering*, *105*, 43−50. Available from https://doi.org/10.1016/j.petrol.2013.03.028.

Santos, E. F., Teixeira, M. F. S., Converti, A., Porto, A. L. F., & Sarubbo, L. A. (2019). Production of a new lipoprotein biosurfactant by Streptomyces sp. DPUA1566 isolated from lichens collected in the Brazilian Amazon using agroindustry wastes. *Biocatalysis and Agricultural Biotechnology*, *17*, 142−150. Available from https://doi.org/10.1016/j.bcab.2018.10.014.

Sarubbo, L. A., Rocha, R. B., Luna, J. M., Rufino, R. D., Santos, V. A., & Banat, I. M. (2015). Some aspects of heavy metals contamination remediation and role of biosurfactants. *Chemistry and Ecology*, *31*(8), 707−723. Available from https://doi.org/10.1080/02757540.2015.1095293.

Savary, S., Willocquet, L., Pethybridge, S. J., Esker, P., McRoberts, N., & Nelson, A. (2019). The global burden of pathogens and pests on major food crops. *Nature Ecology & Evolution*, *3*(3), 430−439. Available from https://doi.org/10.1038/s41559-018-0793-y.

Schultz, J., & Rosado, A. S. (2020). Extreme environments: A source of biosurfactants for biotechnological applications. *Extremophiles*, *24*(2), 189−206. Available from https://doi.org/10.1007/s00792-019-01151-2.

Shalini, D., Benson, A., Gomathi, R., John Henry, A., Jerritta, S., & Melvin Joe, M. (2017). Isolation, characterization of glycolipid type biosurfactant from endophytic Acinetobacter sp. ACMS25 and evaluation of its biocontrol efficiency against *Xanthomonas oryzae*. *Biocatalysis and Agricultural Biotechnology*, *11*, 252−258. Available from https://doi.org/10.1016/j.bcab.2017.07.013.

Sibila, M. A., Garrido, M. C., Perales, J. A., & Quiroga, J. M. (2008). Ecotoxicity and biodegradability of an alkyl ethoxysulphate surfactant in coastal waters. *Science of the Total Environment*, *394*(2−3), 265−274. Available from https://doi.org/10.1016/j.scitotenv.2008.01.043.

Silva, E. J., Rocha e Silva, N. M. P., Rufino, R. D., Luna, J. M., Silva, R. O., & Sarubbo, L. A. (2014). Characterization of a biosurfactant produced by *Pseudomonas cepacia* CCT6659 in the presence of industrial wastes and its application in the biodegradation of hydrophobic compounds in soil. *Colloids and Surfaces B: Biointerfaces*, *117*, 36−41. Available from https://doi.org/10.1016/j.colsurfb.2014.02.012.

Singh, P., & Rale, V. (2022). *Applications of microbial biosurfactants in biocontrol management. Biocontrol mechanisms of endophytic microorganisms* (pp. 217–237). Elsevier. Available from https://doi.org/10.1016/B978-0-323-88478-5.00009-2.

Singh, P., Patil, Y., & Rale, V. (2019). Biosurfactant production: Emerging trends and promising strategies. *Journal of Applied Microbiology, 126*(1), 2–13. Available from https://doi.org/10.1111/jam.14057.

Sütterlin, H., Alexy, R., & Kümmerer, K. (2008). The toxicity of the quaternary ammonium compound benzalkonium chloride alone and in mixtures with other anionic compounds to bacteria in test systems with Vibrio fischeri and *Pseudomonas putida. Ecotoxicology and Environmental Safety, 71*(2), 498–505. Available from https://doi.org/10.1016/j.ecoenv.2007.12.015.

Tabohashi, T., Ninomiya, R., & Imori, Y. (1998). A novel amino acid derivative for hair care products. *Fragrance Journal, 26*, 58–63.

Thorsteinsson, T., Másson, M., Kristinsson, K. G., Hjálmarsdóttir, M. A., Hilmarsson, H., & Loftsson, T. (2003). Soft antimicrobial agents: Synthesis and activity of labile environmentally friendly long chain quaternary ammonium compounds. *Journal of Medicinal Chemistry, 46*(19), 4173–4181. Available from https://doi.org/10.1021/jm030829z.

Utsunomiya, A., Watanuki, T., Matsushita, K., Nishina, M., & Tomita, I. (1997). Assessment of the toxicity of linear alkylbenzene sulfonate and quaternary alkylammonium chloride by measuring 13C-glycerol in Dunaliella sp. *Chemosphere, 35*(11), 2479–2490. Available from https://doi.org/10.1016/S0045-6535(97)00316-0.

Varjani, S., & Upasani, V. N. (2019). Evaluation of rhamnolipid production by a halotolerant novel strain of *Pseudomonas aeruginosa. Bioresource Technology, 288*. Available from https://doi.org/10.1016/j.biortech.2019.121577.

Vecino, X., Rodríguez-López, L., Gudiña, E. J., Cruz, J. M., Moldes, A. B., & Rodrigues, L. R. (2017). Vineyard pruning waste as an alternative carbon source to produce novel biosurfactants by *Lactobacillus paracasei. Journal of Industrial and Engineering Chemistry, 55*, 40–49. Available from https://doi.org/10.1016/j.jiec.2017.06.014.

Vollenbroich, D., Pauli, G., Ozel, M., & Vater, J. (1997). Antimycoplasma properties and application in cell culture of surfactin, a lipopeptide antibiotic from *Bacillus subtilis. Applied and Environmental Microbiology, 63*(1), 44–49. Available from https://doi.org/10.1128/aem.63.1.44-49.1997.

Wang, S., & Mulligan, C. N. (2004). Rhamnolipid foam enhanced remediation of cadmium and nickel contaminated soil. *Water, Air, and Soil Pollution, 157*(1–4), 315–330. Available from https://doi.org/10.1023/B:WATE.0000038904.91977.f0.

Yañez-Ocampo, G., Somoza-Coutiño, G., Blanco-González, C., & Wong-Villarreal, A. (2017). Utilization of agroindustrial waste for biosurfactant production by native bacteria from Chiapas. *Open Agriculture, 2*(1), 341–349. Available from https://doi.org/10.1515/opag-2017-0038.

Ying, G. G. (2006). Fate, behavior and effects of surfactants and their degradation products in the environment. *Environment International, 32*(3), 417–431. Available from https://doi.org/10.1016/j.envint.2005.07.004.

CHAPTER 19

Biosurfactants: sustainable alternatives to chemical surfactants

Arif Nissar Zargar and Preeti Srivastava
Department of Biochemical Engineering and Biotechnology, Indian Institute of Technology Delhi, Hauz Khas, New Delhi, India

19.1 Introduction

Significant environmental changes in recent years have prompted an assessment of the role and impact of human activities on it (Manga et al., 2021). Life cycle sustainability assessment (LCA) is one of the common approaches used recently to quantify the impact of human activities on the environment (Kloepffer, 2008). These assessments usually aim at evaluating the economic, social, and environmental consequences of these activities (Zamagni, 2012). The accumulation of toxic waste in the environment and the inevitable consequences of climate change on all life forms on this planet are two major forces currently driving the global sustainability agendas (Halsnaes, 2021).

One of the most extensively researched fields in terms of sustainability is the manufacture of synthetic materials and chemicals, which is considered to have economic, social, and environmental implications (Wang & Hellweg, 2021; Zuin et al., 2021). The key areas that are the focus of lifecycle sustainability assessment include the extraction and processing of raw materials involved in the production, the nature of the waste generated in the production process, the nature of the synthesized product, safe disposal, and degradation of the synthesized product and waste recycling.

One such type of compound that is synthesized chemically and has been investigated from the standpoint of sustainability is the production of surfactants (Jimoh & Lin, 2019; Manga et al., 2021). Surfactants are amphipathic molecules that reduce the surface tension of a liquid and the interfacial tension between two phases (Zargar et al., 2022a). Today, surfactants have become indispensable due to their widespread applications in the detergent, cosmetic, agricultural, pharmaceutical, and petroleum industries (Zargar et al., 2022b). Chemical surfactants are generally produced from fossil fuels (Levison, 2009). The use of fossils for the production of surfactants has been linked to an increase in the level of environmental pollution (Jimoh & Lin, 2019). The limited availability of fossil fuels in nature and the increase in environmental pollution due to the use of petrochemicals has escalated the concerns about the sustainable production of surfactants (Jimoh & Lin, 2019). This has created a demand for greener alternatives that are environmentally friendly. Biosurfactants, that is, surfactants of plant and microbial origin, have been recognized as efficient alternatives to synthetic surfactants (Marchant & Banat, 2012). Microbial biosurfactants have been explored more than those obtained from plants due to the possibility of their large-scale production and commercialization. Biosurfactant production from sustainable sources has been reported to result in an 8% reduction in lifetime CO_2 emissions by preventing the release of 1.5 million tons of CO_2 (Farias et al., 2021; Rocha e Silva et al., 2019). Biosurfactants are generally considered to have lower toxicity and higher biodegradability than synthetic surfactants (Sahnoun et al., 2014; Varvaresou & Iakovou, 2015). This ensures their safe use and disposal in agricultural, petroleum, and pharmaceutical industries. Due to their biodegradable nature, they are considered environmentally friendly, which is a key advantage in these industries (Adu et al., 2020; Fracchia et al., 2019).

The sustainable production of biosurfactants and their advantages over chemical surfactants have resulted in a continuous increase in global demand. The worldwide biosurfactant market was worth more than $3.66 billion in 2020 and is projected to reach $5.71 billion by the end of 2028, expanding at a CAGR of 5.4%. This has attracted several companies, such as AGAE Technologies, Jeniel, and Rhamnolipid Inc., Ecover and BASF-Cognis, for large-scale and commercial biosurfactant manufacturing.

This chapter aims to provide insight into the production of surfactants from a sustainability point of view. The chapter discusses the sustainability-driven shift from synthetic surfactants to biosurfactants and gives an idea of the methods involved in the sustainable production of biosurfactants.

19.2 Drive for global sustainability

Sustainability is the capacity to constantly maintain or support a process over a long term (Wilkinson et al., 2001). Sustainable drives aim to meet the current needs without compromising the needs of future generations. Sustainability drives ensure that the actions performed today will have a beneficial impact on human life and the entire Earth's habitat (Portney, 2015). A key idea of sustainability is to have a balance between economic prosperity, environmental preservation, and social responsibility (Manga et al., 2021) (Fig. 19.1). This means evaluating different product or process alternatives not only for the cost of products or processes developed but also their social impact and overall effect on the environment. Sustainable technologies will likely lie inside the zone where these three notions overlap.

Shift from nonrenewable to renewable resources for all inputs along the value chain of a product is a major theme of the sustainable development goals (Marcelino et al., 2020). Nonrenewable resources are found in limited quantities within the Earth, raising concerns about their long-term acquisition as raw materials for developing any process

Figure 19.1 Key aspects of sustainability.

or product. In additon, using fossil fuels as a source of raw materials for any process or product development has also been linked to a rise in carbon emissions, which are further linked to climate change (Wuebbles & Jain, 2001). The production of greenhouse gases due to the usage of fossil fuels has been established to have a considerable impact on the environment, with the potential to affect all forms of life on this planet. This has sparked serious concerns about the usage of fossil fuels in present sustainable development scenarios, creating a demand for greener alternatives to be considered.

Various international climate change treaties reflect global efforts to limit greenhouse gas emissions. Environmental sustainability worries extend beyond climate change to include pollution and the depletion of natural resources (Olasanmi & Thring, 2018).

Recent environmental changes have prompted an examination of the role and impact of human activity on it, which has sparked the drive for sustainability (Manga et al., 2021). Sustainability is a relatively recent concept proposed to reconcile economic expansion with emerging social and environmental issues (Purvis et al., 2019). Various attempts have been made to accurately quantify the effects of human activities on the environment. However, inconsistencies have been observed in different methods, leading to inaccurate conclusions. Life cycle assessment (LCA) is standardized tool used to accurately measure the impact of a process or product on the environment (Ögmundarson et al., 2020). The key framework of LCA includes goal and scope definition, inventory analysis, impact assessment, and interpretation. LCA considers all the stages involved in the process, including acquisition of raw materials, manufacturing process, packaging, transportation, consumption, and end disposal (Guinee et al., 2011; Rydberg, 2010).

19.3 Chemical surfactants and their production

In 2019, the global surfactant market was valued at 41.3 billion USD. The wide applications of surfactants have led to a continuous increase in global demand, and the market is expected to reach 51.5 billion USD by 2027 (Roelants & Soetaert, 2022). The need for personal care products in the cosmetics sectors and home care products, such as detergents and disinfectants, has increased, which has fueled the surfactant market (Moldes et al., 2021). The need for surfactants has also expanded due to other applications in the agricultural, pharmaceutical, petroleum, and food and beverage industries (Fracchia et al., 2019). Due to their application in the above-mentioned industries, especially in home care and personal care industries, surfactants have become indispensable commodities (Adu et al., 2020).

Surfactants are primarily synthesized chemically from petroleum, and the primary steps in the surfactant supply chain include raw materials processing, feedstock processing, and chemical production, surfactant conversion, product formulation, and distribution (Levison, 2009). The first step involved in the production of surfactants is the processing of crude petroleum into petrochemicals like petroleum oil distillates (Scheibel, 2009). The process converts crude petroleum into various chemicals such as paraffins, benzene, and other aromatic derivatives. Similarly, oleochemical processors extract and purify seed oils from various sources to produce triglyceride oils with variable chain lengths.

The above-mentioned basic raw materials are transformed into a variety of derivatives by feedstock producers, including alkylbenzenes, alkylphenols, polyalkylenes, olefins, and their derivatives (Ziegler or oxo alcohols), which are used to make surfactants (Canselier, 2008). Petrochemical feedstock producers are often integrated with crude petroleum processors due to economic and logistic advantages in producing their products. Triglyceride oils are processed through splitting, transesterification, hydrogenolysis, and hydrogenation and converted into fatty acids, methyl esters, and natural alcohols.

The final step in synthesizing a chemical surfactant is the attachment of a hydrophilic group to the hydrophobic compound synthesized in the earlier steps. For this, surfactant converters add highly reactive chemicals to the hydrophobic chains that affix or create a hydrophilic head group on them. Examples of the highly reactive chemicals supplied by diversified chemical producers include sulfur trioxide (SO_3), phosphorous pentoxide (P_2O_5), ethylene oxide, propylene oxide, dimethyl sulfate, hydrogen peroxide, epichlorohydrin, monochloroacetic acid, and methyl chloride (Levison, 2009).

Surfactant converters, as discussed earlier, affix or create a highly water-soluble functional group on water-insoluble feedstock. The process is usually carried out by one of the following chemical reactions: sulfonation, sulfation, amidation, alkoxylation, esterification, amination, phosphation, and quaternization (Texter, 2001). The resulting molecule is a surfactant which based on the charge of the head group, can fall into one of the four broad categories: anionic, cationic, nonionic, and amphoteric (Porter, 2013). Anionic and nonionic surfactants, due to their low price, accessibility, and broad-range application in home care, dominated the surfactant market in 2019, accounting for almost 80% of the total surfactant market. Cationic and amphoteric surfactants account for substantially lesser market shares.

Globally, surfactants are used in a wide variety of consumer and industrial product formulations, and their active concentrations range from nearly 100% in some cleaning products to just a few parts per million in high-performance applications such as drug delivery systems, precision optics coatings, and electronics manufacturing.

19.4 Sustainability assessment of chemical surfactants

Life cycle sustainability assessments for synthetic surfactant production have established that the production and usage of chemical surfactants do not match the sustainability requirements stated in current scenarios (Liu et al., 2020). Raw materials processing, feedstock processing, and chemical manufacturing, surfactant conversion, product formulation, and distribution are all examples of procedures that fail to fulfill sustainability requirements.

The primary raw material utilized in the synthesis of chemical surfactants is crude oil. The world has 1.65 trillion barrels of known oil reserves, of which the world consumes 35,442,913,090 barrels of oil annually, or 97,103,871 barrels per day. At the current oil consumption rates, there will be enough oil for around 47 years. This limited availability of oil reserves has raised severe concerns about their long-term sustainability for producing various chemicals, including biosurfactants.

The first step in the synthesis of surfactants is the conversion of crude petroleum into petrochemicals and petroleum oil distillates such as paraffins, benzene, and other aromatic compounds. This step is typically carried out in an oil refinery, which is generally considered as a major source of environmental pollution. Refineries have long been associated with contamination of local air, land, and water (Adebiyi, 2022; Thorat & Sonwani, 2022; Varjani et al., 2017). Petroleum refineries are a major source of toxic and harmful air pollutants, especially BTEX chemicals (benzene, toluene, ethylbenzene, and xylene) (Bustillo-Lecompte et al., 2018). They are also a significant producer of sulfur dioxide, nitrogen oxides, carbon monoxide, hydrogen sulfide, and particle matter. Most of these chemicals are known to be harmful to living things, have been identified as probable carcinogens, and have been associated with complications of the reproductive, respiratory, and developmental systems (Varjani et al., 2017). The wastewater produced within the refineries is often disposed of via deep injection wells, and some of these pollutants reach aquifers and groundwater, resulting in their contamination. Soil pollution, although not as significant as air and water pollution, is another problem with petroleum refineries and often associated with oil spills. Although economically viable, the production of petrochemicals by crude oil refining has environmental and social implications and, therefore, does not meet the sustainable standards.

The utilization of highly reactive chemicals in the last step of the surfactant synthesis that does not meet the established criteria. The chemicals used in this step are highly toxic and categorized in "right to know hazardous" and "special health hazard substance" lists. Sulfur trioxide is a carcinogen, can cause pulmonary edema, and is highly corrosive and explosive (Muller, 2000). Prolonged exposure to phosphorous pentoxide has been associated with mouth ulceration and damage to teeth, mucous membranes, conjunctivitis, and organ damage, especially lungs, and may affect the biochemical systems. Apart from these social implications, these highly reactive chemicals also affect the environment, e.g., sulfur trioxide is a constituent of acid rain.

Overall, this clearly establishes that the surfactant production process has immense social and environmental implications and does not meet the current sustainability standards. The toxicity and nonbiodegradability of chemical surfactants are two characteristics that again raise concerns about their sustainability. These characteristics allow them to stay in the environment in the long term and have hazardous consequences. The toxic effects of surfactants on the microbial world, aquatic systems, and their effect on higher vertebrates have been well described (Rebello et al., 2014). Due to the poor sustainable characteristics of synthetic surfactants, the demand for greener alternatives (biosurfactants) has been increasing.

19.5 Biosurfactants: sustainable alternatives to chemical surfactants

Biosurfactants are surfactants derived from biological sources. They are typically produced by plants, animals, or microorganisms, and their production is related to various applications in these organisms (Xu et al., 2011). Antimicrobial properties of biosurfactants generated in plants and animals play a significant part in their immunity (Gomaa, 2013). Microbes, on the other hand, produce biosurfactants to get better access to hydrophobic carbon sources. Besides nutrient assimilation, they aid in cellular growth, biofilm formation, cellular communication, and motility (Gomaa, 2013; Sharma et al., 2021). Microbial biosurfactants have received greater attention than plant biosurfactants due to the prospect of large-scale production and commercialization.

Microbes produce biosurfactants in a growth-associated or nongrowth associated manner. In growth-associated biosurfactant production, biosurfactants start accumulating in the culture broth as soon as the microorganism starts to grow and continues till the culture reaches the stationary phase (Zargar et al., 2022a). While in nongrowth-associated production, accumulation of biosurfactant continues in the stationary phase. During their production, biosurfactants are secreted out by the cells into the extracellular broth, where they start accumulating with time. After the fermentation run, the culture broth is taken, cells are removed, and biosurfactant is extracted from the cell-free broth by precipitation, followed by solvent extraction.

LCA studies show that biosurfactants are green and sustainable alternatives to chemical surfactants (Fig. 19.2). Biosurfactants have been reported to play an important role in reducing carbon dioxide emissions from the atmosphere (Karlapudi et al., 2018; Rahman & Gakpe, 2008). According to estimates, switching from synthetic to biological surfactants will reduce CO_2 emissions by 8% over the course of a person's lifetime and prevent 1.5 million tonnes of carbon dioxide from entering the atmosphere (Farias et al., 2021).

One of the major differences between synthetic and bio-based surfactants that plays a key role in determining sustainability is the source and availability of raw materials.

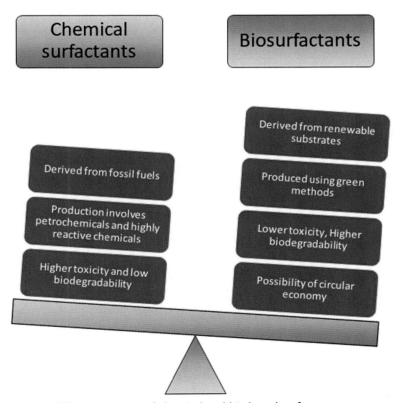

Figure 19.2 Sustainability assessment of chemical and bio-based surfactants.

Compared to synthetic surfactants derived from petrochemicals, biosurfactants are derived from low-cost renewable substrates (Makkar et al., 2011). The substrates used for the production of biosurfactants include agro-industrial wastes, crop residues, and by-products of the food and dairy industry (Gudiña et al., 2016; Magalhães et al., 2018; Olasanmi & Thring, 2018). These resources are relatively cheap but highly rich in carbohydrate and lipid content and do not compete with the food requirements (Verma et al., 2020). The use of inexpensive substrates has long-term benefits in terms of sustainability. Furthermore, using industrial waste streams and by-products as substrates promotes economic and environmental sustainability by reducing the possible contamination hazards associated with their otherwise disposal (Dahiya et al., 2020; Sellami et al., 2021). The use of these wastes for biosurfactant synthesis also reduces the costs associated with their treatment if they were disposed into the environment.

Another important factor that distinguishes biosurfactants from synthetic surfactants is the method involved in their synthesis. Chemical surfactant synthesis involves processes and chemicals that do not meet the social and environmental sustainability requirements. Biosurfactants, on the other hand, are synthesized using fermentation of sugars used for the growth of the microbial culture (Zargar et al., 2022a). Fermentation is completely green and does not involve any hazardous chemicals. The residual waste from the fermentation is also safe for the environment, and no social concerns associated with it. This makes the entire biosurfactant production process sustainable in terms of economic, social, and environmental aspects (Elias et al., 2021).

One of the issues with biosurfactant production is downstream processing for biosurfactant extraction. Extraction is generally done using expensive and hazardous chemical solvents such as chloroform. Since chloroform is toxic, its use poses social and environmental concerns. However, other ways of extracting biosurfactants that do not involve the use of harmful chemicals have also been reported, for example, gravity separation and foam fractionation. The use of chemical-free methods enhances sustainable biosurfactant production (Chen et al., 2006; Makkar & Cameotra, 1999; Santos et al., 2016; Winterburn & Martin, 2012).

The use of psychrophiles for the production of biosurfactants offers additional advantages in terms of the cost involved in the production. Perfumo et al. highlighted the production of biosurfactants by psychrophiles as a newly developing area with prospective energy-saving and sustainable products and processes (Perfumo et al., 2018). According to the authors, biosurfactants may be produced without heat, which promotes the development of biotechnological processes with minimal energy requirements.

The chemical nature of the biosurfactants also gives them an advantage over synthetic surfactants because biosurfactants have higher structural diversity, lower toxicity, higher biodegradability, lower CMC and are stable under extreme environmental conditions (Banat et al., 2014; Sahnoun et al., 2014; Varvaresou & Iakovou, 2015). Biosurfactants are a structurally quite diverse group of compounds whose structure is primarily dictated by the genetic makeup of the producer microorganism and the substrate utilized for biosurfactant production. The structural diversity makes them specific for a particular application. Studies have shown that rhamnolipid biosurfactants are 10 times less toxic than synthetic surfactants (Poremba et al., 1991). Another study reported that a synthetic surfactant (Marlon A-350) is more hazardous and mutagenic than a biosurfactant produced by *Pseudomonas aeruginosa* (Flasz et al., 1998). Thus, biosurfactants are less toxic than the synthetic surfactants, making them more compatible with cosmetic, pharmaceutical, and food applications. Another advantage of biosurfactants over chemical surfactants is that they have higher biodegradability than synthetic surfactants. This makes them better choices for agricultural, petroleum, and environmental applications such as bioremediation.

Despite being appealing substitutes for synthetic surfactants, commercialization and large-scale production of biosurfactants have been challenging. This is primarily due to the higher economics of the process attributed to higher costs in downstream processing of biosurfactants and lower yields of the fermentation process (Zargar et al., 2023). Various approaches can be used to overcome these limitations and enhance biosurfactant production.

19.6 Concluding remarks and future outlook

Global sustainability agendas, public awareness of environmental pollution, and safety concerns have resulted in assessments of processes and products used in modern life. Surfactants have become indispensable commodities due to their widespread use in home and personal care industries. However, sustainability drives have raised questions about the impact of chemical surfactants and the processes involved in their production on the environment and humans. This has created a demand for greener alternatives that fulfill the criteria of present-day sustainability. Biosurfactants have emerged as sustainable alternatives to chemical surfactants. Biosurfactants have been documented to have lower toxicity and higher biodegradability than chemical surfactants, which promotes their use in personal care, home care, pharmaceutical, and agricultural industries. LCA studies have concluded that the raw materials and steps involved in biosurfactant production meet the requirements of social and environmental sustainability. A major concern, however, is the economic sustainability of biosurfactants. In the present scenario, biosurfactants are economically costlier than chemical surfactants, which are valued at 2 USD per kg. The higher cost of biosurfactant production has been attributed to the cost involved in downstream processing and low yields. This limits the large-scale production of biosurfactants. Various efforts can be made to improve the yield of biosurfactants and reduce the cost involved in their production and extraction. Such approaches are expected to make biosurfactants economically viable options and will determine their large-scale production and commercialization. Successful replacement of synthetic surfactants by biosurfactants, although visualized as a sustainably viable option, will only be possible if they are priced equally or less than their chemical counterparts. The future of biosurfactants is, therefore, dependent on the process innovations that make their production cost—effective.

References

Adebiyi, F. M. (2022). Air quality and management in petroleum refining industry: A review. *Environmental Chemistry and Ecotoxicology, 4*, 89—96.

Adu, S. A., Naughton, P. J., Marchant, R., & Banat, I. M. (2020). Microbial biosurfactants in cosmetic and personal skincare pharmaceutical formulations. *Pharmaceutics, 12*, 1099.

Banat, I. M., De Rienzo, M. A. D., & Quinn, G. A. (2014). Microbial biofilms: Biosurfactants as antibiofilm agents. *Applied Microbiology and Biotechnology, 98*, 9915—9929.

Bustillo-Lecompte, C. F., Kakar, D., & Mehrvar, M. (2018). Photochemical treatment of benzene, toluene, ethylbenzene, and xylenes (BTEX) in aqueous solutions using advanced oxidation processes: towards a cleaner production in the petroleum refining and petrochemical industries. *Journal of Cleaner Production, 186*, 609—617.

Canselier, J. P. (2008). *Production of alkanesulfonates and related compounds (high-molecular-weight sulfonates)*. CRC Press.

Chen, C. Y., Baker, S. C., & Darton, R. C. (2006). Continuous production of biosurfactant with foam fractionation. *Journal of Chemical Technology & Biotechnology, 81*, 1915—1922.

Dahiya, S., Katakojwala, R., Ramakrishna, S., & Mohan, S. V. (2020). Biobased products and life cycle assessment in the context of circular economy and sustainability. *Materials Circular Economy, 2*, 1—28.

Elias, A. M., Longati, A. A., Ellamla, H. R., Furlan, F. F., Ribeiro, M. P., Marcelino, P. R., dos Santos, J. C., da Silva, S. S., & Giordano, R. C. (2021). Techno-economic-environmental analysis of sophorolipid biosurfactant production from sugarcane bagasse. *Industrial & Engineering Chemistry Research, 60*, 9833—9850.

Farias, C. B. B., Almeida, F. C., Silva, I. A., Souza, T. C., Meira, H. M., Rita de Cássia, F., Luna, J. M., Santos, V. A., Converti, A., & Banat, I. M. (2021). Production of green surfactants: Market prospects. *Electronic Journal of Biotechnology, 51*, 28—39.

Flasz, A., Rocha, C., Mosquera, B., & Sajo, C. (1998). A comparative study of the toxicity of a synthetic surfactant and one produced by *Pseudomonas aeruginosa* ATCC 55925. *Medical Science Research, 26*, 181—185.

Fracchia, L., Ceresa, C., & Banat, I. M. (2019). Biosurfactants in cosmetic, biomedical and pharmaceutical industry. In I. M. Banat, & R. Thavasi (Eds.), *Microbial Biosurfactants and Their Environmental and Industrial Applications* (pp. 258—288). CRC Press.

Gomaa, E. Z. (2013). Antimicrobial activity of a biosurfactant produced by *Bacillus licheniformis* strain M104 grown on whey. *Brazilian Archives of Biology and Technology, 56*, 259—268.

Gudiña, E. J., Rodrigues, A. I., de Freitas, V., Azevedo, Z., Teixeira, J. A., & Rodrigues, L. R. (2016). Valorization of agro-industrial wastes towards the production of rhamnolipids. *Bioresource Technology, 212*, 144—150.

Guinee, J. B., Heijungs, R., Huppes, G., Zamagni, A., Masoni, P., Buonamici, R., Ekvall, T., & Rydberg, T. (2011). *Life cycle assessment: Past, present, and future*. ACS Publications.

Halsnaes, K. (2021). *A review of the literature on climate change and sustainable development*. Routledge.

Jimoh, A. A., & Lin, J. (2019). Biosurfactant: A new frontier for greener technology and environmental sustainability. *Ecotoxicology and Environmental Safety, 184*, 109607.

Karlapudi, A. P., Venkateswarulu, T., Tammineedi, J., Kanumuri, L., Ravuru, B. K., Ramu Dirisala, V., & Kodali, V. P. (2018). Role of biosurfactants in bioremediation of oil pollution-a review. *Petroleum, 4*, 241—249.

Kloepffer, W. (2008). Life cycle sustainability assessment of products. *The International Journal of Life Cycle Assessment*, *13*, 89–95.

Levison, M. (2009). Surfactant production: Present realities and future perspectives. *Handbook of Detergents Part F: Production*, *142*, 1–38.

Liu, Y., Lu, S., Yan, X., Gao, S., Cui, X., & Cui, Z. (2020). Life cycle assessment of petroleum refining process: A case study in China. *Journal of Cleaner Production*, *256*, 120422.

Magalhães, E. R. B., Silva, F. L., Sousa, M. A. D. S. B., & Dos Santos, E. S. (2018). Use of different agroindustrial waste and produced water for biosurfactant production. *Biosciences Biotechnology Research Asia*, *15*, 17–26.

Makkar, R. S., & Cameotra, S. S. (1999). Biosurfactant production by microorganisms on unconventional carbon sources. *Journal of Surfactants and Detergents*, *2*, 237–241.

Makkar, R. S., Cameotra, S. S., & Banat, I. M. (2011). Advances in utilization of renewable substrates for biosurfactant production. *AMB Express*, *1*, 1–19.

Manga, E. B., Celik, P. A., Cabuk, A., & Banat, I. M. (2021). Biosurfactants: Opportunities for the development of a sustainable future. *Current Opinion in Colloid & Interface Science*, *56*, 101514.

Marcelino, P. R. F., Gonçalves, F., Jimenez, I. M., Carneiro, B. C., Santos, B. B., & da Silva, S. S. (2020). *Sustainable production of biosurfactants and their applications. Lignocellulosic biorefining technologies* (pp. 159–183). Wiley.

Marchant, R., & Banat, I. M. (2012). Biosurfactants: a sustainable replacement for chemical surfactants? *Biotechnology Letters*, *34*, 1597–1605.

Moldes, A. B., Rodríguez-López, L., Rincón-Fontán, M., López-Prieto, A., Vecino, X., & Cruz, J. M. (2021). Synthetic and bio-derived surfactants versus microbial biosurfactants in the cosmetic industry: An overview. *International Journal of Molecular Sciences*, *22*, 2371.

Muller, T. L. (2000). *Sulfuric acid and sulfur trioxide. Kirk-othmer encyclopedia of chemical technology*. Wiley.

Ögmundarson, Ó., Herrgård, M. J., Forster, J., Hauschild, M. Z., & Fantke, P. (2020). Addressing environmental sustainability of biochemicals. *Nature Sustainability*, *3*, 167–174.

Olasanmi, I. O., & Thring, R. W. (2018). The role of biosurfactants in the continued drive for environmental sustainability. *Sustainability*, *10*, 4817.

Perfumo, A., Banat, I. M., & Marchant, R. (2018). Going green and cold: Biosurfactants from low-temperature environments to biotechnology applications. *Trends in Biotechnology*, *36*, 277–289.

Poremba, K., Gunkel, W., Lang, S., & Wagner, F. (1991). Toxicity testing of synthetic and biogenic surfactants on marine microorganisms. *Environmental Toxicology and Water Quality*, *6*, 157–163.

Porter, M. R. (2013). *Handbook of surfactants*. Springer.

Portney, K. E. (2015). *Sustainability*. MIT Press.

Purvis, B., Mao, Y., & Robinson, D. (2019). Three pillars of sustainability: in search of conceptual origins. *Sustainability Science*, *14*(3), 681–695.

Rahman, P. K., & Gakpe, E. (2008). Production, characterisation and applications of biosurfactants-Review. *Biotechnology*, *7*(2), 360–370.

Rebello, S., Asok, A. K., Mundayoor, S., & Jisha, M. (2014). Surfactants: Toxicity, remediation and green surfactants. *Environmental Chemistry Letters*, *12*, 275–287.

Rocha e Silva, N. M. P., Meira, H. M., Almeida, F. C. G., Soares da Silva, Rd. C. F., Almeida, D. G., Luna, J. M., Rufino, R. D., Santos, V. A., & Sarubbo, L. A. (2019). Natural surfactants and their applications for heavy oil removal in industry. *Separation & Purification Reviews*, *48*, 267–281.

Roelants, S. L., & Soetaert, W. (2022). *Industrial perspectives for (microbial) biosurfactants*. Springer.

Rydberg, T. (2010). Life cycle assessment: past, present, and future. *Environmental Science & Technology*, *45*, 90–96.

Sahnoun, R., Mnif, I., Fetoui, H., Gdoura, R., Chaabouni, K., Makni-Ayadi, F., Kallel, C., Ellouze-Chaabouni, S., & Ghribi, D. (2014). Evaluation of *Bacillus subtilis* SPB1 lipopeptide biosurfactant toxicity towards mice. *International Journal of Peptide Research and Therapeutics*, *20*, 333–340.

Santos, D. K. F., Rufino, R. D., Luna, J. M., Santos, V. A., & Sarubbo, L. A. (2016). Biosurfactants: Multifunctional biomolecules of the 21st century. *International Journal of Molecular Sciences*, *17*, 401.

Scheibel, J. J. (2009). Production of alcohols and alcohol sulfates. *Handbook of detergents Part F: Production* (pp. 117–135). Florida: CRC Press Taylor & Francis Group.

Sellami, M., Khlifi, A., Frikha, F., Miled, N., Belbahri, L., & Rebah, F. B. (2021). Agro-industrial waste based growth media optimization for biosurfactant production by *Aneurinibacillus migulanus*. *Journal of Microbiology, Biotechnology and Food Sciences, 2021*, 578–583.

Sharma, J., Sundar, D., & Srivastava, P. (2021). Biosurfactants: potential agents for controlling cellular communication, motility and antagonism. *Frontiers in Molecular Biosciences, 893*.

Texter, J., 2001. Reactions and synthesis in surfactant systems.

Thorat, B. N., & Sonwani, R. K. (2022). Current technologies and future perspectives for the treatment of complex petroleum refinery wastewater: A review. *Bioresource Technology, 355*, 127263.

Varjani, S. J., Gnansounou, E., & Pandey, A. (2017). Comprehensive review on toxicity of persistent organic pollutants from petroleum refinery waste and their degradation by microorganisms. *Chemosphere, 188*, 280–291.

Varvaresou, A., & Iakovou, K. (2015). Biosurfactants in cosmetics and biopharmaceuticals. *Letters in Applied Microbiology, 61*, 214–223.

Verma, R., Sharma, S., Kundu, L. M., & Pandey, L. M. (2020). Experimental investigation of molasses as a sole nutrient for the production of an alternative metabolite biosurfactant. *Journal of Water Process Engineering, 38*, 101632.

Wang, Z., & Hellweg, S. (2021). First steps toward sustainable circular uses of chemicals: advancing the assessment and management paradigm. *ACS Sustainable Chemistry & Engineering, 9*, 6939–6951.

Wilkinson, A., Hill, M., & Gollan, P. (2001). The sustainability debate. *International Journal of Operations & Production Management, 21*, 1492–1502.

Winterburn, J., & Martin, P. (2012). Foam mitigation and exploitation in biosurfactant production. *Biotechnology Letters, 34*, 187–195.

Wuebbles, D. J., & Jain, A. K. (2001). Concerns about climate change and the role of fossil fuel use. *Fuel Processing Technology, 71*, 99–119.

Xu, Q., Nakajima, M., Liu, Z., & Shiina, T. (2011). Biosurfactants for microbubble preparation and application. *International Journal of Molecular Sciences, 12*, 462–475.

Zamagni, A. (2012). *Life cycle sustainability assessment* (pp. 373–376). Springer.

Zargar, A. N., Kumar, M., & Srivastava, P. (2023). Biosurfactants: Challenges and future outlooks. In *Advancements in Biosurfactants Research* (pp. 551–576). Cham: Springer International Publishing.

Zargar, A. N., Lymperatou, A., Skiadas, I., Kumar, M., & Srivastava, P. (2022a). Structural and functional characterization of a novel biosurfactant from Bacillus sp. IITD106. *Journal of Hazardous Materials, 423*, 127201.

Zargar, A. N., Mishra, S., Kumar, M., & Srivastava, P. (2022b). Isolation and chemical characterization of the biosurfactant produced by Gordonia sp. IITR100. *PLoS One, 17*, e0264202.

Zuin, V. G., Eilks, I., Elschami, M., & Kümmerer, K. (2021). Education in green chemistry and in sustainable chemistry: Perspectives towards sustainability. *Green Chemistry, 23*, 1594–1608.

CHAPTER 20

Biosurfactants for sustainability

Oluwaseun Ruth Alara[1,2], Nour Hamid Abdurahman[1,2] and Hassan Alsaggaf Ali[3]
[1]Centre for Research in Advanced Fluid and Processes (Fluid Centre), Universiti Malaysia Pahang, Gambang, Pahang, Malaysia
[2]Faculty of Chemical and Process Engineering Technology, Universiti Malaysia Pahang, Gambang, Pahang, Malaysia
[3]Eastern Unity Technology, Kuala Lumpur, Malaysia

20.1 Introduction

The ongoing challenges facing the environment about burning fossil fuels such as coal and petrol have resulted in increasing realization that encourages the exploration of sustainable processes. As a matter of urgency, various countries are now adopting certain measures with the application of sustainable developments to reduce global environmental and climatic challenges. The adoption of sustainable bio-based products as an alternative to synthetic products is a major solution (Ng et al., 2020). The continuous progress in science and technology has propelled humanity into the exploration of natural resources. These have opened diverse activities, including exploration of crude oil; digging of fossil fuels; utilization of crude-oil-related products such as petrol, diesel, and kerosene; and continuous use of chemicals in the pharmaceutical and agricultural products that have ameliorated the global way of life. However, the introduction of chemicals to ease lifestyle has also had some side effects on human beings and their environment as heavy metals, synthetic chemicals, materials, and solvents are needed for these developments (Wilton et al., 2018).

Moreover, the effect of using synthetic surfactants has increased concerns about the environment due to the side effect, leading to a growing body of research on biosurfactants. They are endured with reduced toxicity, minimal effects on the ecosystem, and biodegradability (Sajna et al., 2013). Because of these important properties of biosurfactants, they are being applied in diverse industrial applications, including cosmetics, food and beverage, environment, and pharmaceuticals (Varjani et al., 2021). In the global market, biosurfactant production is correspondingly low, mainly because of higher downstream processing and feedstock costs (Adesra et al., 2021). The costs of feedstocks in the production of biosurfactants amount to more than half of the overall production costs. The continuous growth has accumulated a compound annual growth rate from of 4.3% from 2014 to 2020 (Gaur et al., 2022). In 2020, about USD 1.8 billion in sales were generated as revenue from the biosurfactant market, and this revenue is forecasted to exceed USD 2.6 billion by 2027. Europe and the

Table 20.1 Some of the commercialized biosurfactants according to country of manufacturing and their applications.

S/N	Country	Biosurfactants	Applications
	Belgium	Sophorolipids	Antibiofilm, enhanced oil recovery (EOR), detergent action, oil recovery, and processing
	China	Green surfactant alkyl polyglucoside, rhamnolipids	Cosmetics, cleaning of hard surfaces, institutional and industrial surface cleaning, detergents for homecare, EOR, high-tech services, oil recovery from storage tanks, and personal care
	France	Lipopeptides	Cosmetics and agriculture
	Germany	Cellobiose lipid biosurfactants and glycolipid	Dishwashing liquids, cleaning products, and bioactive characteristics in pharmaceutical products
	Germany/USA	Green surfactant alkyl polyglucoside	Cosmetics, cleaning of hard surfaces, institutional and industrial surface cleaning
	Japan	Sodium surfactin and sophorolipids	Cosmetics and cleaning products
	South Africa	Surfactin	Cleaning products
	South Korea	Sophorolipids (sopholine)	Personal care products
	USA	Rhamnolipids	Agriculture, cleaners for households, cleaning products, cosmetics, detergents for homecare, EOR, high-tech services, oil recovery from storage tanks, and personal care

United States of America are the emerging largest markets (Table 20.1). Of the different biosurfactants, sophorolipids are mostly used in the detergent industry.

In the biosurfactant market, the major players are Jeneil Biotech Inc. (USA), Ecover (Belgium), Akzo Nobel, BASF Cognis, Evonik Industry, Saraya, and Urumqi Unite (Germany) (Singh et al., 2019). The use of renewable and cheaper materials, including agro-industrial wastes, agricultural residues, and wastes from food industries, are being investigated as the alternative sources for the production of biosurfactants to overcome the economic issues associated with biosurfactant production (Sharma et al., 2020). The global market has experienced a greater boost as various entrepreneurs are interested in the use of biosurfactants because they are eco-friendly. Thus, this chapter focuses on the sustainability of biosurfactants.

20.2 Production of biosurfactants from wastes and renewable materials for sustainability

During metabolism and growth, various microorganisms including fungi, yeasts, and bacteria can synthesize surface-active biomolecules. The synthesis of biosurfactants mostly occurs through microbial cell systems. Biologically, biosurfactants are produced using different substrates, including hydrophobic mixtures, hydrocarbons, solvents, chemicals, vegetable oils, dairy products, wastes from cooking oils, and others (Abbot et al., 2022; Gayathiri et al., 2022). Currently, the utilization of raw materials at reduced costs is studied to resolve the challenges associated with a higher cost of biosurfactant production. Various wastes are being generated from the industries; these include molasses, starch wastes, soap stock, corn steep liquor, and animal fats; these low-cost raw materials can be explored in biosurfactant production (Banat et al., 2014). These wastes are sent into the environment as by-products of coproducts, thus, polluting the environment.

Wastes used to produce biosurfactants are generated from different sources, including animal wastes, food wastes, brewery wastes, dairy and bakery wastes, lignocellulosic wastes, starch-rich wastes, and wastes from cooking oils. These wastes are generated in abundance across the globe because most of these wastes are produced daily. Controlling the accumulation of these wastes is essential for a safe environment. Another importance of using industrial wastes as substrates of raw materials is that they are readily available in large quantities to produce biosurfactants on a larger scale for commercialization. Rhamnolipid was produced by the strain of *Pseudomonas aeruginosa* LBI using substrates of soybean oil waste, whey, molasses, and cassava flour (Nitschke et al., 2010). Surfactin was produced by *Bacillus safensis* J2 utilizing bagasse (Das & Kumar, 2018). In another study, biosurfactants were produced by *Candida* sp., *Pseudomonas* sp., and *Bacillus* sp., using wheat straw, barley bran, cassava flour waste, rice paddy, potato wastes, and de-oiled cakes as potential substrates (Rajasimman et al., 2021). The peels from pineapple were used as substrates to prepare the culture media in the production of biosurfactants; this food waste can be used as a substitute for glucose and salt nutrients (Vieira et al., 2021).

Various vegetable cooking oils are consistently used as substrates to produce biosurfactants. For instance, phenylalanine and olive oil were employed alongside *Nocardiopsis* sp. B4 in the production of biosurfactants (Khopade et al., 2012). A study utilized vegetable oil and different cheap renewable substrates to replace the usual carbon sources through *Pseudomonas aeruginosa* PB3A (Saravanan & Vijayakumar, 2014). In dairy, different substrates, including whey, lactic whey, curd whey, whey wastes, and cheese whey, are generated excessively every day; these have been used as substrates in biosurfactant production (Jimoh & Lin, 2019). Agricultural products are other sources that generate abundant waste; these include molasses, sugarcane straws, rice straws, beet molasses, bran, sugarcane bagasse, soy hulls, corn hulls, cassava flours, and their wastewater.

All these wastes and many more are being studied to produce biosurfactants (Jimoh & Lin, 2019; Marcelino et al., 2019; Nitschke et al., 2010; Samad et al., 2017).

Different kinds of organic compounds, chemicals, hydrophobic mixtures, and organic compounds are extensively used by microorganisms in the form of energy and carbon sources. Therefore, the solubilization of these compounds is enhanced by biosurfactants at concentrations that improve their utilization, microbial uptake, and bioavailability (Patowary et al., 2017). Lipopeptide was produced using the strains of *Paenibacillus* sp. D9 and *Paenibacillus dendritiformis* CN5 by enhancing pyrene biodegradation, motor oil, and polycyclic aromatic hydrocarbon (PAH) (Jimoh & Lin, 2019).

Efforts are being made through different studies to replace synthetic surfactants; however, the cost of implementing biosurfactants is the bone of contention. The higher costs of raw materials and low productivity are affecting the economic viability of biosurfactant production through scale-up (Henkel et al., 2012). Thus, the alternative of using free or low-cost raw materials is essential to overcome the problem associated with the economic process. Another important parameter in the production of biosurfactants to consider is the recovery and purification costs, which are high (Banat et al., 2014). This issue can be resolved using low-cost substrates in the production of biosurfactants because these substrates can improve the quantity and quality of the produced biosurfactants. Moreover, the cost of producing biosurfactants with high-quality products should be reduced. This can only be achieved using by-products rich in oil and viable in the large production of biosurfactants (Partovi et al., 2013). The choice of cheaper substrates depends on the availability as this varies from nation to nation. For instance, the cost of procuring corn steep liquor will be cheaper in a country where the corn is largely cultivated.

20.3 Methods for enhancing sustainability in biosurfactant production

Biosurfactants can be synthesized from different pathogens predominantly attached either extracellularly or intracellularly during growth (Santos et al., 2016). Most biosurfactants secreted by pathogens occur when the nutrient-preventing condition overrides the production media (stationary growth phase). Moreover, biosurfactants can be produced naturally or influenced by varied molecules, stress, temperature, aeration, inoculum size, agitation speed, and diversified pH. In addition, some studies have reported that the production of biosurfactants can be induced through elements including phosphorus, nitrogen, iron, carbon, manganese, and sulfur (Abbasi et al., 2012). Some groups of scholars have reported that the ratios of several elements, including carbon:nitrogen, carbon:phosphorus, carbon:magnesium, or carbon:iron, can be optimized to improve the biosurfactant yields (Jimoh & Lin, 2019). Moreover, several techniques are used to reduce the production cost of biosurfactants. The cost reduction can make biosurfactants economically viable by using inexpensive raw materials, screening of new microorganisms, enhancing production yields, improving cheap processes, media optimization, cost-effective downstream procedures utilizing statistical models, improved yields of biosurfactants using superlative mutants and genetically engineered bacteria, and purifications of end products (Jimoh & Lin, 2019). Some of the techniques used to enhance the production of biosurfactants are based on production economics (Table 20.2).

20.3.1 Nitrogen sources

Nitrogen is regarded as the second prominent vital nutrient required in producing biosurfactants by pathogens. Composites originating from nitrogen are crucial to producing bioactive compounds, development of cell components, and microbial growth. Various organic and inorganic nitrogen sources had shown to affect biosurfactant production. An earlier study stated that amino acids and ammonium nitrates were utilized as a source of nitrogen by *P. aeruginosa*. In the synthesis of 2.8 g/L rhamnolipid through *P. aeruginosa*, an optimum of 0.3% of KNO_3 was reported as the desired nitrogen source (Patil et al., 2014). Similar results have been presented as the optimal nitrogen sources using *Paenibacillus* sp. D9 using ammonium sulfate (Jimoh & Lin, 2019) and *P. aeruginosa* R2 and 0.4% ammonium nitrate (Komal et al., 2012). Another study by Silva et al. (2010) reported that the produced rhamnolipid from *P. aeruginosa* UCP0092 in a production medium composed of 3% glycerol was affected through 0.6% of sodium nitrate. Biosurfactant synthesis increased using urea, NH_4NO_3, $NaNO_3$, or KNO_3, besides ammonium sulfate by some *Bacillus* strains. The variation in choosing inorganic rather than organic nitrogenous origin or vice versa may depend on the microbial strain or medium constituent (Elazzazy et al., 2015; Ghribi & Ellouze-Chaabouni, 2011). Therefore, organic nitrogen sources have a high impact on increasing biosurfactant properties. The application has become vital because of its reduced cost and availability.

An investigation revealed the highest emulsifying effect at 82% and the lowest surface effect at 29.50 mN/m with the use of urea in the form of organic nitrogen origins by *Virgibacillus salaries* KSA-T (Elazzazy et al., 2015). The strain *Vibrio* sp. 3B-2 used organic nitrogenous origins to improve microbial growth and biosurfactant yields with smaller secretion generated from inorganic nitrogen origins. In addition, (Hu et al., 2015) reported that using fungi extract as a source of nitrogen helps to achieve the highest biosurfactant yields from strain 3B-2. Another study indicated a noticeable reduction of surface tension to 29.7 mN/m while utilizing fungi extract as a nitrogen source to produce anaerobic lipopeptide by *Bacillus mojavensis* JF-2 (Liang et al., 2017). A study by Abbasi et al. (2012) reported that during the production of biosurfactants by *P. aeruginosa*, the result indicated that a synergistic relationship between sodium nitrate and fungi extract significantly influenced the production of biosurfactants.

Table 20.2 Techniques used in improving biosurfactant production.

S/N	Techniques	Microorganisms	Biosurfactants	Reference(s)
1.	Hyperproducing strains: Recombination and genetic modification	*Acinetobacter calcoaceticus* A2, *Bacillus* sp., *B. licheniformis*, *B. subtilis*, *B. subtilis* SK320, *B. licheniformis*, *Paenibacillus* sp., *Rhodococus erythropolis* SB-1A	Biodispersan, glycolipids, lichenysin, lipopeptide, surfactin	Anburajan et al. (2015), Cai et al. (2016), Jung et al. (2012), Kanna et al. (2014), Qiu et al. (2014), Shekhar et al. (2015)
2.	Optimization of downstream processes	*B. subtilis*, *P. aeruginosa* (MTCC 2453)	Rhamnolipid, surfactin	Jadhav et al. (2018), Joshi and Desai (2013), Soares da Silva et al. (2019)
3.	Statistical model optimization of enviromental factors and media components	*Aspergillus ustus* (MSF3), *B. subtilis* SPB1, *Brevibacterium aureum* MSA13, *P. aeruginosa* (MTCC 2297), *Paenibacillus alvei* ARN63, *Serratia nbidaea* SNAU02	Lipopeptide, mannosylerythritol lipids, sophorolipid	Joshi and Desai. (2013), Nalini and Parthasarathi (2014), Niu et al. (2019)
4.	Upscaling processes	*B. subtilis*, *B. licheniformis* R2, *Candida bombicola* ATCC 22214, *K. pneumoniae* WMF02, *P. cepacia* CCT6659	Lipopeptide, phospholipid, sophorolipid	Joshi and Desai (2013), Soares da Silva et al. (2019)
5.	Using cheap substrates	*Actinomycetes nocardiopsis* A17, *B. subtilis*, *B. subtilis* B20, *B. subtilis* B30, *B. subtilis* ICA56, *Brevibacterium aureum* MSA13, *Corynebacterium* sp., *C. aquaticum*, *C. xerosis* NS5, *Candida lipolytica* IA 1055, *C. lipolytica* UCP0988, *Klebsiella pneumoniae* WMF02, *Marinobacter hydrocarbonoclasticus* SdK644, *P. aeruginosa*, *P. aeruginosa* DS10−129, *P. aeruginosa* GS9−119, *P. aeruginosa*, *P. aeruginosa* MA01, *P. alcaligenes*, *Pseudomonas* SWP-4, *Serratia marcescens* UCP 1549	Glycolipid, lipopeptide, phospholipid, rhamnolipid, sophorolipids,	Abbasi et al. (2012), Al-Bahry et al. (2013), Al-Wahaibi et al. (2014), Araújo et al. (2019), Chakraborty et al. (2015), Dalili et al. (2015), De França et al. (2015), Nogueira Felix et al. (2019), Jamal et al. (2012), Jung et al. (2012), Seghal Kiran et al. (2010), Lan et al. (2015), Li et al. (2016), Martins and Martins (2018), Sobri et al. (2018), Zenati et al. (2018)

Martínez-Trujillo et al. (2015) revealed that a combination of fungi extract-$NaNO_3$ demonstrated a higher impact on the activity and biosynthesis of bacterial biosurfactant than the singular inclusion of sodium nitrate.

20.3.2 Carbon sources

It has been shown that different kinds of carbon sources significantly influence the production of biosurfactants in quantity and quality. These carbon sources can vary from low-cost cheap substrates, including oils, starchy substrates, petroleum effluents, distillery wastes, animal fat, lactic whey, plant-based oils, soap stock, olive oil mill effluent, molasses, vegetable oils, oil wastes, hydrocarbons (n-dodecane, n-tetradecane, n-hexadecane, pyrene), hydrophobic mixtures (kerosene, paraffin, crude oil, motor oil, diesel) with a high possibility of improving the production of biosurfactants (Jimoh & Lin, 2019). Abdel-Mawgoud et al. (2010) stated that the application of hydrophobic carbon sources was followed by solid biosurfactant production. A similar study reported optimization of factors influencing various hydrocarbon substrates, which produced a 3.6-fold biosurfactant produced through *Sphingobacterium detergens* (Burgos-Díaz et al., 2013). Furthermore, this was reported to induce biosurfactant on every investigated hydrocarbon on *P. aeruginosa* PBSC1; the best result was achieved with 4.99 g/L of biosurfactant using motor oil to statistically equate with the diesel oil of 4.11 g/L (Jimoh & Lin, 2019) and n-hexadecane of 4.76 g/L (Joice & Parthasarathi, 2014). Hamzah et al. (2013) employed low-molecular-weight carbohydrates to increase biosurfactants productions. Glycerol can be regarded as a C-3 compound and precursor of fatty acid, which makes it highly soluble in biosurfactant production medium. An optimum biosurfactant was produced by *P. aeruginosa* UKMP14T using a mineral salt medium consisting of 1% glycerol. Chakraborty et al. (2015) reported that the highest quantities of lipid-derived biosurfactant were derived in a medium consisting of glycerol.

Furthermore, several studies have reported the use of glycerol and sucrose as the main sources of carbon to increase biosurfactants by various *Bacillus* strains (Pereira et al., 2013). An investigation stated that after 96 h, about 2.11 g/L lipopeptide produced by *Bacillus clausii* 5B was found in a medium containing 1% (w/v) glucose as a carbon source (Aparna et al., 2012); however, this was compared with *Kocuria turfanesis* BS-J and *P. aeruginosa* BS-P when cultured on a media containing a mixture of the distillery and industrial wastes to produce 0.967 and 1.976 g/L yields, respectively (Dubey et al., 2012).

20.3.3 Growth conditions

The output of biosurfactant production can be influenced by various growth conditions such as pH, temperature, incubation time, aeration, and agitation speed. The high dependence of many pathogens on pH for cell growth and secondary metabolite production can be a vital feature. Several studies reported that an increase in production was within the alkaline condition range of pH ≥ 8.0 as given therein. An investigation by Patil et al. (2014) showed the optimal yield of glycolipid biosurfactants produced by *P. aeruginosa* strain F23 at pH 8. Other studies suggested comparative results, which were obtained from *P. aeruginosa* RS29 and WJ-1 with higher biosurfactants at a pH range of 7–8 and 6–8, respectively (Saikia et al., 2012). A study by Kanna et al. (2014) stated that an optimum of 2.5 g/L of lipopeptide biosurfactant was synthesized by *P. putida* MTCC at pH 8.0. Another investigation by Hamzah et al. (2013) reported that the pH of the medium was at an optimum pH of 9.0. Hence, biosurfactant production increased from the pH range of 6.0–9.0 and pH 10, after which the biosurfactant synthesis began to reduce. To some extent, it has been reported that neutral pH of 7.0 and acidic pH conditions (pH ≤ 6.0) can increase biosurfactant synthesis to support its competitiveness on a large scale. For example, rhamnolipid biosurfactant was produced by *Pseudomonas* sp. at its highest pH range of 6–6.5 and with a drastic decline over the pH of 7 (Kannahi & Sherley, 2012). A study revealed that *Paenibacillus alvei* can synthesize lipopeptide biosurfactant within a pH range of 6–8, and the highest biosurfactant yield was achieved at a pH of 6.89 (Najafi et al., 2011). Several investigations have revealed the highest yield of biosurfactants attained at a neutral pH of 7 (Chakraborty et al., 2015). Nevertheless, biosurfactant yield by *Actinomycetes nocardiosis* A17 was achieved at the highest pH of 6.8, although biosurfactant property remained at lower and higher pH (Chakraborty et al., 2015).

An insignificant difference in temperature can be another essential factor influencing various biosurfactant production processes. Patil et al. (2014) stated that the maximum yield of rhamnolipid biosurfactants produced by *P. aeruginosa* strain F23 was observed at 30°C. Another study by Chakraborty et al. (2015) on *A. nocardiosis* A17 stated that the optimal temperature at 28°C was highly effective in producing biosurfactants. However, the majority of these strains such as *B. subtilis* MTCC441 (Chander et al., 2012), *P. aeruginosa* RS29 (Saikia et al., 2012), *P. aeruginosa* WJ-1 (Xia et al., 2012), and *P. putida* (Kannahi & Sherley, 2012) can produce the highest biosurfactant at 37°C. Aeration and agitation speed can be the other main factors influencing the synthesis of biosurfactants by various pathogens. These two factors, which influence the secretion of biosurfactants, could also facilitate the exchange of oxygen from the gaseous to the aqueous phase. Investigation done by Zarur Coelho et al. (2010) examined the impact of agitation speed and aeration on biosurfactant produced by *Yarrowia* sp. The results from the batch fermentation showed that the biosurfactant effect was impacted when the agitation speed rose from 160 to 250 rpm.

Another study showed that the alterations in agitation speed help to increase biosurfactant produced by *P. aeruginosa* UCP 0992 in a medium containing glycerol (Silva et al., 2010). Oliveira et al. (2009) have also pointed out that varying agitation speed between 50 and 200 rpm significantly affects the biosurfactant synthesized by *P. alcaligenes* cultured in oil palm. The results indicated that an increase in the velocity of rotation favors surface tension reduction to 27.6 mN/m in cell-free broth. Jamal et al. (2012) discovered that the highest concentration of the rhamnolipid biosurfactant produced by *P. aeruginosa* was found when the agitation speed remained at a limit between 140 and 160 rpm with varying revolutions between 100 and 200 rpm. The effectiveness of the mass transfer in the medium constituents and oxygen molecules can be influenced by the agitation speed.

It was estimated that the highest surfactin concentration at 4.7 g/L after 21 h in fed-batch culture with an agitation rate of 150 rpm agitation speed and 1 volume of air per medium unit per time aeration rate. A study reported another vital factor that can have important consequences on biosurfactant production is the size of the inoculum (Asif et al., 2013), which can also play an important role in the biomass production and development of products because several physiological processes depend on cell density. An uninterrupted association was found between biomass growth rate and product development of several bacterial products. A study on *B. subtilis* SPB1 indicated that the density and age of inoculum significantly impact the yield and total production cost process. Moreover, a study by Mnif et al. (2013) indicated that after optimizing the conditions of the inoculum, the yield of lipopeptide biosurfactant produced by *B. subtilis* SPB1 increased to about 3.4 g/L. Nalini & Parthasarathi (2018) reported that the highest biosurfactant yield was achieved at 2.4 mL inoculum of strain SNAU02; the inoculum size shows a significant impact on the production of the biosurfactant. Similarly, an increase in the inoculum size can upturn microbial growth to a particular extent, and any further increase may cause a decrease in microbial properties because of insufficient nutrients (Nalini & Parthasarathi, 2018). There can be a lack of oxygen and nutrient because of a large inoculum ratio, while a small inoculum size could lead to significant reduction in the cell number in the production medium. The smaller inoculum size requires considerable time to achieve the maximum increase and use substrates to obtain the yield required for the product. For example, Asif et al. (2013) stated that the use of 1 mL inoculum size could reduce the biosurfactant yield by 2.74 g/L. The results indicated several variations in the production of biosurfactants under various conditions of inoculum densities and bacterial growth.

20.3.4 Carbon/nitrogen ratio

During fermentation, the carbon-to-nitrogen ratio can be regarded as one of the important characteristics that can affect microbial impact and increase in metabolites in biosurfactant production (Xia et al., 2012). Reduction in nitrogen concentrations can prevent bacterial growth, hence, promoting cellular metabolism for producing these biosurfactants. Xia et al. (2012) stated that the relative quantities of multiple elements to carbon, such as the carbon-to-nitrogen ratio, could influence biosurfactant yield in the medium compositions. Several studies have reported an increase in the synthesis of biosurfactants due to nitrogen insufficiency conditions. In addition, it had been revealed that an optimized carbon-to-nitrogen ratio of 7:1 was used as the optimum condition to increase biosurfactant produced by *P. aeruginosa* strain F23. Many investigations have shown that *P. aeruginosa* produced rhamnolipid biosurfactants in the presence of reduced nitrogen sources (Patil et al., 2014; Xia et al., 2012). A study reported that the presence of diesel fuel to $(NH_4)_2SO_4$ ratio of 3:1 could lead to the highest production of biosurfactant at 3.79 g/L with 32.0 mN/m reductions in surface tension and dry cell weight of 0.35 g/L. The results revealed that a higher carbon-to-nitrogen ratio reduced bacterial growth and favored cellular metabolism toward the synthesis of metabolites (Jimoh & Lin, 2019).

On the other hand, the highest biosurfactant yield was by *P. nitroreducens* at the carbon-to-nitrogen ratio of 22:1 with glucose and sodium nitrate, respectively (Onwosi & Odibo, 2013). An investigation by Elazzazy et al. (2015) reported that the minimum surface tension (29 mN/m) and the highest emulsifying effect (85%) were attained at a carbon-to-nitrogen ratio of 30:1. However, any abundance in proportion showed no significant influence on biosurfactant produced by of *Virgibacillus salaries* KSA-T. Lan et al. (2015) showed that the complete carbon-to-nitrogen ratio (waste cooking oil and $NaNO_3$) was 10:1, which led to a drastic reduction in surface tension to 27.5 mN/m and biosurfactant yield to 5 g/L.

20.3.5 Use of optimization tools

The improvement in the development of biosurfactant yield has been further investigated by the introduction of an adequate innovative statistical strategy that recognized the relationship between several factors behind the bioprocess. Response surface methodology (RSM) is composed of a set of statistical techniques used for experimental designs and model building while determining the unique impacts on the factors and establishing optimal conditions. It is important to examine the influence of small independent factors on the system reaction without predetermining the relationship between the objective and the variable's function (Najafi et al., 2011). RSM employed many regression studies using quantitative information from suitably controlled studies to recognize multivalent conditions simultaneously (Seghal Kiran et al., 2010; Najafi et al., 2011). Several studies have effectively used RSM to improve biosurfactant yield by reducing production costs through the selection of balanced proportions of culture medium ingredients and culture optimization conditions (Farias et al., 2021; Najafi et al., 2011).

20.3.6 Using recombinant DNA technology

Recently, genetic enhancement has been suggested to produce biosurfactants. This can be because of the comparatively low production yields, which later reduce the biotechnological uses of these biosurfactants (Jimoh & Lin, 2019). Most importantly, pathogens can reduce and utilize various substrates when synthesizing biosurfactants. However, studies have reported that little information was available about the cloning, molecular properties, and functional characterization of their reduction and biosurfactant processes (Aliakbari et al., 2014). The investigation by Sekhon et al. (2012) revealed that in the presence of hydrocarbons and other carbon sources, biosurfactants produced by pathogens may be attributed to the introduction of particular genes. Because of its biotechnological and industrial importance, a new biosurfactant with the ability to modify the genetic material of pathogens using recombinant DNA technology has been suggested. Soares da Silva et al. (2019) suggested that one way to minimize costs and improve biosurfactant properties is to develop hyperproducing pathogens. It is good to establish recombinant microbial strains that can increase biosurfactant production yields at a reduced cost of production or with the capacity to synthesize powerful congeners that are closely related to bio-products (Bachmann et al., 2014). For biotechnological and petroleum uses, biosurfactant producers must be engineered to withstand harsh process conditions. However, Soares da Silva et al. (2019) proposed a way to find novel genetic modifications from severe conditions, such as pH, temperatures, and high salt concentration. Nevertheless, biotechnological applications of cloned hyperproducing strains have not been adequately attempted, but hyperbiosurfactant producers have been demonstrated (Satpute et al., 2010).

20.4 Conclusion

Biosurfactant production can be a competitive as well as a profitable industry. This is because of its usage in diverse sectors, including food, petroleum, cosmetics, and medicine. The importance of biosurfactants is overwhelming when compared with chemical-based surfactants. Nevertheless, the sustainability of biosurfactants is essential. Less expensive raw materials are being studied to resolve the challenges associated with a higher cost of biosurfactant production. These include molasses, starch wastes, soap stock, corn steep liquor, animal fats, animal wastes, food wastes, brewery wastes, dairy and bakery wastes, lignocellulosic wastes, starch-rich wastes, and wastes from cooking oils. Using cheaper raw materials to produce substrates for biosurfactants is an innovative means of sustaining the production of biosurfactants. Moreover, selecting appropriate techniques can boost and increase the production of biosurfactants. These include screening of new microorganisms, media optimization, improved yields of biosurfactants using superlative mutants and genetically engineered bacteria, and purifications of end products. Thus, the sustainability of biosurfactants can be achieved if appropriate cheap raw materials, especially wastes and renewable materials, and techniques are employed.

Acknowledgments

The authors thankfully acknowledge Eastern Unity Technology, Malaysia, for the financial support through the grant (UIC190806).

References

Abbasi, H., Hamedi, M. M., Lotfabad, T. B., Zahiri, H. S., Sharafi, H., Masoomi, F., Moosavi-Movahedi, A. A., Ortiz, A., Amanlou, M., & Noghabi, K. A. (2012). Biosurfactant-producing bacterium, *Pseudomonas aeruginosa* MA01 isolated from spoiled apples: Physicochemical and structural characteristics of isolated biosurfactant. *Journal of Bioscience and Bioengineering*, *113*(2), 211–219. Available from https://doi.org/10.1016/j.jbiosc.2011.10.002.

Abbot, V., Paliwal, D., Sharma, A., & Sharma, P. (2022). A review on the physicochemical and biological applications of biosurfactants in biotechnology and pharmaceuticals. *Heliyon*, *8*(8), e10149. Available from https://doi.org/10.1016/j.heliyon.2022.e10149.

Abdel-Mawgoud, A. M., Hausmann, R., Lépine, F., Müller, M. M., & Déziel, E. (2010). Rhamnolipids: Detection, analysis, biosynthesis, genetic regulation, and bioengineering of production (pp. 13–55). Springer Science and Business Media LLC. Available from https://doi.org/10.1007/978-3-642-14490-5_2.

Adesra, A., Srivastava, V. K., & Varjani, S. (2021). Valorization of dairy wastes: Integrative approaches for value added products. *Indian Journal of Microbiology*, *61*(3), 270–278. Available from https://doi.org/10.1007/s12088-021-00943-5.

Al-Bahry, S. N., Al-Wahaibi, Y. M., Elshafie, A. E., Al-Bemani, A. S., Joshi, S. J., Al-Makhmari, H. S., & Al-Sulaimani, H. S. (2013). Biosurfactant production by *Bacillus subtilis* B20 using date molasses and its possible application in enhanced oil recovery. *International Biodeterioration and Biodegradation*, *81*, 141–146. Available from https://doi.org/10.1016/j.ibiod.2012.01.006.

Aliakbari, E., Tebyanian, H., Hassanshahian, M., & Kariminik, A. (2014). Degradation of alkanes in contaminated sites. *International Journal of Advanced Biological and Biomedical Research*, *2*, 1620–1637.

Al-Wahaibi, Y., Joshi, S., Al-Bahry, S., Elshafie, A., Al-Bemani, A., & Shibulal, B. (2014). Biosurfactant production by *Bacillus subtilis* B30 and its application in enhancing oil recovery. *Colloids and Surfaces B: Biointerfaces*, *114*, 324–333. Available from https://doi.org/10.1016/j.colsurfb.2013.09.022.

Anburajan, L., Meena, B., Raghavan, R. V., Shridhar, D., Joseph, T. C., Vinithkumar, N. V., Dharani, G., Dheenan, P. S., & Kirubagaran, R. (2015). Heterologous expression, purification, and phylogenetic analysis of oil-degrading biosurfactant biosynthesis genes from the marine sponge-associated *Bacillus licheniformis* NIOT-06. *Bioprocess and Biosystems Engineering*, *38*(6), 1009–1018. Available from https://doi.org/10.1007/s00449-015-1359-x.

Aparna, A., Srinikethan, G., & Hegde, S. (2012). Isolation, screening and production of biosurfactant by *Bacillus clausii* 5B. *Research Journal of Biotechnology*, *3*.

Araújo, H. W. C., Andrade, R. F. S., Montero-Rodríguez, D., Rubio-Ribeaux, D., Alves Da Silva, C. A., & Campos-Takaki, G. M. (2019). Sustainable biosurfactant produced by *Serratia marcescens* UCP 1549 and its suitability for agricultural and marine bioremediation applications. *Microbial Cell Factories*, *18*(1). Available from https://doi.org/10.1186/s12934-018-1046-0.

Asif, S., Habib, H., Abbasi, M. H., Akhtar, R. M., Waqas, M., Zafar, R. M. S., & Awais, H. (2013). Biosurfactant production by pseudomonas aeruginosa strains on 1mL of inoculum size. *Pakistan Journal of Medical and Health Sciences*, *7*(2). Available from http://pjmhsonline.com/apriltojune2013/biosurfactant_production_by_pseudomonas.htm.

Bachmann, R. T., Johnson, A. C., & Edyvean, R. G. J. (2014). Biotechnology in the petroleum industry: An overview. *International Biodeterioration and Biodegradation*, *86*, 225–237. Available from https://doi.org/10.1016/j.ibiod.2013.09.011.

Banat, I. M., De Rienzo, M. A. D., & Quinn, G. A. (2014). Microbial biofilms: Biosurfactants as antibiofilm agents. *Applied Microbiology and Biotechnology*, *98*(24), 9915–9929. Available from https://doi.org/10.1007/s00253-014-6169-6.

Burgos-Díaz, C., Pons, R., Teruel, J. A., Aranda, F. J., Ortiz, A., Manresa, A., & Marqués, A. M. (2013). The production and physicochemical properties of a biosurfactant mixture obtained from Sphingobacterium detergens. *Journal of Colloid and Interface Science*, *394*(1), 368–379. Available from https://doi.org/10.1016/j.jcis.2012.12.017.

Cai, Q., Zhang, B., Chen, B., Cao, T., & Lv, Z. (2016). Biosurfactant produced by a rhodococcus erythropolis mutant as an oil spill response agent. *Water Quality Research Journal of Canada*, *51*(2), 97–105. Available from https://doi.org/10.2166/wqrjc.2016.025.

Chakraborty, S., Ghosh, M., Chakraborti, S., Jana, S., Sen, K. K., Kokare, C., & Zhang, L. (2015). Biosurfactant produced from Actinomycetes nocardiopsis A17: Characterization and its biological evaluation. *International Journal of Biological Macromolecules*, *79*, 405–412. Available from https://doi.org/10.1016/j.ijbiomac.2015.04.068.

Chander, C., Lohitnath, T., Kumar., & Kalaichelvan, P. (2012). Production and characterization of biosurfactant from *Bacillus subtilis* MTCC441 and its evaluation to use as bioemulsifier for food biopreservative. *Advances in Applied Science Research*, *3*, 1827–1831.

Dalili, D., Amini, M., Faramarzi, M. A., Fazeli, M. R., Khoshayand, M. R., & Samadi, N. (2015). Isolation and structural characterization of Coryxin, a novel cyclic lipopeptide from Corynebacterium xerosis NS5 having emulsifying and anti-biofilm activity. *Colloids and Surfaces B: Biointerfaces*, *135*, 425–432. Available from https://doi.org/10.1016/j.colsurfb.2015.07.005.

Das, A. J., & Kumar, R. (2018). Utilization of agro-industrial waste for biosurfactant production under submerged fermentation and its application in oil recovery from sand matrix. *Bioresource Technology*, *260*, 233–240. Available from https://doi.org/10.1016/j.biortech.2018.03.093.

De França, Í. W. L., Lima, A. P., Lemos, J. A. M., Lemos, C. G. F., Melo, V. M. M., De Sant'ana, H. B., & Gonçalves, L. R. B. (2015). Production of a biosurfactant by *Bacillus subtilis* ICA56 aiming bioremediation of impacted soils. *Catalysis Today*, *255*, 10–15. Available from https://doi.org/10.1016/j.cattod.2015.01.046.

Dubey, K. V., Charde, P. N., Meshram, U. S., Yadav, M., Singh, S. K., Juwarkar, S., & Juwarkar, A. A. (2012). Potential of new microbial isolates for biosurfactant production using combinations of distillery waste with other industrial wastes. *Journal of Petroleum & Environmental Biotechnology*, *4*.

Elazzazy, A. M., Abdelmoneim, T. S., & Almaghrabi, O. A. (2015). Isolation and characterization of biosurfactant production under extreme environmental conditions by alkali-halo-thermophilic bacteria from Saudi Arabia. *Saudi Journal of Biological Sciences*, *22*(4), 466–475. Available from https://doi.org/10.1016/j.sjbs.2014.11.018.

Farias, C. B. B., Almeida, F. C. G., Silva, I. A., Souza, T. C., Meira, H. M., Soares da Silva, R. d C. F., Luna, J. M., Santos, V. A., Converti, A., Banat, I. M., & Sarubbo, L. A. (2021). Production of green surfactants: Market prospects. *Electronic Journal of Biotechnology*, *51*, 28–39. Available from https://doi.org/10.1016/j.ejbt.2021.02.002.

Gaur, V. K., Sharma, P., Sirohi, R., Varjani, S., Taherzadeh, M. J., Chang, J. S., Yong Ng, H., Wong, J. W. C., & Kim, S. H. (2022). Production of biosurfactants from agro-industrial waste and waste cooking oil in a circular bioeconomy: An overview. *Bioresource Technology*, *343*. Available from https://doi.org/10.1016/j.biortech.2021.126059.

Gayathiri, E., Prakash, P., Karmegam, N., Varjani, S., Awasthi, M. K., & Ravindran, B. (2022). Biosurfactants: Potential and eco-friendly material for sustainable agriculture and environmental safety—A review. *Agronomy*, *12*(3). Available from https://doi.org/10.3390/agronomy12030662.

Ghribi, D., & Ellouze-Chaabouni, S. (2011). Enhancement of *Bacillus subtilis* lipopeptide biosurfactants production through optimization of medium composition and adequate control of aeration. *Biotechnology Research International*, *2011*, 1–6. Available from https://doi.org/10.4061/2011/653654.

Hamzah, A., Sabturani, N., & Radiman, S. (2013). Screening and optimization of biosurfactant production by the hydrocarbon-degrading bacteria. *Sains Malaysiana*, *42*(5), 615–623. Available from http://www.ukm.my/jsm/pdf_files/SM-PDF-42-5-2013/08%20Ainon%20Hamzah.pdf.

Henkel, M., Müller, M. M., Kügler, J. H., Lovaglio, R. B., Contiero, J., Syldatk, C., & Hausmann, R. (2012). Rhamnolipids as biosurfactants from renewable resources: Concepts for next-generation rhamnolipid production. *Process Biochemistry*, *47*(8), 1207–1219. Available from https://doi.org/10.1016/j.procbio.2012.04.018.

Hu, X., Wang, C., & Wang, P. (2015). Optimization and characterization of biosurfactant production from marine Vibrio sp. strain 3B-2. *Frontiers in Microbiology*, *6*. Available from https://doi.org/10.3389/fmicb.2015.00976.

Jadhav, J., Dutta, S., Kale, S., & Pratap, A. (2018). Fermentative production of rhamnolipid and purification by adsorption chromatography. *Preparative Biochemistry and Biotechnology*, *48*(3), 234−241. Available from https://doi.org/10.1080/10826068.2017.1421967.

Jamal, P., Nawawi, W. M. F., & Alam, M. Z. (2012). Optimum medium components for biosurfactant production by klebsiella pneumoniae WMF02 utilizing sludge palm oil as a substrate. *Australian Journal of Basic and Applied Sciences*, *6*(1), 100−108. Available from http://www.insipub.net/ajbas/2012/January/100-108.pdf.

Jimoh, A. A., & Lin, J. (2019). Biosurfactant: A new frontier for greener technology and environmental sustainability. *Ecotoxicology and Environmental Safety*, *184*, 109607. Available from https://doi.org/10.1016/j.ecoenv.2019.109607.

Joice, P. A., & Parthasarathi, R. (2014). Optimization of biosurfactant production from *Pseudomonas aeruginosa* PBSC1. *International Journal of Current Microbiology and Applied Sciences*, *3*, 140−151.

Joshi, S. J., & Desai, A. J. (2013). Bench-scale production of biosurfactants and their potential in ex-situ MEOR application. *Soil and Sediment Contamination*, *22*(6), 701−715. Available from https://doi.org/10.1080/15320383.2013.756450.

Jung, J., Yu, K. O., Ramzi, A. B., Choe, S. H., Kim, S. W., & Han, S. O. (2012). Improvement of surfactin production in *Bacillus subtilis* using synthetic wastewater by overexpression of specific extracellular signaling peptides, comX and phrC. *Biotechnology and Bioengineering*, *109*(9), 2349−2356. Available from https://doi.org/10.1002/bit.24524.

Kanna, R., Gummadi, S. N., & Kumar, G. S. (2014). Production and characterization of biosurfactant by *Pseudomonas putida* MTCC 2467. *Journal of Biological Sciences*, *14*(6), 436−445. Available from https://doi.org/10.3923/jbs.2014.436.445.

Kannahi, M., & Sherley, M. (2012). Biosurfactant production by Pseudomonas putida and Aspergillus niger from oil contaminated site. *International Journal of Chemistry and Pharmaceutical Sciences*, *3*, 37−42.

Khopade, A., Biao, R., Liu, X., Mahadik, K., Zhang, L., & Kokare, C. (2012). Production and stability studies of the biosurfactant isolated from marine Nocardiopsis sp. B4. *Desalination*, *285*, 198−204. Available from https://doi.org/10.1016/j.desal.2011.10.002.

Komal, K., Anuradha, P., & Aruna, K. (2012). Studies on biosurfactant production by *Pseudomonas aeruginosa* R2 isolated from oil contaminated soil sample. *Asian Journal of Biological Sciences*, *7*, 123−129.

Lan, G., Fan, Q., Liu, Y., Chen, C., Li, G., Liu, Y., & Yin, X. (2015). Rhamnolipid production from waste cooking oil using Pseudomonas SWP-4. *Biochemical Engineering Journal*, *101*, 44−54. Available from https://doi.org/10.1016/j.bej.2015.05.001.

Li, J., Deng, M., Wang, Y., & Chen, W. (2016). Production and characteristics of biosurfactant produced by *Bacillus pseudomycoides* BS6 utilizing soybean oil waste. *International Biodeterioration and Biodegradation*, *112*, 72−79. Available from https://doi.org/10.1016/j.ibiod.2016.05.002.

Liang, X., Shi, R., Radosevich, M., Zhao, F., Zhang, Y., Han, S., & Zhang, Y. (2017). Anaerobic lipopeptide biosurfactant production by an engineered bacterial strain for in situ microbial enhanced oil recovery. *RSC Advances*, *7*(33), 20667−20676. Available from https://doi.org/10.1039/c7ra02453c.

Marcelino, P. R. F., Peres, G. F. D., Terán-Hilares, R., Pagnocca, F. C., Rosa, C. A., Lacerda, T. M., dos Santos, J. C., & da Silva, S. S. (2019). Biosurfactants production by yeasts using sugarcane bagasse hemicellulosic hydrolysate as new sustainable alternative for lignocellulosic biorefineries. *Industrial Crops and Products*, *129*, 212−223. Available from https://doi.org/10.1016/j.indcrop.2018.12.001.

Martínez-Trujillo, M. A., Venegas, I., Vigueras-Carmona, S., Zafra-Jimenez, G., & García-Rivero, M. (2015). Optimización de la producción de un biosurfactante bacteriano. *Revista Mexicana de Ingenieria Quimica*, *14*, 355−362.

Martins, P. C., & Martins, V. G. (2018). Biosurfactant production from industrial wastes with potential remove of insoluble paint. *International Biodeterioration and Biodegradation*, *127*, 10−16. Available from https://doi.org/10.1016/j.ibiod.2017.11.005.

Mnif, I., Ellouze-Chaabouni, S., & Ghribi, D. (2013). Economic production of *Bacillus subtilis* SPB1 biosurfactant using local agro-industrial wastes and its application in enhancing solubility of diesel. *Journal of Chemical Technology & Biotechnology, 88*(5), 779−787. Available from https://doi.org/10.1002/jctb.3894.

Najafi, A. R., Rahimpour, M. R., Jahanmiri, A. H., Roostaazad, R., Arabian, D., Soleimani, M., & Jamshidnejad, Z. (2011). Interactive optimization of biosurfactant production by *Paenibacillus alvei* ARN63 isolated from an Iranian oil well. *Colloids and Surfaces B: Biointerfaces, 82*(1), 33−39. Available from https://doi.org/10.1016/j.colsurfb.2010.08.010.

Nalini, S., & Parthasarathi, R. (2018). Optimization of rhamnolipid biosurfactant production from *Serratia rubidaea* SNAU02 under solid-state fermentation and its biocontrol efficacy against Fusarium wilt of eggplant. *Annals of Agrarian Science, 16*(2), 108−115. Available from https://doi.org/10.1016/j.aasci.2017.11.002.

Nalini, S., & Parthasarathi, R. (2014). Production and characterization of rhamnolipids produced by *Serratia rubidaea* SNAU02 under solid-state fermentation and its application as biocontrol agent. *Bioresource Technology, 173*, 231−238. Available from https://doi.org/10.1016/j.biortech.2014.09.051.

Ng, H. S., Kee, P. E., Yim, H. S., Chen, P. T., Wei, Y. H., & Chi-Wei Lan, J. (2020). Recent advances on the sustainable approaches for conversion and reutilization of food wastes to valuable bioproducts. *Bioresource Technology, 302*. Available from https://doi.org/10.1016/j.biortech.2020.122889.

Nitschke, M., Costa, S. G. V. A. O., & Contiero, J. (2010). Structure and applications of a rhamnolipid surfactant produced in soybean oil waste. *Applied Biochemistry and Biotechnology, 160*(7), 2066−2074. Available from https://doi.org/10.1007/s12010-009-8707-8.

Niu, Y., Wu, J., Wang, W., & Chen, Q. (2019). Production and characterization of a new glycolipid, mannosylerythritol lipid, from waste cooking oil biotransformation by *Pseudozyma aphidis* ZJUDM34. *Food Science & Nutrition, 7*(3), 937−948. Available from https://doi.org/10.1002/fsn3.880.

Nogueira Felix, A. K., Martins, J. J. L., Lima Almeida, J. G., Giro, M. E. A., Cavalcante, K. F., Maciel Melo, V. M., Loiola Pessoa, O. D., Ponte Rocha, M. V., Rocha Barros Gonçalves, L., & Saraiva de Santiago Aguiar, R. (2019). Purification and characterization of a biosurfactant produced by *Bacillus subtilis* in cashew apple juice and its application in the remediation of oil-contaminated soil. *Colloids and Surfaces B: Biointerfaces, 175*, 256−263. Available from https://doi.org/10.1016/j.colsurfb.2018.11.062.

Oliveira, F. J. S., Vazquez, L., de Campos, N. P., & de França, F. P. (2009). Production of rhamnolipids by a *Pseudomonas alcaligenes* strain. *Process Biochemistry, 44*(4), 383−389. Available from https://doi.org/10.1016/j.procbio.2008.11.014.

Onwosi, C., & Odibo, F. (2013). Use of response surface design in the optimization of starter cultures for enhanced rhamnolipid production by *Pseudomonas nitroreducens*. *African Journal of Biotechnology, 12*.

Partovi, M., Lotfabad, T. B., Roostaazad, R., Bahmaei, M., & Tayyebi, S. (2013). Management of soybean oil refinery wastes through recycling them for producing biosurfactant using *Pseudomonas aeruginosa* MR01. *World Journal of Microbiology and Biotechnology, 29*(6), 1039−1047. Available from https://doi.org/10.1007/s11274-013-1267-7.

Patil, S., Pendse, A., & Aruna, K. (2014). Studies on optimization of biosurfactant production by *Pseudomonas aeruginosa* F23 isolated from oil contaminated soil sample. *International Journal of Current Biotechnology, 2*, 20−30.

Patowary, K., Patowary, R., Kalita, M. C., & Deka, S. (2017). Characterization of biosurfactant produced during degradation of hydrocarbons using crude oil as sole source of carbon. *Frontiers in Microbiology, 8*. Available from https://doi.org/10.3389/fmicb.2017.00279.

Pereira, J. F. B., Gudiña, E. J., Costa, R., Vitorino, R., Teixeira, J. A., Coutinho, J. A. P., & Rodrigues, L. R. (2013). Optimization and characterization of biosurfactant production by *Bacillus subtilis* isolates towards microbial enhanced oil recovery applications. *Fuel, 111*, 259−268. Available from https://doi.org/10.1016/j.fuel.2013.04.040.

Qiu, Y., Xiao, F., Wei, X., Wen, Z., & Chen, S. (2014). Improvement of lichenysin production in Bacillus licheniformis by replacement of native promoter of lichenysin biosynthesis operon and medium optimization. *Applied Microbiology and Biotechnology, 98*(21), 8895−8903. Available from https://doi.org/10.1007/s00253-014-5978-y.

Rajasimman, M., Suganya, A., Manivannan, P., & Pandian, A. M. K. (2021). Utilization of agroindustrial wastes with a high content of protein, carbohydrates, and fatty acid used for mass production of biosurfactant. *Green sustainable process for chemical and environmental engineering and science: Microbially-derived biosurfactants for improving sustainability in industry* (pp. 127−146). Elsevier. Available from https://doi.org/10.1016/B978-0-12-823380-1.00007-1.

Saikia, R. R., Deka, S., Deka, M., & Sarma, H. (2012). Optimization of environmental factors for improved production of rhamnolipid biosurfactant by *Pseudomonas aeruginosa* RS29 on glycerol. *Journal of Basic Microbiology, 52*(4), 446−457. Available from https://doi.org/10.1002/jobm.201100228.

Sajna, K. V., Sukumaran, R. K., Jayamurthy, H., Reddy, K. K., Kanjilal, S., Prasad, R. B. N., & Pandey, A. (2013). Studies on biosurfactants from Pseudozyma sp. NII 08165 and their potential application as laundry detergent additives. *Biochemical Engineering Journal, 78*, 85−92. Available from https://doi.org/10.1016/j.bej.2012.12.014.

Samad, A., Zhang, J., Chen, D., Chen, X., Tucker, M., & Liang, Y. (2017). Sweet sorghum bagasse and corn stover serving as substrates for producing sophorolipids. *Journal of Industrial Microbiology and Biotechnology, 44*(3), 353−362. Available from https://doi.org/10.1007/s10295-016-1891-y.

Santos, D. K. F., Rufino, R. D., Luna, J. M., Santos, V. A., & Sarubbo, L. A. (2016). Biosurfactants: Multifunctional biomolecules of the 21st century. *International Journal of Molecular Sciences, 17*(3). Available from https://doi.org/10.3390/ijms17030401.

Saravanan, V., & Vijayakumar, S. (2014). Production of biosurfactant by *Pseudomonas aeruginosa* PB3A using agro-industrial wastes as a carbon source. *Malaysian Journal of Microbiology, 10*(1), 57−62. Available from http://web.usm.my/mjm/issues/vol10no1/Research%208.pdf.

Satpute, S. K., Banpurkar, A. G., Dhakephalkar, P. K., Banat, I. M., & Chopade, B. A. (2010). Methods for investigating biosurfactants and bioemulsifiers: A review. *Critical Reviews in Biotechnology, 30*(2), 127−144. Available from https://doi.org/10.3109/07388550903427280.

Seghal Kiran, G., Anto Thomas, T., Selvin, J., Sabarathnam, B., & Lipton, A. P. (2010). Optimization and characterization of a new lipopeptide biosurfactant produced by marine *Brevibacterium aureum* MSA13 in solid state culture. *Bioresource Technology, 101*(7), 2389−2396. Available from https://doi.org/10.1016/j.biortech.2009.11.023.

Sekhon, K. K., Khanna, S., & Cameotra, S. S. (2012). Biosurfactant production and potential correlation with esterase activity. *Journal of Petroleum & Environmental Biotechnology, 3*.

Sharma, P., Gaur, V. K., Kim, S.-H., & Pandey, A. (2020). Microbial strategies for biotransforming food waste into resources. *Bioresource Technology, 299*.

Shekhar, S., Sundaramanickam, A., & Balasubramanian, T. (2015). Biosurfactant producing microbes and their potential applications: A review. *Critical Reviews in Environmental Science and Technology, 45*(14), 1522−1554. Available from https://doi.org/10.1080/10643389.2014.955631.

Silva, S. N. R. L., Farias, C. B. B., Rufino, R. D., Luna, J. M., & Sarubbo, L. A. (2010). Glycerol as substrate for the production of biosurfactant by *Pseudomonas aeruginosa* UCP0992. *Colloids and Surfaces B: Biointerfaces, 79*(1), 174−183. Available from https://doi.org/10.1016/j.colsurfb.2010.03.050.

Singh, P., Patil, Y., & Rale, V. (2019). Biosurfactant production: Emerging trends and promising strategies. *Journal of Applied Microbiology, 126*(1), 2−13. Available from https://doi.org/10.1111/jam.14057.

Soares da Silva, R. d C. F., de Almeida, D. G., Brasileiro, P. P. F., Rufino, R. D., de Luna, J. M., & Sarubbo, L. A. (2019). Production, formulation and cost estimation of a commercial biosurfactant. *Biodegradation, 30*(4), 191−201. Available from https://doi.org/10.1007/s10532-018-9830-4.

Sobri, I. M., Halim, M., Lai, O.-M., Lajis, A. F., Yusof, M. T., Halmi, M. I. E., Johari, W. L. W., & Wasoh, H. (2018). Emulsification characteristics of rhamnolipids by *Pseudomonas aeruginosa* using coconut oil as carbon source. *Journal of Environmental Microbiology and Toxicology, 6*(1), 7−12. Available from https://doi.org/10.54987/jemat.v6i1.400.

Varjani, S., Rakholiya, P., Yong Ng, H., Taherzadeh, M. J., Hao Ngo, H., Chang, J. S., Wong, J. W. C., You, S., Teixeira, J. A., & Bui, X. T. (2021). Bio-based rhamnolipids production and recovery from waste streams: Status and perspectives. *Bioresource Technology, 319*. Available from https://doi.org/10.1016/j.biortech.2020.124213.

Vieira, I. M. M., Santos, B. L. P., Silva, L. S., Ramos, L. C., de Souza, R. R., Ruzene, D. S., & Silva, D. P. (2021). Potential of pineapple peel in the alternative composition of culture media for biosurfactant production. *Environmental Science and Pollution Research, 28*(48), 68957−68971. Available from https://doi.org/10.1007/s11356-021-15393-1.

Wilton, N., Lyon-Marion, B. A., Kamath, R., McVey, K., Pennell, K. D., & Robbat, A. (2018). Remediation of heavy hydrocarbon impacted soil using biopolymer and polystyrene foam beads. *Journal of Hazardous Materials*, *349*, 153–159. Available from https://doi.org/10.1016/j.jhazmat.2018.01.041.

Xia, W. J., Luo, Z. B., Dong, H. P., Yu, L., Cui, Q. F., & Bi, Y. Q. (2012). Synthesis, characterization, and oil recovery application of biosurfactant produced by indigenous *Pseudomonas aeruginosa* WJ-1 using waste vegetable oils. *Applied Biochemistry and Biotechnology*, *166*(5), 1148–1166. Available from https://doi.org/10.1007/s12010-011-9501-y.

Zarur Coelho, M. A., Fontes, G. C., Fonseca Amaral, P. F., & Nele, M. (2010). Factorial design to optimize biosurfactant production by *Yarrowia lipolytica*. *Journal of Biomedicine and Biotechnology*, *2010*. Available from https://doi.org/10.1155/2010/821306.

Zenati, B., Chebbi, A., Badis, A., Eddouaouda, K., Boutoumi, H., El Hattab, M., Hentati, D., Chelbi, M., Sayadi, S., Chamkha, M., & Franzetti, A. (2018). A non-toxic microbial surfactant from Marinobacter hydrocarbonoclasticus SdK644 for crude oil solubilization enhancement. *Ecotoxicology and Environmental Safety*, *154*, 100–107. Available from https://doi.org/10.1016/j.ecoenv.2018.02.032.

Index

Note: Page numbers followed by "*f*" and "*t*" refer to figures and tables, respectively.

A

Achromobacter sp., 415
    A. xylosoxidans, 126–127
Acid precipitation, 97–98
Acid-producing bacteria (APB), 292–293
Acinetobacter sp., 13–14, 26, 46, 203–204, 413
    A. baumannii, 13, 309
    A. calcoaceticus, 7
    A. radioresistens, 13
    Acinetobacter sp. Y2, 209
Actinobacillus sp., 413
Actinomycetes nocardiosis A17, 443–444
Active coatings, 364
Active pharmaceutical ingredients (APIs), 251–252
    chemical and physical properties of, 252–253
    recalcitrance of, 253–254
Active pharmaceutical ingredients, 255–256
Adhesion, 355–358
Adsorption, 249–250, 263–268
Aedes, 347–348
    A. aegypti, 348
    A. albopictus, 348–350
Aeration, 88–89
Aeromonas, 159–161
AES. *See* Alkylpolyoxy-ethylene sulfates (AES)
AFM. *See* Atomic force microscopy (AFM)
Agitation, 88–89
Agric-food wastes, 66–67
Agricultural/agriculture, 342, 373
    applications of DDT, 134
    biosurfactants as agricultural biopesticides, 342–345
    byproducts, 94–95
    formulations/chemicals, 123
    pests, 343
    products, 439–440
Agro-industrial sector, 43–46
Agro-industrial wastes, 156, 193–194, 407
Air–liquid interfaces, 30
alanA gene, 13
alanB gene, 13

alanC gene, 13
Alasan, 13
Albumin, 358
Alcaligenes piechaudii srs PTA-5580, 170–172
Alcanivorax sp. BIC1A5, 210–211
Alcohol, 39
Algal biosurfactants, 261
Alkyl BDMAC. *See* Alkyl benzyl dimethyl ammonium chloride (Alkyl BDMAC)
Alkyl benzyl dimethyl ammonium chloride (Alkyl BDMAC), 411
Alkyl ethoxysulphate surfactant, 411
Alkyl sugars, 278–279
Alkyl sulfates (AS), 411
Alkyl TMAC. *See* Alkyl trimethyl ammonium chloride (Alkyl TMAC)
Alkyl trimethyl ammonium chloride (Alkyl TMAC), 411
Alkylbenzene sulphonates (LAS), 411
Alkylpolyoxy-ethylene sulfates (AES), 411
α-alkyl β-hydroxy fatty acid, 39
Altererythrobacter, 209
Alzheimer's disease, 154–155
Amino acids, 39, 83–84, 177, 202–203, 395–396
Amphipathic molecules, 371–372
Amphiphilic self-assembly, 30
Amphoteric biosurfactants, 343–345
Amphoteric surfactants, 412
Amphoterics, 26–27
AMR bacteria. *See* Antimicrobial-resistant bacteria (AMR bacteria)
Animal fat, 378
Animal slaughter, 377
Animal wastes from food industry, 69–70
Anionic biosurfactants, 343–345
Anionic surfactants, 411–412
ANN. *See* Artificial Neural Network (ANN)
ANN-GA. *See* Artificial Neural Network combined with Genetic Algorithm (ANN-GA)
Anomalocardia brasiliana, 381

Anopheles, 347–348
  A. culcifacies, 350
  A. stephensi, 350
Anoxic packed-bed biofilm reactor (AnPBR), 175–176
AnPBR. *See* Anoxic packed-bed biofilm reactor (AnPBR)
Anti-adhesive agents, 261
Anti-biofilm property
  combinations of complex biosurfactants with, 328
  current industrial and medical applications and commercialization of biosurfactant compounds with, 329–331
  fengycin-like lipopeptides as potential anti-biofilm agent, 325
  fungi-based biosurfactants with, 328
  glycolipid biosurfactants with, 326–327
  lipopeptide biosurfactants with, 321–326
  polymyxin as potential anti-biofilm agents, 324
  pseudofactin as potential antibiofilm agent, 325–326
  putisolvin as potential anti-biofilm agent, 325
  surface-active secretions mammals with, 328
  surfactin as potential anti-biofilm agent, 325
Antiadhesive biological coating
  antiadhesive coating property of biosurfactants, 359–360
  antiadhesive property of
    glycolipids biosurfactants, 360–362
    lipopeptides biosurfactants, 362–363
  challenges and future perspective, 365
  microbial adhesion and biofilms, 358–359
  patents, 364–365
  production of antiadhesive and antiineffective biomaterials, 363–364
  properties of different biosurfactants, 356*t*
Antiadhesive biomaterials, 363–364
Antiadhesive property
  of glycolipids biosurfactants, 360–362
  of lipopeptides biosurfactants, 362–363
Antibiofilm agent, biosurfactants as, 321–328, 322*t*
Antiineffective biomaterials, 363–364
Antimicrobial activity of surfactin, 12
Antimicrobial agents, 357–358
  biosurfactants as, 309–314
    glycolipid biosurfactants as, 310–312
    glycoprotein biosurfactants as, 313–314
    lipopeptide biosurfactants as, 312–313

Antimicrobial-resistant bacteria (AMR bacteria), 357
APB. *See* Acid-producing bacteria (APB)
Aphis fabae, 347
APIs. *See* Active pharmaceutical ingredients (APIs)
Aquatic ecosystem, 152–154
Aquatic environments, 198–199
  damage caused by presence of surfactants in aquatic environment, 409*f*
  effect of synthetic surfactants on, 408–412
  sources and occurrence of HC/PD in, 195
Aqueous two-phase systems (ATPS), 97–98
Aromatic hydrocarbons, 111
Artemia
  A. franciscana, 411
  A. salina, 43–46, 68, 411
Arthrobacter spp, 261, 279–282, 359
  A. calcoaceticus, 375
Arthrobaterprotophormiae, 260
Arthrofactin, 359
Artificial Neural Network (ANN), 262
Artificial Neural Network combined with Genetic Algorithm (ANN-GA), 94
AS. *See* Alkyl sulfates (AS)
Ascomycetes, 47–48
Aspergillus, 26, 47–48, 107, 170–172
  A. flavus, 5–6
  A. ustus, 376
Astrocaryum aculeatum, 67
Atomic force microscopy (AFM), 296
ATPS. *See* Aqueous two-phase systems (ATPS)
Azotobacter chrococcum, 31–34

B

Bacillus sp., 12, 26, 41, 67–69, 81, 126–127, 202–203, 262, 279–282, 341, 346, 379, 413, 439
  B. amyloliquefaciens, 43–46
  B. aryabhattai, 164–165
  B. asahii, 87–88
  B. cereus, 210–211
  B. clausii, 262
  B. coagulans, 97–98
  B. detrensis, 87–88
  B. fluorescens, 40
  B. licheniformis, 5–6, 40, 67, 86–88, 93, 126–127, 259–260, 263–268, 380
  B. maegaterium, 311, 418

B. mallei, 7
B. methylotrophicus DCS1, 176–177
B. mojavensis JF-2, 441–443
B. polymyxa, 40
B. pseudomallei, 7
B. pumilus, 96, 415
B. safensis J2, 439
B. stearothermophilus, 257
B. subtilis, 3, 5–6, 11–12, 40, 43–46, 84–90, 135–137, 158, 258, 260–261, 291–292, 296, 311–312, 314–318, 321, 325, 375–376, 378, 380, 398–399, 413
B. thailandensis, 7
B. thuringiensis, 135, 346–347
B. velezensis
Bacteria, 43–47, 208–209, 342
Bacterial biosurfactants, 261
Bacterial cellulose (BC), 292
Bacteriocins, 357–358
Bactris gasipaes, 67
BBD. *See* Box Behnken Design (BBD)
BC. *See* Bacterial cellulose (BC)
BCS. *See* Biopharmaceutic drugs (BCS)
BE. *See* Bio-emulsifiers (BE)
Benzene, 429–430
Benzene, toluene, ethylbenzene, and xylene chemicals (BTEX chemicals), 429–430
Benzenesulfonic acid (BSA), 410
Benzodiazepines, 251–252
BEPSs. *See* Bound EPSs (BEPSs)
Beta-1,2-linked glucose units, 202
Beta-sit-G, 229–230
β-lactamase inhibitors, 254
β-lactams, 254
Bifidobacterium, 357–358
Bio-based substances (BS), 43–46
Bio-based surfactants, 155–174, 372
  sustainability assessment of, 431f
Bio-emulsifiers (BE), 26, 43, 47, 107, 380, 395–396
Bio-interfaces
  interaction of biosurfactants with, 226–229
  saponin membrane interaction and delivery, 228f
Biochemical oxygen demand (BOD), 70
Biocorrosion, 292–293
Biodegradability, 51, 60, 260
  and low toxicity, 380–381

Biodegradable polymeric coatings, 364
Biodegradation process, 131–132, 198
Biodiesel, 43–46
Biodispersan, 13
Bioelectrokinetic remediation, 175–176
Bioemulsifying compounds, 158–159
Bioethanol, 377
Biofilm, 4, 127–128, 308
  biofilm-related diseases, 320
  development, 355–357
  formation, 355
  on medical instruments, 357–358
  microbial formation of, 319–321
    problems associated with resistant bacterial biofilms, 320–321
    stages of biofilm formation, 320
Biogeochemical cycles, 121–122
Bioinformatics, 151–152
Biological coating, 364–365
Biological methods, 357
Biological molecules interact, 226–227
Biomolecules, 151–152
Biopharmaceutic drugs (BCS), 233–236
Bioprocessing
  factors of biosurfactant production, 88–90
    aeration and agitation, 88–89
    incubation time, 89
    inoculum concentration, 89–90
  methods, 42
Bioreactors, 376–377
Bioremediation, 151–152, 170–172, 432–433
  plant-based surfactants and role in, 137–139
  process, 414
  technology, 155
Biosulfuric acid, 107–108
Biosurfactants (BS), 1, 3, 25, 30, 39–40, 43, 59–60, 79–80, 107, 121–122, 126–127, 155–174, 193–194, 217, 220–222, 249–250, 268–270, 277, 307, 341, 343–345, 355, 357–359, 373, 395–396, 414–415, 425–426, 430–433, 440–441. *See also* Chemical surfactants
  activity, 179
  adjuvant for decontamination of soil associated with petroleum derivatives, 170–172
  as agricultural biopesticides, 342–345
  algal, 261
  anti-adhesive agents, 261

Biosurfactants (BS) (*Continued*)
  antibacterial properties of, 314–318
  as antibiofilm agent and mechanisms of action, 321–328
    combinations of complex biosurfactants with anti-biofilm agent property, 328
    fungi-based biosurfactants with anti-biofilm properties, 328
    glycolipid biosurfactants with anti-biofilm property, 326–327
    lipopeptide biosurfactants with anti-biofilm property, 321–326
    surface-active secretions mammals with anti-biofilm property, 328
  anticancer property of, 319
  antifungal properties of, 318
  as antimicrobial agent and mechanisms of action, 309–314
  antimicrobial properties of, 314–319, 315t
  antiviral properties of, 318–319
  applications, 5–6, 6f, 262
  bacterial, 261
  beneficial effects on plants, 415–418
  as biocontrol for organic agriculture, 345–350
    Bacillus biosurfactant as insecticidal, 345–347
    as environmentally safe insecticides, 345f
    as green pesticide for vectors of diseases, 347–350
    used for insect vectors of diseases, 349t
  biodegradability, 260
  biosurfactant-based bioremediation, 151–152
  biosurfactant-facilitated recovery from crude oil tank bottom oil sludge, 165–170
    impact of biosurfactant-mediated HC/PD removal on microbial community, 208–209
  biosurfactant-producing microbes, 151–152
  biosurfactant-producing microorganisms, 92–93
  biosurfactant–biointerface interaction, 226–227
  biosurfactants-mediated biodegradation of PAHs, 125–128
  biosurfactants-mediated degradation of polychlorinated biphenyls, 130–132
  biosurfactants-producing microbes, 4
  biosynthesis of, 161–162, 161f
  characteristics of, 259
  characterization methods for, 296–299
    analysis of inhibition effects, 297–298
    component characterization techniques, 297
    surface analysis techniques, 296–297
    theoretical calculation techniques, 298–299
  chemical classification of, 373–374
  classification, 7–14, 26–27, 109–111, 110f, 156
    based on molecular composition, 257–259
    based on molecular weight, 257
    fatty acids, phospholipids, and neutral lipids, 14
    fengycin, 12
    glycolipids, 7
    LPs, 11
    particulate biosurfactants, 13–14
    polymeric microbial surfactants, 13
    RLs, 7–9
    SLs, 9–10
    surfactin, 11–12
    TLs, 10–11
  in cleaning of crude oil storage tanks, 114–116
    mechanism of action, 114–116
    in tank clean-up, 114–116
  CMC values of, 219t
  combinations of complex biosurfactants with anti-biofilm agent property, 328
  corrosion inhibition
    mechanism of, 286–291
    properties of, 280t
  critical micelle concentration and micelle formation, 2f
  current industrial and medical applications and commercialization of, 329–331
  as delivery carriers for DNA-/RNA-based drug vehicles, 229–231, 230t
  from different food wastes, 60–71, 61t
  diverse biosurfactants with anticancer activity, 235t
  as drug delivery agents, 234t
  ecological significance of, 4
  elucidating biosurfactant drug adsorption properties and mechanisms, 262–263
    electrostatic and electron–exchange interaction, 263
    increasing bioavailability of hydrophobic PPCPs, 263
    promoting hydrophobic partitioning/interaction, 262
    surface charge exchange, 262

environmental effect of synthetic surfactants, 408–412
factors affecting biosurfactant production, 30–34, 51, 108
   carbon sources for production of biosurfactants by different microorganisms, 32t
   nitrogen sources for production of biosurfactants by different microorganisms, 33t
factors affecting scale up of, 397–398
fundaments aspects of, 1–3
fungal, 261
glycolipids, 309
for HC/PD removal, 201–204
   high-molecular-mass biosurfactants, 203–204
   low-molecular-mass biosurfactants, 201–203
for hydrophobic drug bioavailability, 221–226
hydrophobic pollutants/petroleum derivatives
   mechanism behind biosurfactants on removal of, 162–163
   strategies involved in application of biosurfactants on remediation of, 164–165
as insects biocontrol, 342f
interaction of biosurfactants with bio-interfaces, 226–229
limitations preventing for extensive application of biosurfactant for drug removal, 268
main advantages of biosurfactants compared to synthetic surfactants, 408f
management of marine oil spills, 173–174
market, 438
metabolic engineering approaches, 177–178
micelle, 220–221
microbial biosurfactants and HCPD, 166t
microbial biosurfactants-based formulations, 330t
monomers forming micelles at interface, 219f
natural role of, 123
new biosurfactants with improved drug target efficiency, 231–236
   niosomes, 232–233
   NPs, 233–236
omics approaches for, 174–175
patents, 138t, 236–238, 399–403
petroleum hydrocarbons waste, 159–162
physiochemical properties of, 27–30, 28t
practical application of, 291–296
   commercial applications of, 295–296

   environments, 294–295
   marine environment, 291–292
   oil and gas pipelines, 292–293
production, 89
   alternative substrates as raw materials, 94–95
   bioprocessing factors, 88–90
   economic constraints, 90
   efficient fermentation design, 95–97
   factors affecting, 80–90, 81f
   factors that affect large-scale production and commercialization of, 90–91, 90f
   improving downstream processing, 97–98
   methods for enhancing sustainability in, 440–447
   microbial strain, 81–82
   new or engineered microbial strains, 92–93
   nutritional parameters, 82–85
   optimizing composition of culture medium, 94
   physicochemical factors, 85–88
   possible approaches to improve, 91–98
   potential of hydrogen, 85
   safety concerns, 91
   salinity, 87–88
   scale-up studies of, 398–399
   technical constraints, 91
   temperature, 85–87
   using wastes from food industry, 66f
production, 4–5
   of biosurfactants from wastes and renewable materials for sustainability, 439–440
   of biosurfactants using microorganisms, 43–50, 44t
   methods, 108–109, 109f
   strategy, 107–108
properties, 108f, 156–159
   CMC, 156–157
   emulsification, 158–159
   low toxicity, 159
   self-assembly, 159
   stability, 158
   surface and interface activity, 158
recent advancements in
   biosurfactant-aided adsorption technologies for removal of drugs from environment, 263–268
   enhancing specificity and functional properties of, 175–178

Biosurfactants (BS) (*Continued*)
  on removal of hydrophobic pollutants/petroleum derivatives, 178–180
    economic feasibility, 179–180
    factors affecting remediation of hydrophobic pollutants, 179
    removal mechanisms of HC/PD in presence of, 204–208, 204f
  removal of petroleum derivatives from petroleum refineries and petrochemical waste effluent stream, 172–173
  role of biosurfactants in environmental pollution, 412–415
  self-assembly, 259
  solubilization, 260
  sources and classification of, 261, 279–286
    fatty acids, 285–286
    glycolipids, 279–283
    lipopeptides, 283
    phospholipids, 285
    polymer class, 283–284
  sources of production of, 41, 42f
  structural diversity of microbial surfactants, 6–7
  suitable carbon and energy source for microbial production of, 159–162
  surface and interface activity, 260
  surfactants—micelle formation mechanism, 157f
  and sustainability, 415
  targeted hydrophobicity enhancements in, 176–177
    coproduction of biosurfactants with value-added products, 176–177
    inducers, 177
  temperature and pH tolerance, 260
  types, 40–41, 157f, 256–257
    glycolipids, 40
    lipopeptides or lipoprotein, 40
    particulate type, 41
    phospholipids and fatty acids, 40–41
    polymeric surfactants, 41
  unique self-assembly features of biosurfactants for drug adaptation and target improvement, 218–221
  unknown aspects of biosurfactants and future directions, 238–239
Biotechnological processes, 378–379
Biotechnology, 39
BOD. *See* Biochemical oxygen demand (BOD)
Bound EPSs (BEPSs), 283–284
Box Behnken Design (BBD), 262
bphA gene, 131–132
bphB gene, 131–132
bphC gene, 131–132
Brassica oleracea, 68
Brevibacterium, 26
  B. casei NK8, 211
Brewery wastes, 69
Brucella anthropi sp., 130–131
BS. *See* Bio-based substances (BS); Biosurfactants (BS)
BSA. *See* Benzenesulfonic acid (BSA)
BTEX chemicals. *See* Benzene, toluene, ethylbenzene, and xylene chemicals (BTEX chemicals)
Burkholderia, 123–124
  B. cenocepacia, 134
  B. cepacia, 287
  B. pseudomallei, 123–124
By-products, 59

C

C8-hydroxyl acid, 311
Cadmium (Cd), 413
Calgary Biofilm Device, 365
Callosobruchus chinensis, 346
CAM. *See* Contact angle measurement methods (CAM)
Candida sp., 26, 41–42, 67, 107, 439
  C. albicans, 311, 318, 321, 360–363
  C. antarctica, 85, 161–162
  C. apicola, 9–10, 85
  C. bombicola, 9–10, 93, 159, 217, 328, 375–376, 378
  C. glaebosa, 49
  C. lipolytica, 41, 85, 375, 413–414
    C. lipolytica UCP 0988, 378, 413–415
  C. lusitaniae, 311
  C. sphaerica, 173–174, 328
  C. tropicalis, 398
Carbamazepine, 251–252
Carbohydrates, 39, 82–83
Carbon, 42
  atoms, 109–111
  compounds, 4–5
  feedstocks, 60
  sources, 31, 32t, 43, 59, 82–83, 285–286, 443
  substrates, 4–5

Carbon dioxide, 357–358
Carbon/nitrogen ratio, 445–446
Carbon-to-nitrogen ratio, 445–446
Carbonic acid, 128
Carboxylic acid (COOH), 39, 109–111
Cassava processing, 377
Cationic biosurfactants, 343–345
Cationic peptides, 254
Cationic surfactants, 411–412
Cell surface hydrophobicity (CSH), 203
Cell-to-cell communication, 4
Cellobiose lipids (CLs), 311
Cellulomonas cellulans, 327
Cellulosic substances, 70–71
Cellulosic sugar wastes, 70–71
Central nervous system (CNS), 320–321
Cetyl trimethylammonium bromide (CTAB), 411
Chaetoceros gracilis, 411
Chaetonium globosum, 311–312
Characterization methods for biosurfactants, 296–299
Chemical pesticides, 342, 343f
Chemical surfactants, 1, 425–426. *See also* Biosurfactants (BS)
  drive for global sustainability, 426–428
  and production, 428–429
  sustainability assessment of, 429–430, 431f
  sustainable alternatives to, 430–433
  synthesis, 432
Chemisorbed biosurfactants, 286–288
Chikungunya, 348
Chiral drugs, 252–253
Chlorella spp., 261
Chlorinated polycyclic aromatic hydrocarbons (Cl-PAHs), 124–126
Chlorobenzoic acids, 131–132
Chloroform, 432
Chlorpyrifos, 133–135
Chronic prostatitis, 355–357
Chryseobacterium indologenes MUT. 2, 291–292
Citrobacter, 26, 107
Citrus
  C. lambiri, 67
  C. medica, 67
Cl-PAHs. *See* Chlorinated polycyclic aromatic hydrocarbons (Cl-PAHs)

Cleaning products, 373
Clofibric acid, 252–253
Clostridium sp., 26, 279–282
CLs. *See* Cellobiose lipids (CLs)
CMC. *See* Critical micelle concentration (CMC)
CNS. *See* Central nervous system (CNS)
Coal, 437
Coatings, 363
Cobetia, 86
Cold-rolled steel (CRS), 291–292
Colistin, 324
Collagen, 358
Colletotrichum, 5–6
  C. gloeosporioides, 318
  C. musae, 318
Commercialization of biosurfactants
  challenges and future outlook, 404
  factors affecting scale up of biosurfactants, 397–398
  global biosurfactant market and impact on COVID-19 pandemic, 396–397
  patents related to biosurfactants, 399–403
  production, 90–91
  scale-up studies of biosurfactant production, 398–399
Conidiophores, 47–48
Contact angle measurement methods (CAM), 296
Continuous phase, 221
Cooking oils, wastes from, 68–69
Copper oxide, 358
Cord factors, 202
Core–shell type nanocapsules, 231–232
Corrosion inhibition mechanism of biosurfactants, 286–291
  chemisorbed type, 286–288
  electrochemical reaction mechanism on carbon steel, 288f
  electrostatic attraction type, 288–289
  film-forming mixed type inhibitor, 289–291
Corynebacteria, 10–11
Corynebacterium spp., 26, 41, 107, 202
  C. xerosis, 310
Corynomucolic acids, 40–41
Cosmetics, 5–6, 373, 381
Cost, 268–270
Cost–effective downstream procedures, 440–441
Covalent forces, 220

COVID-19, 318–319
    global biosurfactant market and impact on COVID-19 pandemic, 396–397
CPP. *See* Critical packing parameter (CPP)
Crescentia cujete. *See* Novel tropical fruit (*Crescentia cujete*)
Crescentiacujete, 255–256
Critical micelle concentration (CMC), 1–2, 26, 125–126, 156–157, 194, 218–219, 219t, 371–372
Critical packing parameter (CPP), 221–222
Crop protection in organic agriculture, 345
CRS. *See* Cold-rolled steel (CRS)
Crude biosurfactant, 413–414
Crude oil, 111–112
    crude-oil-related products, 437
Crude oil storage tank clean-up, 112–114
    automated cleaning, 113–114
    biosurfactants production strategy, 107–108
    challenges and future outlook, 116–117
    classification of biosurfactants, 109–111, 110f
    factors affecting biosurfactants production, 108
    manual method, 113
    oil storage tanks in industry, 111–114
    production methods of biosurfactants production, 108–109, 109f
    storage tanks, 112
    use of biosurfactants in cleaning of crude oil storage tanks, 114–116
Cryptococcus
    C. curvatus, 70
    C. neoformans, 311
Crystal molecules, 255–256
CSH. *See* Cell surface hydrophobicity (CSH)
CTAB. *See* Cetyl trimethylammonium bromide (CTAB)
Culex, 347–348
    C. quinquefasciatus, 350
Culture media, 31–34
Culture medium, 82
    optimizing composition of, 94
Curcumin, 236
Curd whey, 70
Cutaneotrichosporon mucoides UFMG-CM-Y6148, 70–71
Cyclic esters, 109–111
Cyclic lipopeptides, 283
Cyclic peptide, 39
Cystic fibrosis, 355–357

D

DADMAC. *See* Dialkyl dimethyl ammonium chloride (DADMAC)
Dairy industries, 377
Dairy whey, 378
Daphnia magna, 410
DDSs. *See* Drug delivery systems (DDSs)
DDT. *See* Dichlorodiphenyltrichloroethane (DDT)
Degradation process, 127–128
Degradation rate, 126–127
Delivery carriers for DNA-/RNA-based drug vehicles, biosurfactants as, 229–231
Dengue, 348
Desulfurization, 111–112
Di-rhamnolipid, 358
Dialkyl dimethyl ammonium chloride (DADMAC), 411
2,4-diaminobutyric acid (DAB), 324
Dichlorodiphenyltrichloroethane (DDT), 133–134, 195
Diesel, 437
    fuel, 152–154
    oil, 375
1,2-dimyristoyl-sn-glycro-3-phosphocholine (DMPC), 238
Dip-coating method, 364
Diseases
    biosurfactants as green pesticide for vectors of, 347–350
    in organic agriculture, 345
DNA-based drug vehicles, biosurfactants as delivery carriers for, 229–231
Dodecyl trimethylammonium bromide (DTAB), 411
Domestic waste, 123
Downstream processing, improving, 97–98
DOX. *See* Doxorubicin (DOX)
Doxorubicin (DOX), 233–236
Drug adsorption, 238
    advantages and disadvantages of microbial biosurfactants used in, 269t
    from aqueous medium, 265t
    elucidating biosurfactant drug adsorption properties and mechanisms, 262–263
    future prospects, 268–270
    insight on properties pertaining to ecotoxicological impact of pharmaceutical drugs, 250–256
    limitations preventing for extensive application of biosurfactant for drug removal, 268

recent advancements in biosurfactant-aided adsorption technologies for removal of drugs from environment, 263–268
Drug delivery, 220–221
Drug delivery systems (DDSs), 217, 229–230, 331
Drug encapsulation, 220
Drug removal, limitations preventing for extensive application of biosurfactant for, 268
Drug target efficiency, new biosurfactants with improved, 231–236
DTAB. *See* Dodecyl trimethylammonium bromide (DTAB)
Dunaliella sp., 411
  D. salina, 261, 411

E

Economic constraints of biosurfactant production, 90
EcoPiling, 170–172
Ecosystem
  agriculture formulations/chemicals, 123
  automotive and transportation, 123
  domestic waste, 123
  industrial manufacturing, 123
  pollution and impact sources on, 122–123
Ecotoxicity of commercial surfactants, 411
Ecotoxicological effects by pharmaceutical drugs, 250f
Ectomielois ceratoniae, 346
EDX. *See* Energy dispersive X–ray (EDX)
Electrochemical method, 298
Electron–exchange interaction, 263
Electrostatic attraction type, 288–289
Electrostatic–exchange interaction, 263
ELM. *See* Emulsion liquid membrane (ELM)
*emt*1 gene, 161–162
Emulsan, 41, 380, 395–396
Emulsification, 158–159, 380
  HC/PD, 207
  index, 207
Emulsion formation by biosurfactants for hydrophobic drug bioavailability, 221–226
Emulsion liquid membrane (ELM), 262
Enantioselectivity, 252–253
Endocarditis, 355–357
Endocrine-active pharmaceutical hormone, 251–252

Energy, 42
  source for microbial production of biosurfactants, 159–162
Energy dispersive X–ray (EDX), 289
Engineered microbial strains, 92–93
Engineering of biosurfactants for removal of hydrophobic pollutants/petroleum derivatives, 181–182
Enterobacter, 26, 107
  E. aerogenes, 311–312
  E. cloacae, 5–6
Enterococcus, 364–365
  E. faecalis, 357
Environment, 425
  fate of HC/PD in, 195–198
Environmental pollution, role of biosurfactants in, 412–415
Environmental sustainability, 407
Environmentally safe insecticides, biosurfactant as, 345f
Enzymatic catalysis, 277
Ephestia kuehniella, 346
1,2-epoxyethane, 26–27
EPS. *See* Exopolysaccharides (EPS)
EPSs. *See* Extracellular polymeric substances (EPSs)
Erysipelas sp., 279–282
Escherichia coli, 10, 309, 357, 362–363
Ethanol, 357–358
Ethinylsterol, 251–252
Ex situ strategies, 414
Exiguobacterium profundum, 68–69
Exogenous minerals, 67
Exopolysaccharides (EPS), 107, 293, 357–358
Extracellular polymeric substances (EPSs), 283–284, 319–320, 326–327

F

FAO. *See* Food and Agriculture Organization (FAO)
Fats, 415
Fatty acids, 14, 40–41, 80, 109–111, 177, 278, 285–286, 373, 374t
  typical structures of, 286f
Fed-batch operation, 96
Feedstock, waste as, 42–43
Fengycin, 12, 307
  fengycin-like lipopeptides as potential anti-biofilm agent, 325

Fermentation, 445–446
  efficient fermentation design, 95–97
  process, 42, 397–398
FE–SEM. *See* Field-emission scanning electron microscopy (FE–SEM)
Fibrogen, 358
Fictibacillus barbaricus, 87–88
Field-emission scanning electron microscopy (FE–SEM), 289
Film-forming mixed type inhibitor, 289–291
Fish wastes, 69
Flocculosin, 311
Foam formation, 384
Food, 5–6, 373
  industry
    animal wastes from, 69–70
    biosurfactant production using wastes from, 66f
  products, 59
  wastes, 59, 67
    agric-food wastes, 66–67
    animal wastes from food industry, 69–70
    biosurfactants from different, 60–71, 61t
    BOD, 70
    brewery wastes, 69
    lignocellulosic wastes, 70–71
    starch-rich wastes, 71
    wastes from cooking oils, 68–69
Food and Agriculture Organization (FAO), 68
Fossil fuels, 124–125, 426–427, 437
Fourier transform infrared spectroscopy (FTIR), 49, 287, 296
FTIR. *See* Fourier transform infrared spectroscopy (FTIR)
Fungal biosurfactants, 261
Fungi, 47–48, 208–209, 342
  fungi-based biosurfactants with anti-biofilm properties, 328
  strains, 357
  used to produce biosurfactants, 48t
Fusarium, 47–48
  F. moniliforme, 318
  F. oxysporum, 5–6, 318
  F. solani, 318

G

GA. *See* Genetic Algorithm (GA)
Galvanostatic charge-discharge test, 263–268

γ–polyglutamate, 288
Gas chromatography (GC), 205
Gas chromatography flame ionization detection (GC-FID), 202
Gas chromatography–mass spectrometry (GC-MS), 27, 296
Gas pipelines, 292–293
GC. *See* Gas chromatography (GC)
GC-FID. *See* Gas chromatography flame ionization detection (GC-FID)
GC-MS. *See* Gas chromatography–mass spectrometry (GC-MS)
Generally Regarded as Safe (GRAS), 357–358
Genetic algorithm (GA), 94
Genetic engineering, 151–152
Geobacillus sp., 279–282
Geobacillus thermodenitrificans NG80-2, 380
Gibbs free energy, 1–2
Gliocladium virens, 311–312
Global biosurfactant market and impact on COVID-19 pandemic, 396–397, 397f
Glomerella cingulata, 312
GLs. *See* Glycolipids (GLs)
Glucose, 41, 60, 82–83, 375
Glycerol, 375
Glycerophospholipids, 285
Glycoglycerolipids, 80
Glycolipid biosurfactants, 258, 279–282
  with anti-biofilm property, 326–327
    rhamnolipids as potential anti-biofilm agent, 327
    sophorolipids as potential anti-biofilm agent, 326–327
  antiadhesive property of, 360–362
  as antimicrobial agents, 310–312
    cellobiose lipids, 311
    MELs, 312
    oligosaccharide lipids, 311
    RLs, 311–312
    SLs, 310
    trehalose lipids and succinyl trehalose lipids, 312
    xylolipids, 311
Glycolipids (GLs), 3, 6–7, 40, 60, 80, 107, 201–202, 257, 278–283, 307–308, 355, 359, 373, 374t
Glycolipopeptides, 83–84
Glycolipoproteins, 80

Glycopeptides, 80, 254
Glycoprotein biosurfactants as antimicrobial agent, 313–314
Glycylcyclines, 254
Glycyrrhizic acid, 227
GNB. *See* Gram-negative bacteria (GNB)
Gold nanoparticles, surfactin enhancing skin penetration of, 237f
GPB. *See* Gram-positive bacteria (GPB)
Gram-negative bacteria (GNB), 290, 313, 357
Gram-positive bacteria (GPB), 313, 357
GRAS. *See* Generally Regarded as Safe (GRAS)
Grease, 404
Green chemistry, 348
Green pesticide for vectors of diseases, biosurfactants as, 347–350
Green process, 408
Green surfactants, 30, 155–174
Groundwater pollution, 121–122
Growing media, 41

H

Hair gel, 381
Haloarcula
    H. argentinensis, 87–88
    H. japonica, 87–88
    H. vallismortis, 87–88
Halomonas sp., 47, 86, 210–211
Halophilic microorganisms, 87–88
Halorubrum tebenquichense, 87–88
HAS. *See* Human serum albumin (HAS)
HC/PD. *See* Hydrophobic contaminant/petroleum derivate/derivatives (HC/PD)
Health and Safety Executive (HSE), 113
Heavy metals, 412–413
Hemicellulose hydrolysate, 70–71
Herpes simplex virus (HSV), 318
High molecular weight, 257
High-molecular-mass biosurfactants, 203–204
    polymeric and particulate surfactants, 203–204
High-molecular-weight biosurfactants, 279
High-molecular-weight compounds, 26, 373
High-performance liquid chromatography (HPLC), 27, 49
Higher carbon-to-nitrogen ratio, 83–84
HLB. *See* Hydrophilic–lipophilic balance (HLB)
HOCs. *See* Hydrophobic organic compounds (HOCs)

Homogenization, 221–222
HPLC. *See* High-performance liquid chromatography (HPLC)
HSE. *See* Health and Safety Executive (HSE)
HSV. *See* Herpes simplex virus (HSV)
Human serum albumin (HAS), 411
Humicola, 209
Hydrocarbon-derived epoxides, 154–155
Hydrocarbons, 13–14, 25, 82–83, 122, 259, 395–396, 415
Hydrogen, 85
Hydrogen peroxide, 357–358
Hydrogen sulfide, 357–358
Hydrophilic component of BioS, 109–111
Hydrophilic groups, 39
Hydrophilic moieties, 39, 307
Hydrophilic–lipophilic balance (HLB), 3, 221–222
Hydrophobic contaminant/petroleum derivate/derivatives (HC/PD), 166t, 194
    impact of biosurfactant-mediated HC/PD removal on microbial community, 208–209
    biosurfactants for HC/PD removal, 201–204
    conventional treatment technologies for mitigation of, 155
    conventions and regulations for environmental prevention of, 196t
    environmental prevalence of, 152–154
    influence HC/PD removal by biosurfactant, 199f
    removal mechanisms of HC/PD in presence of biosurfactants, 204–208
        HC/PD emulsification, 207
        HC/PD mobilization, 207–208
        HC/PD solubilization, 205–206
    sources, occurrence, fate, and implications, 195–201, 200f
        ecological and health implications of HC/PD, 198–201
        fate of HC/PD in environment, 195–198
        sources and occurrence of HC/PD in aquatic and terrestrial environments, 195
Hydrophobic drugs, 222–226
    solubility and emulsion formation by biosurfactants for hydrophobic drug bioavailability, 221–226
Hydrophobic groups, 39

Hydrophobic mixtures, 440
Hydrophobic moieties, 307
Hydrophobic organic compounds (HOCs), 30, 193
Hydrophobic partitioning/interaction, promoting, 262
Hydrophobic pollutants using biosurfactants, factors affecting remediation of, 179
Hydrophobic pollutants/petroleum derivatives
  bottlenecks in real-time application of biosurfactants on removal of, 178–180
  challenges and future prospects, 182–183
  engineering of biosurfactants for removal of, 181–182
  filed level applications toward remediation of, 174–175
  green surfactants, 155–174
  HCPD
    conventional treatment technologies for mitigation of, 155
    environmental prevalence of, 152–154
    hydrophobic pollutants environmental prevalence of, 153f
    mechanism behind biosurfactants on removal of, 162–163
  omics approaches for biosurfactants, 174–175
  recent advancements in enhancing specificity and functional properties of biosurfactants, 175–178
  strategies involved in application of biosurfactants on remediation of, 164–165
  toxic impacts of petroleum hydrocarbons, 154–155
Hydrophobic pollution, 195
Hydrophobic portion, 6–7
Hydrophobic PPCPs, increasing bioavailability of, 263
Hydrophobicity, 262, 289
3-(3-hydroxyalkanoyloxy) alkanoic acids (HAAs), 176–177, 201–202
Hydroxyapatite-coated implants, 364
Hydroxyl fatty acid, 39
3-hydroxyl-1,3-methyl-tetradecanoic acid, 258

I

IARC. *See* International Agency for Research on Cancer (IARC)
IB. *See* Iron bacteria (IB)
Immobilization, 207
Implants, 355–357
In situ strategies, 414
Incubation time, 89, 376
Induced systemic resistance (ISR), 415
Industrial applications, 10
Industrial byproducts, 94–95
Industrial chemicals, 277
Industrial feedstock, 41
Industrial manufacturing, 123
Industrial pharmaceutical production, 249–250
Industrial processing of vegetables and fruits, 66
Industrial sector, 395–396
Industrial wastes, 404
Inoculum concentration, 89–90
Inorganic ions, 84–85
Inorganic nitrogen sources, 83–84
Insect control, 341
Insecticidal, *Bacillus* biosurfactant as, 345–347
Insecticidal activity, 341, 346–347
Insecticidal potential of biosurfactants
  biosurfactants as
    agricultural biopesticides, 342–345
    biocontrol for organic agriculture, 345–350
    insects biocontrol, 342f
  chemical pesticides, 342
Insecticides, 343–345
Insects, 342
Interface activity, 158, 260
Interfacial interactions, 227
Interfacial tension, 4, 8–9, 25, 121–122, 158
International Agency for Research on Cancer (IARC), 165–170
International climate change treaties, 427
Ionic biosurfactants, 222
Ionic strength tolerance, 381–382
Iron, 358
Iron bacteria (IB), 292–293
Isochrysis galbana, 411
ISR. *See* Induced systemic resistance (ISR)

J

JE1058BS, 115

K

Kerosene, 437
Ketolides, 254
Kitchen waste oil (KWO), 382–383

Klebsiella sp., 135–137
  K. pneumonia, 311–312, 314–318
    IVN51, 285
    MUT.1, 291–292
    KOD36, 158
Kocuria turfanesis, 70
  K. turfanesis BS-J, 443
Kukuria marina BS15, 87–88
Kurtzmanomyces, 49
KWO. *See* Kitchen waste oil (KWO)

L

L-amino acids, 42
LAB. *See* Lactic acid bacteria (LAB)
Lactam antibiotics, 236
Lactic acid bacteria (LAB), 357–358
Lactobacilli, 309
Lactobacillus, 26, 107, 307–308, 313–314, 357–358
  L. acidophilus, 357
  L. casei, 360
  L. cellobiosus TM1, 313–314
  L. delbrueckii N2, 313–314
  L. fermentum, 357
  L. paracasei, 381, 415
  L. pentosus, 262, 414–415
Lactococcus, 357–358
  L. lactis, 311
Lactones, 109–111
Large-scale production of biosurfactant production, 90–91
Larvicidal activity, 350
LAS. *See* Alkylbenzene sulphonates (LAS); Linear alkylbenzene sulfonate (LAS); Linear anionic surfactant alkylbenzene sulfonate (LAS)
LB-EPS. *See* Loosely bound EPS (LB-EPS)
LC-Ms. *See* Liquid chromatography-mass spectrometry/spectroscopy (LC-Ms)
LCA. *See* Life cycle assessment (LCA)
LCC. *See* Life cycle costing (LCC)
Lead (Pb), 413
Leucococcus enterica, 293
Leuconostoc, 26, 107, 357–358
Lichenysin, 259
Life cycle assessment (LCA), 425, 427
  studies, 431
Life cycle costing (LCC), 428

Life cycle sustainability assessments, 429
Lignocellulosic wastes, 70–71
Linear alkylbenzene sulfonate (LAS), 411
Linear anionic surfactant alkylbenzene sulfonate (LAS), 410
Linear lipopeptides, 283
Lipids, 40, 109–111, 227, 395–396
Lipopeptides (LPs), 3, 6–7, 11, 40, 43–46, 60, 80, 83–84, 107, 123–124, 164–165, 202–203, 258, 278, 283, 307, 309, 355, 359, 373, 374*t*, 381, 440
  biosurfactants, 413
    with anti-biofilm property, 321–326
    antiadhesive property of, 362–363
    as antimicrobial agents, 312–313
  typical structure of, 284*f*
Lipopolysaccharides (LPS), 80, 204–205, 313, 355
Lipopolysaccharides–protein complexes, 6–7
Lipoproteins, 6–7, 40, 60, 80, 258, 261, 355, 373
Liposomes, 220, 229–232
Liquid chromatography-mass spectrometry/spectroscopy (LC-Ms), 27, 49
Liquid fermentation, 108–109
Liquid surface tension, 79
Liquid wastes, 121–122
Liquid–liquid interfaces, 1–2, 30
Listeria monocytogenes, 314–318
Long-chain fatty acid, 39
Loosely bound EPS (LB-EPS), 283–284
Low molecular weight, 257
Low toxicity, 159
Low-cost byproducts, 42–43
Low-cost sources, 25–26
Low-molecular-mass biosurfactants, 201–203
  glycolipids, 201–202
  lipopeptides, 202–203
Low-molecular-weight biosurfactants, 279, 373
Low-molecular-weight compounds, 26
LPs. *See* Lipopeptides (LPs)
LPS. *See* Lipopolysaccharides (LPS)
Lysinibaciilus sp., 130–131
Lysobacter, 209

M

*mac*1 gene, 161–162
Maconellicoccus hirsutus, 346
Macronutrients, 4–5, 84–85
Macrophomina phaseolina, 418

Madin-Darby canine kidney (MDCK), 412
MALDI-TOF. See Matrix-assisted laser desorption/ionization−time of flight mass spectroscopy (MALDI−TOF)
Mammalian cells, 229−230
Manganese-oxidizing bacteria (MOB), 292−293
Mannosyl erythritol lipids (MELs), 68−69, 161−162, 307, 312, 348−350
Mannosylerythritol, 312
Mannosylerythritol lipids-A (MEL-A), 229−230
Marine algae, 43−46
Marine environment, 291−292
Marine inhibitory bacterium (*Vibrio neocaledonicus* sp.), 292
Marine oil spills, management of, 173−174
Mariniluteicoccus flavus, 314−318
Marinobacter sp., 128
Marinomonas, 86
Mass spectrometry, 201−202
*mat*1 gene, 161−162
Matrix-assisted laser desorption/ionization−time of flight mass spectroscopy (MALDI−TOF), 27
MDCK. See Madin-Darby canine kidney (MDCK)
MDDS. See Microemulsion based drug delivery systems (MDDS)
MDR. See Multidrug resistance (MDR)
MDS. See Molecular dynamics simulation (MDS)
Mechanisms of action, biosurfactants as, 309−314, 321−328
Medical devices, 329, 361
Medical implants, 355−357
Medicine, 373
MEL-A. See Mannosylerythritol lipids-A (MEL-A)
Melanaphis sacchari, 347
MELs. See Mannosyl erythritol lipids (MELs)
MEOR. See Microbial enhanced oil recovery (MEOR); Microbial-enhanced oil recovery (MEOR)
Mercury (Hg), 413
Metabolic engineering approaches, 177−178
Metabolite rhamnolipids, 287
Metabolomics, 174−175
Metagenomics, 175
Metal ions, 375−376
Metallic salts, 84−85
Metaproteomics, 175
Metatranscriptomics, 175
Methanobacterium thermoautotrophium, 257
Methicillin-resistant *Staphylococcus aureus* (MRSA), 308
MIC. See Minimum inhibition concentration (MIC)
Micellar assemblies, 221−222
Micelles, 1−3, 30, 125−126, 218f, 220, 371−372
Microbes, 80, 85−86, 127−128, 131−132, 159−161, 307, 430−431
Microbial adhesion, 355−357, 364
  and biofilms, 358−359
Microbial biosurfactants, 80, 166t, 341, 355, 359, 372−373, 425−426
  classes/subclasses, 374t
  production, 375−377
Microbial biotechnology, 348
Microbial cells, 123−124, 173
  microbial cell-bound biosurfactants, 375
Microbial cellular metabolism, 277
Microbial community, impact of biosurfactant-mediated HC/PD removal on, 208−209
Microbial corrosion, 295
Microbial enhanced oil recovery (MEOR), 380
Microbial growth, 85−86
Microbial metabolites, 309, 342−343, 345
Microbial resources
  challenges and future research directions, 51
  factors affecting biosurfactants production, 51
  fermentation process, 42
  fungi, 47−48
  low-cost byproducts and waste as feedstock, 42−43
  production of biosurfactants using microorganisms, 43−50
  sources of production of biosurfactants, 41
  types of biosurfactants, 40−41
  yeast, 49−50
Microbial strain, 81−82, 86
Microbial surfactants, 1, 372
  structural diversity of, 6−7
Microbial-based bioremediation, 164−165
Microbial-based biosurfactants, 308
Microbial-enhanced oil recovery (MEOR), 414
Microbiologically influenced corrosion, 292−293
Microemulsion based drug delivery systems (MDDS), 221−222, 223f

Microemulsions, 207
  difference between micro-and nanoemulsion droplets, 225f
  as drug delivery agents, 234t
  formation, 3
  formulation, 221–222
  properties of, 224t
Microorganisms, 3, 6–7, 26, 31, 39–41, 60, 80–81, 107–108, 124–125, 126f, 128, 208–209, 283–284, 289–290, 415, 430
  bacteria, 43–47
  to HC/PD, 204–205
  production of biosurfactants using, 43–50, 44t
Minimum inhibition concentration (MIC), 311–312
MOB. *See* Manganese-oxidizing bacteria (MOB)
Mobilization, HC/PD, 207–208
Mode of action, 251–252
Moesziomyces antarcticus, 86
Molasses, 378
Molecular dynamics simulation (MDS), 298
Mono-rhamnolipids, 205, 358
MRSA. *See* Methicillin-resistant *Staphylococcus aureus* (MRSA)
Multidrug resistance (MDR), 229–230, 231f, 357
Multifunctional di-rhamnolipid, 207
Musa paradisiaca, 67
Mycobacteria, 10–11
Mycobacterium, 202, 359
Mycolic acids, 40–41
Myerozyma sp., 4–5, 9–10
Myzus persicae, 346

N

Nannochloropsis gaditana, 411
Nanoemulsions, 222–226
  application of, 226f
  difference between micro-and nanoemulsion droplets, 225f
  properties of, 224t
Nanomaterials, 358
Nanoparticles (NPs), 231–236
Nanotechnology, 151–152, 178–179, 181–182
Naphthalene sulfonic acid (NSA), 410
Natural biological extraction, 277
Natural non-ionic compounds, 137–139
Natural sources, 122

Natural surfactants, 25, 27, 373
  glycolipids, 27
Neosporin, 324
Nerium oleander, 328
Neutral acids, 80
Neutral lipids, 14, 374t
New microbial strains, 92–93
Nickel (Ni), 413
Niosomes, 231–233, 233f
Nitrates, 83–84
Nitrogen, 42, 375–376
  sources, 31, 33t, 83–84, 441–443
NMR. *See* Nuclear magnetic resonance (NMR)
Nocardia, 10–11, 202
Nocardioides, 327
Non-polar groups, 278
Non-*Pseudomonas* RL producers, 7
Noncovalent forces, 220
Noncovalent interactions, 218–219
Nonionic surfactants, 26–27, 411
Nonpolar end, 25
Nonrenewable raw materials, 373
Nonrenewable resources, 426–427
Nonribosomal peptide synthetase (NRPS), 11, 123–124
Novel drug delivery systems, 218
Novel tropical fruit (*Crescentia cujete*), 263–268
NPs. *See* Nanoparticles (NPs)
NRPS. *See* Nonribosomal peptide synthetase (NRPS)
NSA. *See* Naphthalene sulfonic acid (NSA)
Nuclear magnetic resonance (NMR), 27
Nucleic acids, 84–85
Nutrients, 31
  nutrient-limiting circumstances, 107–108
Nutritional parameters, 82–85
  carbon source, 82–83
  inorganic ions, 84–85
  nitrogen source, 83–84

O

O/W system. *See* Oil-in-water system (O/W system)
Ochrobactrum intermedium, 164–165
Octapeptin A and B, 313
OECD. *See* Organization for Economic Cooperation and Development (OECD)
Oil-in-water system (O/W system), 221

Oils, 404, 415
  industry, 414
  pipelines, 292–293
  reservoir, 111–112
  sludge, 165–170
  spill, 173–174, 174f
  spreading, 208
  storage tanks, 111–112
    crude oil, 111–112
    in industry, 111–114, 111f
  wastes, 377–379
    animal fat, 378
    dairy whey, 378
    molasses, 378
    soap stock, 378–379
    starchy substrates, 378
Oily sludge hydrocarbons, 116
Oily wastes, 156
Oily wastewater, 172–173
Oleic acid, 377
Oligosaccharide lipids, 311
Olive oil mill effluent (OOME), 159–161, 377
Omics approaches, 151–152, 178–179
  for biosurfactants, 174–175
  metabolomics, 174–175
  metagenomics, 175
  metaproteomics, 175
  metatranscriptomics, 175
OOME. See Olive oil mill effluent (OOME)
OPs. See Organopesticides (OPs)
Optimization
  of oxygen transfer, 376
  tools, 446
Organic acids, 357–358
Organic agriculture, biosurfactant as biocontrol for, 345–350
Organic carbon, 70
Organic compounds, 440
Organic contaminants, 122
Organic matter, 198, 278–279
Organic nitrogen sources, 83–84
Organic pollutants, 255–256
Organization for Economic Cooperation and Development (OECD), 68
Organochlorine pesticides, 195
Organopesticides (OPs), 132–133
  organochlorinated pesticides, 136t
  surfactants enhanced degradation of, 133–137
    chlorpyrifos, 134–135
    dichlorodiphenyltrichloroethane, 134
    PHE, 135–137
    quinalphos, 135
Organosilicon surfactants, 412
Otitis media, 355–357
OTR. See Oxygen transfer rate (OTR)
OUR. See Oxygen uptake rate (OUR)
Oxazolidinones, 254
Oxide, 358
Oxygen, 88, 290
Oxygen transfer rate (OTR), 88
Oxygen uptake rate (OUR), 88

P

Paclitaxel (PTX), 233–236
Paenibacillus, 26
  P. alvei, 443–444
  P. dendritiformis CN5, 440
  P. polymyxa, 362
Paenisporosarcina indica, 87–88
PAHs. See Petroleum hydrocarbons (PAHs); Polyaromatic hydrocarbons (PAHs); Polycyclic aromatic hydrocarbons (PAHs)
Palate, lung, nasal epithelium clone (PLUNC), 328
Palm oil, 377
Pantoea stewartii, 7
Pantothenic acids, 70
Paper industries, 413
Paraffins, 429–430
Parkinson's disease, 154–155
Particulate biosurfactants, 13–14, 259
Particulate surfactants, 373, 374t
Patents, 364–365
  related to biosurfactants, 399–403, 400t
Pathogenic biofilm, 359
Pathogens, 355
PCBs. See Polychlorinated biphenyls (PCBs)
pDNA. See Plasmid DNA (pDNA)
Pectate lyase (PEL), 176
Pectin lyase (PNL), 176
PEL. See Pectate lyase (PEL)
Penicillium, 26, 47–48, 107
Peptide component, 40
Perfluoroalkyl carboxylic acids (PFCAs), 195
Periodontitis, 355–357
Pest control, 342
Pest management in organic agriculture, 345

Pesticides, 122, 195, 413
Petrochemical sector, 79
Petrochemical wastewater, 172–173
Petroleum, 5–6, 25, 151–152, 373, 375
 derivatives, 413
  adjuvant for decontamination of soil associated with, 170–172
  conventional treatment technologies for mitigation of, 155
  environmental prevalence of, 152–154
  from petroleum refineries and petrochemical waste effluent stream, 172–173
  industries, 152–154
  oil distillates, 429–430
  refineries, 429–430
  sludge, 112–113, 159–161
Petroleum hydrocarbons (PAHs), 47, 162–163
 biosynthesis of biosurfactants, 161–162
 toxic impacts of, 154–155
 waste, 159–162
Petroleum-based surfactants, 373
 challenges limit production of biosurfactants, 383–384
  foam formation, 384
  price and low productivity, 383–384
  recovery and purification, 384
 chemical classification of biosurfactants, 373–374
 microbial biosurfactant production, 375–377
 production using renewable raw materials, 382–383
 properties and advantages of biosurfactants, 379–382
  surface and interface activity, 379–382
 renewable natural resources used in biosurfactant production, 377–379
PFCAs. *See* Perfluoroalkyl carboxylic acids (PFCAs)
PG. *See* Polygalacturonase (PG)
pH, 381–382
pH. *See* Potential of hydrogen (pH)
Phaeodactylum tricornutum, 411
Pharmaceutical drugs, 249–250
 APIs, 251–252
  chemical and physical properties of, 252–253
  recalcitrance of, 253–254
 insight on properties pertaining to ecotoxicological impact of, 250–256, 250f
 plausible biosurfactant-mediated treatment approach, 255–256
Pharmaceuticals, 81–82, 251–252, 251f, 262, 373
 compounds, 252–254
 industry, 5–6, 217, 249–250
  absorption patterns, 238f
  emulsifying power of surfactin, 237f
  patents concerning applications of biosurfactants for, 236–238
  surfactin enhancing skin penetration of gold nanoparticles, 237f
PHE. *See* Phenanthrene (PHE)
Phenanthrene (PHE), 135–137
Phosphates, 39, 84–85
Phospholipid fatty acid (PLFA), 208–209
Phospholipids, 14, 40–41, 80, 84–85, 161–162, 238, 278, 285, 285f, 373, 374t, 395–396
Phosphorous pentoxide (P_2O_5), 428
Photobacterium phosphoreum, 260, 410
Phyllanthus emblica, 293
Physa acuta Draparnaud, 411
Pichia sp., 289
 P. guilliermondii, 165–170
 P. lynferdii, 165–170
 P. sydowiorum, 165–170
Planomicrobium okeanokoites, 205
Plant-based surfactants
 biosurfactant patents, 138t
 and role in bioremediation, 137–139
Plants, beneficial effects on, 415–418
Plasma membrane, 220–221
Plasmid DNA (pDNA), 229–230
Plastics, 195
PLFA. *See* Phospholipid fatty acid (PLFA)
PLUNC. *See* Palate, lung, nasal epithelium clone (PLUNC)
PNL. *See* Pectin lyase (PNL)
Poecilia vivipara, 210–211, 381
Pollutants, 155, 255–256, 277
Pollution, 412
Pollution mitigation utilizing biosurfactants
 biosurfactant-producing strains and potential applications in pesticide degradation, 129t
 biosurfactants-mediated degradation of polychlorinated biphenyls, 130–132
 natural role of biosurfactants, 123
 OPs, 132–133
  surfactants enhanced degradation of OPs, 133–137
 PAHs, 124–125

Pollution mitigation utilizing biosurfactants (*Continued*)
    biosurfactants-mediated biodegradation of, 125–128
    PCBs, 128–130
    plant-based surfactants and role in bioremediation, 137–139
    pollution and impact sources on ecosystem, 122–123
Polyaromatic hydrocarbons (PAHs), 152–155
Polyaspartate, 288
Polybrominated diphenyl ethers, 195
Polychlorinated aromatic hydrocarbons, 124–125
Polychlorinated biphenyls (PCBs), 122, 128–130, 131*f*, 133*f*, 195
    biosurfactants-mediated degradation of, 130–132
Polycyclic aromatic hydrocarbons (PAHs), 122, 124–125, 130*f*, 195, 440
    biosurfactants-mediated biodegradation of, 125–128
Polyfluoroalkyl substances, 195
Polygalacturonase (PG), 176
Polymer class, 283–284
Polymeric biosurfactants, 13, 60, 259, 374*t*
Polymeric microbial surfactants, 13
Polymeric surfactants, 41, 278, 373
Polymyxins, 313
    as potential anti-biofilm agents, 324
Polyoxyethylene alkyl ethers (polyoxyethylene AE), 411
Polyoxylethylene alkyl-phenyl ethers (polyoxylethylene APE), 411
Polypeptide chain, 40
Polysaccharides, 41, 228–229
Polysaccharides–protein–fatty acid complexes, 6–7
Polystyrene surface (PS), 360
Population explosion, 132–133
Porphyridium cruentum, 261
Potato, 71
Potential of hydrogen (pH), 85
Practical application of biosurfactants, 291–296
Prays oleae, 346
Primary hydrocarbons, 111
Probiotics, 357–358
Propionibacterium
    P. acnes, 310
    P. freudenreichii, 362

Proteins, 228–229, 355
    biosurfactant, 41
Proteus, 357
PS. *See* Polystyrene surface (PS)
Pseudoalteromonas, 86
    P. lipolytica, 291–292
Pseudofactin as potential antibiofilm agent, 325–326
Pseudokirchneriella subcapitata, 411
Pseudomonas sp., 26, 42–43, 67, 81, 86, 126–127, 152–154, 207, 279–282, 359, 375, 439, 443–444
    P. aeruginosa, 3–5, 83–84, 116, 123–124, 134–135, 159–161, 201–202, 255–256, 260, 282–283, 287, 321, 324, 327, 348, 361, 375–376, 432–433, 441, 445–446
    P. alcaligenes, 137–139, 376, 444–445
    P. aphidis ZJUDM34, 68–69
    P. azotoformans AJ15, 67
    P. cepacia
    P. chlororaphis, 123–124
    P. fluorescens, 43, 203
    P. guariconensis LE3, 418
    P. marginalis, 134–135
    P. mosselii F01, 287, 293
    P. nitroreducens, 446
    P. putida, 93, 170–172, 410, 413
    P. putis strain PCL 1445, 278–279
    P. stutzeri NA3, 350
Pseudostelium sp., 279–282
Pseudozyma sp., 49, 312
    P. aphidis, 165–170
    P. flocculosa, 311
    P. fusiformata, 311
Psychrophilic microbes, 86
PTX. *See* Paclitaxel (PTX)
Putisolvin as potential anti-biofilm agent, 325

Q
QS. *See* Quorum Sensing (QS)
Quantum chemical computing, 298
Quaternary alkylammonium chloride, 411
Quillaja saponaria, 137–139
Quinalphos, 133–135
Quinolone-related agents, 254
Quorum Sensing (QS), 4

R

Ralstonia pickettii srs PTA-5579, 170–172
Raphidocelis subcapitata, 411
Raw materials, 90
    alternative substrates as, 94–95
RBCs. *See* Rhamnolipid biosurfactants (RBCs)
Recombinant DNA technology, 446–447
Red algae, 261
Renewable by-products for production of biosurfactant, use of, 416t
Renewable feedstock, 193
Renewable materials for sustainability, production of biosurfactants from, 439–440
Renewable natural resources used in biosurfactant production, 377–379
    oil wastes, 377–379
Renewable raw materials, production using, 382–383
Renewable sources, 25–26, 249–250
Renewable substrates, 80
Renibacterium salmoninarum, 327
Residual oil, 81–82
Response surface methodology (RSM), 4–5, 446
Response Surface Methodology-Box Behnken Design (RSM-BBD), 262
Rhamnolipid biosurfactants (RBCs), 293, 299, 329
Rhamnolipids (RLs), 7–9, 8f, 82, 109–111, 123–124, 126–127, 201–202, 205, 258–259, 282, 287, 295–296, 298, 307, 309–312, 327, 355, 358, 360–361. *See also* Sophorolipids (SLs)
    anti-adhesive and antibiofilm properties, 358
    as potential anti-biofilm agent, 327
    production, 83–84
    rhamnolipid A, 236–238
    rhamnolipid B, 236–238
    schematic structures of, 282f
Rhamnols, 395–396
Rhamnose, 201–202
Rhamnosyltransferase 1 A (rhlA), 175–176
Rhamnosyltransferase 1B (rhlB), 175–176
Rhizobacterium *Pseudomonas putida* strain PCL1445, 325
rhlA. *See* Rhamnosyltransferase 1 A (rhlA)
rhlB. *See* Rhamnosyltransferase 1B (rhlB)
rhlC gene, 201–202

Rhodococcus sp., 10–11, 26, 47, 86, 107, 116, 130–131, 375–376
    R. erythropolis, 5–6, 47, 135–137, 261, 311–312
    R. fascians, 47
    R. opacus, 47
    R. rhodochrous, 47
    R. ruber, 47
Rhophalosiphum maidis, 347
Riboflavin, 70
RLs. *See* Rhamnolipids (RLs)
RNA interference (RNAi), 229–230
RNA-based drug vehicles, biosurfactants as delivery carriers for, 229–231
RNAi. *See* RNA interference (RNAi)
Robinia pseudoscscia, 328
RSM. *See* Response surface methodology (RSM)
RSM-BBD. *See* Response Surface Methodology-Box Behnken Design (RSM-BBD)

S

Saccharomyces, 26, 107
    S. cerevisiae, 31–34, 176–177
Salicola sp., 87–88
Salinibacter ruber, 87–88
Salinity, 87–88
Salmonella
    S. enterica, 357
    S. tyhimurium, 357
SAMB. *See* Self assembling marine biofilm (SAMB)
Sapindus mukorossi, 137–139, 262
Saponins, 137–139
    membrane interaction and delivery, 228f
    saponins–cholesterol interaction, 227
SARS-CoV-2 infection, 5–6
Scale–up studies of biosurfactant production, 398–399
Scanning electron microscopy (SEM), 289
Scenedesmus subspicatus, 411
SCWO technology. *See* Supercritical water oxidation technology (SCWO technology)
SDS. *See* Sodium dodecyl sulfate (SDS)
Self assembling marine biofilm (SAMB), 324
Self-assembly, 159, 218–219
    in aqueous solutions, 30
    of biosurfactant, 259

Self-micro-emulsifying drug delivery systems (SMEDDS), 221–222
SEM. *See* Scanning electron microscopy (SEM)
SEPSs. *See* Soluble EPSs (SEPSs)
Serratia sp., 181–182
    S. marcescens, 203
    S. rubidaea SNAU02, 68
Silver oxide, 358
siRNA. *See* Small interfering RNA (siRNA)
sit-G. *See* Sitosterol beta-D-glucoside (sit-G)
Sitosterol beta-D-glucoside (sit-G), 229–230
Skeletonema costatum, 411
SLs. *See* Sophorolipids (SLs)
Sludge waste, 413
Small interfering RNA (siRNA), 229–230
SMEDDS. *See* Self-micro-emulsifying drug delivery systems (SMEDDS)
SmF. *See* Submerged fermentation (SmF)
Soap stock, 378–379
Social LCA techniques, 428
Sodium dodecyl sulfate (SDS), 309–310, 381, 410–411
Soils, 198
    adjuvant for decontamination of soil associated with petroleum derivatives, 170–172
    pollution, 121–122, 128–130, 429–430
    soil-dwelling microorganisms, 123–124
    soil/sediment matrix, 194
Solid wastes, 42–43, 121–122
    management, 42–43, 51
Solid-state fermentation (SSF), 42, 96, 108–109
Solid–liquid interfaces, 30
Solubility formation by biosurfactants for hydrophobic drug bioavailability, 221–226
Solubilization, 162–163, 205, 260
    HC/PD, 205–206
Soluble EPSs (SEPSs), 283–284
Sophorolipids (SLs), 9–10, 10*f*, 49, 70–71, 109–111, 161–162, 202, 217, 258, 282–283, 307, 309–310, 326, 355
    biosurfactant, 318–319
    main structures of, 283*f*
    as potential anti-biofilm agent, 326–327
Sources organic carbon, 70–71
Sphingobacterium detergens, 443
Sphingomgelins, 285
Sphingomonas sp., 135–137, 209
Spiculisporic acid, 14

Spider mite, 347
Spodoptera littoralis, 346
Sporisorium scitamineum, 318
SRB. *See* Sulfate-reducing bacteria (SRB)
SrfAA module, 177
SrfAB module, 177
SrfAC module, 177
SSF. *See* Solid-state fermentation (SSF)
Stability, 158
Staphylococcus
    S. aureus, 308–309
    S. epidermidis, 311–312, 327
    S. sciuri subsp. rodentium strain SE I, 278–279
Starch, 404
Starch-rich wastes, 71
Starchy by-products, 378
Starchy substrates, 378
Starmerella bombicola, 9–10, 49, 93, 165–170, 314, 375
Stern layer, 158
STF. *See* Surface tension (STF)
Storage tanks, 112
Streptococcus, 357–358
    S. faecium, 310
    S. mitis, 360
    S. mutans, 361–363
    S. thermophiles A and L. lactis 53, 359
Streptogramins, 254
Streptomyces, 134
Structure–function relationships, 3
Submerged fermentation (SmF), 108–109
Succinyl trehalose lipids, 312
Sucrose, 375
Sugar cane, 404
Sugars, 82–83
    molecules, 257
Sulfate-reducing bacteria (SRB), 292–293
Sulfur trioxide (SO_3), 428, 430
Sunflower seed oil, 377
Supercritical water oxidation technology (SCWO technology), 155
Surface activity, 158, 260
Surface analysis techniques, 296–297
Surface tension (STF), 1–2, 4, 25, 59, 121–122, 380
    of biosurfactants, 126–127
Surface-active agents, 25–26
Surface-active molecules, 47

Surface-active secretions mammals with anti-biofilm property, 328
Surface-bound anti-infective agents, 363–364
Surfactants, 1, 26–27, 39, 79, 277, 371–372, 372f, 395, 409–410, 412, 425–426, 428
  converters, 429
  production process, 430
Surfactin, 11–12, 177, 202–203, 236, 258–259, 307, 313, 346, 379
  emulsifying power of, 237f
  as potential anti-biofilm agent, 325
  production, 85–86
Sustainability, 426–427
  assessment of
    chemical and bio-based surfactants, 431f
    chemical surfactants, 429–430
  biosurfactants and, 415, 438t
  key aspects of, 427f
  methods for enhancing sustainability in biosurfactant production, 440–447
    carbon sources, 443
    carbon/nitrogen ratio, 445–446
    growth conditions, 443–445
    nitrogen sources, 441–443
    using recombinant DNA technology, 446–447
    techniques used in improving biosurfactant production, 442t
    use of optimization tools, 446
  production of biosurfactants from wastes and renewable materials for, 439–440
  use of renewable by-products for production of biosurfactant, 416t
Sustainable alternatives to chemical surfactants, 430–433
Synthetic pesticides, 342
Synthetic surfactants, 1–2, 25, 27, 193, 343, 372–373, 407
  environmental effect of, 408–412

T

t-RNase synthetase inhibitors, 254
Talaromyces trachyspermus, 14
Tank bottom petroleum oil sludge (TBOS), 181–182
Tank clean-up, biosurfactants in, 114–116
TB-EPS. *See* Tightly bound EPS (TB-EPS)
TBOS. *See* Tank bottom petroleum oil sludge (TBOS)

Technical constraints of biosurfactant production, 91
Techniques, 440–441
Temperature, 85–87, 260, 381–382
Tenacibaculum litoreum W–4, 289
Tenacibaculum mesophilum D–6, 289
Terrestrial environments, 198–199
  effect of synthetic surfactants on, 411–412
  sources and occurrence of HC/PD in, 195
Tetragenococcus koreensis, 327
Tetranychus urticae, 347
Tetraselmis chuii, 411
Theobroma grandiflorum, 67
Thermophilic microbes, 86
Thermophilic microorganisms, 86–87
Thermus thermophilus HB8, 377
Thin-layer chromatography (TLC), 27, 49, 201–202
Thiobacillus, 26, 107
Thodococcus ruber IEGM 231, 360
Tightly bound EPS (TB-EPS), 283–284
Time of flight mass spectroscopy (TOF mass spectroscopy), 27
TLC. *See* Thin-layer chromatography (TLC)
TLs. *See* Trehalolipids (TLs)
TOF mass spectroscopy. *See* Time of flight mass spectroscopy (TOF mass spectroscopy)
Toothpaste, 381
Torulopsis
  T. bombicola, 375
  T. petrophilum, 49
Toxic compounds, 413
Toxic refractory organic pharmaceutical compounds, removal of, 262
Traditional plasma spraying method, 364
Trehalolipids (TLs), 10–11, 258, 307
Trehalose lipids, 202, 312
Trichophyton metaggrophytes, 327
Trichosporon asahii, 311
Tricoderma atroviride, 318
Triglyceride oils, 428
Triton X-100, 410, 413
Tsukamurella DM 44370, 311–312
Tuta absoluta, 346

U

UN. *See* United Nations (UN)
UN-SDGs. *See* United Nations Sustainable Development Goals (UN-SDGs)

United Nations (UN), 209–210
United Nations Sustainable Development Goals (UN-SDGs), 194, 209–211
   eco-sustainable biosurfactant-based HC/PD removal approach toward achieving selected, 209–211
United States Environmental Protection Agency (USEPA), 152–154
USEPA. *See* United States Environmental Protection Agency (USEPA)
Ustilago, 26, 107, 312
   U. maydis, 161–162

V

Value-added products, coproduction of biosurfactants with, 176–177
VDS. *See* Vesicular drug systems (VDS)
Vector-borne diseases, 347–348
Vegetable cooking oils, 439–440
Vehicle exhausts, 123
Vesicular drug systems (VDS), 331
Vibrio
   V. alginolyticus, 292
   V. cholera, 314–318
   V. fischeri, 410
   V. Neocaledonicus sp., 287, 294–295
   V. parahaemolyticus, 292
Virgibacillus salaries KSA-T, 441–443, 446
VOCs. *See* Volatile organic compounds (VOCs)
Volatile organic compounds (VOCs), 154–155

W

Wastes
   from cooking oils, 68–69
   as feedstock, 42–43
   generation, 59
   production of biosurfactants from wastes for sustainability, 439–440
   use strategy, 408
   vegetable oils, 377
Water-in-oil type of emulsion, 158–159
Water-insoluble carbon substrates, 415
Weissella cibaria PN3, 193–194
Whey, 378
Wickerhamomyces anomalus, 165–170
Winsor-R ratio, 221–222
Woody biomass, 123

X

XPS. *See* X–ray photoelectron spectroscopy (XPS)
X–ray photoelectron spectroscopy (XPS), 296
Xylolipids, 311

Y

Yarrowia lipolytica, 31–34, 85, 87–88, 380
Yeast, 49–50, 50t, 208–209
Yellow fever, 348

Z

Zika viruses, 348
Zinc oxide, 358

Printed in the United States
by Baker & Taylor Publisher Services